Urban Evolutionary Biology

Urban Evolutionary Biology

EDITED BY

Marta Szulkin
Centre of New Technologies, University of Warsaw, Poland

Jason Munshi-South
Louis Calder Center-Biological Field Station, Fordham University, Armonk, NY, USA

Anne Charmantier
Anne Charmantier Centre d'Ecologie Fonctionnelle et Evolutive, Univ. Montpellier, CNRS, EPHE, IRD, Univ. Paul Valery Montpellier 3, Montpellier, France

OXFORD
UNIVERSITY PRESS

OXFORD
UNIVERSITY PRESS

Great Clarendon Street, Oxford, OX2 6DP,
United Kingdom

Oxford University Press is a department of the University of Oxford.
It furthers the University's objective of excellence in research, scholarship,
and education by publishing worldwide. Oxford is a registered trade mark of
Oxford University Press in the UK and in certain other countries

© Oxford University Press 2020

The moral rights of the authors have been asserted

First Edition published in 2020

Published in the United States of America by Oxford University Press
198 Madison Avenue, New York, NY 10016, United States of America

British Library Cataloguing in Publication Data
Data available

Library of Congress Control Number: 2019947644

ISBN 978–0–19–883684–1 (hbk.)
ISBN 978–0–19–883685–8 (pbk.)

DOI: 10.1093/oso/9780198836841.001.0001

Printed and bound by
CPI Group (UK) Ltd, Croydon, CR0 4YY

Links to third party websites are provided by Oxford in good faith and
for information only. Oxford disclaims any responsibility for the materials
contained in any third party website referenced in this work.

To our urban children - Bruno, Helena, Kiran, Zulekha, Vincent and Laetitia.

...and to all urban dwellers rediscovering the urban realm after COVID-19

All illustrative drawings (above and at the start of each chapter) by Allison Fitzmorris

Foreword

Evolutionary biology is faced with perhaps one of the greatest challenges since Darwin's formulation of the theory of evolution: the emergence of cities as novel agents of evolutionary change.

In his historical voyage to the Galapagos Islands, Charles Darwin was astonished by the diversity of life he found in this unique place on Earth; but more than the diversity, what most surprised Darwin was that many of the birds, which varied from island to island, were indeed originally from one single species. Each bird population had adapted to its island's unique environmental niche, and, over time, the populations varied significantly from each other. Darwin was witnessing evolution by natural selection and the origins of new species.

Almost two centuries later, today we are witnessing the emergence of a new potential source of natural selection caused by a major planetary transition. Rapid urbanization is reshaping fundamental ecological and evolutionary processes, both locally and globally, with significant consequences for Earth's biodiversity. Today's laboratory for evolutionary studies is not on a far-off island somewhere in the Pacific Ocean but rather it is much closer to home. In fact, this laboratory *is* home for many of us. Studies have shown that cities are driving evolutionary changes in many species. At the time Darwin visited the Galapagos (1835), there were just over 1.1 billion people on the planet and only about 5.4 per cent (about 60 million) were living in urban settlements (United Nations 1976). Today, of the 7.5 billion people on the planet, more than half of them, 4.2 billion, live in cities; and it is expected that by 2050 nearly 70 per cent of the population will (United Nations 2019). Urbanization is drastically changing habitat and biotic interactions. It is also driving systematic socioeconomic transi-

tions toward an increased pace of life, expanding the impact of cities on natural resources beyond their boundaries.

During the last decade, a number of studies have provided evidence of urban-driven evolutionary changes and divergent evolution in species of birds, fish, plants, mammals, and invertebrates. Since evolutionary biologists began to pay increasing attention to the urban environment, a new frontier has opened, prompting new scientific questions and challenges. One of the biggest challenges is to understand the mechanisms by which human settlements affect all evolutionary processes, including natural selection, mutation, gene flow, and genetic drift.

Urban ecosystems are particularly complex because a variety of new selective pressures alter the structure and function of communities and ecosystems in unexpected ways. Despite the dominant rhetoric of humans' special place in nature, we are not the only species that modifies its environment and the environments of other organisms, thereby altering the selection pressures to which they and other organisms must adapt. It is well known that microbes, earthworms, beavers, and many other organisms have changed and continue to alter natural processes that maintain life on Earth. Perhaps what makes human agency unique is our *extraordinary capacity* to transform the Earth through rapid advancement in technology. However, like other organisms, we engage in this transformation well beyond our understanding of the power we have. As David Grinspoon states in his book *Earth in Human Hands*, 'We're at the controls on planet Earth, but we're not in control' (Grinspoon 2017, p. ix).

Cities have fascinated me since an early age. Perhaps one of my most inspiring encounters with cities and city building was my first visit as a child to Rome's ancient aqueduct, Aqua Appia. Built in 312 BC, it was the first of eleven Roman aqueducts—sophisticated systems of underground tunnels and built structures. By the third century CE, the aqueducts totaled 315 miles in length and were used to convey water to the rapidly expanding city and to ensure its drinking quality to a million inhabitants. It is then when I first said aloud: 'When I grow up, I want to be an urban planner.' It is perhaps my Italian homeland, combined with many opportunities to visit beautiful cities across the world, that has inspired (and continues to inspire) my endless curiosity for cities, which I consider the greatest expression of human civilization.

Many natural scientists attribute the cause of the current environmental crisis to urbanization and consider cities a major threat to the planet. In most natural science and biology books, cities are still portrayed as the end of nature. But the increasing attention that (primarily young) natural scholars are placing on the human built environment is starting to reveal a new picture. The city is a distinct biome created by human action. Urbanization may represent a turning point in the evolution of our planet and the interaction between humans and the natural environment.

What captured my attention visiting the Roman aqueduct became much clearer later in my career as an urban ecologist. I saw a distinct parallel between humans as ecosystem engineers, who modify their environment to optimize environmental conditions for their survival (niche construction), and the action of other organisms, who do the same. I wanted to know if a new understanding of humans as ecosystem engineers could inform a new convergence of urban biology/ecosystem science with city design and planning practices. Just as all biological systems, at various scales of biological organization, develop the close integration of numerous subunits to function (West et al. 1997), humans build complex infrastructure to support human functions and complex networks that supply resources, process waste, and regulate system functioning (West and Brown 2005). Along the same lines of West and his colleagues' search for similarities between scaling laws in biology and human systems, I was asking: Are there universal laws governing the ways that human societies self-organize and construct their niche similar to those that govern other biological systems? And

how can we build cities that facilitate a new cooperation between humans and other organisms?

When I set out to study urbanism at the University of Architecture in Venice (itself a stunning example of the complex relationships between the built and natural environments), I was already dreaming about city planning that was grounded in the principles of ecology and natural systems. It was quite clear to me that I had to study both urban planning and biology. I was lucky to spend the next few years pursuing my doctoral and post-doctoral studies in urban studies at Massachusetts Institute of Technology and in biology at Stanford University. But I soon realized that not much was known about city planning grounded in biology, and not much was even known about *how* to achieve such understanding. My experience at two of the best schools in their respective fields was that urban planners and biologists were not even talking to each other, let alone speaking a common language.

The evolution of cities as part of nature is not new. It dates back at least to 1915 when Sir Patrick Geddes of Scotland published his book *Cities in Evolution* (Geddes 1915). Geddes' focus was not evolution. Rather he was inspired by evolutionary theory and applied evolutionary concepts to the city. Anne Spirn (1985) traces the roots of ecological urbanism all the way back to the writings of Hippocrates (ca. fifth century BCE), Vitruvius (ca. first century BCE), and Leon Battista Alberti (1485). During the last century, the idea took form and evolved in areas of study in various disciplines including sociology (Park et al. 1925), geography (Berry 1964; Zimmerer 1994), ecology (Odum 1953; Wolman 1971; Sukopp 1990; McDonnell et al. 1993), anthropology (Vayda and Rappaport 1968), history (Cronon 1991), and urban design and planning (Lynch 1961; McHarg 1969; Spirn 1984). But it is only recently that urban ecology has emerged as a field of study of the science of cities (Grimm et al. 2008; McPhearson et al. 2016).

During the last three decades, empirical research in Europe (Sukopp 2002) and the establishment of urban sites in the Long-Term Ecological Research Network, ecological studies funded by the US National Science Foundation (McDonnell 2011; Forman 2014), were instrumental in setting the groundwork for building the bridge between ecology and urban planning. These teams have articulated the unique interactions between patterns and processes and their human and ecological functions, and they have developed new conceptual frameworks and approaches to study cities as ecosystems. Lessons learned by the urban ecology community will be invaluable to urban evolutionary biologists.

Despite the significant progress made in the past three decades in understanding urban ecosystem dynamics and the functional implications of urbanization, the roles of evolutionary change in urban community dynamics have received little attention. Understanding the complex coupled interactions between human activity and eco-evolutionary processes requires new conceptual frameworks, theories, models, and methodologies that transcend disciplinary approaches. This book is a remarkable effort to bring to the attention of both evolutionary biologists and ecologists the implications that a rapid urbanizing world have for the evolution of species traits and biodiversity. The editors lay out clearly both the challenges and the opportunities of studying evolution in urban environments. Among the foremost challenges faced by scientists studying cities is to define a common framework and set of metrics to quantify urbanization. But to do that, we will need to answer a fundamental question: What is urban? Ask any scholar of urban systems across different disciplines and you will get a different answer. The definitions even vary within each field. An additional challenge altogether is *measuring* urbanization. At the dawn of such a promising discipline, Marta Szulkin and colleagues offer a productive strategy to tackle the characterization of urban systems through space and time, providing a tool that many scholars of urban evolution can test in their own cities and study systems.

Another challenge of studying the evolutionary consequences of urbanization is the marked diversity of cities across the world. Despite the many commonalities that characterize urban environments, cities are highly diverse. This has been one of the central challenges underpinning the study of cities, and perhaps it is one of the key pillars for designing a successful strategy for interacting with other fields. It is encouraging to see that many authors here, who are tackling different questions, acknowledge that different types of urban environments may generate different selective processes across and within cities, and they recognize the importance of urban heterogeneity in understanding urban evolutionary dynamics.

The book's editors, Marta Szulkin, Jason Munshi-South, and Anne Charmantier, have put together an excellent selection of contributions that represent emerging scientific questions and relevant terrestrial and aquatic study systems. Each chapter provides a systematic account of some of the key themes starting to emerge. Scholars of urban evolution will learn through the numerous examples here how major selective pressure caused by urbanization affects organismal morphology, physiology, behaviour, and life-history traits in many plant and animal populations. Cities not only intensify anthropogenic pressures such as fragmentation of natural habitats and pollution, but also create new ones. Impermeable surfaces, including roads, pavements, and buildings, trap heat and increase temperature, driving changes in biochemical processes and altering biological systems, as Sarah Diamond and Ryan Martin illustrate. Organisms living in urban environments are simultaneously exposed to multiple pressures such as increased air pollution, artificial night light, noise and other disturbances, as Caroline Isaksson and Frances Bonier investigate. And some of the effects may be unexpected due to the complexity of the urban matrix and its impact on gene flow, as, for example, Jason Munshi-South and Jonathan Richardson show.

Urban evolution is more than a missing chapter in the study of evolution. Investigating evolution in urban environments provides unique opportunities to examine parallel evolution in natural populations on a large scale. As author James Santangelo and colleagues point out, urban environments represent globally replicated, large-scale disturbances to the landscape. Understanding the shared features of urban environments and exploring parallel species responses to rapid urban development will provide new insight in evolutionary biology.

Among the greatest contributions that urban evolution studies can provide to ecology is the opportunity to study evolving metacommunities—interconnected populations where evolutionary processes and community ecological processes interact in a landscape context, responding to anthropogenic disturbances. Despite the increasing number of studies showing genetic divergence of individual species to urban environments, Brans and colleagues highlight that an integrated approach that combines community data with data on genetic responses is still lacking. Another important contribution of urban evolution studies is understanding the effect of the urban context on mutualism, an aspect often overlooked when studying biotic interactions and here reviewed by Irwin and colleagues. Mutualisms are important in maintaining the health and biodiversity of ecosystems. And the ecological dynamics in urban landscapes may affect the stability of mutualisms among species with important consequences for the ecosystem dynamic. Furthermore, as several contributors point out, the recent rapid advances in genomic methods allows exploration of previously intractable questions, such as disentangling the contributions to adaptation of genetic change and phenotypic plasticity.

The book also should appeal to conservation biologists and others involved in the practice of managing and maintaining biodiversity. Understanding the implications of urban development on evolutionary dynamics will provide novel insights for urban conservation planning and for designing strategies that will help species adapt to urbanization, improve the forecasting of population declines, and help develop appropriate conservation and management plans.

This book comes at an exciting time as new collaborations among evolutionary biologists, ecosystem ecologists, and urban planners are converging in pursuit of the ambitious goal of establishing a shared vision for the emerging field of urban evolutionary biology.

Marina Alberti
University of Washington

References

Berry, B.J. (1964). Cities as systems within systems of cities. *Papers in Regional Science*, 13(1), 146–63.

Cronon, W. (1991). The significance of significance in American history. In: Etulain, R.W. (ed.) *Writing Western History: Essays on Major Western Historians*, p. 73. University of New Mexico Press, Albuquerque.

Forman, R.T. (2014). *Urban Ecology: Science of Cities*. Cambridge University Press, Cambridge.

Geddes, P. (1915). *Cities in Evolution*. Williams & Norgate, London.

Grimm, N.B., Faeth, S.H., Golubiewski, N.E., et al. (2008). Global change and the ecology of cities. *Science*, 319(5864), 756–60.

Grinspoon, D. (2017). *Earth in Human Hands: Shaping our Planet's Future*. Grand Central Publishing, New York.

Kemp, M.C. (1969). *Pure Theory of International Trade and Investment*. Prentice Hall, Upper Saddle River, NJ.

Lynch, K. (1961). The pattern of the metropolis. *Daedalus*, 90(1), 79–98.

McDonnell, M.J. (2011). The history of urban ecology. *Urban Ecology*, 9, 5–17.

McDonnell, M.J., Pickett, S.T., and Pouyat, R.V. (1993). The application of the ecological gradient paradigm to the study of urban effects. In: McDonnell, M.J. and Pickett, T.A. (eds) *Humans as Components of Ecosystems*, pp. 175–89. Springer, New York.

McHarg, I.L. (1969). *Design with Nature*. University of Pennsylvania, New York.

McPhearson, T., Pickett, S.T., Grimm, N.B., et al. (2016). Advancing urban ecology toward a science of cities. *BioScience*, 66(3), 198–212.

Odum, P.E. (1953). *Fundamentals of Ecology*. W.B. Saunders, Philadelphia.

Park, R.E., Burgess, E.W., and McKenzie, R.D. (1925). The growth of the city: an introduction to a research project. In: Park, R.E. and Burgess, E.W. (eds) *The City*, pp. 85–97. University of Chicago Press, Chicago.

Spirn, A.W. (1984). *The Granite Garden: Urban Nature and Human Design*. Basic Books: New York.

Spirn, A.W. (1985). Urban nature and human design: renewing the great tradition. *Journal of Planning Education and Research*, 5(1), 39–51.

Sukopp, H. (1990). Urban ecology and its application in Europe. *Urban Ecology: Plants and Plant Communities in Urban Environments*, 1(2).

Sukopp, H. (2002). On the early history of urban ecology in Europe. *Preslia*, 74, 373–93.

United Nations (1976). Orders of Magnitude of the World's Urban Population in History. E/CN.9/XIX/CRP. United Nations, New York.

United Nations (2019). World Urbanization Prospects. The 2018 Revision. United Nations, New York.

Vayda, A.P. and Rappaport, R.A. (1968). *Ecology, Cultural and Noncultural*, pp. 476–98. California Indian Library Collections, Berkeley.

West, G.B. and Brown, J.H. (2005). The origin of allometric scaling laws in biology from genomes to ecosystems. *Journal of Experimental Biology*, 208, 1575–92.

West, G.B., Brown, J.H., and Enquist, B.J. (1997). A general model for the origin of allometric scaling laws in biology. *Science*, 276(5309), 122–6.

Wolman, M.G. (1971). The nation's rivers. *Science*, 174(4012), 905–18.

Zimmerer, K.S. (1994). Human geography and the 'new ecology': the prospect and promise of integration. *Annals of the Association of American Geographers*, 84(1), 108–25.

Contents

List of Contributors   xiii

1 Introduction   1
Marta Szulkin, Jason Munshi-South, and Anne Charmantier

2 How to Quantify Urbanization When Testing for Urban Evolution?   13
Marta Szulkin, Colin J. Garroway, Michela Corsini, Andrzej Z. Kotarba, and Davide Dominoni

3 Urban Environments as a Framework to Study Parallel Evolution   36
James S. Santangelo, Lindsay S. Miles, Sophie T. Breitbart, David Murray-Stoker, L. Ruth Rivkin,
Marc T.J. Johnson, and Rob W. Ness

**4 Landscape Genetic Approaches to Understanding Movement and
Gene Flow in Cities**   54
Jason Munshi-South and Jonathan L. Richardson

5 Adaptation Genomics in Urban Environments   74
Charles Perrier, Aude Caizergues, and Anne Charmantier

6 Evolutionary Consequences of the Urban Heat Island   91
Sarah E. Diamond and Ryan A. Martin

7 The Evolutionary Ecology of Mutualisms in Urban Landscapes   111
Rebecca E. Irwin, Elsa Youngsteadt, Paige S. Warren, and Judith L. Bronstein

8 Sidewalk Plants as a Model for Studying Adaptation to Urban Environments   130
Pierre Olivier Cheptou and Susan C. Lambrecht

9 Adaptive Evolution of Plant Life History in Urban Environments   142
Amanda J. Gorton, Liana T. Burghardt, and Peter Tiffin

10 Urbanization and Evolution in Aquatic Environments   157
R. Brian Langerhans and Elizabeth M.A. Kern

11 Evolutionary Dynamics of Metacommunities in Urbanized Landscapes   175
Kristien I. Brans, Lynn Govaert, and Luc De Meester

12 Terrestrial Locomotor Evolution in Urban Environments   197
Kristin M. Winchell, Andrew C. Battles, and Talia Y. Moore

13 Urban Evolutionary Physiology   217
Caroline Isaksson and Frances Bonier

14 Urban Sexual Selection   234
Tuul Sepp, Kevin J. McGraw, and Mathieu Giraudeau

15 Cognition and Adaptation to Urban Environments   253
Daniel Sol, Oriol Lapiedra, and Simon Ducatez

16 Selection on Humans in Cities   268
Emmanuel Milot and Stephen C. Stearns

List of Glossary Terms Definition   289
Index   295

List of Contributors

Andrew C. Battles Department of Biology, University of Rhode Island, Kingston, RI, USA

Frances Bonier Department of Biology, Queen's University, Kingston, ON, Canada

Kristien I. Brans Laboratory of Aquatic Ecology, Evolution and Conservation, KU Leuven, Belgium

Sophie T. Breitbart Department of Biology, University of Toronto Mississauga; Department of Ecology and Evolutionary Biology, University of Toronto; and Centre for Urban Environments, University of Toronto Mississauga, ON, Canada

Judith L. Bronstein Department of Ecology and Evolutionary Biology, University of Arizona, Tucson, AZ, USA

Liana T. Burghardt Department of Plant and Microbial Biology, University of Minnesota, St. Paul, MN, USA

Aude Caizergues Centre d'Ecologie Fonctionnelle et Evolutive, Univ. Montpellier, CNRS, EPHE, IRD, Univ. Paul Valery Montpellier 3, Montpellier, France

Anne Charmantier Centre d'Ecologie Fonctionnelle et Evolutive, Univ. Montpellier, CNRS, EPHE, IRD, Univ. Paul Valery Montpellier 3, Montpellier, France

Pierre Olivier Cheptou Centre d'Ecologie Fonctionnelle et Evolutive, Univ. Montpellier, CNRS, EPHE, IRD, Univ. Paul Valery Montpellier 3, Montpellier, France

Michela Corsini Centre of New Technologies, University of Warsaw, Poland

Luc De Meester Laboratory of Aquatic Ecology, Evolution and Conservation, KU Leuven, Belgium

Sarah E. Diamond Department of Biology, Case Western Reserve University, Cleveland, OH, USA

Davide Dominoni Animal Health and Comparative Medicine, Institute of Biodiversity, University of Glasgow, UK; and Department of Animal Ecology, Netherlands Institute of Ecology, Wageningen, The Netherlands

Simon Ducatez CREAF-CSIC (Centre for Ecological Research and Applied Forestries), Cerdanyola del Vallès Catalonia, Spain; and Department of Biology, McGill University, Montréal, QC, Canada

Colin J. Garroway Department of Biological Sciences, University of Manitoba, Winnipeg, MB, Canada

Amanda J. Gorton Graduate Program in Ecology, Evolution, and Behavior, University of Minnesota, St. Paul, MN, USA

Mathieu Giraudeau Centre de Recherche en Écologie et Évolution de la Santé (CREES), CREEC/MIVEGEC UMR5290 (CNRS – IRD – Université de Montpellier)

Lynn Govaert Department of Evolutionary Biology and Environmental Studies, University of Zürich, Switzerland; and Department of Aquatic Ecology, Eawag: Swiss Federal Institute of Aquatic Science and Technology, Dübendorf, Switzerland

Rebecca E. Irwin Department of Applied Ecology, North Carolina State University, Raleigh, NC, USA

Caroline Isaksson Department of Biology, Lund University, Sweden

Marc T.J. Johnson Department of Biology, University of Toronto Mississauga; Department of Ecology and Evolutionary Biology, University of Toronto; and Centre for Urban Environments, University of Toronto Mississauga, ON, Canada

Elizabeth M.A. Kern Department of Life Science, Ewha Womans University, Seoul, South Korea

Andrzej Z. Kotarba Centrum Badań Kosmicznych Polskiej Akademii Nauk (CBK PAN), Warsaw, Poland

Susan C. Lambrecht Department of Biological Sciences, San Jose University, CA, USA

R. Brian Langerhans Department of Biological Sciences and W.M. Keck Center for Behavioral Biology, North Carolina State University, Raleigh, NC, USA

Oriol Lapiedra CREAF-CSIC (Centre for Ecological Research and Applied Forestries), Cerdanyola del Vallès, Catalonia, Spain

Ryan A. Martin Department of Biology, Case Western Reserve University, Cleveland, OH, USA

Kevin J. McGraw School of Life Sciences, Arizona State University, Tempe, AZ, USA

Lindsay S. Miles Department of Biology, University of Toronto Mississauga; and Centre for Urban Environments, University of Toronto Mississauga, ON, Canada

Emmanuel Milot Department of Chemistry, Biochemistry and Physics, Université du Québec à Trois-Rivières, QC, Canada

Talia Y. Moore Robotics Institute, Department of Ecology and Evolutionary Biology, Museum of Zoology, University of Michigan, Ann Arbor, MI, USA

Jason Munshi-South Louis Calder Center-Biological Field Station, Fordham University, Armonk, NY, USA

David Murray-Stoker Department of Biology, University of Toronto Mississauga; Department of Ecology and Evolutionary Biology, University of Toronto; and Centre for Urban Environments, University of Toronto Mississauga, ON, Canada

Rob W. Ness Department of Biology, University of Toronto Mississauga; Department of Ecology and Evolutionary Biology, University of Toronto; and Centre for Urban Environments, University of Toronto Mississauga, ON, Canada

Charles Perrier Centre de Biologie pour la Gestion des Populations. UMR CBGP, INRAE, CIRAD, IRD, Montpellier SupAgro, Univ Montpellier, Montpellier, France

Jonathan L. Richardson Department of Biology, University of Richmond, VA, USA

L. Ruth Rivkin Department of Biology, University of Toronto Mississauga; Department of Ecology and Evolutionary Biology, University of Toronto; and Centre for Urban Environments, University of Toronto Mississauga, ON, Canada

James S. Santangelo Department of Biology, University of Toronto Mississauga; Department of Ecology and Evolutionary Biology, University of Toronto; and Centre for Urban Environments, University of Toronto Mississauga, ON, Canada

Tuul Sepp Department of Zoology, University of Tartu, Estonia

Daniel Sol CREAF-CSIC (Centre for Ecological Research and Applied Forestries), Cerdanyola del Vallès, Catalonia, Spain

Stephen C. Stearns Department of Ecology and Evolutionary Biology, Yale University, New Haven, CT, USA

Marta Szulkin Centre of New Technologies, University of Warsaw, Poland

Peter Tiffin Department of Plant and Microbial Biology, University of Minnesota, St. Paul, MN, USA

Paige S. Warren Department of Environmental Conservation, University of Massachusetts, Amherst, MA, USA

Kristin M. Winchell Department of Biology, Washington University, St. Louis, MO, USA

Elsa Youngsteadt Department of Applied Ecology, North Carolina State University, Raleigh, NC, USA

CHAPTER 1

Introduction

Marta Szulkin, Jason Munshi-South, and Anne Charmantier

Szulkin, M., Munshi-South, J. and Charmantier, A., *Introduction* In: *Urban Evolutionary Biology*. Edited by Marta Szulkin, Jason Munshi-South and Anne Charmantier, Oxford University Press (2020). © Oxford University Press.
DOI: 10.1093/oso/9780198836841.003.0001

1.1 Urban evolutionary biology

'*A bustling city*'—a catchphrase often used to describe cities and towns, one that refers to the busyness of people and to the crowd we create by living *en masse* in realms of concrete and glass. But as we slow down for a minute, we start noticing a myriad of life pulsing throughout the urban space—in air and water, under the ground, and in remnants of vegetation. There is still some way to go before dusty, sterile grounds representing human settlements in post-apocalyptic movies such as *Blade Runner 2049* become the norm. Our emerging biological knowledge of the urban space, if applied wisely, may allow us to steer clear from such extreme forms of urbanization. As urban areas grow exponentially worldwide, scientists, government authorities, and urban planners are becoming increasingly aware that biological life needs to be preserved in cities. To provide an in-depth understanding of the dynamics of biological variation and survival in the urban space, the following questions need to be addressed: How do we quantify the urban environment so that it allows for pertinent evolutionary testing? To what extent are cities sinks or sources of life? What are the sources of urban adaptation? And crucially—given the extreme environmental alterations induced by cities—can life evolve in urban environments, and do cities trigger parallel evolution worldwide?

Urban evolutionary biology is the study of evolutionary change in populations as a response to human-built environments and all that high densities of human population entail. Put succinctly, *urban evolution* is the study of how features of cities drive allele, genotype, and phenotype frequency changes in populations, across generations. Because urbanization has increased rapidly in the last century, urban evolution largely focuses on microevolutionary change, including the processes of mutation, selection, gene flow, and genetic drift. At the same time, building an accurate picture of urban evolution also requires understanding the respective roles of phenotypic plasticity and epigenetic processes in producing evolutionary change in cities, all of which are discussed in multiple contexts throughout this volume. The pace of evolutionary change in this context can span a few to many hundreds of generations. Because it is the evolu-

tionary response to environmental change that is of interest, urban evolutionary biology relies on urban ecology knowledge—we want to understand how organismal interactions with biotic and abiotic urban environments promote and constrain evolutionary processes. Urban ecology is now a mature field of biological research (Niemelä et al. 2011; Douglas and James 2014), and we thus have a well-established body of work describing the changes in ecological interactions that occur with urbanization. Urban evolutionary biologists can thus use this existing ecological knowledge as a context for the testing of evolutionary hypotheses in an urban environment.

With its conspicuously altered ecological dynamics, the urban environment stands in stark contrast to the natural environment as the latter was used as research ground for virtually all long-term studies of vertebrates investigated in the wild and cornerstone in evolutionary ecology research (Clutton-Brock and Sheldon 2010). There is thus a considerable gap in our understanding of the dynamics of organismal trait variation, selection, and response to selection in an urban setting. In other words, a greater understanding of evolutionary biology in urban environments is needed. One of the first commentaries explicitly recognizing the great potential of cities in the study of evolution was written by Jared Diamond (1986), who encouraged evolutionary biologists to use cities as fascinating opportunities to study natural selection in action. Fast forward 30 years and this idea has indeed started to permeate evolutionary biology: we are now witnessing a rapidly growing body of evidence quantifying the effect of urbanization on phenotypes and genotypes worldwide. In a comprehensive review published by Johnson and Munshi-South (2017), 192 studies on 134 species document evolutionary responses to urbanization, with over 50 per cent of these published in the past 5 years. In this volume, 155 species from across the tree of life contribute to our understanding of urban evolutionary biology (Figure 1.1). Recent attention to urban evolutionary biology is also fuelled by the recognition that anthropogenic activities are often associated with the fastest rates of evolutionary change: in a meta-analysis of ca. 1600 phenotypic changes recorded worldwide, Alberti et al. (2017) report that rates of phenotypic change are greater

Figure 1.1 Representation of distinct species discussed in this volume, pooled into taxonomic classes from the tree of life. Only species whose exact Latin names were given are included. Each species was counted only once per chapter, and more than once if it was mentioned in more than one chapter. Species were pooled into taxonomic classes represented here in the following order: birds, plants, insects, mammals, reptiles, fish, amphibians, branchiopods, gastropods. Taxonomic classes representing ≤ 5 per cent of all species representatives (arachnids, bacteria, malacostraca) are not included in the figure. Height of each taxonomic class (relative to the bird class) relates to the number of times species representatives of this class are cited in the book (based on the above described criteria). Total number of distinct species listed in the volume: N = 155. Cumulated number of species cited across chapters: N = 233. (Photo credit: Copyright-free PhyloPic silhouettes.)

in urbanizing systems compared with natural and non-urban anthropogenic systems. The growing rate of urbanization and the maturation of urban study systems worldwide leave little doubt that interest in the urban environment as an agent of evolutionary change will grow considerably in the years to come.

It is thus clear that we are currently witnessing the start of a nascent, yet fast-growing field of research in urban evolution. This development is much needed: despite the fact that cities occupy ca. 3 per cent of the total surface of the Earth's habitable land area (Center for International Earth Science Information Network - Columbia University et al. 2011)—a value that will inexorably grow in the next decades—the urban environment has been a largely neglected study site among evolutionary biologists. But we can benefit from considerable urban ecology knowledge, a much more mature field of biological research, and use it as a valuable experimental context. Urbanization can thus offer a great range of opportunities to test for rapid evolutionary processes as a consequence of human activity, both because of replicate contexts for hypothesis testing and because cities are characterized by an array of easily quantifiable environmental axes of variation (Chapter 2) and thus testable agents of selection. Thanks to a possible wide breadth of inference in terms of taxa that can be studied (from bacteria to wild vertebrates) and analytical methods (from gene flow to traits, from signatures of selection to patterns of gene expression, from correlative

to experimental approaches), urban evolution has the potential to stand at a fascinating multidisciplinary crossroad, enriching the field of evolutionary biology with emergent yet incredibly potent new research themes where the urban habitat is key.

There is no doubt that urban evolutionary biology can—and should—have a societal impact, and has a role in strengthening the drive towards more sustainable cities. These themes are discussed in the following sections. We further present an overview of chapters reporting core evidence of up-to-date urban evolutionary biology research. We continue with a brief discussion of challenges faced by the field, as well as an outline of some inspiring future directions awaiting this nascent area of research.

1.2 Societal impact of urban evolutionary biology

1.2.1 Education and outreach

Dobzhansky's thought that 'nothing in biology makes sense, except in the light of evolution' (Dobzhansky 1973) is pervasive in biology, yet it is far from obvious to the general public (Miller et al. 2006). We believe that studying evolutionary biology in a city context is a unique opportunity to disseminate evolutionary knowledge to a large audience. Educating people about evolution is of fundamental importance for several reasons that we outline here (see also https://evokeproject.org:

Evolutionary Knowledge for Everyone): first, because it is fundamental for humans to recognize how we came to be. Knowledge of our evolutionary history helps us understand the roots of some of our habits, our characteristics, and our limitations, as well as to position ourselves in the biodiversity network. Second, evolution is essential knowledge in our everyday lives and decisions, even more so in a world where *CRISPR/Cas9* opens opportunities for humans to direct evolution (Chapter 16). It is necessary to grasp the fundamentals of evolution to understand developments in medicine (Stearns and Medzhitov 2015), but also in agriculture and of course in the conservation and management of the environment. Third, urban residents who are educated about evolutionary biology will be more likely to take part in citizen science (also called 'community science') projects on urban evolution. City dwellers thus represent an unprecedented task force to combine data collection and science education. This reasoning is what brought the Evolution MegaLab initiative to set up the first European-scale citizen science case study. While introducing participants to Darwinian evolution, the Evolution MegaLab website coordinated a volunteer survey of shell polymorphism in the banded snails *Cepaea nemoralis* and *Cepaea hortensis* (Worthington et al. 2012). In 2009, the 6461 volunteers registered on the website collected half a million snails in 2990 populations. This was compared to a historical dataset of 6515 sampled populations to test for evolutionary changes in shell albedo in the context of climate change (Silvertown et al. 2011). A telling difference between the historical scientific survey and the volunteer survey was that the latter relied much more heavily on records in urban environments. In this case, it is thus the broader community that opened the eyes of the scientists regarding the opportunity of studying evolution in cities, while at the same time becoming more educated about evolutionary ecology.

1.2.2 Sustainable cities

Interest in building sustainable and resilient cities has increased dramatically in recent years as urban dwellers have faced serious challenges from severe weather, disease, aging infrastructure, social unrest, and myriad other pressures. These challenges, combined with recent developments in new information gathering technologies, have led to the concept of 'smart cities', whereby new technologies are used to manage cities and resources in a more sustainable manner (see de Jong et al. 2015; Marvasti and Rees (accessed 5 January 2020) for definitions of terms surrounding the sustainable city literature). However, few of these efforts have sought to incorporate evolutionary principles into urban planning. Cities are habitats constructed by and for humans, and thus the needs of non-human populations will typically be lower priority. The time lag necessary to observe evolutionary change even when pressures are strong and responses are rapid also makes it difficult to account for evolution over typical political timescales. Perhaps most importantly, evolution of a single species occurs in a broader ecological context, and thus evolutionary responses may be difficult to predict or control in ways that will be useful to improving city life for humans.

Take for example the case of landscape connectivity for native and invasive rodent species in New York City, USA, described in Chapter 4 of this book. A commensal pest such as the brown rat (*Rattus norvegicus*) is largely deemed undesirable, and research into their connectivity across urban landscapes could be quite useful for their control. Identification of the landscape features that facilitate gene flow of rats could be incorporated into municipal pest control efforts to create cities that are increasingly hostile to their movement and survival. However, native species such as the white-footed mouse (*Peromyscus leucopus*) present much more complex issues than pests. Evolutionary analyses have identified the types and spatial extent of urban land cover needed for this species to move through the urban landscape (Chapter 4; Munshi-South 2012; Munshi-South et al. 2016), but it is not entirely clear persistence and gene flow of this rodent should be an urban environmental goal. White-footed mice are an important component of forest ecosystems across eastern North America, and their presence would seem desirable if we wish to promote healthy forest ecosystems within cities in this region. However, the white-footed mouse is also implicated in infection cycles of several tick-borne pathogens, particularly the bacteria that cause Lyme disease, and thus in some sense their presence in cities is as undesirable as the rat. There will thus be unavoidable tradeoffs when making decisions about how to influence the environment and evolutionary processes in cities.

Efforts to incorporate evolutionary processes into urban planning will need to carefully weigh the impacts on humans, wildlife, and habitats. Leveraging evolutionary data to make decisions about urban sustainability will also lead to value judgements about different evolutionary outcomes. For example, is it generally good, bad, or neutral that some species adapt to urban conditions? In general, one might be tempted to indicate that it is good for species that would otherwise be excluded from urban areas (i.e., resistance to pollution, higher temperatures, or sidewalk habitats), or at least represents a neutral evolutionary outcome. However, some selection pressures, such as those that may cause species to adapt to persisting on human waste, may be seen as less desirable. While some trends in urban evolutionary outcomes are apparent from the work presented in this book, there will undoubtedly be surprises in the future that will prompt scientists, planners, and other stakeholders to reconsider principles upon which their decision-making is made. For example, to what extent should humans promote 'good' adaptations in other organisms and prevent or counteract 'bad' adaptations?

One area of urban sustainability that most stakeholders can agree on may be the value of native habitats and species in urban areas. The general well-being of people living in cities is well known to be improved by access to green spaces and biodiversity (e.g., Jorgensen and Gobster 2010), so urban planning that incorporates conservation of wildlife habitats (in many cases, novel habitats that support unique combinations of native and non-native species) should be part of every city's planning efforts. Although only a small percentage of global species occur in cities (Aronson et al. 2014), urban habitats may act as refugia for some of them. Declining species may find new life in cities where diseases, competitors, and predators are less prevalent, and in some cases may exhibit local adaptation to these conditions (Hall et al. 2017). That said, conservation biology in urban areas has historically largely ignored evolutionary processes. When optimizing the space preserved for natural habitats in cities, conservation strategies focus largely on economic and sociocultural values of biodiversity (e.g.,

Mace 2014; Whitehead et al. 2014). However, the idea of evolutionary-focused conservation efforts (or 'evocentric conservation'; see Sarrazin and Lecomte 2016) has recently emerged. This evocentric approach to solve the acute environmental crises faced today would not only improve the strategies to preserve biodiversity by taking into account species evolutionary potential and evolutionary trajectories, but also redefine human responsibility and identity (Sarrazin and Lecomte 2016). We believe that the city is an ideal place to stimulate a dialogue between social scientists, ecologists, and evolutionary biologists and face the challenge of combining knowledge from these fields to find a new and sustainable equilibrium between humans and non-human life.

1.3 Overview of chapters

In this book, we reveal how evolutionary biology applied to urban ecosystems has many promising avenues both to increase our evolutionary knowledge and understanding, as well as to contribute to preserving wild populations living in cities. Ultimately, such initiatives would allow us (1) to confirm whether a fascinating, yet largely ignored evolutionary process is occurring literally on our doorstep, across the globe, and in each city independently, but also (2) improve our understanding of how evolutionary biology can contribute to strengthening the resilience of wildlife in urban landscapes.

Urban Evolutionary Biology consists of sixteen chapters, and draws on biological data from a large array of representatives from the tree of life (Figure 1.1). A glossary of important terms from across chapters is located at the end of the book. While the majority of chapters feature an exhaustive and comprehensive quantitative review that summarizes distinct aspects of urban evolutionary biology, others focus on development in theory or methods, or are centred on empirical examples. *Chapter 2* highlights the fact that cities are a mosaic of contrasting environments. It outlines study frameworks, methodology, and study design suggestions to maximize robust and informative evolutionary hypothesis testing in an urban landscape. *Chapter 3* examines the fantastic experimental attribute of cities in that they are globally replicated, large-scale disturbances to the landscape. The extent to which species display parallel evolutionary responses across independent urban environments is discussed. *Chapter 4* shows that urban heterogeneity can lead to complex movement patterns in wildlife, which can be inferred with population genetic analyses. These can unravel spatial patterns of genetic variation and signatures of adaptive evolution across the genome with increasing resolution. *Chapter 5* ties in overarching scientific fields—quantitative genetics, population genetics, and the recently emerging epigenetic frameworks—with natural population monitoring and field experiments to further our understanding of urban-driven adaptation genomics. *Chapter 6* discusses contemporary evolutionary change associated with urban heat islands and the relative roles of phenotypic plasticity vs evolution of thermally sensitive traits. *Chapter 7* reviews evidence for urbanization affecting interactions such as transportation, protection, and nutritional mutualisms, and asks whether cities can alter the evolutionary outcomes of mutualisms. *Chapter 8* highlights the fact that plants in cities are often relegated to small, isolated patches of soil in a concrete matrix. It uses the case study of an annual weed (*Crepis sancta*) to show how urbanization has led to contemporary local adaptation in a large suite of plant traits. *Chapter 9* reviews current evidence to lay a foundation for understanding the adaptation of plant life histories to urban environments and points to directions for further research. *Chapter 10* looks at urban evolution in aquatic environments, and examines the evolutionary consequences of major types of urban-induced changes to ecosystems: biotic interactions, habitat connectivity and hydrology, temperature, and pollution. *Chapter 11* calls for taking an integrated approach combining community data with data on genetic responses of focal taxa to understand the evolution of metacommunities in urban environments, which is also of direct relevance to our comprehension of eco-evolutionary feedbacks and urban ecosystem functions and services. *Chapter 12* explores the possible adaptive responses of arboreal species to the spatial structure and properties of climbing substrates relevant to terrestrial and climbing locomotion in urban environments. *Chapter 13* highlights how three central physiological components of coping responses

to environmental challenges—detoxification, endocrine and metabolic systems—are the likely targets of selection imposed by the challenges confronting urban populations. *Chapter 14* reviews evidence for urban-driven changes in sexual selection pressures and sexually selected traits, with particular attention to the potential rapid adaptive and plastic shifts in traits of signallers and receivers. *Chapter 15* discusses how cognition may play a major role in facilitating evolutionary adaptation of animals to the urban environment by (1) allowing individuals to choose the habitats and resources that better match their phenotypes and (2) helping animals to construct learned responses to challenges they have never or rarely experienced. Finally, *Chapter 16* looks at the effects of urbanization on the species that is itself the cause of this radical habitat modification—*Homo sapiens*. Elements needed to understand the evolutionary potential of humans living in cities, particularly focusing on traits affecting health, are presented.

1.4 Challenges and emerging topics

Each chapter in this book reports on important findings on rapid evolution in city environments. At the same time, all chapters also agree with the fact that research on evolutionary processes in urban habitats is still nascent. Many of the chapters therefore have a great value in suggesting testable hypotheses or most promising experimental, empirical, or theoretical approaches that should be explored in the future. As well as a review of what has been done, this book is very much about the exciting future of a nascent field. It also discusses challenges this field of research needs to address to enable informative and prolific development. In this section, we cover only a small range of the emerging topics that will keep urban evolutionary biologists busy in the future, listing in particular some challenges that are not covered in the following chapters. For some of these, our insights are kept short because they are developed further in specific chapters.

1.4.1 Challenges

Some attributes of twenty-first-century biological data collection can hinder endeavours for efficient urban evolution inference, and we urge the community of evolutionary biologists as well as their funding bodies to take the following two points under consideration:

(1) **Data replication**: The recommendation of replicated experimental design for urban evolutionary biology inference involves running data collection on multiple cities—this is logistically challenging, particularly when phenotypic, genetic, and fitness data are collected. While such inference is a gold standard for evolutionary ecology research (Barrett and Hoekstra 2011), it can be difficult to achieve in practice, particularly when multiple countries are involved. This situation is further exacerbated in the case of protected species where different legislations are involved regarding, for example, blood sampling, tagging, or ringing. 'Bottom-up' initiatives, involving the participation of citizens worldwide, can sometimes yield spectacular results (see the MegaLab initiative on banded snails; Worthington et al. 2012), although these will usually remain limited to instances where the species is not protected by law and/or to experimental setups where fitness data are not collected. Ideally, funding agencies supporting urban evolutionary endeavours should take these limitations into account and facilitate collaborative projects with efficient sharing of financial and administrative resources between collaborators.

(2) **Long-term inference**: The evolutionary consequences of urbanization occur not only in space, but also in time (see discussion in Chapter 2). This ideally requires the long-term monitoring of populations and experimental setups. Yet short-term funding sources are the challenging staple upon which many long-term studies of natural populations rely (Clutton-Brock and Sheldon 2010). Funding enabling the core functioning of long-term projects should thus be available to support these projects. While inspiring examples for such initiatives exist (see the Long-term Ecological Research networks financed by the National Science Foundation in the USA, providing stable platforms for the ecological sciences (Callahan 1984)), similar schemes should be promoted worldwide,

specifically monitoring long-term trends in the urban space, in contrasted climatic zones, and across different continents.

1.4.2 Are urban environments genetic sources or sinks?

Neutral genetic variation is profoundly influenced by drift and gene flow in urban environments. Loss of genetic variation within and increased differentiation between populations was the most common result identified in a recent review of 192 studies in urban environments (Johnson and Munshi-South 2017; Miles et al. 2019). Such patterns result from restricted gene flow between isolated populations occupying patchy habitats in cities (Van Rossum 2008; Gortat et al. 2015), or more subtle effects of urbanization on structural and functional connectivity (Beninde et al. 2016; Richardson et al. 2017; Perrier et al. 2018). New advances in spatial population genetics (Wang and Bradburd 2014) and the study of urban landscape connectivity (LaPoint et al. 2015) have greatly improved our ability to understand the influence of drift and gene flow in cities. Coalescent modelling approaches have also provided insights into the neutral demographic histories of urban populations (Harris et al. 2016; Lourenço et al. 2017). Moreover, the rapid development of genotype-by-environment analyses (Rellstab et al. 2015) has improved our ability to detect selection on genes that are relevant to fitness in urbanized environments (Harris and Munshi-South 2017), potentially even when gene flow is ongoing. Drift and gene flow in urban populations are closely linked to population demography and dispersal patterns, but these relationships are as yet poorly understood. If some urban populations act as sources of migrants (Björklund et al. 2010), then urban areas—or specific areas within cities—may be important components of regional metapopulations. But urban habitats may also function as sinks, ecological traps (Demeyrier et al. 2016), or evolutionary traps (Robertson and Chalfoun 2016) for some species, and in these cases one would predict reductions in genetic diversity and other negative effects of drift in urban populations. Finally, urban populations may be relatively isolated but large enough to avoid demographic or genetic decline, have the potential to adapt to local conditions, or experience evolutionary rescue or the spread of advantageous alleles from surrounding urban or non-urban populations (Bourne et al. 2014). The respective role of these three possible scenarios of evolutionary change in the urban space needs further research across taxa and replicated urban gradients.

1.4.3 What are the sources of urban adaptation?

Similar to other types of studies exploring phenotypic variation and responses to rapid environmental change, one of the main unresolved challenges for urban evolutionary ecology studies is to distinguish between plastic and microevolutionary mechanisms of trait divergence between city and natural habitats, as well as across urbanization gradients (Chapters 5 and 6; Diamond et al. 2017). In the face of the very wide range of studies describing urban-specific phenotypes, only a handful have so far provided evidence for one of the two adaptive mechanisms (Donihue and Lambert 2015), using experimental approaches (e.g., LaZerte et al. 2016) and in particular common garden experiments (Miranda et al. 2013), genomic studies (Mueller et al. 2013; see also Schell 2018), or replicated urbanization gradients (Johnson et al. 2018). In a recent case study based on common garden experiments on urban and rural acorn ants (*Temnothorax* sp.), shifts in thermal tolerance under urbanization were attributed to both evolutionary change and phenotypic plasticity whereby urban ants display higher thermal tolerance (Chapter 6; Diamond et al. 2017, 2018). These studies also highlighted the possibility of an evolution of plasticity itself (Crispo et al. 2010), since urban acorn ants exhibited greater plasticity in lower thermal tolerance. The authors also note that maternal and/or epigenetic effects could partly explain the observed shift in thermal tolerance interpreted as evolutionary change. Such cautionary notes are important since growing evidence suggests that epigenetic mechanisms such as DNA methylation could explain a substantial part of the phenotypic responses observed in cases of rapid adaptation to novel environments (Chapter 5; Angers et al. 2010; Day and Bonduriansky 2011). While the role of epigenetic variation in urban adaptation remains largely unknown, a small num-

ber of studies have recently shown epigenetic differences between urban and rural populations (in Darwin finches: McNew et al. (2017); in mice: Yauk et al. (2008); and in humans: Fagny et al. (2015)). We gauge that future explorations of epigenetic variation in urban environments will soon offer a substantial contribution to our understanding of environmentally induced epimutations and their role in adaptation to new habitats.

1.4.4 Urbanization and mutation rates

Darwinian evolution requires the presence of additive genetic variance for natural selection, sexual selection, or genetic drift to induce genetic evolution, or a change in allelic frequencies. This genetic variance across individuals can come either from pre-existing standing genetic variation, or from *de novo* mutations. Mutation rates can vary across time, space, organisms, and along the genome. Since the rate of evolutionary response in a given trait depends on the amount of genetic variance in that trait, factors that influence the rate of genetic mutation (either advantageous or deleterious for fitness in a given context) can play an important role in evolutionary adaptation to urbanization.

Experiments on mice (*Mus musculus*) in laboratories (Somers et al. 2002, 2004) have provided evidence that pollution via airborne contaminants can increase the rate of genetic mutation. Likewise, herring gulls (*Larus argentatus*) nesting close to steel mills in the American Great Lakes region had higher germline mutation rates than those at rural sites (Yauk and Quinn 1996). Overall, in several human and non-human studies, exposure to air pollutants has been related to DNA mutation, but also to epigenetic alterations (Chapter 5). Note that mutation rate and epigenetic processes are tightly linked, since some epigenetic marks (in particular DNA methylation) are often mutagens (Danchin et al. 2019). Overall, these results suggest that air, water, and soil pollution, along with other stressors associated with urban life, are likely to result in increased rates of mutations and epimutations. Further research is needed to explore the fitness consequences of such increased (epi)mutation rates in urban dwellers, and to determine whether these environmentally induced changes in the (epi)genome are transmitted across generations.

1.4.5 Domesticated species as case studies of microevolution

Domestication has provided some of the best examples of rapid evolution induced by humans, and has been previously instrumental in developing some of the methodologies in studying evolution, such as quantitative genetics. Although this book does not include a specific chapter focused on domestication, the evolution of domesticated species (as well as commensals) is closely related to the history of human settlement and urbanization (Vigne 2011), and it remains a promising area to explore urban-driven evolutionary processes. In particular, the recent development of paleo-genomics (and paleo-epigenomics) harnessed to reconstruct evolutionary pathways and trajectories (Lindqvist and Rajora 2019) has provided some emblematic case studies of micro-evolution, such as the evolution from wolf to dog in *Canis lupus* (e.g., adaptation to a starch-rich diet in dogs (Ollivier et al. 2016); see also similar process in avian commensals such as house sparrows *Passer domesticus* (Ravinet et al. 2018)). Future perspectives in genomic studies of species that have co-evolved with the development of human settlements (e.g., cats and dogs) include explorations of genes involved in cognitive abilities, behaviour, or communication—that is, characteristics targeted by urban-induced selection. Reconstructing the early evolutionary history of domestication linked to human settlements may also require a multidisciplinary approach, combining geometric morphometrics, archaeology, paleogenetics, and stable isotopes (Vigne 2015).

1.4.6 The gut microbiome

Evolutionary change mediated though changes in DNA sequence is likely to be only one part of the answer of organismal response to rapid environmental change (Alberdi et al. 2016). This is particularly true for vertebrates with considerably longer generation times relative to organisms reproducing at a rapid rate such as bacteria. Instead, phenotypic plasticity is the alternative mechanism through which rapid adaptation at the phenotypic level can occur. Phenotypic plasticity in evolutionary ecology studies is to date largely viewed independently of the fact that every animal is in fact host to a complex gut microbial community (its gut microbiota and underlying microbiome), and that microbial genes may outnumber a host's genes by orders of magnitude (Hooper and Gordon 2001; Qin et al. 2010). Recently, the gut microbiome has been identified as central to animal fitness as it is thought to help allow species to adapt to new environments (Alberdi et al. 2016). However, this assumption is hard to test in practice and there are currently few data, particularly in urban environments (but see Teyssier et al. 2018). Specifically, little is known about the extent to which the gut microbiota is a host 'trait', determined (at least in part) by the host genome, or a contributor to the host hologenome that itself shapes phenotypic responses.

In recent years, the study of the gut microbiome has driven a paradigm shift in which individuals cannot truly be considered independent of the bacterial communities they host, as these sustain multiple essential functions for the hosts (Zilber-Rosenberg and Rosenberg 2008; McFall-Ngai et al. 2013). The possession of the gut microbiome offers a vertebrate host the potential for rapid adaptation to environmental change that they could not achieve with their own genome (Alberdi et al. 2016). Their contribution to host plasticity in the context of rapid urbanization is thus of particular interest. However, the feedback loops between host genomics, gut microbial composition, associated microbiome metagenomics and the vertebrate phenome in a context of environmental change have yet to be fully understood, not least in a gradient of urbanization. Recently, Teyssier et al. (2018) reported that relative to natural sites, the gut microbiota of house sparrows from urban sites was characterized by shifts in taxonomic composition, community structure, and functional composition. Available evidence from natural populations, however, does not disentangle host phenotypic quality from the gut microbiota community it is hosting, and low individual fitness can be confounded by poor environment and associated poor diet. Consequently, urban-driven cause-and-effect relationships between host genome, gut microbiome, and host phenotype and fitness need to be addressed in relevant cross-fostering or common garden contexts.

1.5 Conclusions

Urban evolution is coming to a turning point: a large body of urban ecology knowledge is now available, and emerging efforts targeting large-scale sampling across multiple urbanization gradients will soon allow us to efficiently test increasingly robust evolutionary hypotheses. This is our opportunity, as evolutionary biologists, to contribute towards a better understanding of evolution in the urban space, and to suggest avenues aimed at strengthening effective population sizes of wild populations in the urban environment. To achieve these aims, we emphasize that stable sources of funding, capable of sustaining long-term (1) population monitoring, (2) common garden, and (3) reciprocal transplant experiments, are essential to allow robust replicated inference of evolutionary dynamics in gradients of urbanization.

Acknowledgements

Marta Szulkin and Anne Charmantier respectively thank the Polish National Science Centre (NCN) and the European Research Council (ERC) for providing them with 5 years of stable funding, without which this book would not have come to light. MS also thanks members of the Wild Urban Evolution & Ecology Lab and her fellow editors, Anne Charmantier and Jason Munshi-South, for extensive discussions of the urban realm. We all thank Menno Schilthuizen for an inspiring read of *Darwin Comes to Town*. M.S. was funded by Sonata BIS and Opus grants from the Polish National Science Centre (NCN2014/14/E/NZ8/00386, NCN2016/21/B/NZ8/03082). A.C. is thankful for discussions with colleagues from the Centre d'Ecologie Fonctionnelle (CEFE) and the scientific committee of the Fondation pour la Recherche sur la Biodiversité (FRB), in particular François Sarrazin and Jean-Denis Vigne. A.C. was supported by the European Research Council (Starting grant ERC-2013-StG-337365-SHE). J.M.-S. thanks members of his lab for fruitful discussions. J.M.-S. was supported by National Science Foundation Grant DEB 1457523.

References

Alberdi, A., Aizpurua O., Bohmann K., and Zepeda-Mendoza, M.L. (2016). Do vertebrate gut metagenomes confer rapid ecological adaptation? *Trends in Ecology & Evolution*, 31(9), 689–99.

Alberti, M., Correa, C., Marzluff, J.M., et al. (2017). Global urban signatures of phenotypic change in animal and plant populations. *Proceedings of the National Academy of Sciences of the United States of America*, 114(34), 8951–6.

Angers, B., Castonguay, E., and Massicotte, R. (2010). Environmentally induced phenotypes and DNA methylation: how to deal with unpredictable conditions until the next generation and after. *Molecular Ecology*, 19(7), 1283–95.

Aronson, M.F., La Sorte F.A., Nilon C.H., et al. (2014). A global analysis of the impacts of urbanization on bird and plant diversity reveals key anthropogenic drivers. *Proceedings of the Royal Society B: Biological Sciences*, 281(1780), 20133330.

Barrett, R.D.H. and Hoekstra, H.E. (2011). Molecular spandrels: tests of adaptation at the genetic level. *Nature Reviews Genetics*, 12(11), 767–80.

Beninde, J., Feldmeier S., Werner, M., et al. (2016). Cityscape genetics: structural vs. functional connectivity of an urban lizard population. *Molecular Ecology*, 25(20), 4984–5000.

Björklund, M., Ruiz, I., and Senar, J.C. (2010). Genetic differentiation in the urban habitat: the great tits (*Parus major*) of the parks of Barcelona city. *Biological Journal of the Linnean Society*, 99(1), 9–19.

Bourne, E.C., Bocedi G., Travis, J.M.J., et al. (2014). Between migration load and evolutionary rescue: dispersal, adaptation and the response of spatially structured populations to environmental change. *Proceedings of the Royal Society B: Biological Sciences*, 281(1778), 20132795.

Callahan, J.T. (1984). Long-term ecological research. *BioScience*, 34(6), 363–7.

Center for International Earth Science Information Network - Columbia University, International Food Policy Research Institute, The World Bank, and Centro Internacional de Agricultura Tropical (2011). Global Rural–Urban Mapping Project, Version 1 (GRUMPv1): Urban Extents Grid. NASA Socioeconomic Data and Applications Center (SEDAC), Palisades, NY. https://doi.org/10.7927/H4GH9FVG.

Clutton-Brock, T. and Sheldon, B.C. (2010). Individuals and populations: the role of long-term, individual-based studies of animals in ecology and evolutionary biology. *Trends in Ecology & Evolution* (Special Issue: Long-Term Ecological Research), 25(10), 562–73.

Crispo, E., DiBattista, J.D., Correa, C., et al. (2010). The evolution of phenotypic plasticity in response to anthropogenic disturbance. *Evolutionary Ecology Resarch*, 12(1), 47–66.

Danchin, E., Pocheville A., Rey O., Pujol B., and Blanchet, S. (2019). Epigenetically facilitated mutational assimilation: epigenetics as a hub within the inclusive evolutionary synthesis. *Biological Reviews*, 94(1), 259–82.

Day, T. and Bonduriansky, R. (2011). A unified approach to the evolutionary consequences of genetic and non-genetic inheritance. *The American Naturalist*, 178(2), E18–36.

de Jong, M., Joss S., Schraven, D., Zhan, C., and Weijnen, M. (2015). Sustainable–smart–resilient–low carbon–eco–knowledge cities; making sense of a multitude of concepts promoting sustainable urbanization. *Journal of Cleaner Production* (Special Issue: Toward a Regenerative Sustainability Paradigm for the Built Environment: From Vision to Reality), 109, 25–38.

Demeyrier, V., Lambrechts, M.M., Perret, P., and Arnaud, G. (2016). Experimental demonstration of an ecological trap for a wild bird in a human-transformed environment. *Animal Behaviour*, 118, 181–90.

Diamond, J.M. (1986). Natural selection: rapid evolution of urban birds. *Nature*, 324, 107.

Diamond, S.E., Chick, L., Perez, A., Strickler, S.A., and Martin, R.A. (2017). Rapid evolution of ant thermal tolerance across an urban–rural temperature cline. *Biological Journal of the Linnean Society*, 121(2), 248–57.

Diamond, S.E., Chick, L., Perez, A., Stricker, S.A., and Martin, R.A. (2018). Evolution of thermal tolerance and its fitness consequences: parallel and non-parallel responses to urban heat islands across three cities. *Proceedings of the Royal Society B: Biological Sciences*, 285(1882), 20180036.

Dobzhansky, T. (1973). Nothing in biology makes sense except in the light of evolution. *American Biology Teacher*, 35(3), 125–9.

Donihue, C.M. and Lambert, M.R. (2015). Adaptive evolution in urban ecosystems. *AMBIO*, 44(3), 194–203.

Douglas, I. and James, J. (2014). *Urban Ecology: An Introduction*. Routledge, Abingdon.

Fagny, M., Patin, E., MacIsaac, J.L., et al. (2015). The epigenomic landscape of African rainforest hunter-gatherers and farmers. *Nature Communications*, 6, 10047.

Gortat, T., Rutkowski, R., Gryczyńska, A., et al. (2015). Anthropopressure gradients and the population genetic structure of *Apodemus agrarius*. *Conservation Genetics*, 16(3), 649–59.

Hall, D.M., Camilo, G.R. Tonietto, R.K. (2017). The city as a refuge for insect pollinators. *Conservation Biology*, 31(1), 24–9.

Harris, S.E. and Munshi-South, J. (2017). Signatures of positive selection and local adaptation to urbanization in white-footed mice (*Peromyscus leucopus*). *Molecular Ecology*, 26(22), 6336–50.

Harris, S.E., Xue, A.T., Alvarado-Serrano, D., et al. (2016). Urbanization shapes the demographic history of a native rodent (the white-footed mouse, *Peromyscus leucopus*) in New York City. *Biology Letters*, 12(4).

Hooper, L.V. and Gordon, J.I. (2001). Commensal host-bacterial relationships in the gut. *Science (New York, NY)*, 292(5519), 1115–18.

Johnson, M.T.J. and Munshi-South, J. (2017). Evolution of life in urban environments. *Science*, 358(6363), eaam8327.

Johnson, M.T.J., Prashad, C.M., Lavoignat, M., and Saini, H.S. (2018). Contrasting the effects of natural selection, genetic drift and gene flow on urban evolution in white clover (*Trifolium repens*). *Proceedings of the Royal Society B: Biological Sciences*, 285(1883), 20181019.

Jorgensen, A. and Gobster, P.H. (2010). Shades of green: measuring the ecology of urban green space in the context of human health and well-being. *Nature and Culture*, 5(3), 338–63.

LaPoint, S., Balkenhol, N., Hale, J., Sadler, J., and van der Ree, R. (2015). Ecological connectivity research in urban areas. *Functional Ecology*, 29(7), 868–78.

LaZerte, S.E., Slabbekoorn, H., and Otter, K.A. (2016). Learning to cope: vocal adjustment to urban noise is correlated with prior experience in black-capped chickadees. *Proceedings of the Royal Society B: Biological Sciences*, 283(1833), 20161058.

Lindqvist, C. and Rajora, O.P. (eds) (2019). *Paleogenomics: Genome-Scale Analysis of Ancient DNA*. Springer, New York.

Lourenço, A., Álvarez, D., Wang, I.J., and Velo-Antón, G. (2017). Trapped within the city: integrating demography, time since isolation and population-specific traits to assess the genetic effects of urbanization. *Molecular Ecology*, 26(6), 1498–514.

Mace, G.M. (2014). Whose conservation? *Science*, 345(6204), 1558–60.

Marvasti, R. and Rees, A. (accessed 5 January 2020). *Smart Cities: Efficient, Sustainable, Digitised Living*. RESET editorial. https://reset.org/node/27044.

McFall-Ngai, M., Hadfield, M.G., Bosch, T.C.G., et al. (2013). Animals in a bacterial world, a new imperative for the life sciences. *Proceedings of the National Academy of Sciences of the United States of America*, 110(9), 3229–36.

McNew, S.M., Beck, D., Sadler-Riggleman, I., et al. (2017). Epigenetic variation between urban and rural populations of Darwin's finches. *BMC Evolutionary Biology*, 17(1), 183.

Miles, L.S., Rivkin, L.R., Johnson, M.T.J., Munshi-South, J., and Verrelli, B. (2019). Gene flow and genetic drift in urban environments. *Molecular Ecology*, 28(18), 4138-4151.

Miller, J.D., Scott, E.C., and Okamoto, S. (2006). Public acceptance of evolution. *Science*, 313(5788), 765–6.

Miranda, A.C., Schielzeth, H., Sonntag, T., and Partecke, J. (2013). Urbanization and its effects on personality traits: a result of microevolution or phenotypic plasticity? *Global Change Biology*, 19(9), 2634–44.

Mueller, J.C., Kuhl, H., Boerno, S., Tella, J.L., Carrete, M., and Kempenaers, B. (2013). Candidate gene polymorphisms for behavioural adaptations during urbanization in blackbirds. *Molecular Ecology*, 22(13), 3629–37.

Munshi-South, J. (2012). Urban landscape genetics: canopy cover predicts gene flow between white-footed mouse (*Peromyscus leucopus*) populations in New York City. *Molecular Ecology*, 21(6), 1360–78.

Munshi-South, J., Zolnik, C.P., and Harris, S.E. (2016). Population genomics of the Anthropocene: urbanization is negatively associated with genome-wide variation in white-footed mouse populations. *Evolutionary Applications*, 9(4), 546–64.

Niemelä, J., Breuste, J.H., Guntenspergen, G., McIntyre, N.E., Elmqvist, T., and James, P. (eds) (2011). *Urban Ecology: Patterns, Processes, and Applications*. Oxford University Press, Oxford.

Ollivier, M., Tresset, A., and Bastian, F. (2016). Amy2B copy number variation reveals starch diet adaptations in ancient European dogs. *Royal Society Open Science*, 3(11), 160449.

Perrier, C., Lozano del Campo, A., Szulkin, M., et al. (2018). Great tits and the city: distribution of genomic diversity and gene–environment associations along an urbanization gradient. *Evolutionary Applications*, 11(5), 593–613.

Qin, J., Li, R., Raes, J., et al. (2010). A human gut microbial gene catalogue established by metagenomic sequencing. *Nature*, 464(7285), 59–65.

Ravinet, M., Elgvin, T.O., Trier, C., et al. (2018). Signatures of human-commensalism in the house sparrow genome. *Proceedings of the Royal Society B: Biological Sciences*, 285(1884), 20181246.

Rellstab, C., Gugerli, F., Eckert, A.J., Hancock, A.M., and Holdregger, R. (2015). A practical guide to environmental association analysis in landscape genomics. *Molecular Ecology*, 24(17), 4348–70.

Richardson, J.L., Burak, M.K., Hernandez, C., et al. (2017). Using fine-scale spatial genetics of Norway rats to improve control efforts and reduce leptospirosis risk in urban slum environments. *Evolutionary Applications*, 10(4), 323–37.

Robertson, B.A. and Chalfoun, A.D. (2016). Evolutionary traps as keys to understanding behavioral maladaptation. *Current Opinion in Behavioral Sciences* (Behavioral Ecology), 12, 12–17.

Sarrazin, F. and Lecomte, J. (2016). Evolution in the Anthropocene. *Science*, 351(6276), 922–3.

Schell, C.J. (2018). Urban evolutionary ecology and the potential benefits of implementing genomics. *Journal of Heredity*, 109(2), 138–51.

Silvertown, J., Cook, L., Cameron, R., et al. (2011). Citizen science reveals unexpected continental-scale evolutionary change in a model organism. *PLOS ONE*, 6(4), e18927.

Somers, C.M., Yauk, C.L., White, P.A., Parfett, C.L.J., and Quinn, J.S. (2002). Air pollution induces heritable DNA mutations. *Proceedings of the National Academy of Sciences of the United States of America*, 99(25), 15904–7.

Somers, C.M., McCarry, B.E., Malek, F., and Quinn, J.S. (2004). Reduction of particulate air pollution lowers the risk of heritable mutations in mice. *Science (New York, NY)*, 304(5673), 1008–10.

Stearns, S.C. and Medzhitov, R. (2015). *Evolutionary Medicine*. Oxford University Press, Oxford.

Teyssier, A., Rouffaer, L.O., Saleh, H.N., et al. (2018). Inside the guts of the city: urban-induced alterations of the gut microbiota in a wild passerine. *Science of the Total Environment*, 612, 1276–86.

Van Rossum, F. (2008). Conservation of long-lived perennial forest herbs in an urban context: *Primula elatior* as study case. *Conservation Genetics*, 9(1), 119–28.

Vigne, J.-D. (2011). The origins of animal domestication and husbandry: a major change in the history of humanity and the biosphere. *Comptes Rendus Biologies* (On the Trail of Domestications, Migrations and Invasions in Agriculture), 334(3), 171–81.

Vigne, J.-D. (2015). Early domestication and farming: what should we know or do for a better understanding? *Anthropozoologica*, 50(2), 123–50.

Wang, I.J. and Bradburd, G.S. (2014). Isolation by environment. *Molecular Ecology*, 23(23), 5649–62.

Whitehead, A.L., Kujala, H., Yves, C.D., et al. (2014). Integrating biological and social values when prioritizing places for biodiversity conservation. *Conservation Biology*, 28(4), 992–1003.

Worthington, J.P., Silvertown, J., Cook, L., et al. (2012). Evolution MegaLab: a case study in citizen science methods. *Methods in Ecology and Evolution*, 3(2), 303–9.

Yauk, C.L. and Quinn, J.S. (1996). Multilocus DNA fingerprinting reveals high rate of heritable genetic mutation in herring gulls nesting in an industrialized urban site. *Proceedings of the National Academy of Sciences of the United States of America*, 93(22), 12137–41.

Yauk, C., Polyzos, A., and Rowan-Carroll, A. (2008). Germline mutations, DNA damage, and global hypermethylation in mice exposed to particulate air pollution in an urban/industrial location. *Proceedings of the National Academy of Sciences of the United States of America*, 105(2), 605–10.

Zilber-Rosenberg, I. and Rosenberg, E. (2008). Role of microorganisms in the evolution of animals and plants: the hologenome theory of evolution. *FEMS Microbiology Reviews*, 32(5), 723–35.

CHAPTER 2

How to Quantify Urbanization When Testing for Urban Evolution?

Marta Szulkin, Colin J. Garroway, Michela Corsini, Andrzej Z. Kotarba, and Davide Dominoni

Szulkin, M., Garroway, C.J., Corsini, M., Kotarba, A.Z. and Dominoni, D., *How to Quantify Urbanization When Testing for Urban Evolution?* In: *Urban Evolutionary Biology*. Edited by Marta Szulkin, Jason Munshi-South and Anne Charmantier, Oxford University Press (2020). © Oxford University Press.
DOI: 10.1093/oso/9780198836841.003.0002

2.1 Introduction

In just under 12 000 years, humans have transitioned from a hunter-gatherer lifestyle, living largely in forests and bushland, into the most prominent ecosystem engineers on the planet (Malhi 2017), predominantly living in their own built environments. With time, cities made of wood and stone have gradually been replaced with brick, steel, tarmac, glass, and concrete. Currently, more than half of the global human population lives in cities, and forecasts predict that by 2050 that estimate will grow to 70 per cent (United Nations 2018). The increasing rate of human migration towards cities will cause urban areas to steadily encroach on natural habitat and agricultural land. Urban areas currently cover ca. 3 per cent of the total surface of the Earth's land area (Center for International Earth Science Information Network et al. 2011; Liu et al. 2014) and are forecasted to grow several-fold in the next decades (Seto et al. 2012). To advance our understanding of urban evolution, it is essential to both accurately quantify urban-driven environmental changes and take in stride the opportunities and complexities inherent to urban environments that are relevant to urban evolution experimental design and hypothesis testing. Both themes are covered in this chapter.

How do we define urban space? The baseline definition of urban areas used by international or governmental agencies usually involves a threshold of human population density, which varies by country depending on methodology and definitions (McIntyre et al. 2000; United Nations 2018). For example, the European Commission defines urban centres as 'continuous grid cells of 1 km^2 with a population density of at least 1500 inhabitants per km^2 and collectively a minimum population of 50 000 inhabitants after gap filling' (Eurostat 2019). Urban clusters are 'contiguous grid cells of 1 km^2 with a population density of at least 300 inhabitants per km^2 and a minimum population of 5000 inhabitants.' In contrast, the United Nations relies on National Census Offices to provide country-specific thresholds of human density for its definitions of urban areas. Censuses, however, cannot be readily transformed into estimates of urban areal extent. Global urban maps, providing regularly updated estimates of the extent and spatial distribution of urban land, often use a combination of remotely sensed imagery and census data (Potere and Schneider 2007). Urban areas can also be defined by the physical attributes and composition of the Earth's land cover. In this light, the concept of percentage of impervious surface area (ISA) plays a central role in global assessments of urbanization, although threshold values of what is considered urban can vary from values as low as 10 per cent to values of 50 per cent depending on the type of study or dataset used (Potere and Schneider 2007; Liu et al. 2014). Depending on the type of methodology used, urban parks and urban natural reserves can fall short of some 'urban' definitions, yet they contribute to the mosaic of urban environments experienced by organisms in cities.

Urbanization is the ultimate replacement of all natural elements (soil, hydrologic system, vegetation, and fauna) by man-made ones: roadways, sewage network, lighting and heating apparatus, living and working construction (Dansereau 1957). Thus, beyond the arbitrary urbanization thresholds set by humans in a socioeconomic context, urbanization should be thought of as a process of environmental change that can be detected across multiple, non-exclusive, and often correlated environmental axes—such as sound or light pollution, chemical pollution, or percentage of impervious surface or canopy cover (Box 2.1). Regardless of which precise definition we choose to adopt, urbanization is progressively modifying natural habitat towards city cores. While distance from city centre has been an often-used metric to quantify environmental change triggered by urbanization, it is now widely acknowledged that such a transect approach does not capture well variation across an urban gradient. Indeed, urbanization generates a mosaic of interspersed, often heterogeneous habitats, so that more or less transformed environments alternate as patches from urban edges towards the urban centre in often 'complex' environmental gradients (McDonnell and Pickett 1990; McDonnell et al. 2012).

Any characterization of the urban environment needs to withstand the demands of evolutionary hypothesis testing. This requires that biologically relevant urbanization variables are used, quantified, and applied to relevant spatial and temporal scales of evolutionary change, in the context of a statistical framework that best identifies the underlying proximate causes of phenotypic or genetic change. In this chapter, we discuss established and emerging frameworks analysing urbanization in ways that are of relevance for urban evolutionary biology inference (section 2.2). Historically, these are grounded in sociological and ecological research, but evolutionary context is increasingly permeating these general frameworks (Donihue and Lambert 2015; Swaddle et al. 2015; Hulme-Beaman et al. 2016; Johnson and Munshi-South 2017; Ouyang et al. 2018; Rivkin et al. 2019). We further outline methods for detailed quantification of the urban environment, including remote-sensing and ground-based approaches (section 2.3). Finally, we highlight the importance of capturing accurate spatial and temporal scales of urbanization, and discuss study design considerations for urban evolutionary biology hypothesis testing (section 2.4). We end the chapter with conclusions and outlook for the future (section 2.5).

2.2 Frameworks for describing and quantifying urbanization

Urban growth results from the attractiveness of urban areas for employment and residence, and is classically viewed in terms of two types of development: *densification*—that is, the construction of new housing units within existing residential areas, and *expansion*—the development of new residential areas on land that was formerly open (Broitman and Koomen 2015). Historically, urban growth was relatively slow as urban populations fluctuated considerably due to wars, famines, and diseases. In the past two centuries, together with the exponential growth of the human population worldwide, cities have expanded rapidly (United Nations 2018). By examining 25 mid-sized cities from different geographical settings and levels of economic development, Schneider and Woodcock (2008) observed four major types of urban development: (1) low-growth cities with modest rates of infilling; (2) high-growth cities with rapid, fragmented development; (3) expansive-growth cities with extensive dispersion at low population densities; and (4) frantic-growth cities with extraordinary land conversion rates at high population densities. These different types of urban growth occur in contrasted climatic regimes and biomes, and under different socioeconomic,

historical, and cultural legacies, all of which can influence the evolution of urban organisms. Within these cities, humans thus govern urban ecosystem structure and processes by introducing a myriad of deviations from natural systems. Consequently, urban ecosystems function not only in spite of people, but also largely because of human participation (Alberti et al. 2003, 2017; Donihue and Lambert 2015). While some urban features are likely to stay the same worldwide, other urban features are likely to differ (Table 2.1); these similarities and contrasts among cities need to be taken under consideration when attempting to generalize urban-driven evolutionary responses, and when interpreting discrepant results from replicated gradients or studies using the same data collection protocols.

Differences in the history of urbanization, together with intrinsic socioeconomic and ecological differences between cities, make it challenging to develop a unified framework for describing and quantifying urbanization (Table 2.1). But the growing availability of global datasets capturing standardized parameters of the urban environment worldwide (section 2.3) increasingly allows for a standardized quantification of urbanization irrespective of geographical, historical, and cultural heritage.

2.2.1 Classic urban ecology frameworks

Urbanization is shorthand for the ecological processes created by the growth of cities and associated human activities. When measured from the core of the city to its suburban and periurban areas, the gradient paradigm implies changes in environmental conditions. McDonnell and Pickett (1990) emphasize the complexity of patterns underlying such gradients and consequently refer to them as 'complex urban–rural gradients'. Indeed, urban–rural gradients are not necessarily linear: more often than not, they are composed of interspersed patches of heterogeneous habitat contributing to the overall urban mosaic (McDonnell and Pickett 1990; McDonnell et al. 2012). As a consequence, when a particular patch in an urban area is the focus of our investigation, it can be relevant to consider the characteristics of the neighbouring patches (section 2.4.3). This perspective draws from major developments in other urban ecology frameworks, such as that of patch dynamics (Pickett et al. 2001). The patch

Table 2.1 Similarities and contrasts among urban features worldwide; comparisons are made relative to natural habitats.

Urban dimension	Urban feature likely to stay the same worldwide	Urban features likely to differ between cities worldwide
Human population density	Higher in urban areas	Several decades ago, most of the world's largest urban agglomerations were found in the more developed regions, but today's large cities are concentrated in developing countries. The fastest growing urban agglomerations are medium-sized cities and cities with less than 1 million inhabitants located in Asia and Africa (United Nations 2018). There is significant diversity in urbanization levels reached by different world regions. The most urbanized regions include North America (82 per cent living in urban areas in 2014), Latin America and the Caribbean (80 per cent), and Europe (73 per cent). Africa and Asia remain mostly rural, with 40 and 48 per cent of their respective populations living in urban areas. All regions are expected to urbanize further over the coming decades. Africa and Asia are urbanizing faster than the other regions and are projected to become 56 and 64 per cent urban, respectively, by 2050.
Anthropogenic food sources	Greater availability in urban areas	Can vary between countries and continents as a consequence of human diet and socioeconomic variability.
Artificial light at night (ALAN) levels	Higher in urban areas	Light intensity can differ depending on electrical grid capacity. ALAN spectral composition can depend on the time infrastructural lighting was set up or upgraded.
Sound levels	Higher in urban areas	Can differ depending on the specification of distinct sound emission sources (e.g., diesel vs petrol engines, siren sounds) as well as human behaviour (e.g., honking).
Land cover • Impervious surface	• Greater proportion in urban areas	• Amount of impervious surface can vary depending on the age of a city, urban growth rate, geographical location, and socioeconomics.
• Green areas	• Fewer in urban areas	• See above.
• Water bodies	• –	• Influenced by climate and historical context.
Temperature	Higher inside the city due to impervious surfaces and waste heat (urban heat island effect)	Cities differ in their thermal profile depending on the season, topography, and climatic zone they are located in. The strength of the urban heat island (temperature increase relative to surrounding natural habitat) is also likely to vary depending on the climatic zone and presence of water bodies in the city.
Urban topology • Building height	• More tall buildings in urban areas	• Influenced by historical and cultural context.
• City area	• –	• Influenced by age of city, and historical and political context: some cities are centrally planned via a top-down approach (e.g., Brasilia), while others develop bottom-up, where urban development decisions are made locally. All cities can be affected by urban sprawl (unrestricted urban growth with little concern for urban planning), although countries apply different strategies to tackle this process.
Age of cities	–	Old cities are unequally represented across the globe.

dynamics approach can be further extended by considering the flow of energy, nutrients, organisms, and species between different patches, effectively moving towards an urban metapopulation dynamics framework (Marzluff et al. 2001).

The degree of environmental change in space will determine the steepness of the gradient in system structure and function, and possibly in resulting selection gradients acting on organisms. While linear gradients or transects are often used to quantify environmental change at the urban–rural interface, awareness that these are in fact composed of a mosaic of heterogeneous environmental patches is needed. Likewise, it should not be expected that organismal response to environmental change—whether assessed on a transect or in a gradient of change in impervious surface or tree cover—is linear, or that environmental variables are normally distributed. Thus, the scope of biological response to the urban–rural gradient, and the distribution of the underlying environmental variables, should be adequately addressed statistically during data analysis (section 2.4).

When urban ecology emerged as a standalone field of research in the 1970s (Douglas and Goode 2015), it initially worked with data available at the time. These included variables such as distance to the city centre, human population density, and percentage of impervious cover (McDonnell and Hahs 2013). It thus made use of data originally collected to create standards for administrative or urban planning purposes—not to capture biological variation. Such variables are sometimes referred to as 'aggregated variables', as they are surrogate variables to indicate altered environmental conditions. Subsequently, the point was raised that specific hypothesis-driven variables, such as vegetation structure, temperature patterns, the availability of resources such as water, nutrients, nest sites, or food availability, and also noise or light levels should be taken into consideration (McDonnell and Hahs 2013; Swaddle et al. 2015). In addition, there is a growing awareness that, as for any other ecosystem, urban environments are more than the sum of their parts (Alberti et al. 2003), and human preferences and their interactions with demographic and economic factors from the social realm (Grimm et al. 2000; Ouyang et al. 2018) should be acknow-

ledged when analysing urban-driven biological responses.

In practice, it is often difficult to quantify every urban environment with as large a spectrum of ecologically relevant variables as those mentioned above, and the ultimate focus should lie in the ecology of the species under study. Moreover, aggregated variables and ecologically relevant patterns often covary: a case study of a European capital city shows that photosynthetic activity generated by urban vegetation (normalized difference vegetation index, NDVI), or the percentage of tree cover, is highly negatively correlated with the percentage of impervious surface (Box 2.1; Figure 2.1). Depending on the study purpose, it is thus important to establish whether priority should be given to variables that allow straightforward comparisons between cities (such as the percentage of impervious surface) or focus on more ecologically specific, hypothesis-driven measures of the environment (such as temperature at ground level or vegetation composition; McDonnell and Hahs 2013), which may require greater effort in terms of data acquisition.

2.2.2 Time as a missing axis in the study of the evolutionary consequences of urbanization

When measuring the degree of urbanization at specific locations, the temporal dynamics of landscape change can be overlooked (Ramalho and Hobbs 2012), with only the most recent spatial configuration and surrounding land use taken into account. As much as long-term studies in natural habitats have demonstrated their unique value to our understanding of evolution in the wild (Clutton-Brock and Sheldon 2010; Charmantier et al. 2014, 2016; Grant and Grant 2014), a similarly long-term perspective is needed in urban environments to infer evolutionary processes over time. Three key temporal aspects should be taken into account when assessing the evolutionary consequences of urbanization.

First, the age of a city often approximately reflects the time since a species has been exposed to urban environmental factors. Alternatively, the arrival of new species into a city—such as the eastern grey squirrel (*Sciurus carolinensis*) in American towns—

Box 2.1 Case study: Visualizing urban environmental variation in a European capital city

In the city of Warsaw, Poland, Szulkin and colleagues measured nine environmental axes of variation in an urban–rural gradient. Environmental data were collected for 565 coordinates (specifically—nest boxes), located at a minimum distance of 50 m from each other, and measured at nine sites in a gradient of urbanization of 24.3 km. Data were collected both on the ground (human presence, noise pollution, temperature), with a geographic information system (GIS) (distance to paths, distance to roads) and remote sensing tools (NDVI, percentage tree cover, percentage imperviousness, light pollution); see Supplementary Information for details. Covariation between the different environmental axes was visualized with PCA (Figure 2.1).

Figure 2.1 shows that many environmental variables collected in the urban–rural gradient are highly correlated with each other: overall, there is strong overlap in the distribution of loadings on PCA1 in terms of light pollution, imperviousness, temperature and human presence on the one hand, and NDVI (photosynthetic activity) and tree cover on the other (Fig. 2.1 A); this trend is even more noticeable when only sites from within the city's administrative limits are included (Fig. 2.1 B). Note that the negative association between sound or light pollution and distance to roads stems from the fact that small distances between a nestbox and the nearest road covary with high levels of sound and light. While PCA is a useful visualization tool, formal statistical testing of original variables on organismal variation is recommended to enable the comparison of effect sizes across studies.

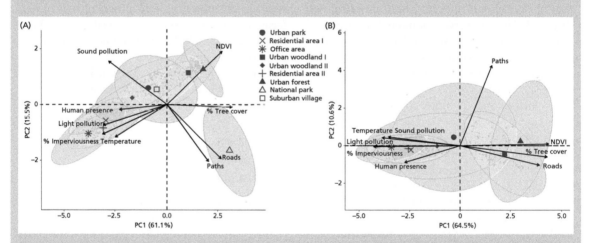

Figure 2.1 Principal Component Analysis based on a correlation matrix of nine environmental axes (vectors). Data was collected on the ground and with remote sensing in a gradient of urbanisation in the city of Warsaw, Poland, for 565 nestboxes (dots) spread across nine study sites (ellipses with centroids corresponding to different habitat types), ordered on the PCA legend by increasing distance to the strict urban centre. Data includes sites close to the urban centre (urban park, office area), two residential areas, two urban woodlands, an urban forest, a suburban village and a national park (coniferous forest). **A.** Data on the full gradient of urbanisation (n = 565 nestboxes, 9 sites). **B.** Data restricted to nestboxes located within the administrative limits of the city (n = 408 nestboxes, 7 sites that do not include the suburban village and the national park sites). A methodological summary of how each environmental variable was derived is available as supplementary information to the chapter.

can often be noteworthy and well documented (Benson 2013). This allows researchers to incorporate the length of exposure to selective environments into analyses, something that is rarely feasible in more natural settings where such data will often be unavailable. The length of exposure might modulate the 'amount' of urban selection pressures that populations have experienced.

Second, the strength of selection exerted by urbanization can be modulated by past characteristics of the urban landscapes, such as the degree of isolation of remnant patches, pollutant levels, and socioeconomic attributes (e.g., population density and build-up density), all of which are likely to change over time. Past landscape attributes can affect species with lasting legacies that persist over long periods of time (Lewis et al. 2006) and can continue to affect natural populations even after other more recent disturbance processes begin operating (Dambrine et al. 2007). A major issue will be to disentangle these various characteristics of the urban setting that can add up or interact together, with often unknown consequences on resulting strength of selection.

Third, species responses to urbanization-driven selection are not likely to be instantaneous, and a time lag is expected between selection pressure and adaptation. This delay can be measured at the species level as a lag in response to selection, and datasets enabling comprehensive testing of the magnitude of such lags are emerging (Alberti et al. 2017; Johnson and Munshi-South 2017). It can also be measured at the community level, where patterns of species richness can be examined as to whether they reflect past or present changes in the urban landscape (Krauss et al. 2010). For example, in old cities, remnant biodiversity might be largely shaped by the disturbance processes originating from the surrounding urban matrix. However, in young and rapidly urbanizing landscapes, communities are likely to still be gradually adjusting to the new environment. During this transient period, biological communities might be better explained by previous rather than current landscape spatial configurations (Krauss et al. 2010; Parker et al. 2010). Taken to its extreme, this process may be compared to the concept of 'extinction debt' (Kussaari et al. 2009), where the consequences of initial habitat loss are delayed, and species that avoided extinction due to direct destruction of their habitat can later become vulnerable to the altered ecosystem dynamics associated with fragmented landscapes (Hahs et al. 2009). By compiling a plant extinction rate dataset from twenty-two cities around the world, Hahs et al. (2009) showed that the legacy of historical urban landscape transformations can last for centuries, suggesting that modern cities could carry an extinction debt. Current urban communities are thus likely to contain a transient number of species and genotypes that are likely to go extinct once the transient period is over (Kuussaari et al. 2009; Krauss et al. 2010).

The current genetic architecture of any plant or animal population will be the result of a series of population demography events and environmental exposures. Luckily in the case of urban-dwelling organisms, data relating to past environments will be more often available than in the case of populations studied in a natural setting. The ability to consider the age of urbanization, land-use legacies, and the history of species colonization of a city can thus considerably enhance our understanding of observed patterns of gene flow and genetic differentiation. More specifically, species measured in different remnant patches in different cities are likely to find themselves at different stages along the course of adaptation to urban life. Thus, if data are interpreted without taking into account the spatial and temporal aspects of environmental and biological processes, this might lead to inconsistent or even contrasting results between different cities, species, or studies. These considerations can be illustrated by three examples. The spatial and temporal characteristics of colonization of urban space can leave distinct genomic signatures resulting from contrasted demographic processes, as follows:

(1) In a study of burrowing owls (*Athene cunicularia*) led by Mueller et al. (2018), entirely sequenced genomes revealed that these colonized each of three South American cities under investigation in an independent fashion. While gene flow occurred between neighbouring urban and rural populations, it was not detected between urban populations of different cities. In line with this finding, all urban populations showed reduced levels of standing genetic variation in rare single nucleotide polymorphisms (SNPs), but different subsets of rare SNPs were found in the different cities. This lowers potential for local adaptation based on rarer variants and decreases potential for detecting genetically parallel, urban-driven signals of selection in the genome (Mueller et al. 2018).

(2) In contrast to independent urban colonization events reported in burrowing owls, several genetic studies of blackbird (*Turdus merula*) suggest that their colonization strategy of urban areas is more compatible with the 'leap-frog' model, where urban-adapted birds colonize other proximal towns and cities (Partecke and Gwinner 2007; Evans et al. 2010).

(3) In a study of tuberculosis and leprosy in humans measured across 17 populations, Barnes et al. (2011) found a high correlation between the frequency of an allele associated with resistance to these diseases and the corresponding age of

regional history of urban settlement. The study demonstrates that populations with a long history of living in towns are better adapted to resist such infections.

2.2.3 Parallel urban evolution framework: replicated insight into urban-driven evolutionary processes

To date, the most compelling evidence for evolutionary repeatability that may mirror urban processes occurring across the globe comes from experimental evolution studies (Blount et al. 2008) on islands with the same or similar combinations of closely related species (Losos et al. 1997). As in the case of islands, cities of varying size now occur on nearly every landmass and in or near every major biome on Earth, with still unfulfilled potential for evolutionary hypothesis testing. Urban homogenization may drive convergent selection across cities as urban areas share common features, to the extent that even distant urban areas can be more environmentally similar than nearer less disturbed areas (McKinney 2006; Groffman et al. 2014). Alternatively, urban heterogeneity can lead to divergent selection within cities or across urbanization gradients. These contrasted evolutionary processes may be taxon-specific, but their respective magnitudes are as yet largely untested.

At a multiple species level, evolutionary biologists can examine the hypothesis that ecological homogenization leads to repeated evolutionary outcomes using highly replicated, global studies of multiple taxa. The set of urban adapters or exploiters (Blair 1996; Isaksson 2015) that survive or thrive in both urban and natural environments (such as great tits (*Parus major*) or lizards) or 'anthrodependent' species (Hulme-Beaman et al. 2016) that occupy nearly every city around the world offer an opportunity for global studies of coordinated evolutionary responses. Similarly, urban-dwelling introduced species are taxonomically diverse, ranging from invertebrates (such as German cockroach (*Blattella germanica*) and bedbugs (*Cimex* sp.)) to vertebrates (rats (*Rattus norvegicus*)) and plants (tree of heaven (*Ailanthus altissima*), and any number of weeds). All of these thus offer great potential to test multifaceted hypotheses about evolution in relation to many urban factors, such as city size, city age, climate, human demographics, and socioeconomic patterns. Similar studies could be designed on a near-continental scale using widely distributed native species, such as the many long-term studies of vertebrates and passerines worldwide (Sanz 1998; Clutton-Brock and Sheldon 2010; Lemoine et al. 2016). For cross-continental comparisons, ecologically similar native species that occur in many cities (such as *Peromyscus* spp. in North America and *Apodemus* spp. in Eurasia) could be employed. Careful choice of sets of species with wide urban and non-urban distributions, well-known natural history, and highly developed genomic resources will facilitate the success of such studies at identifying replicated evolutionary change in the same or similar phenotypes, genetic pathways, or genes.

From a methodological point of view, replication is certainly logistically challenging. In this light, the use of standardized, globally available datasets produced by satellite sensors is invaluable as they capture several relevant environmental dimensions of cities—such as light pollution, impervious surface, or photosynthetic activity—in both space and time (Table 2.2). When study replication is not feasible by single research groups, clearly presenting all spatial data associated with their work is highly encouraged (at a minimum—names and spatial coordinates of sampling sites), as this allows for easier downstream analyses of evolutionary change across studies and taxa. The merit of studying multiple cities, and its potential to identify urban-driven parallel evolution, is discussed in greater detail in Chapter 3.

2.3 Quantifying axes of variation in the urban environment

2.3.1 Urban metrics

Defining urbanization is no simple task, and some datasets, particularly centred on spatial differences in human density, have to rely on national sources, which reflect country-specific thresholds established by national authorities. Many axes of variation in the urban environment have indeed been historically collected on the ground, and are still of great value when specific and/or high-resolution environmental information is needed. However, the repeatability

Table 2.2 Urban metrics: a non-exhaustive list of relevance to urban evolutionary studies. Priority was given to open-access, global datasets of highest resolution—other high-quality resources exist at national and continental levels. Note that maximum resolution may vary depending on inferred timeframes. A review of urban and landscape metrics can also be found in Riiters et al. (1995), McIntyre et al. (2000), and Hahs and McDonnell (2006).

	Axes of variation	Maximum resolution	Earliest coverage	Data source	References
Environmental metrics	Percentage of impervious surface area (ISA)	30 m/pixel	2010	Satellite imagery (Landsat)	De Colstoun et al. (2017)
		500 m/pixel	2001/2002	Satellite imagery (MODIS)	Schneider et al. (2009)
	Percentage tree cover	30 m/pixel	2000	Satellite imagery (Landsat, MODIS)	Hansen et al. (2013); Sexton et al. (2013)
	Percentage grass cover	30 m/pixel	2001	Satellite imagery (Landsat, MODIS)	Ali et al. (2016)
	Land use/cover classes	Vector	Continuously updated	OpenStreetMap (various, crowdsourced)	Goodchild (2007)
	Land use/cover classes	500 m/pixel	2001	Satellite imagery (MODIS)	Friedl et al. (2010); Broxton et al. (2014)
	Distance to nearest road	Point specific	Site specific	Maps/aerial photography/satellite images	Google Earth
	Road density	> 1 m	Project specific	Land-cover maps; aerial/satellite images	Hahs and McDonnell (2006)
	Landscape shape index	Landscape specific	Project specific	Land-cover maps; aerial/satellite images	Hahs and McDonnell (2006)
	Largest patch index	Landscape specific	Project specific	Land-cover maps; aerial/satellite images	Hahs and McDonnell (2006)
	NDVI and EVI (photosynthetic activity)	250 m/pixel	2000	Satellite imagery (MODIS)	Huete et al. (1999); Pettorelli et al. (2005)
		300 m/pixel	2016	Satellite imagery (PROBA-V)	Jacobs et al. (2016)
		4 km/pixel	1981	Satellite imagery (AVHRR)	Beck et al. (2011)
	Water surface	30 m/pixel	2000	Satellite imagery (Landsat)	Feng et al. (2016); Allen and Pavelsky (2018)
	Meteorology (various parameters)	80 km/grid cell	1979	Reanalysis	Dee et al. (2011)
		2.0 deg. × 1.75 deg.	1979	Reanalysis	Kanamitsu et al. (2002)
	Air temperature	Point specific	Project specific	On-ground measurements; weather stations	Villalobos-Jimenez and Hassall (2017)
Pollution metrics	Light—ground light intensity	Point specific	Site specific	On-ground measurements; street lamp censity maps; aerial/satellite images	Bennie et al. (2014); Dominoni et al. (2014)

Table 2.2 Continued

	Axes of variation	Maximum resolution	Earliest coverage	Data source	References
	Light—upward light flux	10–100 m/pixel	2000 (selected locations worldwide)	Satellite imagery/digital photography (International Space Station)	Kyba et al. (2015); Kotarba and Aleksandrowicz (2016)
		1 km/pixel	Daily since 1992; annual and monthly cloud-free mosaics	Satellite imagery (DMSP-OLS; VIIRS—since 2012)	Miller et al. (2013); Hsu et al. (2015)
		1 km	2014	Satellite data (VIIRS); photometers	Falchi et al. (2016)
	Sound	Point specific	Project specific	On-ground measurements; governmental noise maps	Gill et al. (2015); Dominoni et al. (2016)
	Air—NOx level	Point specific	Project specific	On-ground measurements; databases from city councils	Salmon et al. (2018)
	Air—reactive gases (i.e., CO, O_3, NOx)	80 km/pixel	Four times per day, 2003–2012	Reanalysis (satellite data)	Inness et al. (2013)
	Air—particulate matter	Point specific	Project specific	On-ground measurements (AirBeam); databases from city councils	Fusaro et al. (2017)
Temporal metrics	Percentage change in any variable	Landscape specific	Project specific	Land-cover maps	Ramalho and Hobbs (2012)
	Age of city	Landscape specific	Project specific	Historical records	Region-specific historical records
Sociodemographic metrics	Human population density	Country specific	Country specific	National census offices	United Nations (2018)
	Human presence	250 m	2016	Global human settlement layer (GHSL)	European Commission (2016)
		Point-specific	Project specific	Ground measurements	Kight and Swaddle (2007); Corsini et al. (2017, 2019)
	Socioeconomic data	Neighbourhood level	Country specific	National census offices	National census offices

of ground-based quantification can vary depending on protocols, and is cost- and time-consuming. In contrast, the availability of remote sensing data—originally aerial photography and now information largely captured by satellite sensors—is continuously growing. This was initially facilitated by open-access repositories of satellite data granted by the National Aeronautics and Space Administration (NASA) for Landsat and Modis sensors. The European Space Agency (ESA) followed suit by giving open access to Sentinel imagery. Concomitantly, relevant open-source statistical frameworks enabled the analysis of such imagery (e.g., package *Modis* in R, Mattiuzzi et al. (2019)). Generally, free-access satellite sensor data of increasing spatial and temporal resolution are becoming available at a global scale, allowing the derivation of standardized environmental indexes of vegetation, temperature, or impervious surface area worldwide (Table 2.2). This is a particularly welcome development since multiple scales of environmental heterogeneity (section 2.4.3) can be easily extracted from remote sensing data, allowing for the testing of multiple spatial and temporal scales of urban-driven evolutionary processes.

Urban metrics can thus be captured by satellite sensors, but they can also be collected on the ground or through census offices; a non-exhaustive list of these is summarized in Table 2.2. Depending on the study aims, sampling methodology can maximize standardized global spatial coverage, temporal coverage, or fine-scale, field-work-based inference. Choosing relevant sampling span and resolution is likely to be particularly relevant when inferring replicated urban evolution (Chapter 3), the temporal dimension of urbanization, or detailed mechanistic drivers of evolutionary processes. Importantly, the repeatability of urban metrics is likely to vary: it can depend on the environmental variation a given urban metric is capturing (air pollution is likely to be more variable than the amount of impervious surface when measured at two different time points), but also on the underlying method used for environmental quantification. More discussion on the repeatability of environmental metrics is available in Corsini et al. (2019) for human presence, Salmon et al. (2018) for air pollution, and Kotarba and Aleksandrowicz (2016) and Liu et al. (2014) for percentage of impervious surface.

2.3.2 Univariate versus multivariate approaches

As with any ecosystem, the urban environment can be partitioned into distinct (although often correlated) axes of environmental variation—these are discussed in further detail in Hahs and McDonnell (2006), McIntyre et al. (2000), and Riiters et al. (1995). At the same time, the selection pressures organisms are exposed to are likely to reflect multiple axes of environmental variation depending on species-specific biology.

Environmental variation stemming from multiple sources, and particularly the extent to which some environmental variables covary, can be visualized with a principal component analysis (PCA). PCA is a multivariate technique that analyses a data table where observations are described by several correlated quantitative dependent variables (Abdi and Williams 2010). PCA reduces the large number of often collinear variables into a smaller number of orthogonal and independent variables (principal component axes) that can be visualized in two or three-dimensional space. PCA has been used as a means to partition environmental variance into independent, uncorrelated variables in ecological studies (Riiters et al. 1995; Lausch and Herzog 2002; Hahs and McDonnell 2006; Tavernia and Reed 2009) and evolutionary ecology studies (Rodewald et al. 2011; Giraudeau et al. 2014; Perrier et al. 2018). It is undoubtedly useful in visualizing how environmental variables covary in the urban space, as illustrated by Box 2.1: it is shown that while environmental axes capturing the urban environment can have an entirely distinct physical basis, they are often highly correlated with each other—either positively (e.g., light pollution and imperviousness; Figure 2.1) or negatively (e.g., imperviousness and photosynthetic activity (NDVI; Figure 2.1).

While PCA is a valuable visualizing tool, its major drawback lies in the fact that it is difficult to compare effect sizes between studies when PCA axes are used. This is especially problematic in the case of comparative analysis over space or time, or meta-analyses aiming to assess the effect sizes of distinct urban environmental variables. Whenever possible, the reporting of (1) values derived from single variables (rather than synthetic variables such as principal component loadings) and (2) clear methodology of how the environment is quantified are needed to facilitate comparison between studies. Model selection and model averaging can be further used to identify support for specific environmental variables that explain the greatest amount of variance in organismal variation (section 2.4).

2.3.3 How is urbanization quantified in published studies of urban evolution?

To get a general overview of the type of methodology used in studies of urban evolution, we reviewed how urbanization was measured in 192 studies reviewed by Johnson and Munshi-South (2017).

These studies met the criteria for urban evolutionary inference, where either molecular genetic or heritable phenotypic evolutionary change was assessed (e.g., a change in allelic frequency, or a change in the mean value of a heritable trait, respectively), and where an evolutionary process—that is, mutation, gene flow, genetic drift, or selection—was examined. A graphical illustration of review findings is presented in Figure 2.2.

The overall conclusions of this literature review are that:

(1) Few studies of urban evolution replicate the effect of cities in their hypothesis testing: the median number of cities per study is 1 (average:

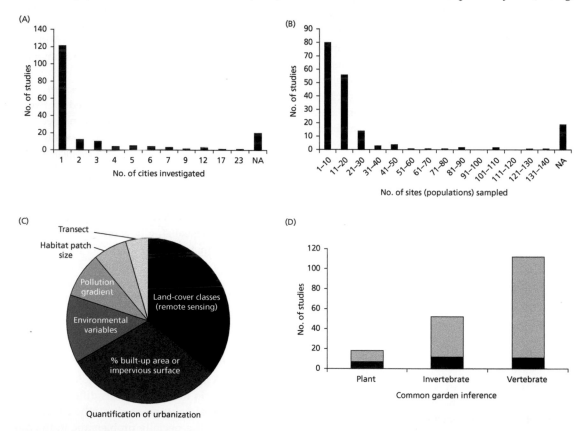

Figure 2.2 An analysis of how urbanization was quantified in studies of urban evolution, defined as urban-driven mutation, genetic drift, gene flow, or selection. Data extracted from 192 studies cited by Johnson and Munshi-South (2017). All studies were published between 1955 and 2017, and examined evolutionary patterns or processes within an urban environment that met one of the following criteria: (1) studies that measured changes in allele frequencies using molecular data; (2) changes in traits with known simple Mendelian inheritance; (3) changes in traits that are shown to be heritable in a common garden; and (4) direct measures of natural selection. (A) Histogram of number of cities investigated in studies of urban evolution. $N_{studies}$ = 185. (B) Histogram of number of sites or populations (urban or other) where biological data were collected. $N_{studies}$ = 165. (C) Summary chart of how urbanization was quantified across studies. $N_{studies}$ = 45 ($N_{studies}$ with non-quantitative definitions of urban areas = 136. Total $N_{studies}$=181). (D) Number of studies with (dark grey) and without (light grey) common garden inference across taxa. $N_{studies}$ = 185.

2.1, $N_{studies}$ =165, Figure 2.2A), while the median number of sites (locations) sampled is 11 (average: 16, $N_{studies}$ = 165, Figure 2.2B).

(2) Seventy-five per cent of studies use qualitative terms in defining urban areas (for example, names of cities; urban/suburban nominal statements), rather than quantitative ones (total $N_{studies}$ = 181). Out of forty-five studies where a quantitative approach to defining urbanization was taken, sixteen quantify urbanization based on remotely sensed land-cover classes, fourteen rely on the percentage of built-up area/extent of impervious surfaces, six use a combination of several environmental variables, four follow light and chemical pollution gradients, three infer habitat patch size, and two take a transect approach running from the city centre towards city edges (Figure 2.2C).

(3) Common garden experiments investigating urban-driven evolutionary processes are proportionally most prevalent in plant studies, followed by invertebrate and vertebrate studies (in decreasing order) (Figure 2.2D).

These results highlight a need for more replicated inference of urbanization processes, quantitative rather than qualitative approaches when inferring urban-driven processes, and experimental approaches allowing for the partitioning of genetic and environmental contributions to traits that are differentially expressed in the urban environment. Urban evolution studies should endeavour to use and publish data collection protocols that allow reproducibility at a global level. Whenever possible, standardized reporting practices of the urban environment and urban evolution testing across several urban–rural gradients are needed (see also Chapter 3). We are aware that such gold standards are computationally and logistically challenging, and often beyond the grasp of a single research group—particularly for vertebrate study systems and/or protected species. We hope that (1) by highlighting the need for standardized protocols capturing urban environmental and organismal variation, (2) with appropriate funding enabling long-term and collaborative endeavours, and (3) the possibility of involving citizen science for some study species (see the Evolution MegaLab project on the recording of banded snails (*Cepaea nemoralis* and *C. hortensis*) across Europe; Worthington et al. 2011) and collaborative science for others (see the Global Urban Evolution Project on the white clover (*Trifolium repens*) www.globalurbanevolution.com), these statistics will rapidly change in favour of more robust urban evolutionary biology hypothesis testing worldwide.

2.4 Study design and statistical approaches for urban evolutionary biology

2.4.1 Model selection and variable fitting

No single set of urbanization metrics is universally applicable (Tavernia and Reed 2009), which underlines the importance of identifying the most relevant dimensions of the urban environment for evolutionary hypothesis testing. In the past decade, multimodel inference and hypothesis ranking based on the evaluation of competing statistical models (e.g., using Akaike Information Criterion (AIC), cross-validation) has been a valuable alternative to traditional null hypothesis testing. With such an approach, several models can be ranked and weighted to provide a quantitative measure of relative support for each competing hypothesis (Grueber et al. 2011). In the context of multiple environmental axes characterizing the urban environment, model ranking, and model averaging based on model weights in particular, are helpful in evaluating relative supports for a given hypothesis over others instead of null hypothesis testing (Burnham et al. 2011). An example of a step-by-step strategy for model averaging is detailed in Marrot et al. (2018), whose study aimed to identify temporal temperature drivers underlying annual selection gradients in the timing of breeding of blue tits (*Cyanistes caeruleus*). While the study was not carried out in an urban setting, it gives valuable inspiration on how a model averaging approach can be used to explore relevant spatial or temporal environmental drivers of organismal variation in an urban context.

Another point to consider is the statistical power associated with different types of categorization of urban environmental predictors: while variables characterizing the urban environment can be fitted as continuous variables (e.g., dwelling density), these can be reclassified into categorical variables, which include ordinal variables (e.g., high, medium, and low density) or nominal variables (urban/rural). By using an AIC approach, Caryl et al. (2014) tested urbanization drivers underlying bat distribution obtained from 36 sites across Melbourne, Australia. The authors found that models fitted with continuous predictors had a better fit and explanatory power in explaining several variables related to bat distribution than those fitted with ordinal predictors due to a loss of statistical power when continuous variables are discretized. That said, models fitted with nominal data explained an equivalent amount of variation as continuous models. This is similar to what is found by Sprau et al. (2016), who tested multiple environmental variables (temperature, humidity, light and noise pollution) in an urban–rural gradient to identify drivers of great tit (*Parus major*) life-history variation. By taking an AIC approach, the authors found that the nominal dichotomy of forest vs city, rather than continuous environmental variables, explained the data best. These results suggest that a priori assumptions of linear covariation between environmental predictors and trait change may be flawed; further work in the field is needed to improve our understanding of the linearity, or non-linearity, between urban environmental variables and phenotypes. Regardless, AIC proves yet again to be an effective statistical framework to identify the best drivers underlying organismal variation.

2.4.2 Controlling for spatial autocorrelation

When replication at the landscape scale is not feasible, it is important to account for pseudoreplication in our study and temper the generality and scale of applicability of our findings. A major issue with single landscape studies is dealing with spatial autocorrelation arising from individuals being sampled close together in space and thus responding to the same environments. This can lead to spurious and inflated estimates of the strength of the evolutionary process—environment relationships. Inflated relationships can arise due to the fact that individuals occupying the same environment will resemble

each other to a greater degree than they otherwise might if exposed to different amounts and configurations of habitats. This issue is particularly salient for species where related individuals stay closer geographically than unrelated individuals, since shared environment will be a confounding factor when exploring genetic effects. There are a great number of statistical approaches for controlling for spatial autocorrelation (Dormann et al. 2007). For example, albeit not in an urban environment, Stopher et al. (2012) found that spatial autocorrelation terms in animal models explained a substantial proportion of the heritability of the traits examined in a population of red deer (*Cervus elaphus*). The authors thus demonstrated that a shared environment and likely limited dispersal inflate the estimation of heritability when not accounted for.

A particularly flexible approach for dealing with spatial autocorrelation when independent replication is not possible is with the use of distance-based Moran's eigenvector maps. These are the eigenvectors of a truncated spatial distance network built with spatial coordinates of sample sites (Borcard and Legendre 2002; Dray et al. 2006). They can be used to summarize spatial patterns in data within multiple statistical frameworks (e.g., because they are orthogonal, they can be directly incorporated into regression analyses as proxy variables accounting for spatial autocorrelation-related processes). Garroway et al. (2013) used distance-based Moran's eigenvector maps to control for spatial patterns in allele frequency variation in a genome–environment association study of a single population of great tits (*Parus major*), resulting in fewer positive associations with putative environmental selection pressures due to accounting for isolation-by-distance directly within the model. These eigenvectors can indeed be used to control for any process (e.g., neutral genetic or ecological) that produces spatial signatures in data regardless of whether the causes are known. This is likely to be particularly relevant in an urban context given the heterogeneity likely to occur in cities.

2.4.3 The problem of scale

We define scale as the spatial or temporal dimension of an evolutionary process of interest. This seemingly simple definition hides a great deal of complexity that can alter how we understand evolution. Organisms experience environments over a limited range of the possible scales of variation, and identifying these most relevant scales is difficult in both space and time (see also section 2.2.2). When our scale of observation for a study is different from the scale at which an environment influences evolutionary processes, we run the risk of poorly estimating or simply not identifying important mechanisms of environmental or evolutionary change (Figure 2.3). Moreover, urban-driven change at one scale of biological organization can alter emergent patterns or mechanisms at another scale (Alberti et al. 2003). Although the importance of considering scale when studying ecological processes was included in early seminal works on the issue (Levin 1992), the choice of scale is rarely explicitly addressed in evolutionary biology studies, and multiscale analyses aimed at identifying the scale of a process are rarer still. It appears likely that the discovery of relationships between evolutionary processes and environments, and the estimation of the strength of such relationships, has been hampered by a lack of explicit consideration of scales of effect relative to the scales of research (e.g., Richardson et al. 2014). For instance,

most studies of local adaptation take place at spatial scales that exceed both the reported spatial grain of environmental variation and the distances moved by study species (Richardson et al. 2014). Thus, with accumulating evidence for fine-scale patterns of local adaptation occurring at or below a species' characteristic dispersal distance (e.g., Garroway et al. 2013; Kaiser et al. 2016; Hämälä et al. 2018), we have likely missed important components of functionally relevant genetic variation.

One approach for the empirical identification of the spatial scale of a process is to test for relationships between the evolutionary response of interest and urban environments across different spatial extents. The use of remotely sensed landscape metrics to characterize environmental variation is particularly amenable to identify which scale of inference is relevant to evolutionary change, as it allows one to model relationships across scales and to use model selection to identify scale with the greatest effect (Figure 2.3; Szulkin et al. 2015). When assessing multiple spatial scales is not possible, we can attempt to select biologically justified scales. Theory (Ricci et al. 2013; Jackson and Fahrig 2015) and empirical (Bowman et al. 2002; Bowman 2003) analyses suggest that traits related to space use (such

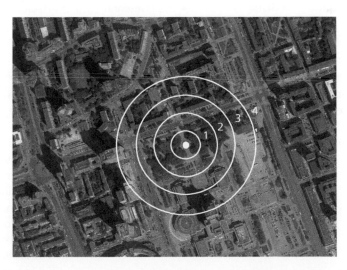

Figure 2.3 Environmental inference across multiple scales of the urban matrix. A hypothetical reproductive event (white dot) can be largely surrounded by green vegetation in perimeter 1. This is decreasingly true when greater perimeters (2, 3, 4) are taken into account. At a population level, the extent to which environment 1, 2, 3, or 4 induces the strongest phenotypic and/or fitness clines can be explicitly tested and compared with model comparison techniques. Ideally, urban evolutionary studies would include multiple and non-overlapping sampling sites across which the configuration of landscapes varies; the above picture sample site would represent a single sampling site. (Source: Google Earth, modified.)

as dispersal or home range size), thus reflecting the scale at which organisms interact with their environments, can help guide our selection of the scale of analysis. It is thus possible that some seemingly variable evolutionary responses to similar environments could in part be explained by the fact that researchers have examined the same question at different and not explicitly articulated spatial scales. Reporting key aspects of spatial data would help determine whether this is the case: such spatial data could include sample coordinates, descriptors of the spatial configuration of study sites and extents, or, better, electronically archived study site maps.

2.5 Conclusions and outlook

The subject matter of evolutionary biology is not new, but studying it in a novel ecological context—cities—is certainly original. Consequently, urban evolutionary biology carries an important set of unique considerations. It is critical to accurately quantify the biological variation stemming from cit-

ies and the organisms that colonize them, thereby enabling the establishment of powerful, spatially replicated evolutionary experiments. Table 2.3 outlines key aspects of urban data collection and analysis to be mindful of when testing evolutionary hypotheses in an urban environment.

Cities are a true environmental mosaic, and a full understanding of the evolutionary consequences of such an environmental patch dynamic has not yet been reached. Variation in city-specific environmental mosaics can generate important variation in findings, and thus can at least partially explain why the same species show different responses in different cities or in different areas within the same city (Ouyang et al. 2018; Rivkin et al. 2019). Thus, an acknowledgement of scale effects, of the different types of urban environments, and of their respective contribution to the urban mosaic should be considered when urban-driven evolutionary hypothesis testing is to be performed. Recognizing that no two cities are the same, we have presented research strategies that have the

Table 2.3 How to optimize urban data collection and analysis for evolutionary hypothesis testing.

	Spatial dimension	Temporal dimension	Data collection	Data analysis
Approaches	• Urban–rural gradient • Landscape	• Static • Urban rate of change • Biological rate of change	• Experimental: common garden/ reciprocal translocations • Observational	• Visualization tools (PCA) • Model selection • Model averaging
What to be mindful of	• Spatial scale of inference • Spatial autocorrelation	• Temporal scale of inference • Acknowledge the generation time of your study species	• Environment/ fitness/ phenotype and genotype need to be quantified whenever possible	• Statistical power resulting from the use of nominal/ discrete/continuous variables • Environmental resolution/ lability needs to be matched to type of trait measured and its own lability • Additionally to differences in means and medians, differences in variance and distributions can be of interest
Take-home message	• Assess repeatability of your environmental axes • Use replicates whenever possible • Report spatial coordinates of sampling locations • Apply standardized protocols and reporting practices across replicated observations			

potential to contribute to our understanding of evolutionary processes in the urban space. Any opportunity for data presentation that facilitates synthesis and spatial replication should be taken whenever possible. With the increasing availability of open datasets coming from populated areas of the world and the ongoing reporting of environmental, climatic, and biological data, including that collected from citizen science programmes, the future of urban evolutionary biology is bright.

Acknowledgements

Marta Szulkin was funded by Sonata BIS and Opus grants from the Polish National Science Centre (NCN2014/14/E/NZ8/00386, NCN2016/21/B/NZ8/03082). Davide Dominoni was funded by the University of Glasgow and a NERC Highlight Topics grant, and Michela Corsini by a Sonata BIS grant awarded to M.S. Colin Garroway was funded by a Natural Sciences and Engineering Research Council of Canada Discovery Grant. We thank Marc-Olivier Beausoleil and Pascal Marrot for enlightening discussions on urban evolution and constructive feedback of the manuscript made at various writing stages. Finally, we sincerely thank Amy Hahs, Sarah Diamond, Jason Munshi-South, and Anne Charmantier for outstandingly constructive comments that substantially improved this chapter.

References

Abdi, H. and Williams, L.J. (2010). Principal component analysis. *Wiley Interdisciplinary Reviews: Computational Statistics*, 2(4), 433–59.

Alberti, M., Marzluff, J.M., Shulenberger, E., et al. (2003). Integrating humans into ecology: opportunities and challenges for studying urban ecosystems. *BioScience*, 53(12), 1169–79.

Alberti, M., Correa, C., Marzluff, J.M., et al. (2017). Global urban signatures of phenotypic change in animal and plant populations. *Proceedings of the National Academy of Sciences of the United States of America*, 114(34), 8951–6.

Ali, I., Cawkwell, F., Dwyer, E., Barrett, B., and Green, S. (2016). Satellite remote sensing of grasslands: from observation to management. *Journal of Plant Ecology*, 9(6), 649–71.

Allen, G.H. and Pavelsky, T.M. (2018). Global extent of rivers and streams. *Science*, 361(6402), 585–8.

Barnes, I., Duda, A., Pybus, O.G., and Thomas, M.G. (2011). Ancient urbanization predicts genetic resistance to tuberculosis. *Evolution: International Journal of Organic Evolution*, 65(3), 842–8.

Beck, H.E., McVicar, T.R., van Dijk, A.I.J.M., et al. (2011). Global evaluation of four AVHRR–NDVI data sets: intercomparison and assessment against Landsat imagery. *Remote Sensing of Environment*, 115(10), 2547–63.

Bennie, J., Davies, T.W., Inger, R., and Gaston, K.J. (2014). Mapping artificial lightscapes for ecological studies. *Methods in Ecology and Evolution*, 5(6), 534–40.

Benson, E. (2013). The urbanization of the eastern gray squirrel in the United States. *Journal of American History*, 100(3), 691–710.

Blair, R.B. (1996). Land use and avian species diversity along an urban gradient. *Ecological Applications*, 6(2), 506–19.

Blount, Z.D., Borland, C.Z., and Lenski, R.E. (2008). Historical contingency and the evolution of a key innovation in an experimental population of *Escherichia coli*. *Proceedings of the National Academy of Sciences of the United States of America*, 105(23), 7899–906.

Borcard, D. and Legendre, P. (2002). All-scale spatial analysis of ecological data by means of principal coordinates of neighbour matrices. *Ecological Modelling*, 153(1–2), 51–68.

Bowman, J. (2003). Is dispersal distance of birds proportional to territory size? *Canadian Journal of Zoology*, 81(2), 195–202.

Bowman, J., Jaeger, J.A.G., and Fahrig, L. (2002). Dispersal distance of mammals is proportional to home range size. *Ecology*, 83(7), 2049–55.

Broitman, D. and Koomen, E. (2015). Residential density change: densification and urban expansion. *Computers, Environment and Urban Systems*, 54, 32–46.

Broxton, P.D., Zeng, X., Sulla-Menashe, D., and Troch, P.A. (2014). A global land cover climatology using MODIS data. *Journal of Applied Meteorology and Climatology*, 53(6), 1593–605.

Burnham, K.P., Anderson, D.R., and Huyvaert, K.P. (2011). AIC model selection and multimodel inference in behavioral ecology: some background, observations, and comparisons. *Behavioral Ecology and Sociobiology*, 65(1), 23–35.

Caryl, F.M., Hahs, A.K., Lumsden, L.F., et al. (2014). Continuous predictors of species distributions support categorically stronger inference than ordinal and nominal classes: an example with urban bats. *Landscape Ecology*, 29(7), 1237–48.

Center for International Earth Science Information Network, International Food Policy Research Institute, The World Bank, and Centro Internacional de Agricultura Tropical (2011). Global Rural–Urban Mapping Project, Version 1 (GRUMPv1): Urban Extents Grid. NASA Socioeconomic Data and Applications Center, Palisades, NY. https://doi.org/10.7927/H4GH9FVG.

Charmantier, A., Garant, D., and Kruuk, L.E.B. (eds) (2014). *Quantitative Genetics in the Wild*. Oxford University Press, Oxford.

Charmantier, A., Doutrelant, C., Dubuc-Messier, G., Fargevieille, A., and Szulkin, M. (2016). Mediterranean blue tits as a case study of local adaptation. *Evolutionary Applications*, 9(1), 135–52.

Clutton-Brock, T. and Sheldon, B.C. (2010). Individuals and populations: the role of long-term, individual-based studies of animals in ecology and evolutionary biology. *Trends in Ecology & Evolution*, 25(10), 562–73.

Corsini, M., Dubiec, A., Marrot, P., and Szulkin, M. (2017). Humans and tits in the city: quantifying the effects of human presence on great tit and blue tit reproductive trait variation. *Frontiers in Ecology and Evolution*, 5(82), 1–12.

Corsini, M., Marrot, P., and Szulkin, M. (2019). Quantifying human presence in a heterogeneous landscape. *Behavioral Ecology*, 30(6), 1632–1641.

Dambrine, E., Dupouey, J.L., Laüt, L., et al. (2007). Present forest biodiversity patterns in France related to former Roman agriculture. *Ecology*, 88(6), 1430–9.

Dansereau, P. (1957). *Biogeography. An Ecological Perspective*. The Ronald Press Company, New York.

De Colstoun, E.C.B., Huang, C., Wang, P., et al. (2017). Documentation for the Global Man-Made Impervious Surface (GMIS) Dataset from Landsat. NASA Socioeconomic Data and Applications Center (SEDAC), Palisades, NY.

Dee, D.P., Uppala, S.M., Simmons, A.J., et al. (2011). The ERA—interim reanalysis: configuration and performance of the data assimilation system. *Quarterly Journal of the Royal Meteorological Society*, 137(656), 553–97.

Dominoni, D.M., Carmona-Wagner, E.O., Hofmann, M., Kranstauber, B., and Partecke, J. (2014). Individual-based measurements of light intensity provide new insights into the effects of artificial light at night on daily rhythms of urban-dwelling songbirds. *Journal of Animal Ecology*, 83(3), 681–92.

Dominoni, D.M., Greif, S., Nemeth, E., and Brumm, H. (2016). Airport noise predicts song timing of European birds. *Ecology and Evolution*, 6(17), 6151–9.

Donihue, C.M. and Lambert, M.R. (2015). Adaptive evolution in urban ecosystems. *Ambio*, 44(3), 194–203.

Dormann, C.F., McPherson, J.M., Araújo, et al. (2007). Methods to account for spatial autocorrelation in the analysis of species distributional data: a review. *Ecography*, 30(5), 609–28.

Dray, S., Legendre, P., and Peres-Neto, P.R. (2006). Spatial modelling: a comprehensive framework for principal coordinate analysis of neighbour matrices (PCNM). *Ecological Modelling*, 196(3–4), 483–93.

Douglas, I. and Goode, D. (2015). Urban natural histories to urban ecologies: the growth of the study of urban nature. In: Douglas, I., Goode, D., Houck, M.C., and Wang R. (eds) *The Routledge Handbook of Urban Ecology*. Routledge, London.

European Commission (2016). GHSL—Global Human Settlement Layer. https://ghsl.jrc.ec.europa.eu.

Eurostat (2019). Glossary: Urban centre. https://ec.europa.eu/eurostat/statistics-explained/index.php/Glossary:Urban_centre.

Evans, K.L., Hatchwell, B.J., Parnell, M., and Gaston, K.J. (2010). A conceptual framework for the colonisation of urban areas: the blackbird *Turdus merula* as a case study. *Biological Reviews*, 85(3), 643–67.

Falchi, F., Cinzano, P., Duriscoe, D., et al. (2016). The new world atlas of artificial night sky brightness. *Science Advances*, 2(6), e1600377.

Feng, M., Sexton, J.O., Channan, S., and Townshend, J.R. (2016). A global, high-resolution (30-m) inland water body dataset for 2000: first results of a topographic–spectral classification algorithm. *International Journal of Digital Earth*, 9(2), 113–33.

Friedl, M.A., Sulla-Menashe, D., Tan, B., et al. (2010). MODIS Collection 5 global land cover: algorithm refinements and characterization of new datasets. *Remote Sensing of Environment*, 114(1), 168–82.

Fusaro, L., Marando, F., Sebastiani, A., et al. (2017). Mapping and assessment of PM10 and O_3 removal by woody vegetation at urban and regional level. *Remote Sensing*, 9(8), 791.

Garroway, C.J., Radersma, R., Sepil, I., et al. (2013). Fine-scale genetic structure in a wild bird population: the role of limited dispersal and environmentally based selection as causal factors. *Evolution*, 67(12), 3488–500.

Gill, S.A., Job, J.R., Myers, K., Naghshineh, K., and Vonhof, M.J. (2015). Toward a broader characterization of anthropogenic noise and its effects on wildlife. *Behavioral Ecology*, 26(2), 328–33.

Giraudeau, M., Nolan, P.M., Black, C.E., et al. (2014). Song characteristics track bill morphology along a gradient of urbanization in house finches (*Haemorhous mexicanus*). *Frontiers in Zoology*, 11(1), 83.

Goodchild, M.F. (2007). Citizens as sensors: the world of volunteered geography. *GeoJournal*, 69(4), 211–21.

Grant, P.R. and Grant, B.R. (2014). *40 years of Evolution: Darwin's Finches on Daphne Major Island*. Princeton University Press, Princeton.

Grimm, N.B., Grove, J.G., Pickett, S.T.A., and Redman, C.L. (2000). Integrated approaches to long-term studies of urban ecological systems: urban ecological systems present multiple challenges to ecologists—pervasive human impact and extreme heterogeneity of cities, and the need to integrate social and ecological approaches, concepts, and theory. *BioScience*, 50(7), 571–84.

Groffman, P.M., Cavender-Bares, J., Bettez, N.D., et al. (2014). Ecological homogenization of urban USA. *Frontiers in Ecology and the Environment*, 12(1), 74–81.

Grueber, C.E., Nakagawa, S., Laws, R.J., and Jamieson, I.G. (2011). Multimodel inference in ecology and evolution: challenges and solutions. *Journal of Evolutionary Biology*, 24(4), 699–711.

Hahs, A.K. and McDonnell, M.J. (2006). Selecting independent measures to quantify Melbourne's urban–rural gradient. *Landscape and Urban Planning*, 78(4), 435–48.

Hahs, A.K., McDonnell, M.J., McCarthy, M.A., et al. (2009). A global synthesis of plant extinction rates in urban areas. *Ecology Letters*, 12(11), 1165–73.

Hämälä, T., Mattila, T.M., and Savolainen, O. (2018). Local adaptation and ecological differentiation under selection, migration, and drift in *Arabidopsis lyrata*. *Evolution*, 72(7), 1373–86.

Hansen, M.C., Potapov, P.V., Moore, R., et al. (2013). High-resolution global maps of 21st-century forest cover change. *Science*, 342(6160), 850–3.

Hsu, F.-C., Baugh, K.E., Ghosh, T., Zhizhin, M., and Elvidge, C.D. (2015). DMSP-OLS radiance calibrated nighttime lights time series with intercalibration. *Remote Sensing*, 7(2), 1855–76.

Huete, A.R., Justice, C., and van Leeuwen, W. (1999). MODIS vegetation index (MOD13)—algorithm theoretical basis document, version 3. https://modis.gsfc.nasa.gov/data/atbd/atbd_mod13.pdf.

Hulme-Beaman, A., Dobney, K., Cucchi, T., and Searle, J.B. (2016). An ecological and evolutionary framework for commensalism in anthropogenic environments. *Trends in Ecology & Evolution*, 31(8), 633–45.

Inness, A., Baier, F., Benedetti, A., et al. (2013). The MACC reanalysis: an 8 yr data set of atmospheric composition. *Atmospheric Chemistry and Physics*, 13(8), 4073–109.

Isaksson, C. (2015). Urbanization, oxidative stress and inflammation: a question of evolving, acclimatizing or coping with urban environmental stress. *Functional Ecology*, 29(7), 913–23.

Jackson, H.B. and Fahrig, L. (2015). Are ecologists conducting research at the optimal scale? *Global Ecology and Biogeography*, 24(1), 52–63.

Jacobs, T., Swinnen, E., Toté, C., and Claes, P. (2016). Product User Manual. Normalized Difference Vegetation Index (NDVI). Collection 300M, Version 1. Report by Gio Global Land Component—Lot I. Issue I 1.11. https://land.copernicus.eu/global/sites/cgls.vito.be/files/products/GIOGL1_PUM_NDVI300m-V1_I1.11.pdf.

Johnson, M.T.J. and Munshi-South, J. (2017). Evolution of life in urban environments. *Science*, 358(6363), eaam8327.

Kaiser, A., Merckx, T., and Van Dyck, H. (2016). The urban heat island and its spatial scale dependent impact on survival and development in butterflies of different thermal sensitivity. *Ecology and Evolution*, 6(12), 4129–40.

Kanamitsu, M., Ebisuzaki, W., Woollen, J., et al. (2002). NCEP–DOE AMIP-II reanalysis (R-2). *Bulletin of the American Meteorological Society*, 83(11), 1631–44.

Kight, C.R. and Swaddle, J.P. (2007). Associations of anthropogenic activity and disturbance with fitness metrics of eastern bluebirds (*Sialia sialis*). *Biological Conservation*, 138(1–2), 189–97.

Kotarba, A.Z. and Aleksandrowicz, S. (2016). Impervious surface detection with nighttime photography from the International Space Station. *Remote Sensing of Environment*, 176, 295–307.

Krauss, J., Bommarco, R., Guardiola, M., et al. (2010). Habitat fragmentation causes immediate and time-delayed biodiversity loss at different trophic levels. *Ecology Letters*, 13(5), 597–605.

Kuussaari, M., Bommarco, R., Heikkinen, R.K., et al. (2009). Extinction debt: a challenge for biodiversity conservation. *Trends in Ecology & Evolution*, 24(10), 564–71.

Kyba, C.C.M., Garz, S., Kuechly, H., et al. (2015). High-resolution imagery of earth at night: new sources, opportunities and challenges. *Remote Sensing*, 7(1), 1–23.

Lausch, A. and Herzog, F. (2002). Applicability of landscape metrics for the monitoring of landscape change: issues of scale, resolution and interpretability. *Ecological Indicators*, 2(1–2), 3–15.

Lemoine, M., Lucek, K., Perrier, C., et al. (2016). Low but contrasting neutral genetic differentiation shaped by winter temperature in European great tits. *Biological Journal of the Linnean Society*, 118(3), 668–85.

Levin, S.A. (1992). The problem of pattern and scale in ecology: the Robert H. MacArthur award lecture. *Ecology*, 73(6), 1943–67.

Lewis, D.B., Kaye, J.P., Gries, C., Kinzig, A.P., and Redman, C.L. (2006). Agrarian legacy in soil nutrient pools of urbanizing arid lands. *Global Change Biology*, 12(4), 703–9.

Liu, Z., He, C., Zhou, Y., and Wu, J. (2014). How much of the world's land has been urbanized, really? A hierarchical framework for avoiding confusion. *Landscape Ecology*, 29(5), 763–71.

Losos, J.B., Warheitt, K.I., and Schoener, T.W. (1997). Adaptive differentiation following experimental island colonization in *Anolis* lizards. *Nature*, 387(6628), 70.

Malhi, Y. (2017). The concept of the Anthropocene. *Annual Review of Environment and Resources*, 42, 77–104.

Marrot, P., Charmantier, A., Blondel, J., and Garant, D. (2018). Current spring warming as a driver of selection on reproductive timing in a wild passerine. *Journal of Animal Ecology*, 87(3), 754–64.

Marzluff, J.M., Bowman, R., and Donnelly, R. (2001). A historical perspective on urban bird research: trends, terms, and approaches. In: *Avian Ecology and Conservation in an Urbanizing World*, pp. 1–17. Springer, Boston, MA.

Mattiuzzi, M., Verbesselt, J., Hengl, T., et al. (accessed 8 January 2020). Package 'MODIS'. https://github.com/MatMatt/MODIS.

McDonnell, M.J. and Hahs, A.K. (2013). The future of urban biodiversity research: moving beyond the 'low-hanging fruit'. *Urban Ecosystems*, 16(3), 397–409.

McDonnell, M.J. and Hahs, A.K. (2015). Adaptation and adaptedness of organisms to urban environments. *Annual Review of Ecology, Evolution and Systematics*, 46, 261–80.

McDonnell, M.J. and Pickett, S.T.A. (1990). Ecosystem structure and function along urban–rural gradients: an unexploited opportunity for ecology. *Ecology*, 71(4), 1232–7.

McDonnell, M.J., Hahs, A.K., and Pickett, S.T.A. (2012). Exposing an urban ecology straw man: critique of Ramalho and Hobbs. *Trends in Ecology & Evolution*, 27(5), 255–6.

McIntyre, N.E., Knowles-Yánez, K., and Hope, D. (2000). Urban ecology as an interdisciplinary field: differences in the use of 'urban' between the social and natural sciences. In: Breuste, J, Feldmann, H, and Uhlmann, O. (eds) *Urban Ecology*, pp. 49–65. Springer, Boston, MA.

McKinney, M.L. (2006). Urbanization as a major cause of biotic homogenization. *Biological Conservation*, 127(3), 247–60.

Miller, S.D., Straka, W., Mills, S.P., et al. (2013). Illuminating the capabilities of the suomi national polar-orbiting partnership (NPP) visible infrared imaging radiometer suite (VIIRS) day/night band. *Remote Sensing*, 5(12), 6717–66.

Mueller, J.C., Kuhl, H., Boerno, S., et al. (2018). Evolution of genomic variation in the burrowing owl in response to recent colonization of urban areas. *Proceeding of the Royal Society B: Biological Sciences*, 285(1878), 20180206.

Ouyang, J.Q., Isaksson, C., Schmidt, C., et al. (2018). A new framework for urban ecology: an integration of proximate and ultimate responses to anthropogenic change. *Integrative and Comparative Biology*, 58(5), 915–28.

Parker, J.D., Richie, L.J., Lind, E.M., and Maloney, K.O. (2010). Land use history alters the relationship between native and exotic plants: the rich don't always get richer. *Biological Invasions*, 12(6), 1557–71.

Partecke, J. and Gwinner, E. (2007). Increased sedentariness in European blackbirds following urbanization: a consequence of local adaptation? *Ecology*, 88(4), 882–90.

Perrier, C., Lozano del Campo, A., Szulkin, M., et al. (2018). Great tits and the city: distribution of genomic diversity and gene–environment associations along an urbanization gradient. *Evolutionary Applications*, 11(5), 593–613.

Pettorelli, N., Vik, J.O., Mysterud, A., et al. (2005). Using the satellite-derived NDVI to assess ecological responses to environmental change. *Trends in Ecology & Evolution*, 20(9), 503–10.

Pickett, S.T.A., Cadenasso, M.L., Grove, J.M., et al. (2001). Urban ecological systems: linking terrestrial ecological, physical, and socioeconomic components of metropolitan areas. *Annual Review of Ecology and Systematics*, 32(1), 127–57.

Potere, D. and Schneider, A. (2007). A critical look at representations of urban areas in global maps. *GeoJournal*, 69(1–2), 55–80.

Ramalho, C.E. and Hobbs, R.J. (2012). Time for a change: dynamic urban ecology. *Trends in Ecology & Evolution*, 27(3), 179–88.

Ricci, B., Franck, P., Valantin-Morison, M., Bohan, D.A., and Lavigne, C. (2013). Do species population parameters and landscape characteristics affect the relationship between local population abundance and surrounding habitat amount? *Ecological Complexity*, 15, 62–70.

Richardson, J.L., Urban, M.C., Bolnick, D.I., and Skelly, D.K. (2014). Microgeographic adaptation and the spatial scale of evolution. *Trends in Ecology & Evolution*, 29(3), 165–76.

Riitters, K.H., O'Neill, R.V., Hunsaker, C.T., et al. (1995). A factor analysis of landscape pattern and structure metrics. *Landscape Ecology*, 10(1), 23–39.

Rivkin, L.R., Santangelo, J.S., Alberti, M., et al. (2019). A roadmap for urban evolutionary ecology. *Evolutionary Applications*, 12(3), 384–98.

Rodewald, A.D., Shustack, D.P., and Jones, T.M. (2011). Dynamic selective environments and evolutionary traps in human-dominated landscapes. *Ecology*, 92(9), 1781–8.

Salmón, P., Watson, H., Nord, A., and Isaksson, C. (2018). Effects of the urban environment on oxidative stress in early life: insights from a cross-fostering experiment. *Integrative and Comparative Biology*, 58(5), 986–94.

Sanz, J.J. (1998). Effects of geographic location and habitat on breeding parameters of great tits. *The Auk*, 115(4), 1034–51.

Schneider A. and Woodcock, C.E. (2008). Compact, dispersed, fragmented, extensive? A comparison of urban growth in twenty-five global cities using remotely sensed data, pattern metrics and census information. *Urban Studies*, 45(3), 659–92.

Schneider, A., Friedl, M.A., and Potere, D. (2009). A new map of global urban extent from MODIS satellite data. *Environmental Research Letters*, 4(4), 044003.

Seto, K.C., Gueneralp, B., and Hutyra, L.R. (2012). Global forecasts of urban expansion to 2030 and direct impacts on biodiversity and carbon pools. *Proceedings of the National Academy of Sciences of the United States of America*, 109(40), 16083–8.

Sexton, J.O., Song, X.-P., Feng, M., et al. (2013). Global, 30-m resolution continuous fields of tree cover: Landsat-based rescaling of MODIS vegetation continuous fields with LiDAR-based estimates of error. *International Journal of Digital Earth*, 6(5), 427–48.

Sprau, P., Mouchet, A., and Dingemanse, N.J. (2016). Multidimensional environmental predictors of variation in avian forest and city life histories. *Behavioral Ecology*, arw130.

Stopher, K.V., Walling, C.A., Morris, A., et al. (2012). Shared spatial effects on quantitative genetic parameters: accounting for spatial autocorrelation and home range overlap reduces estimates of heritability in wild red deer. *Evolution: International Journal of Organic Evolution*, 66(8), 2411–26.

Swaddle, J.P., Francis, C.D., Barber, J.R., et al. (2015). A framework to assess evolutionary responses to anthropogenic light and sound. *Trends in Ecology & Evolution*, 30(9), 550–60.

Szulkin, M., Zelazowski, P., Marrot, P., and Charmantier, A. (2015). Application of high resolution satellite imagery to characterize individual-based environmental hetero-geneity in a wild blue tit population. *Remote Sensing*, 7(10), 13319–36.

Tavernia, B.G. and Reed, J.M. (2009). Spatial extent and habitat context influence the nature and strength of relationships between urbanization measures. *Landscape and Urban Planning*, 92(1), 47–52.

Theobald, D.M. (2004). Placing exurban land-use change in a human modification framework. *Frontiers in Ecology and the Environment*, 2(3), 139–44.

United Nations (2018). World Urbanization Prospects: The 2018 Revision, Online Edition. https://population.un.org/wup/.

Villalobos-Jiménez, G. and Hassall, C. (2017). Effects of the urban heat island on the phenology of Odonata in London, UK. *International Journal of Biometerology*, 61(7), 1337–46.

Worthington, J.P., Silvertown, J., Cook, L., et al. (2011). Evolution MegaLab: a case study in citizen science methods. *Methods in Ecology and Evolution*, 3(2), 303–9.

SUPPLEMENTARY INFORMATION – Chapter 2

Quantification of environmental variation in a heterogeneous urban landscape

Environmental variation was quantified at nestboxes monitored as part of a prospectively long-term project on the ecology and evolution of great tits *Parus major* and blue tits *Cyanistes caeruleus* in Warsaw, Poland. Nine axes of environmental variation were investigated across 9 different urban sites, for a total of 565 specific locations (here: nestboxes). Data was collected on the ground, with the use of GIS and remote sensing using the following methodology:

Variables collected on the ground

1. Human presence

Human presence was quantified by recording all humans (bikers or pedestrians) and dogs (inherently associated with human presence in our study sites) during 30 seconds long counts performed at all fixed locations of interest (here 565 distinct nestboxes). An index of human presence was derived, which can be defined as the average number of humans and dogs detected for any given fixed location (such as a nestbox) across multiple instances of counting. Depending on the site, human presence was counted at each nestbox between 18 and 24 times during two great tit breeding seasons. More details of the methodology (including repeatability) can be found in Corsini et al. (2017, 2019).

2. Temperature (in C°)

Temperature was recorded using Thermocron iButtons DS1921G with a 1 hour sampling frequency. The data loggers were set from 24/04/18 until 30/06/18. The entire recording period resulted in the same number of measurements for each data logger (N = 1632). Depending on study site area size, the number of thermocrones varied between 2 and 3 per site (urban park = 3, residential area I = 2, office area = 2, urban woodland I = 3, urban woodland II = 2, residential area II = 2, urban forest = 3, national park = 3,

suburban village = 2). An average temperature was calculated for the exact same recording timespan for each data logger (total N = 22). Each nestbox was assigned a temperature value corresponding to the nearest Thermocron. (Temperature data were downloaded from the data loggers *via* OpenSource software OneWireViewer).

3. Sound pollution (in Db C)

Sound pollution levels were recorded using hand-held sound level meters with a microphone (model: Digital Sound Level Meter SL-200). The sound level was recorded on DbC scale because the C-curve also detects lower frequencies relative to the often-used A-curve (the latter representing the characteristic hearing curve of the human ear). We selected 4 days of measurements throughout the field season and recorded sound levels three times across each day (at 7:00, at 12:00 and at 17:00, total N = 12 measurements per nestbox). The only exception occurred in the urban woodland I site, where the sound pollution protocol was performed only twice at midday (12:00), due to opening-hours restrictions. During each trial, a fieldworker was assigned to specific tracks within each study site, stopped for 5 seconds at each nestbox, and recorded the highest value of sound pollution displayed while holding the device towards the nestbox entrance.

Variables collected using a GIS approach

4. Distance to closest roads

Distances to closest roads were recorded in meters using the "Measure line" tool of the free and open source Geographic Information System Quantum GIS 2.8.2 "Wien", (version released on the 9th of May 2015) as described in Corsini et al. (2017).

5. Distance to closest paths

Distances to closest paths were recorded in meters using the "Measure line" tool of the free and open source

Geographic Information System Quantum GIS 2.8.2 "Wien", (QGIS, version released on the 9th of May 2015) as described in Corsini et al. (2017)

Variables collected with remote sensing (digital photography, satellite sensors)

6. Light pollution

A map of light pollution for the city of Warsaw of 10m (meters) per pixel resolution was derived from night-time digital photography taken by astronauts at the International Space Station (Kyba et al., 2015). The map was further uploaded in QGIS (v.2.18.25). The relative intensity of light pollution was estimated as average pixel brightness (uncelebrated digital number) within a 100m buffer around each nestbox. The averaged light pollution estimation was computed by using the function *Zonal Statistics* in qGIS.

7. Tree cover

Information on tree cover density (in %) was downloaded from Copernicus Land Monitoring Services (https://land. copernicus.eu/sitemap; Forests/ Tree Cover Density). The basic map of tree cover referred to 2015 and is of 20m per pixel resolution. The averaged value was calculated with a 100m radius buffer analysis as described in point 6 (light pollution).

8. Imperviousness

The percentage of soil sealing and built up areas was downloaded from Copernicus Land Monitoring Services (https://land.copernicus.eu/pan-european/high-resolution-layers/imperviousness/status-maps; Imperviousness / Status maps / Imperviousness Density 2015). The imperviousness data referred to 2015 and is of 20m per pixel resolution. A 100m radius buffer analysis was subsequently performed as described in point 6.

9. NDVI

The Normalised Difference Vegetation Index (NDVI; see Frampton et al., 2013; Pettorelli N et al., 2011) was calculated using satellite images of 10m per pixel resolution from SENTINEL2 (S2MSI2A): selected imagery was downloaded as a TIFF file from Copernicus (source: https://scihub.copernicus.eu/dhus/#/home), which was readily corrected for atmospheric effects (bottom of atmosphere reflectance in cartographic geometry). To calculate the NDVI, two different bands were uploaded in QGIS: 4 and 8 (which respectively represent Red and Near Infrared – RED and NIR- respectively). To extrapolate NDVI, the following formula was computed in QGIS (v.2.18.25) via *Raster calculator* tool:

NIR-RED/NIR+RED   in bands   8 – 4/8 + 4

Once the NDVI was calculated, The averaged NDVI value at the nestbox level was estimated within a 100m radius buffer analysis as described in point 6.

References

Corsini, M., Dubiec, A., Marrot, P. and Szulkin, M. (2017). Humans and tits in the city: quantifying the effects of human presence on Great tit and Blue tit reproductive trait variation. *Frontiers in Ecology and Evolution*, 5(82),1–12.

Corsini, M., Marrot, P. and Szulkin, M. (2019). Quantifying human presence in a heterogeneous urban landscape. *Behavioral Ecology*, 30(6), 1632–1641.

Frampton, W.J., Dash, J., Watmough, G. and Milton, E.J. (2013). Evaluating the capabilities of Sentinel-2 for quantitative estimation of biophysical variables in vegetation. *ISPRS Journal of Photogrammetry and Remote Sensing*, 82, 83–92.

Kyba, C.C.M., Garz, S., Kuechly, H., de Miguel, A.S., Zamorano, J., Fischer, J. and Hölker, F. (2015). High-Resolution Imagery of Earth at Night: New Sources, Opportunities and Challenges. *Remote Sensing*, 7(1), 1–23.

Pettorelli N, Ryan S, Mueller T, Bunnefeld N, Jedrzejewska B, Lima M and Kausrud K. (2011). The Normalized Difference Vegetation Index (NDVI): unforeseen successes in animal ecology. *Climate Research*, 46, 15-27.

CHAPTER 3

Urban Environments as a Framework to Study Parallel Evolution

James S. Santangelo, Lindsay S. Miles, Sophie T. Breitbart,
David Murray-Stoker, L. Ruth Rivkin, Marc T.J. Johnson, and Rob W. Ness

Santangelo, J.S., Miles, L.S., Breitbart, S.T., Murray-Stoker, D., Rivkin, L.R., Johnson, M.T.J. and Ness, R.W., *Urban Environments as a Framework to Study Parallel Evolution* In: *Urban Evolutionary Biology*. Edited by Marta Szulkin, Jason Munshi-South and Anne Charmantier, Oxford University Press (2020).
© Oxford University Press. DOI: 10.1093/oso/9780198836841.003.0003

3.1 Introduction

Parallel evolution—the repeated evolution of the same phenotypes or genotypes in response to similar selection pressures across populations or species—remains at the forefront of research in evolutionary biology (Box 3.1), and has greatly enhanced our understanding of phenotypic and genetic evolution

(Arendt and Reznick 2008; Conte et al. 2012; Oke et al. 2017; Stuart et al. 2017; Bolnick et al. 2018; Langerhans 2018). Quantifying the extent of parallel evolutionary responses provides important insights into the nature of adaptation: high levels of parallelism suggest that adaptation can occur only along a limited number of paths (Gould 2006), thereby increasing our ability to predict future evolutionary

---

Box 3.1 Notes on evolutionary convergence and parallelism

Throughout this chapter, we use the term 'parallel' to refer to two or more populations or species evolving the same phenotype(s) or genotypes(s) in response to similar selective regimes; thus, our definition subsumes what some authors would refer to as 'convergent' evolution (Arendt and Reznick 2008). Traditionally, 'convergence' has been used for cases where similar phenotypes/genotypes evolve in different species or when similar phenotypes evolve via different genetic pathways, whereas 'parallelism' was restricted to cases of common ancestry or evolution from the same genes or pathways. However, these definitions are subjective: it is neither clear what level of relatedness is required to be considered a case of parallelism nor whether genetic changes need to be identical at the sequence level (e.g., does the same gene require that the same nucleotide substitution be selected?) (Losos 2011). For a detailed breakdown of the contentious history surrounding the use of 'convergent' and 'parallel' evolution, we refer the reader to an excellent recent review on the topic (Bolnick et al. 2018). More recently, geometric approaches (e.g., phenotypic change vector analysis; Adams and Collyer 2009) to measure parallel evolution have allowed for a mathematical separation of 'convergence' and 'parallelism', paving a path beyond the inconsistencies surrounding the use of these terms (Bolnick et al. 2018). While such approaches have yet to be applied in an urban context (see section 3.5), we nonetheless refer to the repeated evolution of similar traits or genes as parallelism in the broad sense (Bolnick et al. 2018).

Natural selection is often assumed to be the mechanism driving the repeated adaptation of species or populations to similar environments, although other processes may drive phenotypic or genetic parallelism. For example, high levels of macroevolutionary parallelism can occur under some circumstances when phylogenies are simulated under stochastic processes (i.e., Brownian motion; Stayton 2008). On a microevolutionary scale, spatially restricted gene flow combined with genetic drift can generate repeated clines in

single loci (Vasemägi 2006) or in quantitative traits (Colautti and Lau 2015). Similarly, spatial gradients in the strength of genetic drift are sufficient to drive the repeated evolution of phenotypic clines in traits with non-additive (epistatic) genetic architectures (Santangelo et al. 2018a). When these gradients covary with environmental gradients, they may give the misleading impression that selection is driving parallel spatial clines. Shared genetic correlations and genetic and developmental constraints can additionally lead to the repeated evolution of traits or genes (Losos 2011). Comparing observed data to suitable null models (e.g., through simulations) parameterized using detailed knowledge of the genetic architecture underlying focal traits can help reject neutral processes as important processes driving parallel patterns, giving greater confidence to the role of natural selection in causing parallel evolution (Santangelo et al. 2018a).

The advent of next-generation sequencing technologies has allowed recent work to move beyond the identification of parallel phenotypic responses, to identifying the genomic basis of parallel evolution (Stern 2013). At the genomic level, parallelism may arise due to selection acting on genetic variation already present in populations (Colosimo et al. 2005), or on novel mutations arising independently across populations (Chan et al. 2010), providing insight into the relative roles of standing variation versus *de novo* mutation for adaptation (Barrett and Schluter 2008). Alternatively, parallelism could occur via adaptive introgression of a beneficial allele (Heliconius Genome Consortium 2012). Recent models that distinguish between these three modes of parallelism using patterns of genetic diversity surrounding a selected site (Lee and Coop 2017) promise to shed light on how adaptation proceeds in natural populations. With increasing evidence that urbanization drives repeated phenotypic and genomic responses (see sections 3.2 and 3.5), cities are well suited to test alternative hypotheses about how often and by what mechanisms parallel evolution proceeds in nature.

outcomes in response to similar environmental stressors (Losos 2011). However, parallelism can occur at multiple levels, from phenotypes to amino acids and individual nucleotides, and understanding the factors that facilitate or constrain parallel responses at all levels of biological organization is important for understanding how adaptation occurs in nature.

The frequency with which parallel evolution occurs may in part depend on whether we consider evolution at the phenotypic or genotypic level. For example, because natural selection causes an increase in a population's mean fitness (Fisher 1930), we expect fitness to evolve in parallel across populations that are adapting to their environment, independent of the genotypes and phenotypes being selected (Travisano et al. 1995). However, fitness increases due to selection on traits through their effects on an organism's performance (Arnold 1983); thus, there may be multiple phenotypic solutions to the same environmental challenges (Losos 2011) and/or multiple traits may lead to the same functional outcome (i.e., many-to-one mapping; Thompson et al. 2017). Therefore, numerous trait combinations may lead to similar increases in fitness in a given environment, reducing the likelihood that independent populations will evolve in the same way. Moreover, the probability of parallelism decreases at the genetic level with an increasing number of genes contributing to a given phenotype (Conte et al. 2012; MacPherson and Nuismer 2017). The likelihood that any one gene is involved in parallel adaptation depends on its mutational target size, mutation rate, and the average fitness effects of mutations of the gene (Stern 2013), making highly pleiotropic loci unlikely targets of parallel selection. Overall, we therefore expect levels of parallel evolution to decrease with increasing phenotypic (i.e., many possible traits) or genetic (i.e., many possible genes) complexity.

Studies of phenotypic and genetic parallelism have employed a range of experimental and observational approaches. In the laboratory, experimental studies have propagated replicate populations (e.g., *E. coli*, yeast, *Drosophila*) under the same selective pressures, finding that selection typically targets similar phenotypes and genes under these controlled conditions (Graves et al. 2017; Lenski 2017). While these experiments provide insight into how selection

operates and the frequency with which parallelism can happen, controlled environments contain fewer selective forces than do natural populations. By contrast, evolutionary biologists have investigated parallelism by leveraging 'natural' experiments, such as the repeated colonization of freshwater by marine sticklebacks (*Gasterosteus aculeatus*) (Colosimo et al. 2005; Barrett et al. 2008), Bahamas mosquitofish (*Gambusia hubbsia*) adapting to different predator regimes (Langerhans 2018), insects adapting to their plant hosts (Zhen et al. 2012), and beach mice (*Peromyscus polionotus*) inhabiting different coloured substrates (Steiner et al. 2009). Although the extent of parallel evolutionary responses among populations can be high (e.g., stickleback armour plating), many populations still show imperfect parallelism. For example, morphological adaptation of fish populations in response to similar ecological conditions spans a continuum from non-parallel to parallel evolution (Oke et al. 2017). Thus, a major goal of current research is to disentangle the drivers of parallel and non-parallel evolutionary responses to determine the factors shaping repeated evolution in natural populations.

The spatial replication of cities makes them one of the best replicated, large-scale, and unintended experiments for addressing the extent of parallelism in nature. Cities often show commonalities in many environmental features (see section 3.3) and are associated with changes in land use that may drive parallel evolution of urban populations (Johnson and Munshi-South 2017; Santangelo et al. 2018b; Rivkin et al. 2019). However, no two cities are identical and variation in city age, human population density, infrastructure density, city size, local climatic conditions, socioeconomics, and governmental regulations, among other factors (Chapter 2; Grimm et al. 2008), may all contribute to divergent evolutionary responses across independent urban environments (Johnson and Munshi-South 2017). For example, increased urban temperatures are a known driver of adaptive differentiation between urban and non-urban populations in multiple species (Chapter 6; Brans et al. 2018; Diamond et al. 2018), but larger cities exhibit on average stronger urban heat island effects (Zhou et al. 2017). Thus, larger cities may impose stronger selection for heat tolerance, whereas we may not expect heat tolerance to evolve in smaller cities. In addition, variation in land use and zoning within a city can generate fine-scale heterogeneity in environmental conditions and reduce the amount of parallelism among populations. For example, while highly fragmented urban populations of the plant holy hawksbeard (*Crepis santa*) produce a greater proportion of dispersing seeds to increase the probability of successful establishment, less fragmented urban populations produce approximately the same proportion of dispersing seeds as nearby non-urban populations (Chapter 8; Cheptou et al. 2008; Dubois and Cheptou 2017). The global distribution of urban habitats and the variation they exhibit in ecological and environmental processes make cities well suited to teasing apart the factors contributing to parallel and non-parallel evolutionary responses in nature.

In this chapter, we begin by assessing how often parallel evolutionary responses to urban environments have been observed, both within and across species. We then review the agents responsible for driving parallelism across cities, followed by a consideration of the causes of non-parallel responses to urban environments. Finally, we briefly discuss practical steps for future research.

3.2 How often do species show parallel responses to urbanization?

We qualitatively reviewed the current literature to estimate the frequency of urban-driven parallel evolutionary responses. Studies were identified using two methods: (1) by compiling all multicity studies listed in Johnson and Munshi-South (2017) and Miles et al. (2019); and (2) by searching Google Scholar for studies that included the following combinations of terms: 'parallel' or 'convergence'; and 'selection', 'adaptation', 'population genetics', 'mutation', or 'gene flow'; and 'urban*' or 'city'. We identified 30 studies that together examined the evolutionary responses of 18 species, inferred at the phenotypic or genetic level, across multiple cities (Table 3.1). A third of these studied parallel evolution in arthropod species (N = 6 species), followed by birds (N = 3 species), fish (N = 3 species), reptiles (N = 3 species), plants (N = 2 species), and a single mammal species (Table 3.1). The majority of studies were

conducted in cities in North America (N = 10 species) and Europe (N = 6 species), with the remaining two studies being conducted in South America (Table 3.1).

For a given study, we extracted the following variables: the type of evolutionary mechanism measured (i.e., mutation, natural selection, genetic drift, gene flow), the number of cities measured, the number of sites sampled per city, the average number of individuals per site, and the number of cities that showed the same response. The number of cities sampled per species varied between 2 and 13 (mean = 5 cities), with between 1 and 200 sites sampled per city (mean = 21 sites). For each study, the degree of parallelism was measured as the proportion of cities showing the same direction of evolutionary change in response to urbanization (i.e., 'vote-counting'; Orr 2005; Bolnick et al. 2018). Because random changes in phenotypic traits may sometimes erroneously indicate parallel evolution (Box 3.1), replication of population pairs is essential in distinguishing parallel selection from neutral processes (Bolnick et al. 2018). However, because individual studies sampled only a few cities and studies varied in the traits they measured, we made no formal attempt to test if the observed levels of parallelism exceeded the null expectation; rather, our goal was simply to explore whether species respond in the same way to urban environments.

Parallel evolution in response to cities was generally high, with 78 per cent of species (14/18 species) exhibiting some degree of parallel evolution. However, the extent to which focal taxa responded in parallel across cities was variable. For example, Thompson et al. (2016) and Johnson et al. (2018) identified parallel decreases in cyanogenesis in white clover (*Trifolium repens*) in 16 of the 24 cities sampled, indicating 67 per cent of the cities showed parallelism in this species. Similarly, Winchell et al. (2016, 2018, 2019) identified 80 per cent parallelism (4 of 5 cities sampled) in toepad and limb morphology in Puerto Rican crested anole lizards. In both species the phenotypic changes observed were attributed to similar selection pressures across cities (see section 3.3 for further discussion on selection agents in cities). Additionally, 44 per cent of species (N = 8 species) exhibited parallelism across all cities. For example, the red-tailed bumble bee (*Bombus lapidarius*) exhibited genomic signatures of selection at loci involved in

Table 3.1 List of studies that have examined urban evolutionary patterns across multiple cities. The table includes the species, organism type, and continent; evolutionary mechanisms studied, and degree of parallelism observed; number of cities sampled, number of sites sampled per city, and average number of individuals sampled per sites; and relevant citations. The degree of parallelism was measured as the proportion of cities where the same trend was observed. Where necessary, the degree of parallelism has been broken down into separate values for each mechanism studied. Number of cities is a cumulative value across multiple studies of the same species. Number of sites per city and individuals per site is shown as a range when multiple studies have been combined across species.

Species	Organism type	Continent	Evolutionary mechanism	Degree of parallelism	Number of cities	Sites per city	Average number of individuals per site	References
Abax ater	Arthropod	Europe	Gene flow; genetic drift	0	2	6	185	Desender et al. 2005
Aedes aegypti	Arthropod	North America	Selection	1	6	1	48	Saavedra-Rodriguez et al. 2018
Bombus lapidarius	Arthropod	Europe	Gene flow; genetic drift; selection	1	9	2	198	Theodorou et al. 2018
Latrodectus hesperus	Arthropod	North America	Gene flow	0.30	10	2	10	Miles et al. 2018
Pterostichus madidus	Arthropod	Europe	Gene flow; genetic drift	0	2	6	301	Desender et al. 2005
Temnothorax curvispinosus	Arthropod	North America	Selection	0.67	3	4–6	19 colonies	Diamond et al. 2018
Athene cunicularia	Bird	South America	Gene flow; genetic drift	1	3	3	20	Mueller et al. 2018
Larus argentatus	Bird	North America	Mutation	1	6	1–2	90	Yauk and Quinn 1996; Yauk et al. 2000
Turdus merula	Bird	Europe	Gene flow; genetic drift; mutation; selection	Mutation: 0.83 Selection: 1	13	2	24–30	Partecke and Gwinner, 2007; Evans et al. 2009; Evans et al. 2012; Miranda et al. 2013; Mueller et al. 2013
Fundulus heteroclitus	Fish	North America	Mutation; selection	Genetic drift: 1 Mutation: 0 Selection: 0.75	5	2–63	20–150	McMillan et al. 2006; Whitehead et al. 2010; Whitehead et al. 2012; Reid et al. 2016
Gila orcuttii	Fish	North America	Gene flow; genetic drift	1	6	1–5	19	Benjamin et al. 2016
Semotilus atromaculatus	Fish	North America	Selection	1	2	10–15	21	Kern and Langerhans 2018

Species	Organism type	Continent	Evolutionary mechanism	Degree of parallelism	Number of cities	Sites per city	Average number of individuals per site	References
Rattus norvegicus	Mammal	North America	Gene flow	1	4	NA*	305	Combs et al. 2018
Lepidium virginicum	Plant	North America	Selection; gene flow	Selection: 1 Gene flow: 0.75	5	7–19	6	Yakub and Tiffin 2017
Trifolium repens	Plant	North America	Gene flow; genetic drift; selection	Selection: 0.67 Gene flow/genetic drift: 0.38	4	20–128	20	Thompson et al. 2016; Johnson et al. 2018
Anolis cristatellus	Reptile	South America	Selection	0.8075	5	2–6	17–90	Winchell et al. 2016, 2018, 2019
Podarcis muralis	Reptile	Europe	Gene flow	1	4	200	207	Beninde et al. 2018
Salamandra salamandra	Reptile	Europe	Gene flow; genetic drift	1	2	7–16	10–19	Straub et al. 2015; Lourenço et al. 2017

*In this study, the authors do not report number of sites sampled per city. Instead, they sampled large sections of each city to collect *R. norvegicus* individuals.

oxidative stress in all 9 cities that were sampled (Theodorou et al. 2018). These results suggest that levels of parallelism in cities may overall be quite high.

Our survey indicates that selection imposed by urban environments frequently drives phenotypic and genetic parallelism across cities. For example, all eight of the taxa where phenotypic change across multiple cities was measured showed parallel responses (Table 3.1). Similarly, three out of the four studies that measured adaptive genetic changes across cities found parallel responses. However, few studies have simultaneously examined phenotypic and genetic parallelism, and for those that have, the results are inconsistent across studies. For example, urban populations of European blackbirds (*Turdus merula*) show parallelism at both the phenotypic and genotypic level. Specifically, they have evolved to be less migratory and more cautious than non-urban populations (Evans et al. 2009; Miranda et al. 2013). In 10 out of 12 urban/non-urban pairs, these behavioural shifts corresponded to variation in an exonic microsatellite and increased genetic divergence at the *SERT* locus, a candidate gene for harm avoidance (Mueller et al. 2013). By contrast, Reid et al. (2016) found that four geographically disparate populations of killifish (*Fundulus heteroclitus*) located in polluted estuaries near cities all evolved resistance to toxins, but the genetic architecture of this resistance varied across the populations. Reasons for non-parallelism at the genetic level are discussed in detail in section 3.4.

Studying the effects of urbanization on the genetic diversity of populations can also provide insight into how urbanization affects non-adaptive parallel evolution. The amount of genetic diversity within and between populations is shaped by genetic drift, gene flow, and natural selection. For example, the frequent fragmentation, degradation, and isolation of habitats in urban environments are expected to increase the strength of genetic drift within smaller urban populations and decrease gene flow between populations. In such scenarios we expected lower genetic diversity within populations of many urban-dwelling species and increased genetic differentiation between populations. This reduction of genetic diversity can influence the likelihood of parallel adaptive evolution by decreasing the genetic variation available for natural selection to act upon. For the 12 species in which genetic diver-

sity was measured, 67 per cent ($N = 8$ species) found parallel changes in neutral genetic diversity. However, the direction of change was inconsistent across studies. In some cases, genetic diversity was lower in cities ($N = 4$ species), and urban populations were more differentiated from one another ($N = 3$ species), whereas in other cases, genetic diversity was higher in cities with urban populations less differentiated from each other ($N = 4$ species). While dispersal ability undoubtedly shapes the genetic diversity of populations, parallel shifts in genetic diversity were not explained by any particular dispersal mechanism or taxonomic group in our dataset (Table 3.1). The inconsistent effects of urbanization on patterns of neutral genetic variation and differentiation despite high levels of phenotypic and genetic parallelism observed across studies suggests that cities may not be eroding the genetic variation required for populations to adapt in parallel. This highlights the need to study the effects of urbanization on both adaptive and non-adaptive evolutionary processes simultaneously in the same study system to better understand how variation in non-adaptive evolutionary processes influences the likelihood of repeated phenotypic and genetic responses across cities.

3.3 What agents drive parallel evolution across cities?

Understanding parallel evolution in response to urbanization requires disentangling the agents responsible for adaptive and non-adaptive evolution in cities. Parallel adaptive evolution is most likely to arise when cities consistently alter the biotic and abiotic environmental factors that are important for determining the fitness of urban-dwelling taxa. Alternatively, parallel non-adaptive evolutionary responses (e.g., consistent losses of genetic diversity within populations across cities) are more likely to arise when urbanization affects the physical landscape such that the strength of genetic drift and the amount of gene flow are altered, thereby influencing the evolutionary potential of urban populations. In this section, we review the three most well-documented driving forces behind adaptive and non-adaptive evolutionary responses to urbanization: urban heat islands, pollution, and habitat fragmentation. Importantly, temperature (Bennett et al. 1992), pollution (Klerks and Weis 1987), and fragmentation (Cheptou et al. 2016) are known drivers of evolution in non-urban contexts; however, we are only beginning to understand their role in driving parallel evolutionary responses of urban populations.

3.3.1 Urban heat islands

The urban heat island effect is a phenomenon wherein heat is retained in cities more than it would be in a non-urban area (see also Chapter 6). This effect is often attributed to the ubiquity of impervious and heat-absorbing materials (e.g., asphalt or concrete) in a city, and the relatively low abundance of heat-alleviating vegetated surfaces. Thus, cities often register temperatures higher than the surrounding non-urban areas by several degrees Celsius (Zhou et al. 2017). Additionally, these increased temperatures keep cities warmer throughout the year and have extended warm seasonality (Zhou et al. 2017). Urban heat islands present one of the best replicated 'natural' experiments for assessing phenotypic and genetic responses to increased temperature and the extent to which these responses are similar across cities.

The urban heat island effect has contributed to drastic behavioural and phenotypic changes in animals, plants, and zooplankton. Warmer city temperatures have resulted in reduced migration for multiple bird species including urban populations of the dark-eyed junco (*Junco hyemalis*) and the European robin (*Erithacus rubecula*) (Partecke and Gwinner 2007). Furthermore, some urban European blackbird (*Turdus merula*) populations now overwinter in cities instead of migrating south (Partecke and Gwinner 2007; Evans et al. 2012). Elevated, sustained temperatures are also a probable agent of ecological success in urban populations of Virginia pepperweed (*Lepidium virginicum*). Specifically, urban populations of this species bolt sooner, grow larger, and produce more seeds than their non-urban counterparts in four large North American cities (Yakub and Tiffin 2017). Consequently, the consistent signature of the urban heat island effect has promoted population persistence within cities for organisms that would otherwise not be found.

There is mounting evidence that species are evolving to tolerate the increased temperatures found in cities. For example, researchers have identified evidence of selection on genes associated with regulating environmental stress, including heat stress in urban populations of the red-tailed bumble bee (Theodorou et al. 2018). Acorn ants (*Temnothorax curvispinosus*) in Cleveland, OH, and Knoxville, TN, have evolved reduced cold tolerance in urban populations rather than increased tolerance for heat stress (Diamond et al. 2018). Research investigating water flea (*Daphnia magna*) populations distributed along an urbanization gradient in Belgium suggested the evolution of heat tolerance across several species. Specifically, *D. magna* experienced a faster pace-of-life strategy than its non-urban counterparts, including improved energy metabolism and tolerance to physiological stress induced by elevated water temperatures in the city (Brans et al. 2018). These studies provide evidence for parallel responses within and between species to the altered thermal regimes imposed by urbanization. The response of organisms to urban heat islands offers the opportunity to observe repeated adaption to higher temperatures and may lead to key insights into whether and how organisms will adapt to a warming climate (Youngsteadt et al. 2015).

3.3.2 Pollution

Several forms of pollution pose physiological stresses on urban-dwelling organisms that may impose selection. Light pollution disrupts natural light/dark cycles whereas noise pollution can disrupt animal communication and chemical pollution in the air, water, or soil can influence the physiology and performance of organisms (Isaksson 2015). Although some studies have investigated the evolutionary response to urban pollution in individual cities, chemical pollution is the only form that has been studied across multiple urban areas.

One of the best examples is from the Atlantic coast of North America where industry has released toxic pollutants into many wetlands and aquatic habitats. The release of polychlorinated biphenyls (PCBs) into urban estuaries has created a strong selective pressure on Atlantic killifish native to these sites. As a result, multiple killifish populations have evolved increased tolerance to PCBs via desensitization of the aryl hydrocarbon receptor-based signalling (AHR) pathway (Whitehead et al. 2010, 2012; Reid et al. 2016). While populations varied in the AHR genes underlying pollution tolerance, one particular gene in the AHR pathway, the aryl hydrocarbon receptor-interacting protein (AIP), was consistently differentiated among four sensitive-tolerant population pairs (Reid et al. 2016). A recent reanalysis of this gene identified migration of ancestral variation in three northern tolerant populations and independent *de novo* mutation in the southern population as the most likely mechanism for the repeated evolution of pollution tolerance in killifish (Lee and Coop 2017). The unintentional introduction of chemical pollutants into aquatic environments has generated novel selective pressures and spurred multiple acute impacts on these fish. In addition, the widespread release of chemical pesticides into the environment have indelibly shaped ecosystems and the species that reside within them (Johnson and Munshi-South 2017), leading to possible eco-evolutionary feedbacks that await quantification.

Increased levels of pollution in the environment undoubtedly impose strong selection on populations for increased tolerance. The evolution of resistance is an unintended and undesired consequence of the profuse dissemination of insecticides to control disease vectors. Worldwide, urban populations of the mosquito *Aedes aegypti* have developed resistance to pyrethroid insecticides, which are commonly used to prevent adult populations from spreading dengue, Zika, yellow fever, malaria, and other human diseases. Parallel signatures of selection were found across populations collected from urbanized areas where there were eleven substitutions in the voltage-gate sodium channel gene (*vgsc*) (Saavedra-Rodriguez et al. 2018). The evolution of pesticide Resistance is a common phenomenon in many urban pests including bed bugs (Romero and Anderson 2016), rats (Rost et al. 2009), lice (Koch et al. 2016), and cockroaches (Wada-Katsumata et al. 2013).

3.3.3 Habitat fragmentation

The process of urbanization alters the landscape via expansion of urban development and fragmentation of natural habitats, resulting in increased isolation of natural habitat patches (Grimm et al. 2008). For organisms unable to cross the urban matrix, dispersal between the fragmented patches decreases with greater urban development. In such cases it is expected that population sizes decline, gene flow

decreases, and genetic drift increases, resulting in reduced genetic diversity within patches and increased differentiation between patches (discussed further in section 3.4).

Urban habitat fragmentation affects both land-dwelling and aquatic organisms. For aquatic organisms, urban habitat fragmentation is due to both habitat degradation and redirected waterways through dams, canals, and other forms of altered waterways. Dams fragment streams and prevent migration among upstream and downstream aquatic populations, posing strong barriers to gene flow. Dams are linked to reduced genetic diversity in several species of fish, including the southern California arroyo chub (*Gila orcuttii*) (Benjamin et al. 2016). Specifically, habitat loss caused by extensive damming has resulted in reduced genetic diversity, population bottlenecks, and low effective population sizes in arroyo chub populations (Benjamin et al. 2016). Indeed, although the species previously inhabited much of the Los Angeles River and the Santa Margarita River watersheds, recent surveys by the California Department of Fish and Wildlife reported major declines in the size and distribution of populations. Studies in terrestrial systems have similarly identified losses in genetic diversity and increased differentiation in response to urban habitat fragmentation, although the consequences of fragmentation for parallel adaptive evolution in urban environments have not been explicitly investigated (see section 3.2, Table 3.1).

3.4 Why does parallelism *not* occur?

Although there is evidence supporting parallel evolution in response to urbanization, there are certain evolutionary and ecological mechanisms that may inhibit parallelism. Identifying why parallelism is not observed is important for a complete understanding of urban evolutionary and ecological research. In this section, we review some of the mechanisms that may lead to non-parallel evolutionary responses across cities.

3.4.1 Environmental variation

The role that environmental variation within and between cities plays in inhibiting parallel evolution is often overlooked. Although cities typically have increased impervious surface cover, higher temperatures, increased habitat fragmentation, and greater pollution (Grimm et al. 2008), the magnitude and extent of these anthropogenic alterations varies between cities (Johnson and Munshi-South 2017). For example, variation in the amount of impervious surface cover and the concomitant increased temperatures between cities likely affects variation in the strength of selection on traits related to thermal tolerance across cities. More generally, we expect the likelihood of parallel evolutionary responses to decrease with decreasing similarity in the urban environmental features driving adaptation across cities. Thus, quantifying changes in the direction and magnitude of environmental divergence (e.g., Stuart et al. 2017) across cities will be essential to understanding how quantitative differences in putative selective agents are contributing to parallel or non-parallel evolutionary responses across independent urban environments.

Environmental heterogeneity within a city may additionally constrain parallel evolutionary responses due to variation in the strength and direction of selection along the urbanization gradient. The position of populations in cities can vary with respect to numerous properties like the extent of local vegetation, land-use type, and air temperature, among other factors (Grimm et al. 2008). This variation can arise because urbanization fragments natural habitats, creating isolated patches of natural habitat within the urban matrix. From a landscape planning perspective, there exists heterogeneity in urban land-class types including business districts, subdivisions, urban fringe, and non-urban habitats (see also Chapter 2). These land classes and the heterogeneity they create are intrinsic to cities, and it is crucial to quantify their influence on the selective agents that drive urban evolution to understand how variation in land use across cities influences the likelihood of parallelism.

3.4.2 Gene flow

The extent of parallelism observed across cities is expected to depend strongly on patterns of gene flow between different habitat types (i.e., urban/non-urban) and between the same habitat types across cities. Gene flow between urban and non-urban habitats within a city is expected to constrain

local adaptation to the urban environment (Lenormand 2002). If the magnitude of local adaptation varies across cities due to differences in the landscape barriers to gene flow, then the extent of parallelism among urban–non-urban population pairs will be reduced. For example, smaller cities may have more gene flow between urban and non-urban populations relative to larger cities due to the shorter distance required to cross the urban matrix; therefore, for the same study species, local adaptation may be more constrained in the smaller city, leading to differences in the magnitude of adaptation and reduced parallelism between cities (assuming selection is the same). However, taxa are likely to vary in the extent to which urbanization influences dispersal, and consequently the extent to which gene flow constrains adaptation. For example, human-mediated transport facilitates dispersal in some species (e.g., Booth et al. 2015; Johnson and Munshi-South 2017), such as the western black widow spider (*Latrodectus hesperus*) that has recently colonized non-urban habitats due to human movement (Miles et al. 2018). However, some urban habitats were better connected to non-urban habitats than others, suggesting that the amount of human-mediated dispersal is likely to vary across cities, again leading to variation in the amount of constraint on local adaptation and a reduced likelihood of parallelism. Quantifying how variation across cities (e.g., size, age, location) drives variation in gene flow and local adaptation is essential to our understanding of the causes of (non)parallel (*sensu* Bolnick et al. 2018) evolutionary responses to urbanization and should be a key component of future work in urban evolutionary ecology.

By contrast to gene flow between urban and non-urban populations, gene flow between urban populations across cities can facilitate parallel local adaptation to urbanization by spreading adaptive alleles. In addition to shared variation facilitating pollution tolerance in killifish (section 3.3.2), the spread of a novel transposable element (TE) in the *cortex* gene of peppered moths (*Biston betularia*) facilitated adaptation of moth populations to darkened surfaces caused by soot released from industrial processes across much of Europe (Van't Hof et al. 2016). The potential for human-mediated transport of many species across cities may make adaptive introgression a common mechanism underlying parallel evolutionary responses to urbanization, and there are clear candidate systems worthy of additional exploration. For example, Yakub and Tiffin (2017) found that urban populations of the annual plant *Lepidium virginicum* across four large North American cities were on average more closely related to one another at neutral markers than to nearby non-urban populations and were all phenotypically similarly in a number of traits (section 3.3.1). It is unknown whether this phenotypic parallelism is the result of selection on ancestral variation or adaptive variation shared among cities via gene flow. Importantly, signatures of genomic adaptation from both of these modes can be quite similar and distinguishing them requires detailed knowledge of the demographic history of populations that have adapted in parallel (Lee and Coop 2017).

3.4.3 Genetic drift

Urbanization frequently leads to habitat fragmentation, which isolates populations and, in turn, reduces the effective population sizes within these patches, which increases the strength of genetic drift and reduces genetic diversity (Johnson and Munshi-South 2017). Stronger genetic drift has consequences for urban adaptation because natural selection is more efficient in larger populations (Kimura and Crow 1964). Moreover, stochastic changes in allele frequencies associated with drift decrease the probability that separate populations fix the same alleles or traits (Kimura and Crow 1964; Orr 2005). Given the frequent observation of reduced population sizes in urban environments (e.g., Evans et al. 2009; Benjamin et al. 2016), genetic drift is a likely candidate for reduced parallelism across cities. Moreover, species are likely to vary in the extent to which urbanization results in pronounced demographic changes within populations: invasive species that are tolerant of—and sometime dependent on—environmental disturbance may maintain larger populations and be more resistant to the effects of genetic drift relative to more susceptible native populations (Johnson and Munshi-South 2017; Santangelo et al. 2018a). A major goal of future work in urban evolution concerns understanding how a species'

natural history influences its susceptibility to demographic shifts imposed by urbanization and quantifying how these shifts affect the likelihood of parallel evolution across cities.

3.4.4 Genetic architecture of adaptations

Parallelism can also be limited by genetic architecture and the scale at which parallelism is considered. Polygenic traits or 'many-to-one mapping' of genotype-to-phenotype can allow for parallel phenotypic responses with distinct genetic architectures (Losos 2011). For example, pollution tolerance is a parallel phenotype in killifish, but the genetic mechanism of this tolerance varied among populations; therefore, populations were parallel at the phenotypic level, but due to variation in the genes regulating the response there was incomplete parallelism at the genetic level (Reid et al. 2016). In general, it is probable that phenotypic parallelism will be more common than genetic parallelism and genetic parallelism will be most common when an evolving trait is derived from a single gene (Orr 2005; Conte et al. 2012).

Pleiotropy—when a single gene influences two or more traits—can limit parallel evolutionary responses. Mutations in genes that influence many traits are less likely to have net beneficial fitness effects and persist in natural populations (Stern and Orgogozo 2009; Stern 2013). Thus, replicate populations are less likely to have access to the same genetic variation and the probability of parallel genetic evolution is expected to decrease with increasing pleiotropy (Chevin et al. 2010; Bolnick et al. 2018). However, the fate of mutations in pleiotropic loci depends not only on their effects in influencing trait distributions, but also on the strength and direction of selection on each of the traits, which can vary across environments (Barrett et al. 2009). We still lack an understanding of the traits most often favoured in urban environments and know much less about their genetic architectures. Studies identifying parallel phenotypic responses to urbanization combined with molecular assays identifying the genetic basis underlying local adaptation will facilitate quantifying the contribution of genetic constraints to parallel urban evolutionary responses.

3.5 Recommendations for future studies

Given the growing attention focused on parallel evolution in urban environments, it is useful to outline guidance for the design of future studies. Currently, most urban evolution studies sample populations from only a single city (Figure 2.2), which makes identifying parallelism between cities impossible. Thus, there is a need for studies to collect samples across multiple cities because our confidence that observed parallel responses exceed neutral expectations increases with the number of cities sampled. Additionally, many analyses identifying parallel evolutionary responses rely on pooling samples within habitat types and compare multiple population pairs (Box 3.2), although such analyses have yet to be performed in an urban context.

Therefore, as the number of cities sampled increases, fewer urban and non-urban populations need to be sampled per city, provided at least multiple individuals are sampled to accurately estimate trait and/or allele frequency means and variances (Lotterhos and Whitlock 2015). However, with fewer cities sampled, more populations per habitat type and more individuals per population will be required to increase the precision of trait and allele frequency means and variances, and thus to improve power to detect evolutionary change in response to urbanization. In summary, if the goal is to identify the extent of parallel evolutionary responses to urbanization, we recommend researchers prioritize sampling more cities, followed by increasing population and individual sampling within cities.

Box 3.2 Assessing parallelism across urban environments

We recommend that recently developed statistical tools be employed to assess the extent to which urbanization has driven parallel evolution across cities. For example, variance partitioning (e.g., MANCOVA; Langerhans 2018; Figure 3.1A) can be used to assess the amount of trait divergence among habitat types (i.e., urban/non-urban), cities, and the extent to which urban/non-urban trait divergence is consistent across cities. Models with traits as the response variable can be fit with habitat type, city, and the habitat by city interaction as predictors. In this model, a significant habitat effect corresponds to parallel evolution to urban environments (i.e., consistent effect of cities on mean trait values), and the city term measures variation in mean trait values across different cities (averaged across urban and non-urban populations). Finally, a significant habitat type by city interaction suggests the extent of parallelism varies across cities, justifying future work exploring the many potential causes of this divergence. This approach additionally allows researchers to quantify the effect size for each term in the model (e.g., using partial η^2) and plot these against one another to get a sense of the relative importance of parallel vs divergent evolution (Figure 3.1A); some traits may show primarily parallel responses across cities (light grey point in Figure 3.1A), while others may show principally divergent responses (dark grey point in Figure 3.1A). By plotting the effect sizes for all measured traits, researchers can assess the relative contributions of parallel and divergent evolutionary responses to

cites. However, where divergent responses to cities are found to be common among traits in a given system, variance partitioning does not distinguish between divergence due to differences in the direction (i.e., urban populations have larger trait values than non-urban populations in some cities but smaller values in others) or magnitude (i.e., absolute difference in mean trait values between urban and non-urban populations) of evolutionary change (Bolnick et al. 2018).

In cases where multiple traits have been measured, we additionally recommend the use of phenotypic change vector analysis (PCVA) (Adams and Collyer 2009; Figure 3.1B), which has recently become a popular approach in studies of parallelism (Oke et al. 2017; Stuart et al. 2017; Bolnick et al. 2018) and overcomes the limitations of variance partitioning (see previous paragraph) (Bolnick et al. 2018). For each urban/non-urban population pair, PCVA calculates the mean multivariate trait value (i.e., 'centroid') and connects the centroids for each city by a line forming a vector (Figure 3.1B). The angles (Θ) and differences in lengths (ΔL) among the vectors measure the direction and magnitude of trait divergence across population pairs (i.e., cities), respectively (Bolnick et al. 2018). Because of the geometric approach taken by PCVA, it allows parallel and convergent evolution to be distinguished and separately quantified: strict urban convergence occurs when urban centroids (E_d in Figure 3.1B) are closer in trait space than their nearby non-urban

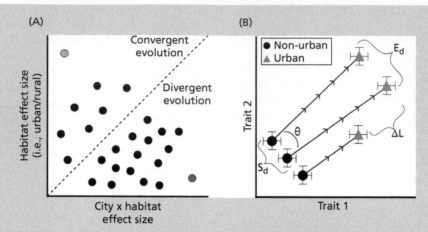

Figure 3.1 Parallel evolution to urban environments can be tested using recently developed statistical frameworks that are common among studies of parallel evolution in well-replicated systems. (A) Variance partitioning allows effect sizes of parallel and divergent evolution to be compared for a large number of traits, thereby quantifying ubiquity of parallel evolution in a given system. Researchers can fit a model with traits as response variables, and habitat type (i.e., urban/non-urban), city, and the city × habitat type interaction as predictors. From this model, researchers can plot habitat effect size (e.g., using partial η^2; Langerhans and DeWitt 2004), which measures parallel evolution, against the city × habitat effect size, which measures unique responses to different cities (i.e., divergence). Examining individual traits (points), some are likely to show primarily parallel responses (light grey point), whereas others may show primarily divergent responses (dark grey point). In this hypothetical example, divergent responses are more common across all traits, suggesting that, on average, different traits respond uniquely to the presence of cities. (B) We additionally recommend use of phenotypic change vector analysis (PCVA) to analyse multivariate trait data. Traits are condensed into 'centroids' (points) in multivariate trait space for both urban (grey triangles) and non-urban (black points) populations. These points are then connected to form a vector, with one vector for each city. Angle (Θ) and differences in length (ΔL) separating all pairwise vectors can be calculated and quantify the direction and magnitude of evolutionary change across cities, respectively. In addition, comparing distance between non-urban populations (S_d, starting distance in Bolnick et al. 2018) and urban populations (E_d, end distance in Bolnick et al. 2018) in multivariate trait space allows parallelism and convergence to be distinguished; strict convergence occurs when $E_d < S_d$, whereas strict parallelism occurs when $S_d = E_d$ and Θ and ΔL are both 0. (Figures modified from Stuart et al. (2017) and Bolnick et al. (2018).)

centroids (i.e., $E_d < S_d$, Figure 3.1B). In other words, urbanization has resulted in populations more similar to one another than their nearby non-urban populations initially were. By contrast, true parallelism occurs when two or more urban/non-urban population pairs have truly parallel divergence vectors (i.e., magnitude and direction of evolutionary change, $E_d = S_d$), despite urbanization changing mean trait values. In this approach, convergent populations need not be parallel and parallel populations need not be convergent. In addition, PCVA can be applied to data at other mechanistic levels (e.g., genome-wide neutral SNPs) (Stuart et al. 2017), allowing for a quantitative estimate of parallelism at

both phenotypic and genotypic scales. This approach can be performed on environmental data as well, allowing the direction and magnitude of phenotypic change to be directly compared to the extent of environmental variation across populations, providing important insight into the drivers of parallel and non-parallel evolutionary responses. While this approach can be extended for continuous sampling (e.g., urban/non-urban gradients), it suffers from more difficult interpretation of the contribution of individual traits to the differences in direction and length between multivariate trait vectors, unclear null hypotheses, and sensitivity to the number of traits included in the analyses (Bolnick et al. 2018).

Assessing the extent to which evolutionary responses are parallel across cities is a crucial step to understanding the ubiquity of urban evolution. However, equally important is understanding the features of cities that lead to non-parallel responses, which includes identifying the contribution of

environmental heterogeneity within urban and non-urban populations to divergent responses (see also Chapters 2 and 4). Research that seeks to quantify the effects of heterogeneity within habitat types will require prioritizing sampling populations and individuals within cities to better identify population-specific responses to urbanization. However, multiple cities would still be required to quantify how population-specific responses constrain or facilitate parallel evolutionary responses across cities. Additionally, this approach would facilitate identifying the landscape-level predictors of gene flow and population structure within and between urban and non-urban populations, something that is not possible with designs that include few populations per habitat type. Furthermore, when addressing urban landscape heterogeneity and landscape-level predictors of gene flow, the quantification of the urban environment is a useful tool (see Chapter 2). For highly mobile species with no clear boundaries between populations, an individual sampling strategy can be adopted to identify urban predictors of gene flow (Beninde et al. 2018), or identify genetic loci under selection across the genome, provided urban and non-urban individuals are pooled. Once phenotypic and genotypic data have been collected, we recommend the use of recently developed statistical tools for identifying parallel evolution in well-replicated systems (Box 3.2, Figure 3.1).

Insight into phenotypic parallelism will be improved if researchers measure multiple traits. The extent of phenotypic parallelism often differs among traits (Oke et al. 2017; Bolnick et al. 2018; Langerhans 2018), and researchers risk missing signatures of parallelism if there is a focus on a limited number of traits. In addition, making detailed measurements of multiple phenotypes will be important to elucidate the traits most commonly targeted by selection in urban habitats and identifying the agents underlying urban evolutionary change. Moreover, identifying the genetic basis underlying adaptive traits in cities (Chapter 5; Reid et al. 2016; Thompson et al. 2016; Van't Hof et al. 2016) will facilitate quantifying the ubiquity of parallelism at multiple mechanistic levels and enable disentangling the factors that drive parallel (or non-parallel) genetic and phenotypic responses. In systems with well-developed genomic resources (e.g., reference genome or transcriptome), target capture-based approaches (Mamanova et al. 2010) or low-coverage whole-genome sequencing (WGS) can facilitate identifying genes under selection through F_{ST} outlier tests and linking these to particular environmental conditions present in urban habitats through gene–environment association analyses (Hoban et al.

2016). This could be especially powerful when the functions of particular genes are known. In such cases, a useful next step would be to leverage patterns of neutral diversity linked to selected sites to distinguish among the multiple processes that can generate parallel genomic responses (i.e., *de novo* mutation, selection on standing variation, and adaptive introgression) (Lee and Coop 2017). In addition to the Atlantic killifish example already given (see section 3.3.2), other candidate study systems for such an approach include the repeated evolution of harm avoidance in European blackbirds at the *SERT* locus (Miranda et al. 2013; Mueller et al. 2013), differentiation in loci putatively involved in heat and oxidative stress across nine urban/non-urban population pairs of red-tailed bumblebees (Theodorou et al. 2018), and selection on *COL11A* (involved in craniofacial formation) and *AMY2A* (involved in starch digestion) among house sparrow (*Passer domesticus*) populations across Eurasia (Ravinet et al. 2018).

In the absence of genomic resources, alternative reduced representation techniques (e.g., ddRAD-seq) (Peterson et al. 2012) can be used to generate genome-wide SNPs for use in outlier or environmental association studies to identify the genomic basis of local adaptation and the extent to which the genetic targets of selection are consistent across cities. Differential gene expression analyses via RNAseq can additionally be used to identify the genetic basis of urban adaptation and assess the extent to which urbanization drives parallelism at multiple mechanistic levels (e.g., phenotype-to-gene).

3.6 Conclusions

Urbanization has proved to be a major evolutionary driver across many species on a global scale. Cities can be exploited as large-scale experiments to address parallelism for both adaptive and non-adaptive evolutionary change. The evolutionary processes and their likelihood of resulting in parallel and non-parallel evolution present a number of varying predictions to test for (see Table 3.2). For example, when increased temperature, pollution, and habitat fragmentation are similar across multiple cities, these factors can drive parallelism both within and across species. Indeed, parallelism has occurred in 75 per cent of the current urban evolution literature that examined multiple cities. However, even when

Table 3.2 Summary of whether the likelihood of parallelism is increased or decreased as a result of various evolutionary or environmental processes discussed throughout the chapter.

Process	Type of trait	Likelihood of parallelism	Notes
Selection	Polygenic	High at the phenotypic level. Lower at the genetic level	Assumes divergent selection of equal magnitude and direction across habitat types and cities
	Oligogenic	High at the phenotypic and genetic level	With fewer loci, there is a greater chance that phenotypes evolve due to changes in the same genes
Gene flow between urban and rural habitats	Any	Low at the phenotypic and genetic level	Gene flow between habitat types should reduce parallelism in so far as it constrains local adaptation
Gene flow within habitat types (urban–urban; rural–rural) across cities	Any	Variable	If gene flow introduces maladaptive alleles, parallelism should be reduced. However, gene flow may facilitate parallel responses through introgression of adaptive variants
Genetic drift	Any	Low at the phenotypic and genetic level	Strong drift would cause different cities to fix different alleles and phenotypes. In addition, it would reduce efficacy of selection through stochastic loss of adaptive variation. Both processes would reduce parallelism
Environmental heterogeneity	Any	Low at the phenotypic and genetic level	High levels of environmental heterogeneity are expected to reduce parallelism by changing the direction and/or magnitude of selection across populations

cities do support similar environments, various demographic processes and heterogeneity in selective environments across urban populations sometimes prevent parallelism from occurring. Future work will require identifying the genomic basis of parallel phenotypic changes across cities and employ recent statistical techniques to distinguish among the multiple processes that may lead to parallel genomic responses to selection. Cities represent a truly unparalleled frontier for investigating the complex interactions governing parallel evolution, enhancing our understanding of how adaptation proceeds in natural populations.

Acknowledgements

We thank the editors (Marta Szulkin, Anne Charmantier, and Jason Munshi-South) for the invitation to contribute this book chapter. Marta Szulkin, Jason Munshi-South, Anne Charmantier, Rebecca Irwin, and Nicolas Bierne provided helpful feedback that greatly improved the chapter. J.S.S. was funded by an NSERC PGS-D, and L.R.R. was funded by an NSERC CGS-D. This work was additionally funded by NSERC grants to M.T.J.J. and R.W.N. and a Canada Research Chair (CRC) to M.T.J.J.

References

Adams, D.C. and Collyer, M.L. (2009). A general framework for the analysis of phenotypic trajectories in evolutionary studies. *Evolution* 63, 1143–54.

Arendt, J. and Reznick, D. (2008). Convergence and parallelism reconsidered: what have we learned about the genetics of adaptation? *Trends in Ecology & Evolution*, 23, 26–32.

Arnold, S.J. (1983). Morphology, performance and fitness. *American Journal of Zoology*, 23, 347–61.

Barrett, R.D.H. and Schluter, D. (2008). Adaptation from standing genetic variation. *Trends in Ecology & Evolution*, 23, 38–44.

Barrett, R.D.H., Rogers, S.M., and Schluter, D. (2008). Natural selection on a major armor gene in threespine stickleback. *Science*, 322, 255–7.

Barrett, R.D.H., Rogers, S.M., and Schluter, D. (2009). Environment specific pleiotropy facilitates divergence at the ectodysplasin locus in threespine stickleback. *Evolution*, 63, 2831–7.

Beninde, J., Feldmeier, S., Veith, M., and Hochkirch, A. (2018). Admixture of hybrid swarms of native and introduced lizards in cities is determined by the cityscape structure and invasion history. *Proceedings of the Royal Society B: Biological Sciences*, 285, 20180143.

Benjamin, A., May, B., Brien, J.O., and Finger, A.J. (2016). Conservation genetics of an urban desert fish, the Arroyo Chub. *Transactions of the American Fisheries Society*, 8487, 277–86.

Bennett, A.F., Lenski, R.E., and Mittler, J.E. (1992). Evolutionary adaptation to temperature. I. Fitness responses of *Escherichia coli* to changes in its thermal environment. *Evolution*, 46, 16–30.

Bolnick, D.I., Barrett, R.D.H., Oke, K.B., Rennison, D.J., and Stuart, Y.E. (2018). (Non)parallel evolution. *Annual Review of Ecology, Evolution and Systematics*, 49, 303–30.

Booth, W., Balvín, O., Vargo, E.L., Vilímová, J., and Schal, C. (2015). Host association drives genetic divergence in the bed bug, *Cimex lectularius. Molecular Ecology*, 24, 980–92.

Brans, K., Stoks, R., and De Meester, L. (2018). Urbanization drives genetic differentiation in physiology and structures the evolution of pace-of-life syndromes in the water flea *Daphnia magna. Proceedings of the Royal Society B: Biological Sciences*, 285, 20180169.

Chan, Y.F., Marks, M.E., Jones, F.C., et al. (2010). Adaptive evolution of pelvic reduction of a *Pitx1* enhancer. *Science*, 327, 302–5.

Cheptou, P.-O., Carrue, O., Rouifed, S., and Cantarel, A. (2008). Rapid evolution of seed dispersal in an urban environment in the weed *Crepis sancta. Proceedings of the National Academy of Sciences of the United States of America*, 105, 3796–9.

Cheptou, P., Hargreaves, A.L., Bonte, D., and Jacquemyn, H. (2016). Adaptation to fragmentation: evolutionary dynamics driven by human influences. *Philosophical Transactions Royal Society B: Biological Sciences*, 372, 20160037.

Chevin, L.M., Martin, G., and Lenormand, T. (2010). Fisher's model and the genomics of adaptation: restricted pleiotropy, heterogenous mutation, and parallel evolution. *Evolution*, 64, 3213–31.

Colautti, R.I. and Lau, J.A. (2015). Contemporary evolution during invasion: evidence for differentiation, natural selection, and local adaptation. *Molecular Ecology*, 24, 1999–2017.

Colosimo, P.F., Hosemann, K.E., Balabhadra, S., et al. (2005). Widespread parallel evolution in stickleback by repeated fixation of *Ectodysplasin* alleles. *Science*, 307, 1928–33.

Combs, M., Byers, K., Ghersi, B.M., et al. (2018). Urban rat races: spatial population genomics of brown rats (*Rattus norvegicus*) compared across multiple cities. *Proceedings of the Royal Society B: Biological Sciences*, 285, 20180245.

Conte, G.L., Arnegard, M.E., Peichel, C.L., and Schluter, D. (2012). The probability of genetic parallelism and convergence in natural populations. *Proceedings of the Royal Society B: Biological Sciences*, 279, 5039–47.

Desender, K., Small, E., Gaublomme, E., and Verdyck, P. (2005). Rural–urban gradients and the population genetic structure of woodland ground beetles. *Conservation Genetics*, 6, 51–62.

Diamond, S.E., Chick, L.D., Perez, A., Strickler, S.A., and Martin, R.A. (2018). Evolution of thermal tolerance and its fitness consequences: parallel and non-parallel responses to urban heat islands across three cities. *Proceedings of the Royal Society B: Biological Sciences*, 285, 20180036.

Dubois, J. and Cheptou, P.-O. (2017). Effects of fragmentation on plant adaptation to urban environments. *Philosophical Transactions of the Royal Society B: Biological Sciences*, 372(1712), 20160038.

Evans, K.L., Gaston, K.J., Frantz, A.C., et al. (2009). Independent colonization of multiple urban centres by a formerly forest specialist bird species. *Proceedings of the Royal Society B: Biological Sciences*, 276(1666), 2403–10.

Evans, K.L., Newton, J., Gaston, K.J., et al. (2012). Colonisation of urban environments is associated with reduced migratory behaviour, facilitating divergence from ancestral populations. *Oikos*, 121, 634–40.

Fisher, R.A. (1930). The nature of inheritance. In: *The Genetical Theory of Natural Selection*, pp. 2–21. Oxford University Press, Oxford.

Gould, S.J. (2006). *Wonderful Life: The Burgess Shale and the Nature of History*. W.W. Norton, New York.

Graves, J.L., Hertweck, K.L., Phillips, M.A., et al. (2017). Genomics of parallel experimental evolution in *Drosophila. Molecular Biology and Evolution*, 34, 831–42.

Grimm, N.B., Faeth, S.H., Golubiewski, N.E., et al. (2008). Global change and the ecology of cities. *Science*, 319, 756–60.

Heliconius Genome Consortium (2012). Butterfly genome reveals promiscuous exchange of mimicry adaptations among species. *Nature*, 487, 94–8.

Hoban, S., Kelley, J.L., Lotterhos, K.E., et al. (2016). Finding the genomic basis of local adaptation: pitfalls, practical solutions, and future directions. *The American Naturalist*, 188, 379–97.

Isaksson, C. (2015). Urbanization, oxidative stress and inflammation: a question of evolving, acclimatizing or coping with urban environmental stress. *Functional Ecology*, 29, 913–23.

Johnson, M.T.J. and Munshi-South, J. (2017). Evolution of life in urban environments. *Science*, 358(6363), eaam8327.

Johnson, M.T.J., Prashad, C., Lavoignat, M., and Saini, H.S. (2018). Contrasting the effects of natural selection, genetic drift and gene flow on urban evolution in white clover (*Trifolium repens*). *Proceedings of the Royal Society B: Biological Sciences*, 285, 20181019.

Kern, E.M.A. and Langerhans, R.B. (2018). Urbanization drives contemporary evolution in stream fish. *Global Change Biology*, 24(8), 3791–803.

Kimura, M. and Crow, J.F. (1964). The number of alleles that can be maintained in a finite population. *Genetics*, 49, 725–38.

Klerks, P.L. and Weis, J.S. (1987). Genetic adaptation to heavy metals in aquatic organisms: a review. *Environmental Pollution*, 45, 173–205.

Koch, E., Clark, J.M., Cohen, B., et al. (2016). Management of head louse infestations in the United States—a literature review. *Pediatric Dermatology*, 33, 466–72.

Langerhans, R.B. (2018). Predictability and parallelism of multitrait adaptation. *Journal of Heredity*, 109, 59–70.

Langerhans, R.B. and DeWitt, T.J. (2004). Shared and unique features of evolutionary diversification. *The American Naturalist*, 164, 335–49.

Lee, K.M. and Coop, G. (2017). Distinguishing among modes of convergent adaptation using population genomic data. *Genetics*, 207, 1591–619.

Lenormand, T. (2002). Gene flow and the limits to natural selection. *Trends in Ecological Evolution*, 17, 183–9.

Lenski, R.E. (2017). Convergence and divergence in a long-term experiment with bacteria. *The American Naturalist*, 190, S57–68.

Losos, J.B. (2011). Convergence, adaptation, and constraint. *Evolution*, 65, 1827–40.

Lotterhos, K.E. and Whitlock, M.C. (2015). The relative power of genome scans to detect local adaptation depends on sampling design and statistical method. *Molecular Ecology*, 24, 1031–46.

Lourenço, A., Álvarez, D., Wang, I.J., and Velo-Antón, G. (2017). Trapped within the city: integrating demography, time since isolation and population-specific traits to assess the genetic effects of urbanization. *Molecular Ecology*, 26, 1498–514.

MacPherson, A. and Nuismer, S.L. (2017). The probability of parallel genetic evolution from standing genetic variation. *Journal of Evolutionary Biology*, 30, 326–37.

Mamanova, L., Coffey, A.J., Scott, C.E., et al. (2010). Target-enrichment strategies for next-generation sequencing. *Nature Methods*, 7, 111–18.

McMillan, A.M., Bagley, M.J., Jackson, S.A., and Nacci, D.E. (2006). Genetic diversity and structure of an estuarine fish (*Fundulus heteroclitus*) indigenous to sites associated with a highly contaminated urban harbor. *Ecotoxicology*, 15, 539–48.

Miles, L.S., Dyer, R.J., and Verrelli, B.C. (2018). Urban hubs of connectivity: contrasting patterns of gene flow within and among cities in the western black widow spider. *Proceedings of the Royal Society B: Biological Sciences*, 285, 20181224.

Miles, L.S., Rivkin, L.R., Johnson, M.T.J., Munshi-South, J., and Verrelli, B.C. (2019). Gene flow and genetic drift in urban environments. *Molecular Ecology*, 28(18), 4138–51.

Miranda, A.C., Schielzeth, H., Sonntag, T., and Partecke, J. (2013). Urbanization and its effects on personality traits: a result of microevolution or phenotypic plasticity? *Global Change Biology*, 19, 2634–44.

Mueller, J.C., Partecke, J., Hatchwell, B.J., Gaston, K.J., and Evans, K.L. (2013). Candidate gene polymorphisms for behavioural adaptations during urbanization in blackbirds. *Molecular Ecology*, 22, 3629–37.

Mueller, J.C., Kuhl, H., Boerno, S., et al. (2018). Evolution of genomic variation in the burrowing owl in response to recent colonization of urban areas. *Proceedings of the Royal Society B: Biological Sciences*, 285, 20180206.

Oke, K.B., Rolshausen, G., LeBlond, C., and Hendry, A.P. (2017). How parallel is parallel evolution? A comparative analysis in fishes. *The American Naturalist*, 190, 1–16.

Orr, A.H. (2005). The probability of parallel evolution. *Evolution*, 59, 216–20.

Partecke, J. and Gwinner, E. (2007). Increased sedentariness in European blackbirds following urbanization: a consequence of local adaptation? *Ecology*, 88, 882–90.

Peterson, B.K., Weber, J.N., Kay, E.H., Fisher, H.S., and Hoekstra, H.E. (2012). Double digest RADseq: an inexpensive method for de novo SNP discovery and genotyping in model and non-model species. *PLOS ONE*, 7, e37135.

Ravinet, M., Elgvin, T., Trier, C., et al. (2018). Signatures of human-commensalism in the house sparrow genome. *Proceedings of the Royal Society B: Biological Sciences*, 285, 20181246.

Reid, N.M., Proestou, D.A., Clark, B.W., et al. (2016). The genomic landscape of rapid repeated evolutionary adaptation to toxic pollution in wild fish. *Science*, 354, 1305–8.

Rivkin, L.R., Santangelo, J.S., Alberti, M., et al. (2019). A roadmap for urban evolutionary ecology. *Evolutionary Applications*, 12, 384–98.

Romero, A. and Anderson, T.D. (2016). High levels of resistance in the common bed bug, *Cimex lectularius* (Hemiptera: Cimicidae), to neonicotinoid insecticides. *Journal of Medical Entomology*, 53, 727–31.

Rost, S., Pelz, H., Menzel, S., et al. (2009). Novel mutations in the *VKORC1* gene of wild rats and mice—a response to 50 years of selection pressure by warfarin? *BMC Genetics*, 9, 1–9.

Saavedra-Rodriguez, K., Maloof, F.V., Corey, L.C., et al. (2018). Parallel evolution of vgsc mutations at domains IS6, IIS6 and IIIS6 in pyrethroid resistant *Aedes aegypti* from Mexico. *Scientific Reports*, 8, 1–9.

Santangelo, J.S., Johnson, M.T.J., and Ness, R.W. (2018a). Modern spandrels: the roles of genetic drift, gene flow and natural selection in the evolution of parallel clines. *Proceedings of the Royal Society B: Biological Sciences*, 285, 20180230.

Santangelo, J.S., Rivkin, L.R., and Johnson, M.T.J. (2018b). The evolution of city life. *Proceedings of the Royal Society B: Biological Sciences*, 285, 20181529.

Stayton, C.T. (2008). Is convergence surprising? An examination of the frequency of convergence in simulated datasets. *Journal of Theoretical Biology*, 252, 1–14.

Steiner, C.C., Römpler, H., Boettger, L.M., Schöneberg, T., and Hoekstra, H.E. (2009). The genetic basis of phenotypic convergence in beach mice: similar pigment patterns but different genes. *Molecular Biology and Evolution*, 26, 35–45.

Stern, D.L. (2013). The genetic causes of convergent evolution. *Nature Reviews Genetics*, 14, 751–64.

Stern, D.L. and Orgogozo, V. (2009). Is genetic evolution predictable? *Science*, 323, 746–52.

Straub, C., Pichlmüller, F., and Helfer, V. (2015). Population genetics of fire salamanders in a pre-Alpine urbanized area. *Salamandra*, 51, 245–51.

Stuart, Y.E., Veen, T., Weber, J.N., et al. (2017). Contrasting effects of environment and genetics generate a continuum of parallel evolution. *Nature Ecology & Evolution*, 1, 1–7.

Theodorou, P., Radzevičiūtė, R., Kahnt, B., et al. (2018). Genome-wide single nucleotide polymorphism scan suggests adaptation to urbanization in an important pollinator, the red-tailed bumblebee (*Bombus lapidarius* L.). *Proceedings of the Royal Society B: Biological Sciences*, 285, 20172806.

Thompson, C.J., Ahmed, N.I., Veen, T., et al. (2017). Many-to-one form-to-function mapping weakens parallel morphological evolution. *Evolution*, 71, 2738–49.

Thompson, K.A., Renaudin, M., and Johnson, M.T.J. (2016). Urbanization drives the evolution of parallel clines in plant populations. *Proceedings of the Royal Society B: Biological Sciences*, 283, 20162180.

Travisano, M., Mongold, J.A., Bennett, A.F., and Lenski, R.E. (1995). Experimental tests of the roles of adaptation, chance, and history in evolution. *Science*, 267, 87–90.

Van't Hof, A.E., Campagne, P., Rigden, D.J., et al. (2016). The industrial melanism mutation in British peppered moths is a transposable element. *Nature*, 534, 102–105.

Vasemägi, A. (2006). The adaptive hypothesis of clinal variation revisited: single-locus clines as a result of spatially restricted gene flow. *Genetics*, 173, 2411–14.

Wada-katsumata, A., Silverman, J., and Schal, C. (2013). Changes in taste neurons support the emergence of an adaptive behavior in cockroaches. *Science*, 340, 972–6.

Whitehead, A., Triant, D.A., Champlin, D., and Nacci, D. (2010). Comparative transcriptomics implicates mechanisms of evolved pollution tolerance in a killifish population. *Molecular Ecology*, 19, 5186–203.

Whitehead, A., Pilcher, W., Champlin, D., and Nacci, D. (2012). Common mechanism underlies repeated evolution of extreme pollution tolerance. *Proceedings of the Royal Society B: Biological Sciences*, 279, 427–33.

Winchell, K.M., Reynolds, R.G., Prado-Irwin, S.R., Puente-Rolón, A.R., and Revell, L.J. (2016). Phenotypic shifts in urban areas in the tropical lizard *Anolis cristatellus*. *Evolution*, 70, 1009–22.

Winchell, K.M., Maayan, I., Fredette, J., and Revell, L.J. (2018). Linking locomotor performance to morphological shifts in urban lizards. *Proceedings of the Royal Society B: Biological Sciences*, 285, 20180229.

Winchell, K.M., Briggs, D., Revell, L.J., et al. (2019). The perils of city life: patterns of injury and fluctuating asymmetry in urban lizards. *Biological Journal of the Linnean Society*, 126, 276–88.

Yakub, M. and Tiffin, P. (2017). Living in the city: urban environments shape the evolution of a native annual plant. *Global Change Biology*, 23, 2082–9.

Yauk, C.L. and Quinn, J.S. (1996). Multilocus DNA fingerprinting reveals high rate of heritable genetic mutation in herring gulls nesting in an industrialized urban site. *Proceedings of the National Academy of Sciences of the United States of America*, 93, 9614–19.

Yauk, C.L., Fox, G.A., Mccarry, B.E., and Quinn, J.S. (2000). Induced minisatellite germline mutations in herring gulls (*Larus argentatus*) living near steel mills. *Mutation Research*, 452, 211–18.

Youngsteadt, E., Dale, A.G., Terando, A.J., Dunn, R.R., and Frank, S.D. (2015). Do cities simulate climate change? A comparison of herbivore response to urban and global warming. *Global Change Biology*, 2, 97–105.

Zhen, Y., Aardema, M.L., Medina, E.M., Schumer, M., and Andolfatto, P. (2012). Parallel molecular evolution in an herbivore community. *Science*, 337, 1634–7.

Zhou, B., Rybski, D., and Kropp, J.P. (2017). The role of city size and urban form in the surface urban heat island. *Scientific Reports*, 7, 1–9.

CHAPTER 4

Landscape Genetic Approaches to Understanding Movement and Gene Flow in Cities

Jason Munshi-South and Jonathan L. Richardson

Munshi-South, J. and Richardson, J.L., *Landscape Genetic Approaches to Understanding Movement and Gene Flow in Cities* In: *Urban Evolutionary Biology*. Edited by Marta Szulkin, Jason Munshi-South and Anne Charmantier, Oxford University Press (2020). © Oxford University Press. DOI: 10.1093/oso/9780198836841.003.0004

4.1 Introduction

As an increasing proportion of the world's human population resides in cities, the footprint of cities expands globally (Seto et al. 2012). The resulting urban habitat differs dramatically from exurban and rural areas, with a more complex matrix of roads, buildings, bare soil, slope, green space, and general infrastructure (Byrne 2007; Ottensmann 2017). The heterogeneity of urban environments has challenged the persistence of some species, but bolstered the prospects of others. Some species found in cities occupy residual native habitat (Wood and Pullin 2002), or are transient individuals dispersing through cities (Benson et al. 2016). On the other end of the spectrum are commensal 'anthrodependent' species, sometimes called urban exploiters, that thrive in cities alongside humans and their infrastructure, waste, and other environs (Blair 2001; Hulme-Beaman et al. 2016).

Species moving within or across cities are also moving their genetic material. Gene flow occurs when dispersers reproduce in their new location. The influences of gene flow include greater connectivity of genetic demes (i.e., groups of individuals that are likely to breed with each other due to proximity), and a reduction in the risks of inbreeding depression through genetic rescue effects when new alleles arrive (Ingvarsson 2001). Alternatively, gene flow can disrupt locally adapted gene pools by introducing maladapted alleles and reducing the fitness of the recipient population, a process known as outbreeding depression (Lenormand 2002; Sambatti and Rice 2006). Regardless of whether it has a positive or negative impact, the role of gene flow is likely to be outsized in cities where selection pressures are intense and diverse at small spatial scales (Richardson et al. 2014; Donihue and Lambert 2015), and where wild populations are likely to be either residual and small (e.g., many native species), or large and targeted by management efforts (e.g., invasive, commensal, and/or pest species).

Estimates of gene flow also provide an important proxy for movement across the urban landscape. Areas where individuals are consistently different genetically from each other indicate parts of the landscape that may restrict the movement of that species. For example, Beninde et al. (2016) found consistent genetic differences among wall lizards (*Podarcis muralis*) separated by a river in Trier, Germany, while Combs et al. (2018b) observed subtle genetic differences between brown rats (*Rattus norvegicus*) in New York City (NYC) separated by an area of low residential density. Other parts of the urban landscape matrix can promote movement and gene flow, as we see when individuals in two different areas share very similar genetic variation. Unfried et al. (2013) found a positive relationship between the amount of forest cover and genetic similarity of song sparrows (*Melospiza melodia*) in Seattle, WA, USA. These indirect genetic proxies of movement complement, but do not replace, direct measures of movement, such as visual surveys, tracking with radio or GPS transmitters, citizen science surveys, or complaint data gathered by municipal governments. Genetic approaches often allow for more individuals to be studied, and over a larger spatial scale, than direct tracking would allow; in many ways, gene flow proxies are very efficient at increasing the scope of a study. Recently, researchers have also relied more on indirect measures of species occurrence using genetic data, including metabarcoding and environmental DNA (eDNA) designed to identify the presence of species through detection of genetic material (Forin-Wiart et al. 2018).

While moving through cities, organisms face a complex matrix of habitats and landscape elements. One of the primary features unique to urban environments is spatial variation in the attributes of human populations. Cities often show drastic socioeconomic heterogeneity at very small spatial scales (Figure 4.1A), also referred to as spatial segregation (Reardon et al. 2008; Hacker et al. 2013). This variation may play an influential role in local urban ecology (Pickett et al. 2001). In many parts of the world, socioeconomics are associated with city services related to sewage treatment, rubbish collection, or pest management (Reis et al. 2008). For example, in NYC there was a negative association between the presence of *R. norvegicus* and median household income (Johnson et al. 2016), as well as education levels (Walsh 2014).

On the other end of the spectrum, higher socioeconomic status often translates to larger properties, lower human densities, and increased vegetation cover. Plant species diversity varies consistently with several socioeconomic variables, including household income and building age (Hope et al.

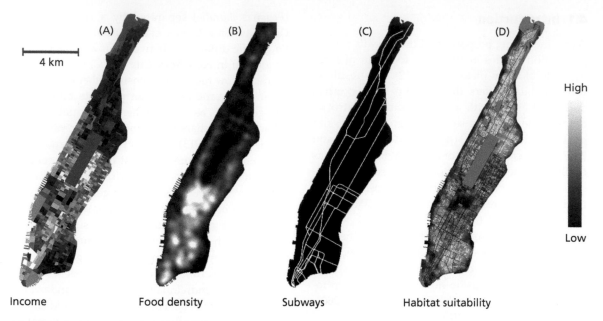

Figure 4.1 Environmental attributes of cities can vary across small spatial scales, illustrated for the island of Manhattan in NYC. (A) shows median family income by census block; (B) represents kernel density of restaurants and food carts (i.e., food density); subway lines are shown in (C); (D) is a map of a habitat suitability model developed for brown rats (*Rattus norvegicus*) based on city inspection data (Richardson, Combs, and Munshi-South, unpublished data). Central Park appears as a homogenously coloured rectangle in the middle of Manhattan due to lack of data on rat presence. In all four panels, lighter colours represent higher values.

2003; Kinzig et al. 2005). Lizard species occupancy also increased with increasing property size and decreasing housing density in Tucson, AZ (Germaine and Wakeling 2001). While few studies have examined this phenomenon, we expect that socioeconomic variation and the attendant variation in habitat will often shape the movement of organisms in and through cities. For example, red foxes (*Vulpes vulpes*) in Vienna are more likely to be encountered in gardens, parks, or low-density residential areas than in industrial areas or urban forests (Walter et al. 2018).

Another important landscape consideration in cities is the high proportion of impervious surface. These surfaces consist of human-made materials that are not porous to water, such as concrete, asphalt roads, and rooftops. The dominance of this one land-cover type negatively affects many native species that rely on vegetated habitat for resources or movement corridors. Other species that can use these surfaces, or the interstitial gaps between impervious surfaces, may not be affected. For example, certain species do quite well taking root or burrowing in small tree planters or even cracks in concrete, including many ants and 'weedy' plant species

(Savage et al. 2015; Salinitro et al. 2018). For such species, the dominance of this one land-cover type may facilitate movement through a process called 'secondary dispersal'. For example, roads in Berlin, Germany were found to facilitate the dispersal of tree of heaven (*Ailanthus altissima*) seeds (Kowarik and von der Lippe 2011). In this study, researchers marked all seeds from a focal tree, tracked their dispersal over 11 days, and created a dispersal kernel. Increased wind speed in the direction of traffic led to longer dispersal averaged along this high-traffic road corridor than had been observed in forest habitat. Another study found that, despite extremely low natural movement ability, passive dispersal of terrestrial snails (*Cornu asperum*) in cities was facilitated by impervious roads and pavements (Balbi et al. 2018).

Green spaces in cities are also highly heterogeneous, and may include parks, cemeteries, golf courses, and green streetscape vegetation managed for human leisure, stormwater, or other goals besides nature conservation. Even though urban green spaces are often dominated by non-native vegetation (Foster and Sandberg 2010), they are consistently associated with increased species diversity compared to spaces dominated by impervious surfaces. Thus, urban green spaces may serve as critical habitat (Sandström et al. 2006; Vergnes et al. 2013), and they have also been identified as important corridors of movement and gene flow (Lepczyk et al. 2017). Munshi-South (2012) found that urban canopy cover within city parks, cemeteries, and other green spaces facilitated the genetic connectivity of white-footed mouse (*Peromyscus leucopus*) populations in NYC. Similarly, Saarikivi et al. (2013) reported that golf courses serve as corridors of genetic connectivity for the common frog (*Rana temporaria*) in Helsinki, Finland. Non-native pest species can also take advantage of public green spaces. Both Walsh (2014) and Johnson et al. (2016) reported a positive association between proximity to public open spaces and brown rat occurrence. Any assessment of movement and gene flow through cities will need to take the range of green spaces into account.

Research into the movement and gene flow of plants and animals across cities is in its infancy (LaPoint et al. 2015). It is clear, however, that urban environments are distinct from non-urban habitats in both their matrix structure and environmental characteristics. Thus, a different framework is required to study movement and gene flow in and through cities than those that have been developed to date for work in non-urban areas. Cities have been neglected study subjects for decades, in part due to the common perception that they are homogenous environments with depauperate floral and faunal communities. In fact, cities are generally a heterogeneous patchwork of diverse environments and microhabitats with surprising species diversity (Figure 4.1). This matrix can lead to more complex movement patterns in urban landscapes, especially if species are habitually avoiding or associating with humans. In this chapter, we highlight the special considerations in studying movement in cities using landscape genetic approaches.

4.2 Analytical approaches for investigating movement and gene flow in urban areas

4.2.1 Choice of molecular markers in urban evolution studies

The influence of urbanization on drift and gene flow is the most common phenomenon studied to date in the urban evolution literature (Johnson and Munshi-South 2017). Estimating rates of gene flow and the effects of drift have a long history in population genetics, and the options for analysis have only increased with the advent of genome-wide markers such as single nucleotide polymorphisms (SNPs).

Early studies of urban gene flow were limited to a few classes of molecular markers and did not directly test the influence of urban barriers or the landscape on patterns of genetic variation. Allozymes were the marker of choice before the discovery and widespread adoption of polymerase chain reaction (PCR), followed by mitochondrial haplotypes due to the higher amount of mtDNA compared to nuclear DNA in eukaryotic cells. However, a large majority of published estimates of gene flow in urban areas have employed microsatellites due to their rapid mutation rate and ease of genotyping (see Table S1 in Johnson and Munshi-South (2017) for a recent list of such studies). Microsatellite-based analyses have largely analysed situations where populations are isolated in urban parks or similar fragmented green spaces. These fragmentation scenarios typically reduce gene flow due to barriers to movement. When gene flow between populations is reduced, populations tend to lose genetic variation and diverge genetically as genetic drift changes allele frequencies in the subdivided groups. Genotyping samples from different areas and calculating metrics like F_{ST} (Weir and Cockerham 1984) provide the most common approach to characterizing gene flow and consequent genetic differentiation. While F_{ST} is a useful metric of genetic differentiation, it can fail dramatically as an indirect measure of gene flow (Whitlock and McCauley 1999) due to violation of one or more assumptions of the underlying model (particularly that populations are at equilibrium between drift and migration). Lastly, reduced gene flow is not the only mechanism that can lead to increased genetic differentiation between groups of organisms. Urban environments often support smaller numbers of individuals within populations, leading to elevated genetic drift and reduced effective population sizes, which can generate similar patterns of elevated genetic differentiation.

Beyond F_{ST}, microsatellite-based studies have sought to use genetic assignment tests or evolutionary clustering analyses to identify recent migrants or their offspring (Munshi-South and Kharchenko 2010). Assignment tests may only succeed at identifying gene flow within the last few generations, and only when (sub)populations are sufficiently genetically isolated. Evolutionary clustering approaches, such as those employed by the widely used STRUCTURE (Pritchard et al. 2000) or ADMIXTURE (Alexander et al. 2009), can produce misleading signatures of recent population mixing due to gene flow when one erroneously assumes that the analysis has identified the true value of distinct evolutionary clusters, K (Lawson et al. 2018). The Bayesian coalescent approach in Migrate-*n* has been used often to estimate equilibrium rates of migration between populations using microsatellites (Beerli 2006) and represented a substantial advance in estimating longer-term gene flow (i.e., over more than a few generations). However, such estimates are rare in the urban literature, with a few exceptions (Munshi-South 2012).

Relatively few studies to date have employed genome-wide SNPs to analyse urban gene flow. However, this gap is closing rapidly with the widespread adoption of reduced representation sequencing approaches (Puritz et al. 2014) such as RAD-Seq (restriction site associated DNA sequencing)/GBS (genotyping-by-sequencing), SNP chips, sequence capture, and low-coverage whole genome resequencing. With SNP data from throughout the genome, researchers can take advantage of more sophisticated demographic modelling approaches based on coalescent theory and the site frequency spectrum (Excoffier et al. 2013). These approaches can estimate migration rates in complicated scenarios of population divergence with ongoing gene flow.

4.2.2 Advances in spatial population genomics and landscape genetics for testing gene flow hypotheses in urban environments

The development of spatially explicit methods for investigating gene flow is also transforming our understanding of fine-scale patterns of genetic variation. Well-known techniques such as spatial autocorrelation analyses (Peakall and Smouse 2006) benefit from the ability to estimate genetic distance

or relatedness more finely with SNPs than with microsatellites. Combining the spatial location of organisms with their genome-wide SNP genotypes has proven powerful in identifying subtle changes across landscapes in gene flow and departures from isolation-by-distance. These new high-resolution markers have also led to renewed interest in the concept of genetic neighbourhoods or 'demes' for understanding gene flow in cities (Wright 1969).

The approach in the spatial genetic diversity (sGD) software package (Shirk and Cushman 2011), as well as the use of Moran's eigenvector maps (MEM) in the MEMGENE package (Galpern et al. 2014), identified an uptown/downtown split in NYC brown rats (Figure 4.2C). The sGD approach estimates genetic diversity after placing populations into overlapping genetic neighbourhoods that reflect the underlying spatial population structure, and the latter identifies spatial neighbourhoods that combine MEM and a regression framework to identify 'cryptic' spatial signals in genetic variation. Another spatial method, estimated effective migration surfaces (EEMS), identifies areas of a landscape where 'effective' migration is low and genetic similarity between individuals decays quickly (Petkova et al. 2016). This model assumes that rates of gene flow vary across the landscape and visualizes departures from isolation-by-distance that result from this variation. These methods are particularly useful for single populations within cities that may exhibit high gene flow, but spatial genetic patterns are still influenced by barriers, socioeconomic attributes of human neighbourhoods, or any other landscape heterogeneity that influences rates of movement (Figure 4.2D; Richardson et al. 2017). If applied widely to a number of populations across multiple cities and species, these new methods have great potential for identifying common influences on gene flow that may be generally applicable to many urbanization scenarios.

While the approaches above identify patterns that may be due to the impacts of landscape heterogeneity on drift and gene flow, they are not designed to explicitly measure the influence of particular landscape factors. Landscape genetics combines landscape ecology, population genetics, and spatial statistics to understand how the structure and composition of landscapes influences gene flow (Manel

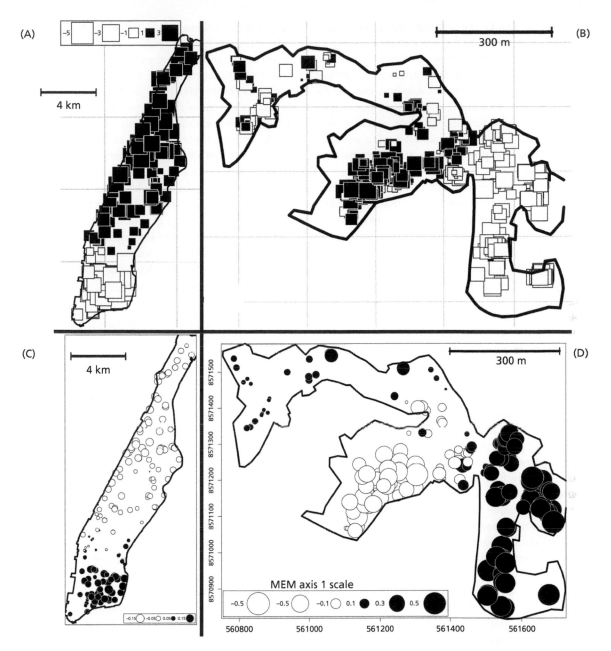

Figure 4.2 Genetic patterns vary at fine spatial scales for brown rats in two large cities, as shown by spatial principal components analysis (sPCA) for (A) the Manhattan section of NYC, U.S.A. and (B) the Pau da Lima section of Salvador, Brazil. The scale is the first principal component score, represented by size and colour of squares, each at the location of a sampled individual. (C) and (D) represent the first axis from MEM analysis for the same two cities as (A) and (B). Scale is based on circle size and colour. Note spatial genetic pattern is very similar for both sPCA and MEM analyses within each city, even though the colour scale (black vs white) is inverted between analyses. (Figures modified from Richardson et al. (2017) and Combs et al. (2018b).)

et al. 2003). Landscape genetics researchers have often focused on the influence that human-caused landscape change has on gene flow (Manel and Holderegger 2013), and thus landscape genetics will likely play an important role in urban evolutionary biology moving forward. Basic population genetic models for gene flow include island models that generally assume equal rates of gene flow between discrete populations at migration–drift equilibrium.

Isolation-by-distance (IBD) is another common model, often observed empirically, that is used as a null hypothesis for spatial genetic studies of natural populations (Jenkins et al. 2010). IBD arises when pairs of individuals or subpopulations that are closer together are more genetically similar than pairs further apart, creating a positive relationship between geographic distance and genetic dissimilarity. However, neither IBD nor island models explicitly deal with the role that spatial dependence plays in patterns of genetic variation (Meirmans 2012). Landscape genetics allows one to test hypotheses about specific landscape features that may restrict or facilitate gene flow or can serve a more exploratory purpose when the relative importance of different landscape attributes is not known a priori. Despite the power of these tools and the focus on gene flow and drift in the urban evolutionary literature, relatively few urban studies have employed landscape genetics to date. The slow adoption is surprising because these methods will be particularly useful for urban landscapes that exhibit greater contrast between land-cover types than seen in natural environments (Jaquiéry et al. 2011).

Landscape genetic approaches based on conceptual models such as isolation-by-resistance (IBR) can explicitly test the influence of particular landscape variables. IBR evaluates the relationship between the hypothetical level of resistance to movement that a focal landscape imposes and genetic differentiation between individuals sampled across that landscape. IBR analyses are often implemented in the software package CIRCUITSCAPE, which represents a substantial advance for landscape genetics because it calculates a metric of overall connectivity between pairs of sites, but also visualizes all hypothetical routes of movement across a landscape (McRae and Beier 2007; Dickson et al. 2019). The main drawback is that a priori information is required to calibrate resistance values between adjacent nodes in the landscape. Thus, these analyses draw upon hypotheses about the variables that restrict gene flow and the magnitude of their influence (often referred to as landscape friction or resistance). Some of these hypotheses about specific landscape influences come from previous empirical work and an understanding of species biology, i.e., 'expert opinion'. While empirical ecological or physiological data are the gold standard for generating hypotheses about landscape effects, they are also rare (Peterman et al. 2014; Nowakowski et al. 2015). Often, landscape genetic analyses use expert opinion in the absence of empirical movement data to parameterize landscape resistance, i.e., the relative costs an organism will encounter when trying to move across the focal landscape.

4.2.3 Analytical challenges to landscape genetic analyses in cities

When extensive empirical data on animal movement through urban landscapes is not available, or very limited, one alternative to researcher opinion is to estimate habitat use based on occurrence data. These 'habitat suitability models' (HSM) or 'ecological niche models' rely on statistical associations between georeferenced presence points for the species in question and geographic data describing climatic, structural, vegetational, and other variation across the landscape (Spear et al. 2010; Milanesi et al. 2016). Occurrence data are often easier to obtain than detailed ecological data, making habitat suitability model development more feasible than new empirical studies to characterize habitat use and preferred environments. Importantly for landscape genetics, using these habitat suitability models can be a more objective way to choose land-cover variables that influence gene flow than expert opinion. Alternatively, optimization procedures such as the approach implemented in ResistanceGA (Peterman 2018) may identify the best combinations of landscape variables and resistance values. Some evidence suggests that optimization is much better at detecting associations between urban landscape heterogeneity and gene flow than habitat suitability models alone (Peterman et al. 2014; Mateo-Sánchez et al. 2015). This optimization approach uses linear mixed modelling and information criteria to identify the combination of variables and their resistance levels that best fit the observed genetic dataset.

Researchers should also consider the temporal resolution possible with landscape genetic approaches, much of which is influenced by the genetic markers used and the length of time urban landscape elements have been in place. For example, if a contiguous population of organisms (i.e., one panmictic gene pool) is suddenly separated by a newly constructed barrier to movement (e.g., a high-volume roadway), some time lag is required before those two gene pools—that are now spatially separated—diverge genetically via drift or selection. That period of time depends on the rate of drift and mutation at those locations on the genome, the strength of selection, and how strongly the new road severs movement and gene flow between the populations. One study using simulated landscapes found that 100–200 generations may be required to detect genetic divergence after the introduction of a barrier (Landguth et al. 2010).

It is particularly important to account for this time lag in cities, dominated by urban infrastructure added within the last few centuries (see also section 2.2 in Chapter 2). Generation time is important to consider as well, since 100 generations will elapse much more quickly for some species than others. Certain markers will also show a signal of genetic divergence sooner than others. Genetic markers have often been selected because they are not under natural selection, and adhere to assumptions inherent to population genetic analyses based on migration–selection–drift balance. Those same markers tend to evolve much more slowly than genomic regions under selection. For example, using > 12 000 SNPs, McCartney-Melstad et al. (2018) found significant genetic divergence between eastern tiger salamander (*Ambystoma tigrinum*) populations in an urbanized part of Long Island, NY, suggesting very limited gene flow between those ponds. In contrast, a study using 12 microsatellite markers looking at the same populations found very little genetic differentiation (Titus et al. 2014). Subsampling of the SNP data indicated that at least 300–400 SNP loci were necessary to detect the genetic structuring observed at the fine spatial scales where these populations occur.

Landscape genetics has been plagued by debates about the validity of commonly used analyses such as partial Mantel tests (Castellano and Balletto 2002). However, new and exciting approaches such as the optimization described above have been steadily developed in recent years. Spatially explicit modelling that can simultaneously detect and account for barriers, IBD, IBR, and/or isolation-by-environment (IBE) would seem to be one of the next methodological leaps forward (Bradburd et al. 2018). IBE is the hypothesis that genetic distances between individuals or populations will increase with environmental differences between the habitats examined. Richardson et al. (2016) also review a number of newer approaches and make recommendations for improving landscape genetics research. In particular, they call for more careful interpretation of the ecological and evolutionary consequences of gene flow, as well as consideration of the scope of inference of any single study.

Replication across cities and species will likely be the most important aspect of future research for understanding how urbanization influences gene flow (Chapter 3; Johnson and Munshi-South 2017). There will be serious challenges in obtaining and analysing data from the same landscape elements for multiple cities (especially for global comparisons). Studies that can test the generality of urban landscape effects on gene flow should be given higher priority than extremely detailed, idiosyncratic models for single species within individual cities.

4.2.4 Landscape genomics approaches to identifying genes under selection in urban environments

The field of landscape *genomics* has developed in parallel with landscape genetics to identify particular genetic variants (usually SNPs) that are statistically associated with environmental variables (Rellstab et al. 2015). When gene flow is restricted, populations will begin to follow separate evolutionary trajectories and may adapt to local selection pressures. Such local adaptation in allopatry can eventually be detected from genomic data as classic signatures of selection, such as elevated F_{ST}, Tajima's D, or extended haplotype homozygosity around selected alleles. In situations with ongoing gene flow, local adaptation may be prevented by ongoing migration of less well-adapted genotypes into the population, lowering the overall fitness of the recipient population

(Kawecki and Ebert 2004). However, when local selection pressures are sufficiently strong, this swamping effect of gene flow may not be enough to prevent adaptation (Richardson et al. 2014). In these cases, landscape genomic approaches such as genotype-by-environment association (GEA) tests can identify outlier candidate loci likely under selection based on their statistical association with ecological variables. GEA has been shown to have high power at detecting loci under selection, and has the added advantage of testing ecological hypotheses by identifying associations with environmental variables that are related to the selection pressure (Forester et al. 2018). Only a few studies to date have used GEA to identify potential local adaptation in urban environments.

4.3 Empirical studies of urban gene flow, drift, and landscape genetics

Evidence of the effects of urbanization on genetic drift and gene flow has steadily accrued over the last few decades. Recently, the pace of research has accelerated in both number of publications and scope: published studies have addressed gene flow and drift within cities, between cities, and/or between cities and their surrounding landscape, and the landscape genomics of local adaptation to urbanization. Mammals in the order Rodentia have been especially popular study subjects. Rodents are the most speciose mammal group, they occur in every city around the world, and most cities contain both native and non-native species. Rodents are generally small-bodied and have fast generation times for a mammal, and thus may exhibit evolutionary changes due to urbanization more quickly than other vertebrates. The urban ecology of rodents is also rich due to their roles as seed predators and dispersers, important prey in urban food webs, and public health threats due to the spread of zoonotic diseases by commensal species (Himsworth et al. 2013). Urban commensals such as brown rats are notorious for their gnawing and the damage it does to urban infrastructure. Municipal governments spend millions of pounds on urban rodent control, often with little success (Parsons et al. 2017). Thus, knowledge of gene flow among urban rodent populations is potentially useful for understanding their

evolutionary dynamics and designing better pest management strategies. Below we review the urban evolution literature that uses population and landscape genetic approaches to understand gene flow, with particular reference to rodents as a focal group. However, we also address work on other taxonomic groups to illustrate the breadth of research effort in urban landscape genetics.

4.3.1 Gene flow, drift, and landscape genetics within cities

Habitat patches within cities are typically highly variable in terms of size, vegetation types, and human uses. These patches are also usually highly fragmented, and thus one may predict low rates of gene flow between patches and strong genetic differentiation due to drift. For example, *P. leucopus* populations in isolated NYC parks rapidly diverged from one another (Munshi-South and Kharchenko 2010). Somewhat surprisingly, these populations did not uniformly show evidence of recent bottlenecks or rapid loss of variation at microsatellite loci, perhaps due to high population densities (Munshi-South and Nagy 2014).

Eurasian mice in the genus *Apodemus* are ecologically similar to North American *Peromyscus* and have been the subject of multiple microsatellite studies of urban gene flow and drift. Striped field mice (*A. agrarius*) sampled from a range of sites in Warsaw, Poland were more genetically differentiated from one another than mice from rural areas (Gortat et al. 2013), although urban populations away from the urban core were able to maintain some gene flow with surrounding areas (Gortat et al. 2015). A more recent colonizer of Warsaw, the yellow-necked mouse (*A. flavicollis*), exhibited stronger genetic structure and reduced gene flow along this urbanization gradient compared to *A. agrarius* (Gortat et al. 2017). Commensal rodents such as *R. norvegicus* exhibit greater rates of gene flow within cities than these native rodents, but even urban rats show some divergence at microsatellite loci due to high philopatry, or landscape features such as waterways (Gardner-Santana et al. 2009), valleys (Kajdacsi et al. 2013), or high-traffic roads (Richardson et al. 2017) that restrict gene flow. These microsatellite studies of rodents all generally point to a negative impact of urbanization on gene flow. While some of the details of these studies may be taxon-specific (such as very short dispersal distances in commensal rats), they are exemplary of broader findings for urban populations of non-rodent taxa (Johnson and Munshi-South 2017).

Recent urban studies have also looked at the spatially explicit relationships between genetic structure and city landscapes. For example, Combs et al. (2018b) found a steep decline in relatedness between pairs of brown rats sampled within a few hundred metres of one another, with relatedness reaching nearly zero by 2 km distance. Other studies have documented that most gene flow in rats occurs over small spatial scales of just a few hundred metres or less (Costa et al. 2016). This limited dispersal and gene flow may partially explain why rat colonies on the small island of Manhattan carry distinct communities of microbes (Angley et al. 2018). Subsequent autocorrelation analyses of two other cities found similar patterns of weak gene flow in New Orleans, U.S.A., and Salvador, Brazil, where the spatial scale extended out several hundred metres farther than in NYC (Combs et al. 2018b). In NYC, spatial PCA detected two clusters of rats in uptown and downtown neighbourhoods (Figure 4.2A and C), whereas traditional principal components analysis (PCA) identified a subtle north–south gradient of genetic variation of rats. These areas are separated by a 'resource desert' in midtown Manhattan with low residential density and high levels of pest control; these socioeconomic and structural differences do not represent a strict physical barrier to gene flow but restrict it enough to create an identifiable pattern in rats across the landscape. The EEMS analysis was also able to detect the same area of midtown Manhattan where rat populations begin to diverge into uptown and downtown populations, indicating its utility for very fine-scale studies of gene flow within individual cities (Combs et al. 2018a). A similar pattern was seen for rats in Salvador, Brazil (Figure 4.2B and D), where a topographic saddle and high-traffic road constricted gene flow on a scale of less than 200 m (Richardson et al. 2017). Spatially explicit methods also detected the influence of canals, major roadways, and neighbourhood attributes on gene flow within rat populations in four different cities (Combs et al. 2018a).

White-footed mice in and around the NYC metropolitan area exhibit high rates of genetic connectivity in rural areas, but increasingly restricted gene flow and diversity as sampling sites become more urban (Figure 4.3; Munshi-South et al. 2016). These SNP-based results provided important context to earlier microsatellite analyses that showed highly restricted gene flow between parks within NYC (Munshi-South 2012).

Similarly, urban burrowing owls (*Athene cunicularia*) exhibit low rates of gene flow between urban and nearby rural areas, but no gene flow between different cities (Mueller et al. 2018). In contrast, red-tailed bumblebees (*Bombus lapidarius*) exhibited little to no effect of urbanization on gene flow: urban and rural populations from multiple cities in Germany showed little genetic structure, and no significant effect of distance classes beyond IBD (Theodorou et al. 2018). In some circumstances, urbanization may facilitate gene flow through human transport on vehicles (Balbi et al. 2018), plants, soil, building materials, or waste, or by organisms tracking resources preferentially located in urban areas (e.g., organisms that feed at rubbish bins). The most detailed example of urban facilitation is black widow spiders (*Latrodectus hesperus*) in western North America. Conditional genetic distances visualized as networks of populations (Dyer and Nason 2004) showed that *L. hesperus* populations in cities are more connected to one another than to rural populations, and some cities act as hubs of connectivity (Miles et al. 2018a). Within-city patterns over finer spatial scales were similar for some cities but not others, indicating that urbanization may not facilitate gene flow in all urban contexts (Miles et al. 2018b).

The case studies and spatial population genomic analyses described above have greatly increased our knowledge of the impacts of urbanization on gene flow. However, these analyses should be considered exploratory and care must be taken to avoid over-interpreting results when the many assumptions of each approach are not strictly met (Whitlock and McCauley 1999; Lawson et al. 2018). Sampling strategies also play an important role, as relatively uniform sampling for continuously distributed populations is needed to identify more subtle effects of urbanization on gene flow (Box 4.1).

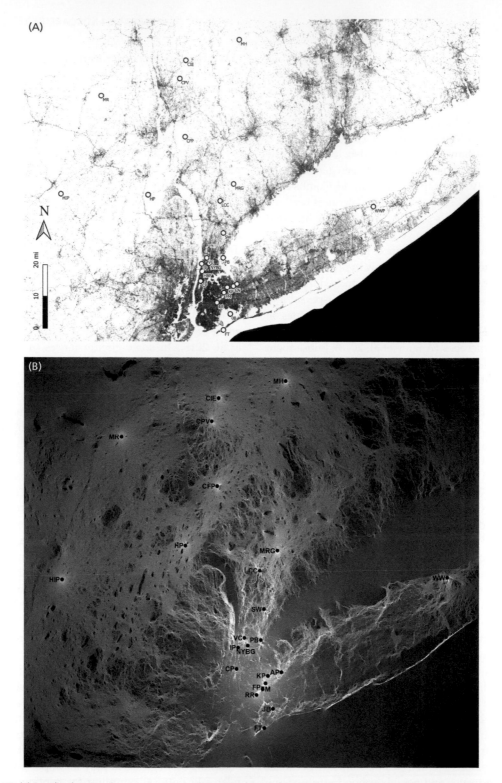

Figure 4.3 (A) Sampling locations for white-footed mice (Peromyscus leucopus) sampled along an urban-to-rural gradient in NYC metropolitan area. Each site is denoted by an open circle with a text label. Darker areas of map indicate greater levels of impervious surface cover. (B) Isolation-by-resistance model based on canopy cover for sampling locations in (A). Lighter areas of map indicate paths of greater connectivity between populations as predicted by Circuitscape. Black areas indicate areas predicted to be barriers to connectivity, such as open water or 10 per cent impervious surface cover. Urban populations are forced through specific paths of high canopy cover, whereas connectivity in rural areas is consistent across most of the landscape. (Figures adapted from Munshi-South et al. (2016).)

Box 4.1 Sampling considerations for studying gene flow in cities

The most compelling studies investigating how the urban landscape shapes gene flow are those that examine multiple cities. Replicated studies are uniquely suited to identify aspects of cities that consistently impede or promote gene flow vs aspects idiosyncratic to each city. Research that includes multiple species further expands the insights into how landscapes shape movement for an urban community of plants and animals. When designing such studies, researchers must make critical decisions about how best to sample individuals. The two main considerations are the geographic distribution of sampling and the numbers of samples to be collected, both of which have been explored previously in the literature. Importantly, if the research team has clear a priori hypotheses about how the urban landscape shapes movement of the focal species, they can design a sampling protocol that specifically targets those aspects of the city environment.

Sampling distribution may fall anywhere along a continuum from uniform sampling along a grid network (Figure 4.4A) to highly clustered (Figure 4.4B). Planning how sampling should be distributed across the cityscape will depend in part on the spatial distribution of the focal organism. For example, commensal species are often distributed more or less continuously throughout a city, and uniformly distributed sampling may be possible. Figure 4.4A represents a hypothetical sampling scheme for a widely distributed species in Vienna, Austria. As shown in Figure 4.4C, researchers were able to sample in this way in Salvador, Brazil, where brown rats were collected along a regular grid that extended SW to NE in order to capture a gradient of socioeconomic variation (J.L. Richardson et al., unpublished data). Dots in Figure 4.4C show individual rat samples collected; lighter background colours represent higher median income values within Salvador.

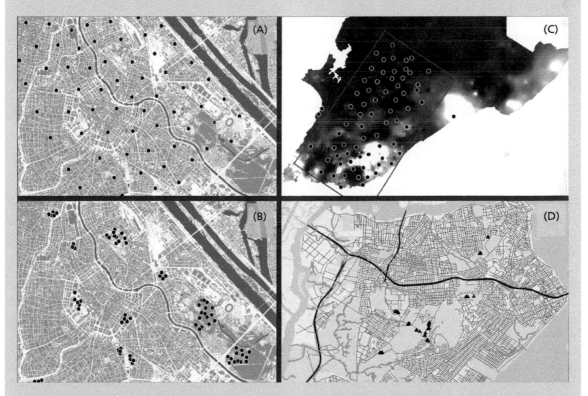

Figure 4.4 Examples of sampling designs. (A) Hypothetical sampling scheme in Vienna, Austria that is nearly uniformly distributed using a grid to select sampling points. (B) The same Vienna landscape but with highly aggregated sampling locations, which would be the case if the focal species is isolated to city parks. (C) and (D) represent empirical sampling schemes, including near-uniform sampling done for rats distributed across Salvador, Brazil (C: from J. Richardson), and clustered sampling of salamanders restricted to isolated streams in the Staten Island section of NYC (D: from N. Fusco). See main text for more information on each panel.

continued

> **Box 4.1 Continued**
>
> However, some species are isolated to patches of resources, which in cities could be green spaces like parks or gardens. This pattern occurs for white-footed mice in NYC (Munshi-South 2012). Figure 4.4B shows a hypothetical sampling scheme for a species restricted to parks in the same Vienna landscape as Figure 4.4A. Similarly, researchers sampling northern two-lined salamanders (*Eurycea bislineata*) found them restricted to the few intact headwater streams and seeps remaining in the Staten Island section of NYC (Figure 4.4D; black triangles are locations of each salamander sampled; N. Fusco et al., unpublished data). With this clustered sampling, they were able to test the hypothesis that high-traffic roads (thick black lines) influence gene flow in this species.
>
> As with sampling distribution, there are no prescriptive standards regarding the number of samples required to identify how movement and gene flow are impacted by urban landscapes. Sample sizes will always be a logistical balance between sampling more separate locations versus more individuals per site, and complex landscapes like cities will often require denser sampling at smaller spatial scales
>
> (Prunier et al. 2013; Richardson et al. 2016). Sample numbers will also depend heavily on the numbers of genetic markers used to detect genetic patterns in the study. For example, analysing thousands of SNPs will provide more information than 10–20 microsatellite loci and will require fewer samples per sampling site to detect patterns (Wang and Bradburd 2014). For species with patchy distributions requiring population-level sampling, rules of thumb ranging between 10 and 50 samples per location have often been used. Yet some studies have found that as few as eight samples per location is sufficient to detect population genomic divergence when thousands of SNP loci are analysed (Nazareno et al. 2017). Individual-based sampling uses genetic dissimilarity among individuals rather than populations to characterize genetic changes across a landscape (Prunier et al. 2013). This approach allows more locations to be sampled, making it attractive for urban landscape genetics studies that require denser sampling or a larger focal area. More detail on sampling considerations can be found in Richardson et al. (2016).

Most importantly, these methods do not directly test hypotheses about the influence of specific attributes of cities on gene flow (Richardson et al. 2016). The rapid development of spatially explicit approaches that incorporate landscape ecology is providing the framework needed for such direct, hypothesis-driven tests.

Recent landscape genetic studies make clear the importance of vegetated habitat for native species persisting in cities. Munshi-South (2012) found that landscape genetic models based on canopy cover were significantly associated with gene flow for *P. leucopus* in NYC parks, even after accounting for IBD. This study used an IBR framework (McRae 2006) to model all possible paths of connectivity between NYC parks; because empirical estimates of dispersal through different levels of canopy cover were not available, reduced levels of canopy cover were modelled as very high resistance to movement based on trapping records that did not record *P. leucopus* outside of closed canopy urban forests. *Peromyscus leucopus* along an urbanization gradient did not exhibit significant IBD but did exhibit IBR based on percentage of impervious surface in the landscape (Figure 4.3B; Munshi-South et al. 2016).

A simple IBE analysis assuming equal variance of allele frequencies for each population along the transect found that being located in NYC compared to surrounding areas was equivalent to an extra 7 km of geographic distance between populations. IBE occurs when gene flow rates are highest between similar habitats, resulting in genetic differentiation between dissimilar areas with high 'environmental distance' between them (Wang and Bradburd 2014). A recent review of published empirical studies indicated that IBE is a more commonly detected empirical pattern than IBD (Sexton et al. 2014). A more complex IBE analysis of *P. leucopus* that accounted for unequal variance in allele frequencies indicated that urban populations deviated from global mean allele frequencies much more than suburban or rural populations (Munshi-South et al. 2016). This latter result suggests that isolation in urban patches may result in strong genetic drift and little to no gene flow for some species.

IBR models that used expert opinion have successfully shown that impervious surfaces associated with urban development restrict yellow-faced bumblebee (*Bombus vosnesenskii*) (Jha and Kremen 2013; Jha 2015) and garden snail gene flow (Balbi et al. 2018).

However, in the latter case of urban snails, roads may increase gene flow over small spatial scales due to transport of snails by humans. This study was also notable because it is one of only two studies (see below) to apply a landscape genetics framework to the same taxon across multiple cities. Similarly, larger buildings and older buildings facilitated movement of invasive urban termites (*Reticulitermes flavipes*) in Paris, France (Baudouin et al. 2018). Landscape genetic modelling has thus found that urbanization may hinder or facilitate gene flow depending on the species and ecological context.

Recent work on urban wall lizards in Germany (mentioned in section 4.1) is exemplary of what urban landscape genetic studies should incorporate moving forward. Across four German cities, Beninde et al. (2018) found that a habitat suitability model developed for this species explained gene flow the best in two cities, while vegetation type and water cover alone were the optimal predictors in the other two cities. Water cover generally reduced gene flow, while semi-natural disturbed vegetation, rocky land cover, and railway tracks facilitated gene flow. Buildings, canopy cover, and roads were less important but still had a negative impact on gene flow. These cities contained complex combinations of native and admixed lineages that responded slightly differently to urban landscapes, indicating the need for large-scale comparative studies to understand gene flow in cities (Beninde et al. 2018). However, it should be appreciated that habitat suitability models complement rather than replace detailed empirical work on local ecology and habitat use. The wall lizard system has made substantial advances to urban landscape genetics by (1) incorporating habitat suitability modelling to identify relevant variables for IBR analyses (Beninde et al. 2016); (2) using an optimization procedure to choose combinations of landscape resistance values that best explain variation in gene flow; and (3) comparing results across multiple cities (Beninde et al. 2018).

4.3.2 Gene flow and drift between urban and rural habitats

Gene flow between urban and rural areas, and between different cities, is likely to be more restricted than gene flow within single cities, although these patterns are likely to be species-dependent. Greater isolation and lower levels of genetic diversity were reported for urban populations of the wood mouse (*A. sylvaticus*) compared to non-urban populations sampled around Dundee, Scotland (Wilson et al. 2016). Similarly, a sigmodontine rodent (*Calomys musculinus*) exhibited higher genetic differentiation and reduced gene flow when inhabiting urban vacant lots compared to populations in agricultural areas in central Argentina (Chiappero et al. 2011).

In some cases, genetic signatures from recent urbanization may coexist with deeper historical demographic signatures. For example, Harris et al. (2016) reported that urban *P. leucopus* in the NYC area are descended from two major lineages that diverged 13–15 000 ybp as mice recolonized the post-glacial landscape. Their demographic modelling also identified multiple recent urban populations that diverged within the last few hundred years when urban parks were created in NYC. Estimates of migration into these parks were variable, but the pattern of urban isolation followed by reduction in effective population size (N_E) was present for all urban populations in the model. Mueller et al. (2018) have also recently used whole genome resequencing to infer that the demographic history of urban and rural burrowing owls was similar until 50–75 years ago, when urban populations experienced a bottleneck that substantially reduced effective population sizes.

Common urban invasive species such as the brown rat are also likely to exhibit complex histories of gene flow between cities and regions. Recent genomic evidence corroborates historic scenarios of expansion of brown rats out of East Asia (particularly northern China), with rats only reaching Europe in the last three to four centuries. Invasion of Africa, the Americas, and Australia was then rapid and relatively recent due to human-assisted movement during the height of European colonialism (Puckett et al. 2016; Puckett and Munshi-South 2019). Most urban brown rat populations are dominated by single lineages that seem to have established and then excluded gene flow from additional lineages (Puckett et al. 2016). Such a result is surprising given the ubiquity of brown rats and their seemingly high capacity for global gene flow.

Before genome-wide SNP data, analyses of the global migration of urban commensals have been limited to inferring relationships between populations from mtDNA haplotype networks, phylogenetic trees, or evolutionary clustering without the ability to directly estimate gene flow or divergence times (Aplin et al. 2011; Song et al. 2014). Such mtDNA studies have revealed that the human transport of house mice (*Mus musculus*) throughout early pre-industrial cities in western Asia and Europe occurred much earlier than movements of *Rattus* species (Jones et al. 2013). Much more recently, blackbirds (*Turdus merula*) arrived in European cities from forest source populations and genetic analyses confirmed that colonization happened independently multiple times (Evans et al. 2009). Blackbirds have since become more sedentary in cities, resulting in accelerated genetic divergence from forest populations as gene flow declined (Evans et al. 2012).

4.3.3 Landscape genomics to identify local adaptation to urbanized environments

While very little landscape genomics work has been done in cities to date, some early studies serve as important examples (see also Chapter 5). Harris and Munshi-South (2017) used a GEA method based on latent factor mixed models (LFMM) (Frichot et al. 2013) to identify 19 candidate genes with SNPs associated with urbanization in six populations of white-footed mice sampled in urban and rural areas. LFMM also accounts for underlying neutral population structure (i.e., differences due to neutral divergence of populations), which was important in this case because the demographic history of these populations was influenced by both recent urbanization and the glacial history of the region (Harris et al. 2016). Many of the outlier genes identified in this study were involved in the same pathways for lipid and carbohydrate metabolism, suggesting that dietary changes are exerting novel selection pressures on urban populations. A recent GEA analysis of bumblebees in Europe identified 201 SNPs associated with land-cover type (urban vs rural), and slightly more than half of these SNPs were also associated with the percentage of impervious surface (Theodorou et al. 2018). Some of these SNPs were in genes associated with metabolism, heat stress, and oxidative stress, and thus may be related to urban selection pressures.

With the widespread adoption of SNPs and low-coverage whole genome sequencing in the coming years, researchers employing GEA tests will likely identify many candidate genes under selection in urban environments. With enough cities and species tested, these studies can robustly test the hypothesis that urbanization drives parallel adaptation in relation to changes in temperature, light, humidity, locomotion on artificial substrates, and other factors (Chapter 3; Johnson and Munshi-South 2017). A main challenge will be the choice of relevant environmental variables that influence local adaptation, as well as accounting for the influence of population structure, demographic history, and gene flow on genetic variation.

4.4 Future directions

To understand the general principles of how urban landscapes impact movement, comparative studies across multiple cities are necessary. While currently rare, our hope is that many more studies will be designed explicitly to include more than one city and compare patterns for the same species and landscape elements (Box 4.1). To our knowledge, such comparisons have only been done for urban rats (Combs et al. 2018a), lizards (Beninde et al. 2018), and land snails (Balbi et al. 2018), all very recent additions to the literature. There will almost certainly be many aspects of urban gene flow that are shaped by attributes specific to a particular city. Yet, the most powerful insights are those that are observed repeatedly across multiple cities, and that can be generalized beyond the focal cities in a predictive manner.

The ability to simulate how movement and gene flow proceed through cities will be an important tool in future studies of landscape genetics and genomics in cities. As urban land cover continues to expand, it will be logistically impossible to study all species and all areas potentially affected. Additionally, different stakeholders can use predictions stemming from simulation models to potentially mitigate the impacts of future development within cities on plant and animal movement. As emphasized in this chapter, strong, data-driven hypotheses are critical in parameterizing simulations and ascribing real-world relevance to the output (Epperson et al. 2010).

Simulations have been used to investigate the impacts of urban landscapes on adaptive trait variation (Santangelo et al. 2018) and population demographic history (Harris et al. 2016). However, few—if any—studies have implemented simulations to predict how city landscapes impact movement and gene flow.

Another interesting research frontier in cities is the use of historical samples to examine occurrence and connectivity in urban areas over time. The field of zooarchaeology has developed to associate animal remnants with human activity and history. Much of the early work in this field has centred on commensal animals (e.g., rats and mice) in cities, and has shed light on the invasion history, ecological niche, and ecosystem impacts of these species (Epperson et al. 2010). In some cases, it is possible to obtain useable DNA from these historical samples, and this genetic material from commensal species has been used in the context of characterizing human settlement patterns (Jones et al. 2013). Depending on the genetic assays used, these ancient DNA samples may also present an untapped resource for landscape genomics in cities, providing a way to investigate long-term patterns of movement and change over time.

4.5 Conclusions

The movement of organisms within and through cities is an important process that we will need to understand better in order to manage wildlife in the rapidly expanding global footprint of cities. In some cases, our management will be geared towards promoting the movement of chosen native species and restricting the movement of other pest species. In either case, genetic analyses will be a valuable tool in characterizing these movements. The approaches developed under the banner of landscape genetics have been used to address specific questions around how the landscape matrix impacts movement in wild species. Generally, urban landscapes will present a much more complex, spatially variable matrix that pushes the bounds of our current analytical approaches. As a result, researchers studying landscape genetics in cities should be incorporating data on socioeconomics, the built environment, and other attributes particular to dense human settlement that may be less relevant outside of the urban context.

Acknowledgements

We thank Joscha Beninde, Anne Charmantier, Charles Perrier, and Marta Szulkin for helpful comments and insights that improved earlier drafts of this chapter. We thank Matthew Combs and Nicole Fusco for sharing data and figures from their research. J. M.-S. was supported by National Science Foundation (NSF) grant DEB 1457523 and J.L.R. by NSF grant OIA 1738789.

References

Alexander, D.H., Novembre, J., and Lange, K. (2009). Fast model-based estimation of ancestry in unrelated individuals. *Genome Research*, 19, 1655–64.

Angley, L.P., Combs, M., Firth, C., et al. (2018). Spatial variation in the parasite communities and genomic structure of urban rats in New York City. *Zoonoses and Public Health*, 65, e113–23.

Aplin, K.P., Suzuki, H., Chinen, A.A., et al. (2011). Multiple geographic origins of commensalism and complex dispersal history of black rats. *PLOS ONE*, 6, e26357.

Balbi, M., Ernoult, A., Poli, P., et al. (2018). Functional connectivity in replicated urban landscapes in the land snail (*Cornu aspersum*). *Molecular Ecology*, 27, 1357–70.

Baudouin, G., Bech, N., Bagnères, A.-G., and Dedeine, F. (2018). Spatial and genetic distribution of a North American termite, *Reticulitermes flavipes*, across the landscape of Paris. *Urban Ecosystems*, 21, 751–64.

Beerli, P. (2006). Comparison of Bayesian and maximum likelihood inference of population genetic parameters. *Bioinformatics*, 22, 341–5.

Beninde, J., Feldmeier, S., Werner, M., et al. (2016). Cityscape genetics: structural vs. functional connectivity of an urban lizard population. *Molecular Ecology*, 25, 4984–5000.

Beninde, J., Feldmeier, S., Veith, M., and Hochkirch, A. (2018). Admixture of hybrid swarms of native and introduced lizards in cities is determined by the cityscape structure and invasion history. *Proceedings of the Royal Society B: Biological Sciences*, 285, 20180143.

Benson, J.F., Mahoney, P.J., Sikich, J.A., et al. (2016). Interactions between demography, genetics, and landscape connectivity increase extinction probability for a small population of large carnivores in a major metropolitan area. *Proceedings of the Royal Society B: Biological Sciences*, 283, 20160957.

Blair, R.B. (2001). Birds and butterflies along urban gradients in two ecoregions of the United States: is urbanization creating a homogeneous fauna?. In: Lockwood, J.L. and McKinney, M.L. (eds) *Biotic Homogenization*, pp. 33–56. Springer, New York.

Bradburd, G.S., Coop, G.M., and Ralph, P.L. (2018). Inferring continuous and discrete population genetic structure across space. *Genetics*, 210(1), 33–52.

Byrne, L.B. (2007). Habitat structure: a fundamental concept and framework for urban soil ecology. *Urban Ecosystems*, 10, 255–74.

Castellano, S., and Balletto, E. (2002). Is the partial Mantel test inadequate? *Evolution*, 56, 1871–3.

Chiappero, M.B., Panzetta-Dutari, G.M., Gómez, D., Castillo, E., Polop, J.J., and Gardenal, C.N. (2011). Contrasting genetic structure of urban and rural populations of the wild rodent *Calomys musculinus* (Cricetidae, Sigmodontinae). *Mammalian Biology—Zeitschrift für Säugetierkunde*, 76, 41–50.

Combs, M., Byers, K.A., Ghersi, B.M., et al. (2018a). Urban rat races: spatial population genomics of brown rats (*Rattus norvegicus*) compared across multiple cities. *Proceedings of the Royal Society B: Biological Sciences*, 285, 20180245.

Combs, M., Puckett, E.E., Richardson, J., Mims, D., and Munshi-South, J. (2018b). Spatial population genomics of the brown rat (*Rattus norvegicus*) in New York City. *Molecular Ecology*, 27, 83–98.

Costa, F., Richardson, J.L., Dion, K., et al. (2016). Multiple paternity in the Norway rat, *Rattus norvegicus*, from urban slums in Salvador, Brazil. *Journal of Heredity*, 107, 181–6.

Dickson, B.G., Albano, C.M., Anantharaman, R., et al. (2019). Circuit-theory applications to connectivity science and conservation. *Conservation Biology*, 33, 239–49.

Donihue, C.M. and Lambert, M.R. (2015). Adaptive evolution in urban ecosystems. *Ambio*, 44, 194–203.

Dyer, R.J. and Nason, J.D. (2004). Population graphs: the graph theoretic shape of genetic structure. *Molecular Ecology*, 13, 1713–27.

Epperson, B.K., Mcrae, B.H., Scribner, K., et al. (2010). Utility of computer simulations in landscape genetics. *Molecular Ecology*, 19, 3549–64.

Evans, K.L., Gaston, K.J., Frantz, A.C., et al. (2009). Independent colonization of multiple urban centres by a formerly forest specialist bird species. *Proceedings of the Royal Society B: Biological Sciences*, 276, 2403–10.

Evans, K.L., Newton, J., Gaston, K.J., Sharp, S.P., McGowan, A., and Hatchwell, B.J. (2012). Colonisation of urban environments is associated with reduced migratory behaviour, facilitating divergence from ancestral populations. *Oikos*, 121, 634–40.

Excoffier, L., Dupanloup, I., Huerta-Sánchez, E., Sousa, V. C., and Foll, M. (2013). Robust Demographic Inference from Genomic and SNP Data. *PLOS Genetics*, 9, e1003905.

Forester, B.R., Lasky, J.R., Wagner, H.H., and Urban, D.L. (2018). Comparing methods for detecting multilocus adaptation with multivariate genotype–environment associations. *Molecular Ecology*, 27, 2215–33.

Forin-Wiart, M.-A., Poulle, M.-L., Piry, S., Cosson, J.-F., Larose, C., and Galan, M. (2018). Evaluating metabar-coding to analyse diet composition of species foraging in anthropogenic landscapes using Ion Torrent and Illumina sequencing. *Scientific Reports*, 8, 17091.

Foster, J. and Sandberg, L.A. (2010). Friends or foe? Invasive species and public green space in Toronto. *Geographical Review*, 94, 178–98.

Frichot, E., Schoville, S.D., Bouchard, G., and François, O. (2013). Testing for associations between loci and envir-onmental gradients using latent factor mixed models. *Molecular Biology and Evolution*, 30, 1687–99.

Galpern, P., Peres-Neto, P., Polfus, J., and Manseau, M. (2014). MEMGENE: spatial pattern detection in genetic distance data. *Methods in Ecology and Evolution*, 5, 1116–20.

Gardner-Santana, L.C., Norris, D.E., Fornadel, C.M., Hinson, E.R., Klein, S.L., and Glass, G.E. (2009). Commensal ecology, urban landscapes, and their influence on the genetic characteristics of city-dwelling Norway rats (*Rattus norvegicus*). *Molecular Ecology*, 18, 2766–78.

Germaine, S.S. and Wakeling, B.F. (2001). Lizard species distributions and habitat occupation along an urban gradient in Tucson, Arizona, USA. *Biological Conservation*, 97, 229–37.

Gortat, T., Rutkowski, R., Gryczynska-Siemiatkowska, A., Kozakiewicz, A., and Kozakiewicz, M. (2013). Genetic structure in urban and rural populations of *Apodemus agrarius* in Poland. *Mammalian Biology—Zeitschrift für Säugetierkunde*, 78, 171–7.

Gortat, T., Rutkowski, R., Gryczyńska, A., Pieniążek, A., Kozakiewicz, A., and Kozakiewicz, M. (2015). Anthropo-pressure gradients and the population genetic structure of *Apodemus agrarius*. *Conservation Genetics*, 16, 649–59.

Gortat, T., Rutkowski, R., Gryczynska, A., Kozakiewicz, A., and Kozakiewicz, M. (2017). The spatial genetic structure of the yellow-necked mouse in an urban envir-onment—a recent invader vs. a closely related perman-ent inhabitant. *Urban Ecosystems*, 20, 581–94.

Guiry, E.J., and Gaulton, B.C. (2016). Inferring human behaviors from isotopic analyses of rat diet: a critical review and historical application. *Journal of Archaeological Method and Theory*, 23, 399–426.

Hacker, K.P., Seto, K.C., Costa, F., et al. (2013). Urban slum structure: integrating socioeconomic and land cover data to model slum evolution in Salvador, Brazil. *International Journal of Health Geographics*, 12, 45.

Harris, S.E. and Munshi-South, J. (2017). Signatures of positive selection and local adaptation to urbanization in white-footed mice (*Peromyscus leucopus*). *Molecular Ecology*, 26, 6336–50.

Harris, S.E., Xue, A.T., Alvarado-Serrano, D., et al. (2016). Urbanization shapes the demographic history of a native rodent (the white-footed mouse, *Peromyscus leucopus*) in New York City. *Biology Letters*, 12, 20150983.

Himsworth, C.G., Parsons, K.L., Jardine, C., and Patrick, D.M. (2013). Rats, cities, people, and pathogens: a systematic review and narrative synthesis of literature regarding the ecology of rat-associated zoonoses in urban centers. *Vector-Borne and Zoonotic Diseases*, 13, 349–59.

Hope, D., Gries, C., Zhu, W., et al. (2003). Socioeconomics drive urban plant diversity. *Proceedings of the National Academy of Sciences of the United States of America*, 100, 8788–92.

Hulme-Beaman, A., Dobney, K., Cucchi, T., and Searle, J.B. (2016). An ecological and evolutionary framework for commensalism in anthropogenic environments. *Trends in Ecology & Evolution*, 31, 633–45.

Ingvarsson, P.K. (2001). Restoration of genetic variation lost—the genetic rescue hypothesis. *Trends in Ecology & Evolution*, 16, 62–3.

Jaquiéry, J., Broquet, T., Hirzel, A.H., Yearsley, J., and Perrin, N. (2011). Inferring landscape effects on dispersal from genetic distances: how far can we go? *Molecular Ecology*, 20, 692–705.

Jenkins, D.G., Carey, M., Czerniewska, J., et al. (2010). A meta-analysis of isolation by distance: relic or reference standard for landscape genetics? *Ecography*, 33, 315–20.

Jha, S. (2015). Contemporary human-altered landscapes and oceanic barriers reduce bumble bee gene flow. *Molecular Ecology*, 24, 993–1006.

Jha, S. and Kremen, C. (2013). Urban land use limits regional bumble bee gene flow. *Molecular Ecology*, 22, 2483–95.

Johnson, M.T.J. and Munshi-South, J. (2017). Evolution of life in urban environments. *Science*, 358, eaam8327.

Johnson, S., Bragdon, C., Olson, C., Merlino, M., and Bonaparte, S. (2016). Characteristics of the built environment and the presence of the Norway rat in New York City: results from a neighborhood rat surveillance program, 2008–2010. *Journal of Environmental Health*, 78, 22–9.

Jones, E.P., Eager, H.M., Gabriel, S.I., Jóhannesdóttir, F., and Searle, J.B. (2013). Genetic tracking of mice and other bioproxies to infer human history. *Trends in Genetics*, 29, 298–308.

Kajdacsi, B., Costa, F., Hyseni, C., et al. (2013). Urban population genetics of slum-dwelling rats (*Rattus norvegicus*) in Salvador, Brazil. *Molecular Ecology*, 22, 5056–70.

Kawecki, T.J. and Ebert, D. (2004). Conceptual issues in local adaptation. *Ecology Letters*, 7, 1225–41.

Kinzig, A.P., Warren, P., Martin, C., Hope, D., and Katti, M. (2005). The effects of human socioeconomic status and cultural characteristics on urban patterns of biodiversity. *Ecology and Society*, 10, 23.

Kowarik, I. and von der Lippe, M. (2011). Secondary wind dispersal enhances long-distance dispersal of an invasive species in urban road corridors. *NeoBiota*, 9, 49–70.

Landguth, E.L., Cushman, S.A., Schwartz, M.K., McKelvey, K.S., Murphy, M., and Luikart, G. (2010). Quantifying the lag time to detect barriers in landscape genetics. *Molecular Ecology*, 19, 4179–91.

LaPoint, S., Balkenhol, N., Hale, J., Sadler, J., and van der Ree, R. (2015). Ecological connectivity research in urban areas. *Functional Ecology*, 29, 868–78.

Lawson, D.J., van Dorp, L., and Falush, D. (2018). A tutorial on how not to over-interpret STRUCTURE and ADMIXTURE bar plots. *Nature Communications*, 9, 3258.

Lenormand, T. (2002). Gene flow and the limits to natural selection. *Trends in Ecology & Evolution*, 17, 183–9.

Lepczyk, C.A., Aronson, M.F.J., Evans, K.L., Goddard, M.A., Lerman, S.B., and MacIvor, J.S. (2017). Biodiversity in the city: fundamental questions for understanding the ecology of urban green spaces for biodiversity conservation. *BioScience*, 67, 799–807.

Manel, S. and Holderegger, R. (2013). Ten years of landscape genetics. *Trends in Ecology & Evolution*, 28, 614–21.

Manel, S., Schwartz, M.K., Luikart, G., and Taberlet, P. (2003). Landscape genetics: combining landscape ecology and population genetics. *Trends in Ecology & Evolution*, 18, 189–97.

Mateo-Sánchez, M.C., Balkenhol, N., Cushman, S., Pérez, T., Domínguez, A., and Saura, S. (2015). A comparative framework to infer landscape effects on population genetic structure: are habitat suitability models effective in explaining gene flow? *Landscape Ecology*, 30, 1405–20.

McCartney-Melstad, E., Vu, J.K., and Shaffer, H.B. (2018). Genomic data recover previously undetectable fragmentation effects in an endangered amphibian. *Molecular Ecology*, 27, 4430–43.

McCormick, M. (2003). Rats, communications, and plague: toward an ecological history. *Journal of Interdisciplinary History*, 34, 1–25.

McRae, B.H. (2006). Isolation by resistance. *Evolution*, 60, 1551–61.

McRae, B.H., and Beier, P. (2007). Circuit theory predicts gene flow in plant and animal populations. *Proceedings of the National Academy of Sciences of the United States of America*, 104, 19885–90.

Meirmans, P.G. (2012). The trouble with isolation by distance. *Molecular Ecology*, 21, 2839–46.

Milanesi, P., Holderegger, R., Caniglia, R., Fabbri, E., and Randi, E. (2016). Different habitat suitability models yield different least-cost path distances for landscape genetic analysis. *Basic and Applied Ecology*, 17, 61–71.

Miles, L.S., Dyer, R.J., and Verrelli, B.C. (2018a). Urban hubs of connectivity: contrasting patterns of gene flow within and among cities in the western black widow spider. *Proceedings of the Royal Society B: Biological Sciences*, 285, 20181224.

Miles, L.S., Johnson, J.C., Dyer, R.J., and Verrelli, B.C. (2018b). Urbanization as a facilitator of gene flow in a human health pest. *Molecular Ecology*, 27, 3219–30.

Mueller, J.C., Kuhl, H., Boerno, S., Tella, J.L., Carrete, M., and Kempenaers, B. (2018). Evolution of genomic variation in the burrowing owl in response to recent colonization of urban areas. *Proceedings of the Royal Society B: Biological Sciences*, 285, 20180206.

Munshi-South, J. (2012). Urban landscape genetics: canopy cover predicts gene flow between white-footed mouse (*Peromyscus leucopus*) populations in New York City. *Molecular Ecology*, 21, 1360–78.

Munshi-South, J. and Kharchenko, K. (2010). Rapid, pervasive genetic differentiation of urban white-footed mouse (*Peromyscus leucopus*) populations in New York City. *Molecular Ecology*, 19, 4242–54.

Munshi-South, J. and Nagy, C. (2014). Urban park characteristics, genetic variation, and historical demography of white-footed mouse (*Peromyscus leucopus*) populations in New York City. *PeerJ*, 2, e310.

Munshi-South, J., Zolnik, C.P., and Harris, S.E. (2016). Population genomics of the Anthropocene: urbanization is negatively associated with genome-wide variation in white-footed mouse populations. *Evolutionary Applications*, 9, 546–64.

Nazareno, A.G., Bemmels, J.B., Dick, C.W., and Lohmann, L.G. (2017). Minimum sample sizes for population genomics: an empirical study from an Amazonian plant species. *Molecular Ecology Resources*, 17, 1136–47.

Nowakowski, A.J., DeWoody, J.A., Fagan, M.E., Willoughby, J.R., and Donnelly, M.A. (2015). Mechanistic insights into landscape genetic structure of two tropical amphibians using field derived resistance surfaces. *Molecular Ecology*, 24, 580–95.

Ottensmann, J.R. (2017). Defining exurban areas for the analysis of urban patterns over time. *SSRN Electronic Journal*. https://papers.ssrn.com/sol3/papers.cfm?abstract_id=2984392.

Parsons, M.H., Banks, P.B., Deutsch, M.A., Corrigan, R.F., and Munshi-South, J. (2017). Trends in urban rat ecology: a framework to define the prevailing knowledge gaps and incentives for academia, pest management professionals (PMPs) and public health agencies to participate. *Journal of Urban Ecology*, 3, jux005.

Peakall, R. and Smouse, P.E. (2006). GENALEX 6: genetic analysis in Excel. Population genetic software for teaching and research. *Molecular Ecology Notes*, 6, 288–95.

Peterman, W.E. (2018). ResistanceGA: an R package for the optimization of resistance surfaces using genetic algorithms. *Methods in Ecology and Evolution*, 9, 1638–47.

Peterman, W.E., Connette, G.M., Semlitsch, R.D., and Eggert, L.S. (2014). Ecological resistance surfaces predict fine-scale genetic differentiation in a terrestrial woodland salamander. *Molecular Ecology*, 23, 2402–13.

Petkova, D., Novembre, J., and Stephens, M. (2016). Visualizing spatial population structure with estimated effective migration surfaces. *Nature Genetics*, 48, 94–100.

Pickett, S.T.A., Cadenasso, M.L., Grove, J.M., et al. (2001). Urban ecological systems: linking terrestrial ecological, physical, and socioeconomic components of metropolitan areas. *Annual Review of Ecology and Systematics*, 32, 127–57.

Pritchard, J.K., Stephens, M., and Donnelly, P. (2000). Inference of population structure using multilocus genotype data. *Genetics*, 155, 945–59.

Prunier, J.G., Kaufmann, B., Fenet, S., et al. (2013). Optimizing the trade-off between spatial and genetic sampling efforts in patchy populations: towards a better assessment of functional connectivity using an individual-based sampling scheme. *Molecular Ecology*, 22, 5516–30.

Puckett, E.E. and Munshi-South, J. (2019). Brown rat demography reveals pre-commensal structure in eastern Asia prior to expansion into Southeast Asia. *Genome Research*, 29, 762–70.

Puckett, E.E., Park, J., Combs, M., et al. (2016). Global population divergence and admixture of the brown rat (*Rattus norvegicus*). *Proceedings of the Royal Society B: Biological Sciences*, 283, 20161762.

Puritz, J.B., Matz, M.V., Toonen, R.J., Weber, J.N., Bolnick, D.I., and Bird, C.E. (2014). Demystifying the RAD fad. *Molecular Ecology*, 23, 5937–42.

Reardon, S.F., Matthews, S.A., O'Sullivan, D., et al. (2008). The geographic scale of metropolitan racial segregation. *Demography*, 45, 489–514.

Reis, R.B., Ribeiro, G.S., Felzemburgh, R.D.M., et al. (2008). Impact of environment and social gradient on Leptospira infection in urban slums. *PLOS Neglected Tropical Diseases*, 2, e228.

Rellstab, C., Gugerli, F., Eckert, A.J., Hancock, A.M., and Holderegger, R. (2015). A practical guide to environmental association analysis in landscape genomics. *Molecular Ecology*, 24, 4348–70.

Richardson, J.L., Urban, M.C., Bolnick, D.I., and Skelly, D.K. (2014). Microgeographic adaptation and the spatial scale of evolution. *Trends in Ecology & Evolution*, 29, 165–76.

Richardson, J.L., Brady, S.P., Wang, I.J., and Spear, S.F. (2016). Navigating the pitfalls and promise of landscape genetics. *Molecular Ecology*, 25, 849–63.

Richardson, J.L., Burak, M.K., Hernandez, C., et al. (2017). Using fine-scale spatial genetics of Norway rats to improve control efforts and reduce leptospirosis risk in urban slum environments. *Evolutionary Applications*, 10, 323–37.

Saarikivi, J., Knopp, T., Granroth, A., and Merilä, J. (2013). The role of golf courses in maintaining genetic connectivity between common frog (*Rana temporaria*) populations in an urban setting. *Conservation Genetics*, 14, 1057–64.

Salinitro, M., Alessandrini, A., Zappi, A., Melucci, D., and Tassoni, A. (2018). Floristic diversity in different urban ecological niches of a southern European city. *Scientific Reports*, 8, 15110.

Sambatti, J.B.M. and Rice, K.J. (2006). Local adaptation, patterns of selection, and gene flow in the California serpentine sunflower (*Helianthus exilis*). *Evolution*, 60, 696–710.

Sandström, U.G., Angelstam, P., and Mikusiński, G. (2006). Ecological diversity of birds in relation to the structure of urban green space. *Landscape and Urban Planning*, 77, 39–53.

Santangelo, J.S., Johnson, M.T.J., and Ness, R.W. (2018). Modern spandrels: the roles of genetic drift, gene flow and natural selection in the evolution of parallel clines. *Proceedings of the Royal Society B: Biological Sciences*, 285, 20180230.

Savage, A.M., Hackett, B., Guénard, B., Youngsteadt, E.K., and Dunn, R.R. (2015). Fine-scale heterogeneity across Manhattan's urban habitat mosaic is associated with variation in ant composition and richness. In: Schonrogge, K. and Orivel, J. (eds) *Insect Conservation and Diversity*, Volume 8, pp. 216–28. John Wiley & Sons, New York.

Seto, K.C., Güneralp, B., and Hutyra, L.R. (2012). Global forecasts of urban expansion to 2030 and direct impacts on biodiversity and carbon pools. *Proceedings of the National Academy of Sciences of the United States of America*, 109, 16083–8.

Sexton, J.P., Hangartner, S.B., and Hoffmann, A.A. (2014). Genetic isolation by environment or distance: which pattern of gene flow is most common? *Evolution*, 68, 1–15.

Shirk, A.J. and Cushman, S.A. (2011). sGD: software for estimating spatially explicit indices of genetic diversity. *Molecular Ecology Resources*, 11, 922–34.

Song, Y., Lan, Z., and Kohn, M.H. (2014). Mitochondrial DNA phylogeography of the Norway rat. *PLOS ONE*, 9, e88425.

Spear, S.F., Balkenhol, N., Fortin, M.J., McRae, B.H., and Scribner, K. (2010). Use of resistance surfaces for landscape genetic studies: considerations for parameterization and analysis. *Molecular Ecology*, 19, 3576–91.

Swift, J.A., Roberts, P., Boivin, N., and Kirch, P.V. (2018). Restructuring of nutrient flows in island ecosystems fol-lowing human colonization evidenced by isotopic analysis of commensal rats. *Proceedings of the National Academy of Sciences of the United States of America*, 115, 6392–7.

Theodorou, P., Radzevičiūtė, R., Kahnt, B., Soro, A., Grosse, I., and Paxton, R.J. (2018). Genome-wide single nucleotide polymorphism scan suggests adaptation to urbanization in an important pollinator, the red-tailed bumblebee (*Bombus lapidarius* L.). *Proceedings of the Royal Society B: Biological Sciences*, 285, 20172806.

Titus, V.R., Bell, R.C., Becker, C.G., and Zamudio, K.R. (2014). Connectivity and gene flow among eastern tiger salamander (*Ambystoma tigrinum*) populations in highly modified anthropogenic landscapes. *Conservation Genetics*, 15, 1447–62.

Unfried, T.M., Hauser, L., and Marzluff, J.M. (2013). Effects of urbanization on song sparrow (*Melospiza melodia*) population connectivity. *Conservation Genetics*, 14, 41–53.

Vergnes, A., Kerbiriou, C., and Clergeau, P. (2013). Ecological corridors also operate in an urban matrix: a test case with garden shrews. *Urban Ecosystems*, 16, 511–25.

Walsh, M.G. (2014). Rat sightings in New York City are associated with neighborhood sociodemographics, housing characteristics, and proximity to open public space. *PeerJ*, 2, e533.

Walter, T., Zink, R., Laaha, G., Zaller, J.G., and Heigl, F. (2018). Fox sightings in a city are related to certain land use classes and sociodemographics: results from a citizen science project. *BMC Ecology*, 18, 50.

Wang, I.J. and Bradburd, G.S. (2014). Isolation by environment. *Molecular Ecology*, 23, 5649–62.

Weir, B.S. and Cockerham, C.C. (1984). Estimating F-statistics for the analysis of population structure. *Evolution*, 38, 1358–70.

Whitlock, M.C. and McCauley, D.E. (1999). Indirect measures of gene flow and migration: FST not equal to 1/(4Nm + 1). *Heredity*, 82, 117–25.

Wilson, A., Fenton, B., Malloch, G., Boag, B., Hubbard, S., and Begg, G. (2016). Urbanisation versus agriculture: a comparison of local genetic diversity and gene flow between wood mouse *Apodemus sylvaticus* populations in human-modified landscapes. *Ecography*, 39, 87–97.

Wood, B.C. and Pullin, A.S. (2002). Persistence of species in a fragmented urban landscape: the importance of dispersal ability and habitat availability for grassland butterflies. *Biodiversity and Conservation*, 11, 1451–68.

Wright, S. (1969). *Evolution and the Genetics of Populations. Volume 2, The Theory of Gene Frequencies*. University of Chicago Press, Chicago.

CHAPTER 5

Adaptation Genomics in
Urban Environments

Charles Perrier, Aude Caizergues, and Anne Charmantier

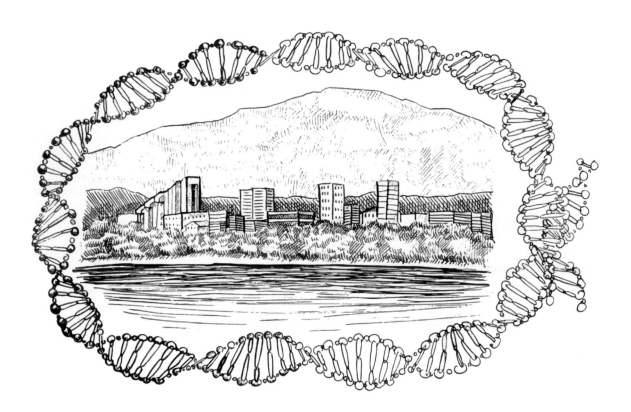

Perrier, C., Caizergues, A. and Charmantier, A., *Adaptation Genomics in Urban Environments* In: *Urban Evolutionary Biology*. Edited by Marta Szulkin, Jason Munshi-South and Anne Charmantier, Oxford University Press (2020). © Oxford University Press.
DOI: 10.1093/oso/9780198836841.003.0005

5.1 Introduction

Urbanization has been escalating for a thousand years and is predicted to continue rising, with up to two-thirds of the global human population expected to live in cities by 2050 (UN 2014). Urbanization is the process whereby natural and rural areas are converted to cities. It drives rapid and tremendous environmental changes, including air, light, noise, and chemical pollution, altered climate, high human and vehicle densities, and modifications of the distribution of organisms' living resources. Wildlife living in the city has to cope with these novel assortments of environmental conditions, which may be often stressful, but also sometimes beneficial in terms of increased survival or reproductive success. For the past four decades, urban ecology has provided ample and pervasive evidence that organisms in cities often differ from those in natural habitats in several key phenotypic characteristics, including both morphological and behavioural traits. The field of urban evolutionary ecology builds on its heritage from the field of urban ecology and aims to document and understand the evolutionary mechanisms underlying microevolution in urban populations (Donihue and Lambert 2015; Johnson and Munshi-South 2017). Studies have revealed striking phenotypic changes in organisms facing human-modified habitats that are much more significant than those seen in populations in natural habitats (Alberti 2015; Alberti et al. 2017). Quantitative genetics and population genetics are of growing importance in this field, respectively deciphering genetic and environmental effects on phenotypic variance and decoding the DNA variation behind heritable variation (Figure 5.1), which may ultimately lead to speciation (Thompson et al. 2018).

Many of the emerging phenotypic differences found between populations from natural and urban environments have been attributed to phenotypic plasticity (Hendry et al. 2008). Indeed, evidence that these trait changes are of genetic origin and are adaptive is still scarce (Pelletier and Coltman 2018), possibly due to a lack of appropriate data. Despite this shortage of data, there is a clear expectation that city populations are exposed to, and might genetically respond to, novel and divergent selection pressures (Alberti 2015). However, maybe counterintuitively, there is also evidence that anthropogenic

disturbances often weaken rather than increase the strength of natural selection (Fugère and Hendry 2018), and therefore the nature and strength of selection in cities may be difficult to predict. Similarly, our overall assessment of the frequency and magnitude of adaptation and maladaptation in cities remains very poor, and the presence of one or the other is often hypothesized without being properly tested (Donihue and Lambert 2015; Johnson and Munshi-South 2017). Hence how often, and by which mechanisms, urbanization results in rapid evolutionary changes, and the nature of these changes, remain open questions. Quantitative genetics can provide answers to these questions, as we review in section 5.2.

Similar to investigations in other ecological contexts, evolution in cities will depend on the strength of natural selection, on the extent of genetic diversity available for selection to act on, on the strength of genetic drift, and on the extent of gene flow (Johnson and Munshi-South 2017). Population genetic studies of urban organisms have begun documenting genetic drift, gene flow, mutations, and genomic footprints of selection in urban populations (Johnson and Munshi-South 2017). Nevertheless, while genetic drift and gene flow have been repeatedly investigated using a wide variety of population genetics methods (see Chapter 4), an understanding of the molecular mechanisms underlying urban adaptive evolution is only beginning to materialize (Johnson and Munshi-South 2017; Schell 2018). As we review in section 5.3, genomic tools made available by recent sequencing developments are instrumental in decoding these molecular mechanisms.

It is known that adaptations result from selection for beneficial DNA variants in a population. However, there is growing evidence that epigenetic variation, this partly heritable variation that does not arise from variation in DNA sequences, can also be implicated in rapid adaptation (Jablonka and Raz 2009). Such epigenetic variation could favour adaptive plasticity and adaptation in new environments. This could happen, for example, by triggering plastic responses towards the local fitness peak or by increasing mutation rates. In section 5.4 we discuss the ways that researchers are taking advantage of new sequencing technologies to investigate the role of epigenetic variation in urban evolution.

In this chapter, we review how quantitative genetics, population genomics, and epigenomics offer the

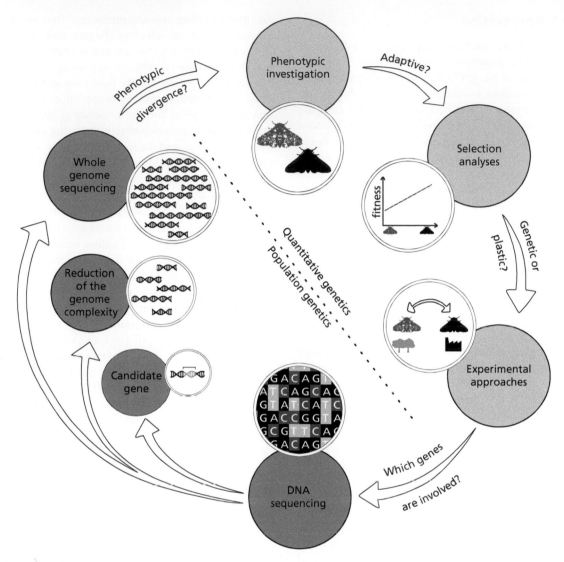

Figure 5.1 Inference of evolutionary processes in urban environments. The genetic or plastic nature of phenotypic and fitness responses in urban environments is investigated using quantitative genetic approaches and experiments. DNA sequencing of candidate genes, of reduced genome complexity or of the entire genome, investigates genes implicated in phenotypic variation, response to selection, and local adaptation in urban environments. Note that other possible arrows can be drawn between the different learning steps.

potential for unprecedented insights into evolution in urban populations. Given the rapidly growing productivity of the urban evolutionary ecology field, this chapter is not exhaustive, but rather presents a diversity of examples illustrating recent advances (see Table 5.1).

We first review how quantitative genetics successfully investigates whether the observed phenotypic variation in urban environments is heritable, adaptive, and results from selection. We then review

how population genomics frameworks offer unique opportunities to pinpoint the genetic variants implicated in evolution in urban environments. We discuss how further use of high-resolution genomic approaches, as well as epigenomics, may improve our understanding of such evolution. We finally conclude on the necessity of integrative studies of quantitative genetics and population genomics for a deeper comprehension of evolution in urban environments.

Table 5.1 Case studies of adaptive evolution in urban environments. For each type of inference illustrated in Figure 5.1, we highlight some emblematic studies that are discussed in this chapter.

Field	Method	Results in several emblematic studies
Quantitative genetics	Phenotypic investigation	Melanism is associated with industrial pollution in the peppered moth (*Biston bettularia*) (Ketterwell 1958)
		Differences in morphology, life history, and behaviour are found in urban populations of great tits (*Parus major*) (Charmantier et al. 2017)
	Selection analysis	Non-dispersing seeds in urban pavements are selected for in the weed *Crepis sancta* (Cheptou et al. 2008)
		Divergence in morphology and life-history traits is not adaptive but probably results from reduced access to resources in the city for great tits (Caizergues et al. 2018)
	Experimentation	Resistance to urban water pollution in the Atlantic killifish (*Fundulus heteroclitus*) (Whitehead et al. 2017)
		A genetic origin of increased heat tolerance is reported in urban acorn ants (*Temnothorax curvispinosus*) (Diamond et al. 2018)
Population genetics	Candidate gene	Candidate genes for personality traits (SERT and DRD4) show SNP variations between urban and rural populations in the blackbird (*Turdus merula*) (Mueller et al. 2013) and black swan (*Cygnus atratus*) (Dongen et al. 2015)
		The SERT gene shows genetic and epigenetic variation associated with urbanization in the great tit (Riyahi et al. 2015)
	Reduced genomic complexity	Multiple SNPs are associated with urbanization in the great tit (Perrier et al. 2018), bumblebee (*Bombus lapidarius*) (Theodorou et al. 2018), and white-footed mice (*Peromyscus leucopus*) (Harris and Munshi-South 2017)
		Multiple differentially expressed genes in the city are found in great tits (Watson et al. 2017)
		Epigenetic variations are found in urban populations of Darwin's finches (*Geospiza fortis* and *Geospiza fuliginosa*) (McNew et al. 2017)
	Whole genome sequencing	A transposable element is identified as responsible for the industrial melanism in British peppered moths (Hof et al. 2011, 2016)
		A few structural variants are associated with resistance to freshwater pollutants in Atlantic killifish (*Fundulus heteroclitus*) (Reid et al. 2016)

5.2 Evolutionary significance of trait variation in an urban context: evidence for genetic adaptation

Evidence of phenotypic divergence between rural and urban populations abounds in the literature across a wide variety of taxa. However, the evolutionary causes and consequences of such divergence in traits remain largely unknown. Urbanization is a growing process driving fast and monumental environmental changes that may require adaptation in order to maintain population persistence (e.g., Alberti 2015; Donihue and Lambert 2015; Pelletier and Coltman 2018). Thus, there is a pressing need to understand how wildlife copes with these changes. Is phenotypic variation across urban and natural populations a response to new environmental pressures induced by the city, i.e., are urban phenotypes adaptive? Are these responses

of a plastic or genetic origin? To address these questions, various methods exist that encompass different subdisciplines of evolutionary biology (Figure 5.1). In this section of the chapter, we briefly outline and discuss some of the methods that do not require the analysis of molecular sequences. In particular, we highlight lessons and results from evolutionary quantitative genetic studies on urban adaptation, i.e., studies of the genetic basis of complex traits and of selection acting on these traits, based on individual repeated measures of these traits and the associated fitness within populations.

5.2.1 Providing quantitative genetic empirical measures of urban-specific selection

Measuring selection in action requires estimating the effects of focal characters on individual fitness.

The evolutionary trajectory of these traits in a given population will depend jointly on the genetic basis of (co)variation for these characters and on the direction and amplitude of selection. The form taken by natural selection and the way it varies across environmental conditions can be quantified using the basic Lande and Arnold (1983) analytical framework to estimate linear or quadratic selection gradients, or via more sophisticated methods that account for statistical complications such as non-normal fitness distributions (e.g., Morrissey and Sakrejda 2013).

Although the claim that urbanization triggers new selection pressures has long been pervasive in the urban evolutionary literature, empirical evidence comparing selection gradients between urban and rural populations is very recent (Fugère and Hendry 2018; see, for example, Gorton et al. 2018 and Caizergues et al. 2018). We believe that this delay is due to the scarcity of individual monitoring programmes, and hence individual fitness estimations, in urban environments. Following 7 years of monitoring of urban and forest great tits (*Parus major*), Caizergues et al. (2018) compared selection gradients in two habitats, revealing that the observed divergence in bird morphology and life history was not adaptive, i.e., it was not aligned with divergence in ongoing selection. For instance, while birds laid their broods on average 3 days earlier in the city, selection for early breeding phenology was found only in the forest and not in the urban habitat. Measuring fitness in the laboratory under different environmental conditions provides a means to bypass long-term programmes while assessing the adaptive nature of urban phenotypes. For instance, a shift towards higher heat tolerance in urban acorn ants (*Temnothorax curvispinosus*) was consistent with results showing positive selection for heat tolerance under laboratory rearing conditions, providing evidence for local adaptation of these ants to the urban heat island (Chapter 6; Diamond et al. 2018).

5.2.2 Testing for plastic versus genetic basis of adaptation

Assessing the potential for an organism to adapt to rapidly changing environmental conditions requires a deep understanding of (1) the standing additive genetic (co)variation in key adaptive traits, and (2) the potential for phenotypic plastic responses in these traits. If selection analyses reveal that an urban phenotypic shift is adaptive, a plastic origin in the shift can be distinguished from a genetic basis by examining whether quantitative trait differences among populations disappear or are maintained in a common garden experiment. This experimental approach has been successfully applied to show that fragmentation and urbanization have driven local adaptation in the annual weed *Crepis sancta* (Asteraceae), following its recent colonization of the city of Montpellier, France. Using a combination of *in situ* and laboratory quantitative genetic experiments, including common garden greenhouse experiments, Cheptou et al. (2008) demonstrated that reduced dispersal observed in this urban weed was the result of rapid contemporary evolution in the proportion of dispersing versus non-dispersing seeds produced, and that this was a response to novel selection (see more details on this exemplary case study in Chapter 8).

In the study mentioned in section 5.2.1 on acorn-dwelling ant species, where selection on thermal tolerance acted in opposite directions between the warmest and coldest rearing conditions (Diamond et al. 2018), a common garden laboratory setup revealed both plastic and genetic processes underlying the higher heat tolerance found in city acorn ants. When raising several colonies of acorn ants sampled in urban and rural sites across three cities, offspring heat tolerance increased with laboratory rearing temperature in all populations, indicative of a plastic response. However, for two of the cities, urban populations displayed greater heat tolerance in warm rearing temperatures, indicative of a genetically based local adaptation.

Compared to common garden experiments in controlled laboratory conditions, reciprocal transplants in field conditions can discriminate between plastic and genetic origins of differentiation, while at the same time assessing the adaptive nature of the differentiation using individual lifetime fitness estimated in the wild. This approach was applied in a recent study on common ragweed (*Ambrosia artemisiifolia*), where seeds collected from multiple urban and rural sites in the Minneapolis–Saint Paul metropolitan area were planted and monitored in two urban and two rural areas (Gorton et al. 2018, and see Chapter 9 for more details on this study). Results from this experimental field study provide evidence for a genetic divergence between urban and rural populations in multiple phenological traits (including flowering time), since strong differences were maintained between plants of urban and rural origin. In terms of adaptive divergence, selection was stronger on the foreign genotype in each habitat, providing support for local adaptation. Additionally, the study reported high spatial heterogeneity for flowering time and fitness across the urban areas, suggesting that local urban adaptation should be investigated at a microgeographic scale in this species (approaches to capturing habitat heterogeneity are discussed in Chapter 2).

Note that while Figure 5.1 presents experimental approaches as the classical first step to differentiate between genetic and plastic origins for urban/rural phenotypic differentiation before implementing genomic approaches, this step can be bypassed. Indeed, field and laboratory projects such as common garden experiments are costly in time and effort, and their feasibility largely relies on the type of species under investigation. In situations where a common garden experiment is a near-impossible task, quantitative genetic approaches based on individual data collected through long-term monitoring programmes can test whether a trait is heritable and whether phenotypic differentiation is an adaptive response to selection (Charmantier et al. 2014). In such a quantitative genetic framework, multivariate models integrate the (genetic) covariation between measured traits involved in adaptation. Building the genetic variance–covariance matrix (G) for different traits allows taking into account multivariate constraints to evolution (Blows 2007). Since evolution in a trait depends on its genetic covariance with other traits as well as on selection acting on these traits, genetic correlations have been identified as one of the main explanations for evolutionary stasis in studies investigating responses to climate change (Merilä 2012). We believe that the diversity and complexity of traits involved in urban adaptation will make it necessary to go beyond a single trait approach in this context as well.

As illustrated in Figure 5.1, once the adaptive and genetic basis of a phenotypic difference have been demonstrated, DNA sequencing offers a range of methods to investigate which genes are involved (see section 5.3). Indeed, conclusive support for local adaptation in a heritable and selected trait requires proof that the trait differentiation is genetically determined. Fortuitously, technological genomic advances in recent years are now providing a collection of tools to identify the genetic variation underlying trait variation, and answer the fundamental questions regarding the nature of adaptation in such cases. These tools, described in section 5.3, can provide evidence for genetic signatures of novel selection in cities and unravel the genetic architecture of urban adaptation (Johnson and Munshi-South 2017; Schell 2018).

5.3 Pinpointing genes implicated in adaptation to urban environments

Population genomics and quantitative genomics offer conceptual and technical frameworks to identify between-individual variation in the genetic code and address how this variation affects phenotype, fitness, and ultimately the evolutionary trajectories of populations in different environments. Many other questions derive from this general topic, particularly those related to the genomic architecture of traits and of local adaptation, such as: Does the appearance of novel variation in traits and the local adaptation of populations to novel environmental conditions, here specifically discussed in the context of urban space, rely on standing genetic variation, *de novo* mutations, or both (Barrett and Schluter 2008)? Do such adaptive processes and trait variation involve a few loci with large effects (Mendelian traits), many loci with small effects (quantitative traits), or both? Among different spatial replicates, is evolution parallel (i.e., repeated) at the level of individual loci or biological functions or does it follow dissimilar paths?

There are two main reasons why urban environments offer an ideal framework to ask these questions and test the associated hypotheses (Donihue and Lambert 2015; Johnson and Munshi-South 2017). First, urban environments often apply new selection pressures compared to more natural habitats (see section 5.2), providing a perfect setting to investigate the genetic loci and mechanisms of rapid and genuine adaptation to environmental change. Second, since cities are distributed in a repeated fashion with often similar characteristics, they are ideal to test for the extent of genetic parallelism and its level (i.e., gene level versus functional level) during repeated adaptation to similar new selection pressures in geographically distant locations (see Chapter 3).

5.3.1 Pioneering use of low-resolution anonymous markers in urban evolution

Before genomic tools applicable to non-model species became widely available, the investigation of the genetic loci underlying trait variation and adaptation in urban environments relied largely on variable anonymous DNA markers. Anonymous markers, such as microsatellites, amplified fragment length polymorphisms, and sequence-specific amplified polymorphisms, have mainly been used to infer diversity, differentiation, effective population size, and gene flow among urban and natural populations (e.g., Chapter 4; Wandeler et al. 2003; Munshi-South et al. 2013). However, they offer relatively little power to identify genes implicated in adaptation to urban environments, since the chance that they are linked to adaptive variants is very small. Nevertheless, a few studies have indirectly used these types of markers to link population genetic parameters to fitness estimates, thereby providing information relative to the putative adaptation or adaptive potential of urban populations compared to rural ones. For instance, Hitchings and Beebee (1997) genotyped common frogs (*Rana temporaria*) from urban and rural ponds using allozyme electrophoresis, showing lower genetic diversity and higher inbreeding, associated with lower fitness, in urban populations than in rural ones. This decrease in fitness could originate from genetic drift in urban habitats, which are generally less connected and smaller than rural ones.

Since anonymous markers can help delineate patterns of gene flow, effective population size, and genetic diversity, they can also help test whether urban and rural populations are genetically isolated due to reduced dispersal and/or reduced gene flow

(see Chapter 4). Theoretically, such genetic isolation would facilitate the establishment of local adaptation (Lenormand 2002) in urban environments, if the effective population size were large enough to retain genetic diversity and allow selection to be effective in the face of genetic drift. Other studies have used similar types of anonymous markers to infer mutagenic effects of chemical pollutions in urban areas (Somers et al. 2002). Air, water, and soil pollution in urban contexts has been related to higher mutation rates in both humans and mice that may be partly heritable (Yauk et al. 2008; Somers and Cooper 2009). While new mutations are mostly deleterious or neutral (Eyre-Walker and Keightley 2007), it has been suggested that increased mutation rates in urban environments, and more generally in stressful conditions, could play a role in adaptive evolution (e.g., Taddei et al. 1997).

5.3.2 Candidate genes

The early documentation of the functional roles of several candidate genes in model species (Zhu and Zhao 2007) allowed ecological studies testing hypotheses relative to the roles of such genes on some phenotypic variations in both natural and urban environments. Among the most famous candidate genes studied in the context of urban adaptation are DRD4 (dopamine receptor D4 gene) and SERT (serotonin transporter), both of which are implicated in behavioural variations in humans (McGeary 2009) and other animals (including birds, detailed below).

These candidate genes have been extensively screened in urban genetic studies in order to investigate their potential effects on large behavioural variation documented in several organisms living in both cities and natural environments. For example, in an inspiring study, Mueller et al. (2013) genotyped either microsatellite or single nucleotide polymorphisms (SNPs) at seven candidate genes, including DRD4 and SERT, in 792 individuals from 12 paired populations of urban and rural blackbirds (*Turdus merula*) across the western Palaearctic. The authors found significant and repeated differentiation at the SERT gene between urban and rural populations and a marginally significant pattern for DRD4. They therefore proposed that behavioural traits related to aggression and associated with the SERT and DRD4 polymorphisms in blackbirds might have experienced parallel selection pressures in urban environments. Further studies in other avian species provided evidence for similar variation at SERT and DRD4 genes underlying behavioural variation between urban and rural environments. DRD4 variation was associated with wariness and urbanization in black swans (*Cygnus atratus*), with urban birds being less wary (Dongen et al. 2015). In great tits, urban and forest birds showed differentiation at SERT (Riyahi et al. 2015), at DRD4, and in behaviour (Riyahi et al. 2016). Many other candidate gene studies in urban environments involve resistance to chemical substances (reviewed by Belfiore and Anderson 2001). Collectively, these studies demonstrate that adaptation in an urban environment can operate via parallel responses to selection on important traits controlled by shared standing genetic variation at a few large-effect genes.

Although genotyping candidate genes identified beforehand can be an entry point to studying adaptation in urban environments, the knowledge gained will remain narrow, will be difficult to generalize, and will have little statistical significance. Indeed, such candidate gene approaches deliver very little to no knowledge about variation in the rest of the genome, offering virtually no chance to detect any other large-effect genes implicated in adaptation. In addition, without such wider information about genetic variation within and between populations, it is difficult to measure the significance of these results obtained for candidate genes. Finally, such a candidate gene approach is well suited for a simple trait with a Mendelian basis, but it is a poor choice for studying complex polygenic traits for which many genetic variants may influence phenotypic variation and adaptation (see Slate 2015 for a detailed discussion of candidate gene studies and the specific case of DRD4). Fortunately, genomic tools facilitate the screening of a genome to identify large-effect genes responsible for simple trait variations and oligogenic local adaptation. They also allow the identification of the numerous genes and biological functions implicated in complex trait variations and polygenic adaptation, while providing a statistically powerful context using many neutral genetic variants.

5.3.3 Urban evolution entering the genomic era: methods used so far

With the rapid development of sequencing technologies and the decrease in their cost, it has become possible to considerably increase the resolution with which one could screen genomes of non-model species to pinpoint evolutionarily relevant genomic variation (Stapley et al. 2010). A large range of methods has been developed, while trying to optimize tradeoffs between genome coverage, number of individuals, and sequencing cost (Schötterer et al. 2014). As the most expensive method, individual whole genome sequencing gives exhaustive information about the entire genome. Moreover, not only SNPs are genotyped but also structural variations (e.g., transposable elements, inversions, copy-number variations, insertions, and deletions). For instance, Reid et al. generated 20 million biallelic variant sites in the genome of the Atlantic killifish (*Fundulus heteroclitus*) using full genome sequencing and identifying genetic determinants of adaptation in urban polluted sites (see later in section 5.3.4 for a detailed description of this study).

Conveniently, it is possible to lower the cost of such genomic investigations, for example by using RAD sequencing and other genotyping by sequencing methods, which enable researchers to genotype fewer SNPs and do not require strong prior genomic resources or knowledge of the species' genetic resources (Narum et al. 2013; Andrew et al. 2016). These latter methods rely on the reduction of the complexity of the genome via enzymatic digestions, multiplexing of barcoded individuals' DNA, amplification of fragments targeted by digestion, sequencing of the entire library of fragments, and downstream bioinformatics (Davey et al. 2011). For example, Theodorou et al. (2018) used a RADseq protocol on a sample of 181 red-tailed bumblebees (*Bombus lapidarius*), calling 110 000 SNPs (discussed in section 5.3.5). While genomic methods—using the entire genomic sequence or taking a reduced genomic complexity approach—have been effective in looking for evolutionarily relevant genomic variants in non-model species sampled in natural and experimental setups, they are only beginning to be implemented in the context of urban evolution (Johnson and Munshi-South 2017). Overall, although genomic tools have been applied in several urban evolutionary contexts oriented towards 'non-adaptive' questions, particularly regarding gene flow and genetic diversity (e.g., Combs et al. 2018; Miles et al. 2018; Mueller et al. 2018; see Chapter 4 for further examples), only a few studies have used these tools in order to investigate footprints of urban adaptation. We will review these studies in the next sections.

5.3.4 Genome-wide sequencing pinpointing oligogenic adaptations in urban environments

In a pioneering genomic study of industrial melanism in the British peppered moth (*Biston betularia*), Van't Hof et al. (2011, 2016) identified the genetic origin of the melanic *carbonaria* morph selected for in industrial areas using whole genome sequencing. This colour polymorphism has long been recognized as a simple trait regulated by a single-locus-dominant allele. However, the location and the nature of this locus were unknown. After unrewarding candidate gene approaches using 16 genes previously implicated in melanization pattern differences in other insects (Van't Hof and Saccheri 2010), Van't Hof et al. (2011) constructed a linkage map of the polymorphism, leading to the discovery of a 200-kb region on chromosome 17. Several years later, using a full genome assembly and fine mapping of the associated region, Van't Hof et al. (2016) showed that the insertion of a large, tandemly repeated, transposable element into the first intron of the gene *cortex* on chromosome 17 was responsible for the *carbonaria* morph. Furthermore, by sequencing a larger set of individuals for this genomic region, the authors inferred that the transposition event occurred around 1819, consistent with the history of the melanic morph records in this species. This textbook example shows how a unique new mutation can control trait variation and adaptive response in a new environment driven by industrial activities.

Reid et al. (2016) studied the genes and genetic mechanisms of resistance to urban/industrial water pollution (polychlorinated biphenyl (PCB) and polycyclic aromatic hydrocarbon (PAH)) in the Atlantic killifish (*Fundulus heteroclitus*). Given the dramatic differences in pollution resistance between individuals from polluted and unpolluted waters, it was known that pollution resistance in this species was likely based on a small number of genes (Whitehead

et al. 2017). Sequencing the entire genomes of 384 individuals with a clever design that implemented both medium and low coverage to lighten the cost, Reid et al. (2016) discovered that several signalling pathways and toxicity-mediating genes were targeted by selection across polluted sites, illustrating parallel evolution at the biological pathway level. They also showed that the populations harboured different footprints of selection at the molecular level (variable deletions and gene copy number), indicating that adaptation took a variety of non-parallel molecular routes, although few genes and similar biological pathways were involved. The authors argued that standing genetic variation was likely the predominant substrate of such adaptation. This study therefore illustrates how a repeated selective pressure in the urban environment triggered divergent molecular evolution underlying a convergent adaptive response at the biological pathway level (see also Chapter 3 for a discussion on parallelism and convergence).

Later on, Oziolor et al. (2019) used similar genome-wide data to investigate adaptive toxicant resistance that recently evolved in the Gulf killifish (*F. grandis*). The authors found that the resistance in the Gulf killifish was due to a few loci that introgressed very recently from the Atlantic killifish. This second study therefore suggests that interspecific gene flow may be an important source of adaptive genetic variation and of adaptation in such a case of rapid and extreme change in selection regime.

These studies illustrate the benefits to be gained from harnessing genomic techniques to identify the genes and genetic mechanisms of adaptation in urban environments. In particular, they show the need for using high-resolution tools that are powerful enough to pinpoint the narrow, and sometimes abundant, footprints of selection associated with adaptation. Whole genome sequencing applied on a few individuals is extremely powerful, and possibly indispensable, for locating narrowly distributed adaptive variants (e.g., the 200-kb region for the peppered moth *carbonaria* large-effect locus).

5.3.5 Polygenic adaptation in urban environments

Although strong selective agents in urban environments sometimes promote rapid evolution of monogenic or oligogenic variation associated with simple trait variation (as seen in section 5.3.4 for melanism in the peppered moth or resistance to pollution in killifish), selection in urban environments may also trigger much more complex adaptive responses. First, the differences in selection pressures between urban and natural environments (e.g., Fugère and Hendry 2018) may target many different traits that are, in turn, influenced by different genes. Second, many traits affected by urban environments might be quantitative, and hence have a polygenic basis (i.e., influenced by many genetic variants of small to intermediate effect size). In this context, genome scans of individuals in urban and natural environments and gene–urbanization level associations can be used to detect as many of the putative loci implicated in adaptation as is possible. In contrast to studying oligogenic adaptation, studying polygenic adaptation requires genotyping hundreds or even thousands of individuals, so that the many variants with small to medium effects expected to contribute to the adaptive response can be identified. A few studies, exemplified in the next paragraph, have tried to identify the potential footprints of divergent selection between urban and rural sites by including relatively large numbers of individuals genotyped at a reduced number of SNPs using RADseq, RNAseq, or SNP arrays in order to decrease sequencing cost.

Harris et al. (2013) and later Harris and Munshi-South (2017) conducted genome scans on white-footed mice (*Peromyscus leucopus*) from NYC that were genotyped first at 31 000 and then at 155 000 SNPs typed using RNA sequencing. They identified hundreds of genetic variants involving metabolic processes that were associated with urbanization. Theodorou et al. (2018) performed genome scans and gene–urbanization associations based on 110 000 SNPs genotyped by RADseq in red-tailed bumble-bees (*Bombus lapidarius*). The authors identified 287 loci potentially under directional selection, some of which were associated with urban land use, and were found in loci most likely implicated in metabolic processes, heat stress, and oxidative stress more often than expected by chance. Perrier et al. (2018) used 50 000 SNPs identified with RAD sequencing to perform gene–urbanization associations and genome scans in great tits sampled across an urbanization gradient. The authors identified several genetic variants that were associated with urbanization. The strengths of these gene–urbanization

associations were, however, relatively low, suggesting a polygenic answer to selection. Perrier et al. (2018) hence used these loci additively to explain differences in response to urbanization between populations or individuals, a method that was conceptually inspired by polygenic risk scores.

These studies illustrate the potential benefits from using cost-effective genomic approaches on subsampled genomes to identify various genes and biological functions implicated in adaptation in urban environments. However, we must recognize that these studies probably did not identify all of the genes implicated in urban adaptation. Indeed, while methods based on a genome with reduced complexity (like RADseq) can be appropriate for species displaying high linkage disequilibrium, they might very well result in insufficient coverage of the genome for species displaying low linkage disequilibrium (Lowry et al. 2016, 2017; but see Catchen et al. 2017 and McKinney et al. 2017). In the latter case, methods based on whole genome sequencing (as used by Reid et al. (2016) and Van't Hof et al. (2016)) should probably be favoured. Nevertheless, identifying the loci implicated in polygenic adaptation may be challenging, even with full genome data, since this task requires both to localize the many loci implicated and many individuals to successfully estimate the small effect size of these loci. It is thus likely that future studies aiming at deciphering polygenic adaptation in urban environments will need to genotype both more loci and more individuals.

5.3.6 Further use of genomics in the field of urban evolution: methodological and taxonomic perspectives

Since sequencing whole genomes requires a large budget, researchers develop methods to decrease the cost, while nevertheless sequencing numerous populations or individuals. One approach is to sequence full genomes at low coverage. For instance, Reid et al. (2016) first sequenced one pair of natural and urbanized-polluted populations at medium coverage (average read depth: 7×) and three additional pairs at low coverage (0.7×). Another way to decrease the cost of sequencing, which has not yet been used in the urban context, is to sequence pools of 50–200 individuals ('pool-seq' or 'pooled sequencing'; Figure 5.2) to capture average allele frequencies (Schlötterer et al. 2014). This method

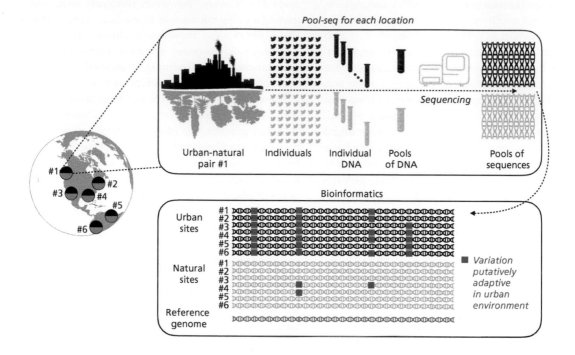

Figure 5.2 Illustration of potential implementation of pool-Seq for population genomics in urban environments. DNA extracts from several individuals are pooled and sequenced for each urban and natural environment pair replicated across space. Bioinformatic analyses align reads to a reference genome, genotyped SNPs, and other structural variants, and reveal putatively adaptive variations in urban environments.

produces cost-effective estimates of differentiation between populations based on large sample sizes (i.e., numerous individuals are generally needed to estimate allele frequency changes between large connected populations).

A strategy at the crossroads of pool-seq and individual genome sequencing is to sequence individual genomes at low coverage and to then pool sequences in a flexible way (e.g., by sex, location, or type of environment; Micheletti and Narum 2018). This integration of low-coverage whole genome sequencing and pool-seq seems very well suited for pinpointing adaptive loci in urban contexts, which typically requires genotyping many individuals but also comparing pairs of several urban and rural sites. However, pool-seq and low-coverage sequencing have limitations, notably the difficulty to estimate linkage disequilibrium and distinguish sequencing errors from low-frequency polymorphisms (Schlötterer et al. 2014).

Several other methodological considerations might very well be important for deciphering the genomic footprints of adaptation in urban environments. First, the mutations fuelling adaptation in the city appear to have diverse origins, since the few studies to date have revealed the roles of SNPs, transposable elements, copy number variations, and insertions/deletions. Therefore, statistical analyses of short reads that have mainly focused on SNP variation could dedicate more effort to detect the structural variation implicated in adaptation, taking advantage of both specialized software and online databases (Albers et al. 2011; Duan et al. 2013). Additionally, while genomic studies often use short-read sequencing, long reads may improve the detection of structural variants (Disdero et Filée 2017).

Other genomic aspects of adaptation in urban environments can be screened using recent genomic advances. Particularly, metagenomics has the power to pinpoint taxonomic diversity and adaptive variation in microbial communities in urban environments. Indeed, while studies of adaptation in urban environments have so far primarily focused on relatively large organisms, microbial species are probably the organisms responding most rapidly to selection in urban environments and could be studied relatively easily using genomics (e.g., Su et al. 2017; Gouliouris et al. 2019). Furthermore, microbiome (e.g., gut and skin microbiomes) diversity and adaptation may be involved in the acclimation and adaptation of their host organisms (Alberdi et al. 2016) in urban environments (e.g., Saxena et al. 2018; Teyssier et al. 2018). Therefore, metagenomics will undoubtedly play an important role in ecological studies aiming at deciphering diverse cases of adaptation in urban environments.

5.4 Epigenetics and the city

Although adaptation is achieved by response to selection via standing additive genetic variation or new genetic mutations, epigenetic variation may also be implicated (Verhoeven et al. 2016). Epigenetics can be defined in several ways, all aiming to understand the heritable molecular variations giving rise to different phenotypes from a similar genetic sequence (Felsenfeld 2014). In the context of studying urban adaptation, the most appropriate definition of epigenetics is probably the study of the sources of heritable (or partly heritable) phenotypic variations that do not arise from nucleotide variation and that are induced by environmental variation, sometimes referred to as transgenerational epigenetic inheritance. In particular, epigenetics examines features and mechanisms contributing to the regulation of gene expression and hereby affecting phenotypes. These features and mechanisms include histone modification, DNA methylation, and non-coding RNAs. The transgenerational stability of epigenetic features induced by transitory environmental variation, in particular DNA methylation, varies greatly from no heritability to stable transmissions over hundreds of generations (reviewed by Jablonka and Raz 2009). This transient epigenetic response may favour adaptation in new environments, because the plasticity resulting from epigenetic regulation may in the short term gain time for adaptation to happen, and because some

epigenetic processes like methylation may in the long term increase mutation rates that could fuel adaptation (Danchin et al. 2018). Taking advantage of the development of sequencing technologies, epigenetic features and mechanisms have been successfully investigated in various contexts, although mainly in humans and other model species (reviewed by Allis and Jenuwein 2016).

In the context of ecological studies, including urban environments, epigenetic studies have focused mainly on the level and distribution of methylation. Indeed, several protocols have been developed to investigate methylation patterns using bisulfite treatment, immunoprecipitation, or methylation-sensitive enzymes, followed by sequencing of a candidate gene, a genome that has been reduced in complexity (i.e., a given proportion of the genome), or the entire genome (reviewed by Jones 2012; Jeremias et al. 2018). We describe here two recent studies of methylation patterns in urban environments in birds.

In our first example, Riyahi et al. (2015) estimated methylation patterns of two famous candidate genes for personality traits, DRD4 and SERT, in great tits from urban and forest environments. They applied bisulphite treatments of the DNA (the treatment of DNA with bisulfite converts cytosine to uracil and leaves 5-methylcytosine unaffected) and strand-specific sequencing to perform the methylation profiling of the CpG dinucleotides in the DRD4 and SERT promoters and in the CpG island overlapping DRD4 exon 3. They also used pyrosequencing to quantify the methylation levels at each CpG. Riyahi et al. found 1–4 per cent higher methylation in urban birds than in forest birds. One CpG dinucleotide located 288 bp from the transcription start site of SERT was related to the exploration score in urban birds. The analyses also revealed that the methylation difference was repeatable in DNA extracted from blood and from brain. Indeed, methylation patterns often differ between different types of cells, notably between brain and blood cells in great tits (Laine et al. 2016). On the genetic side, one SNP in the minimal promoter of SERT was associated with novelty-seeking behaviour in captivity, and the allele favouring novelty seeking was found at higher frequency in urban birds. Riyahi et al. (2015) therefore concluded that both genetic and epigenetic variability in the SERT gene have an important role in shaping potentially adaptive variation in personality traits in great tits from urban and forest environments.

In our second example, McNew et al. (2017) investigated epigenetic variation in Darwin's finches (*Geospiza fortis* and *G. fuliginosa*) in rural and urban environments. Instead of using a candidate gene approach like in the first example, they used methylated DNA immunoprecipitation (MeDIP) in two different tissues to survey genome-wide methylation differences (or differentially methylated regions—DMRs) between urban and rural birds of both species. They identified thousands of DMRs between urban and rural populations, enriched for loci implicated in metabolic pathways, and MAPK and TGFß/BMP signalling. However, very few of these DMRs were shared between the two types of tissues or between the two species, suggesting that such analyses may require more power, gained via higher coverage of the genome and/or more individuals.

Several studies have also documented how urban parameters can alter methylation patterns in lab mice. Yauk et al. (2008) investigated DNA damage, mutation, and DNA methylation in the gametes of male mice experimentally exposed to particulate air pollution in an urban environment. In mice exposed to ambient air, the study showed a 1.6-fold increase in sperm mutation frequency at a particular locus, and their DNA methylation was higher than that for mice breathing filtered air. Hypermethylation remained significantly elevated in mature sperm after removal from the environmental exposure. The authors stressed that mutations and hypermethylation of the germline have the potential to affect the descendants of exposed individuals, particularly in terms of genome stability and regulation of gene expression.

Other epigenetic features can be studied using recently developed protocols (Harris et al. 2010; Schield et al. 2015; Allis and Jenuwein 2016; Verhoeven et al. 2016), and might be of particular importance in the context of ecological studies of rapid adaptation in urban environments. These protocols, like the genomic ones we have discussed, are flexible, and the costs can be alleviated by using pooling strategies or by restricting sequencing to reduced parts

of the genome, such as the GC-rich regions. Indeed, it remains very costly to get access to the entire methylome of many individuals, and the characterization of methylation patterns requires relatively well-annotated genomes. Furthermore, it will be necessary to investigate the functional link between DNA methylation (as measured in McNew et al. 2017, for example) and gene expression (as measured in Watson et al. 2017, for example), as well as to assess the heritability of epigenomic marks.

5.5 Conclusions and summary of the perspectives

Urban evolutionary ecology is a young and dynamic field, with more questions than answers (Johnson and Munshi-South 2017). Opportunistically, this field builds on urban ecology, efficiently using the theory and tools developed in the broader field of evolutionary biology. As highlighted in section 5.2, quantitative genetics uses long-term or experimental data to test whether urban phenotypes are adaptive and whether they are the result of plastic or genetic responses. We have cited some of the numerous success stories in diverse taxa (see Table 5.1), showing either plastic or genetic responses to city life that allow the species to thrive in the urban setting. Now that evolutionary ecologists are firmly considering urbanization as a unique opportunity to study evolution in action, many more case studies with correlative or experimental quantitative genetics are anticipated in the near future. However, limitations to the applications of quantitative genetics in the city will be similar to those for studies in natural environments, notably the difficulty to measure fitness, to obtain pedigrees, and to conduct *in situ* experiments. In particular, the increasingly short-term nature of funding and contracts threaten the sustainability of long-term monitoring projects even though they have proved highly productive in the past (Clutton-Brock and Sheldon 2010).

A valuable feature of research projects in urban environments is that they are often well suited for launching new Citizen Science projects. Involving city-dwellers in data collection can reduce the research costs, and also has many benefits in terms of science dissemination and reconnecting people with nature (Schuttler et al. 2018). While the general public's view of biodiversity is most often inspired by ecological studies and conservation efforts on iconic wildlife, urban evolutionary ecology will most likely involve many good model species that are not often thought about by non-scientists, such as *Daphnia magna* and the weeds *Crepis sancta* and *Trifolium repens*. Citizen science on these models will provide a wider perspective for citizen scientists on the ecology, evolution, and conservation of biodiversity.

As highlighted in section 5.3, genomics has started to allow the identification of genes implicated in evolution in urban populations. Evolutionary responses to selection most likely involve a highly variable number of loci, from the monogenic or oligogenic variation implicated in resistance to pollutants or in the diversification of colour morphs (e.g., Reid et al. 2016; Van't Hof et al. 2016) to the polygenic variation potentially underlying many traits under selection (e.g., Harris and Munshi-South 2017; Perrier et al. 2018; Theodorou et al. 2018). The simultaneous growth of genomics and urban evolutionary ecology will most likely result in many new exciting insights into the genetic mechanisms and genes implicated in urban evolution. Genomic studies implementing whole genome resequencing will likely identify the narrow and potentially numerous footprints of selection (Lowry et al. 2016, 2017; but see McKinney et al. 2017 and Catchen et al. 2017).

However, although sequencing costs have consistently decreased, whole genome resequencing is still expensive, and therefore alternative strategies using pool-seq (Schlötterer et al. 2014; Micheletti and Narum 2018) may be necessary to allow the application of high-resolution tools without prohibitive costs. These methods will also probably be more suitable for designs with replicated urban versus natural comparisons. Other genomics developments, such as measuring selection gradients on individual SNPs (e.g., an example from a non-urban context: Mullen and Hoekstra 2008) or considering the diverse nature of variants implicated in evolution more systematically (e.g., structural variants) and using metagenomics, will also strengthen the understanding of evolution in cities.

As highlighted in section 5.4, epigenetics might be involved in rapid evolution in the city. Epigenomics is offering a growing diversity of tools that can be used to detect the assortment of epigenetic marks and mechanisms that may be involved and to elucidate their potential influence in urban adaptation. These developments will coincide with investigations in the broader field of evolutionary biology aimed at understanding the role of epigenetic mechanisms in evolution in general, and in novel environments more specifically. Pioneering studies have already started to investigate potential epigenomic marks implicated in evolution in urban environments, and there is no doubt that this field will grow quickly in the next decades.

For a comprehensive view of urban evolution, these genetic, genomic, and epigenomic tools will need to be used in an integrative way, along with detailed studies of population demography and of the spatiotemporal structure and intensity of urbanization. Indeed, the age and size of urbanized patches, in combination with species demography, will probably determine the likelihood of adaptation (as discussed in Perrier et al. 2018). For example, older urbanized environments and species with short generation time are more likely to provide case studies on adaptive evolution than long-lived species in recently urbanized areas. Also, species with small foraging and dispersal distances in large urbanized areas will be more likely to display adaptations than species with long dispersal distances or in small urbanized areas. Finally, more intense urbanization should result in stronger selection on adaptive traits, and hence provide more opportunities to witness footprints of urban selection. Overall, the temporal and spatial heterogeneity of the urban environment both within and across cities offers unique possibilities to approach eco-evolutionary processes in a comparative manner (across species and cities), well beyond a single species urban versus rural comparison.

Acknowledgements

This project was funded by the European Research Council (Starting grant ERC-2013-StG-337365-SHE to A.C.). We thank Jon Slate, Emmanuel Milot, Marta Szulkin, and Jason Munshi-South for their constructive comments on previous versions of this manuscript, as well as Carolyn Hall for language corrections.

References

Alberdi, A., Aizpurua, O., Bohmann, K., et al. (2016). Do vertebrate gut metagenomes confer rapid ecological adaptation? *Trends in Ecology & Evolution*, 31, 689–99.

Albers, C.A., Lunter, G., MacArthur, D.G., et al. (2011). Dindel: accurate indel calls from short-read data. *Genome Research*, 21, 961–73.

Alberti, M. (2015). Eco-evolutionary dynamics in an urbanizing planet. *Trends in Ecology & Evolution*, 30(2), 114–26.

Alberti, M., Marzluff, J., and Hunt, V.M. (2017). Urban driven phenotypic changes: empirical observations and theoretical implications for eco-evolutionary feedback. *Philosophical Transactions of the Royal Society B: Biological Sciences*, 372: 20160029.

Allis, C.D. and Jenuwein, T. (2016). The molecular hallmarks of epigenetic control. *Nature Publishing Group*, 17, 487–500.

Andrews, K.R., Good, J.M., Miller, M.R., Luikart, G., and Hohenlohe, P.A. (2016). Harnessing the power of RADseq for ecological and evolutionary genomics. *Nature Publishing Group*, 17, 81–92.

Barrett, R.D.H. and Schluter, D. (2008). Adaptation from standing genetic variation. *Trends in Ecology & Evolution*, 23, 38–44.

Belfiore, N.M. and Anderson, S.L. (2001). Effects of contaminants on genetic patterns in aquatic organisms: a review. *Mutation Research*, 489, 97–122.

Blows, M.W. (2007). A tale of two matrices: multivariate approaches in evolutionary biology. *Journal of Evolutionary Biology*, 20(1), 1–8.

Caizergues, A.E., Gregoire, A., and Charmantier, A. (2018). Urban versus forest ecotypes are not explained by divergent reproductive selection. *Proceedings of the Royal Society B: Biological Sciences*, 285, 20180261.

Catchen, J.M., Hohenlohe, P.A., and Bernatchez, L., et al. (2017). Unbroken: RADseq remains a powerful tool for understanding the genetics of adaptation in natural populations. *Molecular Ecology Resources*, 26, 420–4.

Charmantier, A., Garant, D., and Kruuk, L.E. (2014). *Quantitative Genetics in the Wild*. Oxford University Press, Oxford.

Cheptou, P.O., Carrue, O., Rouifed, S., and Cantarel, A. (2008). Rapid evolution of seed dispersal in an urban environment in the weed *Crepis sancta*. *Proceedings of the National Academy of Sciences of the United States of America*, 105(10), 3796–9.

Clutton-Brock, T. and Sheldon, B.C. (2010). Individuals and populations: the role of long-term, individual-based studies of animals in ecology and evolutionary biology. *Trends in Ecology & Evolution*, 25(10), 562–73.

Combs, M., Byers, K.A., Ghersi, B.M., et al. (2018). Urban rat races: spatial population genomics of brown rats (*Rattus norvegicus*) compared across multiple cities. *Proceedings of the Royal Society B: Biological Sciences*, 285, 20180245.

Danchin, E., Pocheville, A., Rey, O., Pujol, B., and Blanchet, S. (2018). Epigenetically facilitated mutational assimilation: epigenetics as a hub within the inclusive evolutionary synthesis. *Biological Reviews*, 94(1), 259–82.

Davey, J.W., Hohenlohe, P.A., and Etter, P.D., et al. (2011). Genome-wide genetic marker discovery and genotyping using next-generation sequencing. *Nature Publishing Group*, 12, 499–510.

Diamond, S.E., Chick, L.D., Perez, A., Strickler, S.A., and Martin, R.A. (2018). Evolution of thermal tolerance and its fitness consequences: parallel and non-parallel responses to urban heat islands across three cities. *Proceedings of the Royal Society B: Biological Sciences*, 285(1882), 20180036.

Disdero, E. and Filée, J. (2017). LoRTE: detecting transposon-induced genomic variants using low coverage PacBio long read sequences. *Mobile DNA*, 8(5), 1–6.

Dongen, W.F.D., Robinson, R.W., Weston, M.A., Mulder, R.A., and Guay, P.-J. (2015). Variation at the DRD4 locus is associated with wariness and local site selection in urban black swans. *BMC Evolutionary Biology*, 15(253), 1–11.

Donihue, C.M. and Lambert, M.R. (2015). Adaptive evolution in urban ecosystems. *Ambio*, 44(3), 194–203.

Duan, J., Zhang, J.G., Deng, H.W., and Wang, Y.P. (2013). CNV-TV: a robust method to discover copy number variation from short sequencing reads. *BMC Bioinformatics*, 14(1), 150.

Eyre-Walker, A. and Keightley, P.D. (2007). The distribution of fitness effects of new mutations. *Nature Reviews Genetics*, 8, 610–18.

Felsenfeld, G. (2014). A brief history of epigenetics. *Cold Spring Harbor Perspectives in Biology*, 6, a018200.

Fugère, V. and Hendry, A.P. (2018). Human influences on the strength of phenotypic selection. *Proceedings of the National Academy of Sciences of the United States of America*, 115, 10070–75.

Gorton, A.J., Moeller, D.A., and Tiffin, P. (2018). Little plant, big city: a test of adaptation to urban environments in common ragweed (*Ambrosia artemisiifolia*). *Proceedings of the Royal Society B: Biological Sciences*, 285(1881), 20180968.

Gouliouris, T., Raven, K.E., Moradigaravand, D., et al. (2019). Detection of vancomycin-resistant *Enterococcus faecium* hospital-adapted lineages in municipal wastewater treatment plants indicates widespread distribution and release into the environment. *Genome Research*, 29, 626–34.

Harris, R.A., Wang, T., Coarfa, C., et al. (2010). Comparison of sequencing-based methods to profile DNA methylation and identification of monoallelic epigenetic modifications. *Nature Biotechnology*, 28, 1097–105.

Harris, S.E. and Munshi-South, J. (2017). Signatures of positive selection and local adaptation to urbanization in white-footed mice (*Peromyscus leucopus*). *Molecular Ecology*, 26, 6336–50.

Harris, S.E., Munshi-South, J., Obergfell, C., and O'Neill, R. (2013). Signatures of rapid evolution in urban and rural transcriptomes of white-footed mice (*Peromyscus leucopus*) in the New York metropolitan area. *PLOS ONE*, 8(8), e74938.

Hendry, A.P., Farrugia, T.J., and Kinnison, M.T. (2008). Human influences on rates of phenotypic change in wild animal populations. *Molecular Ecology*, 17(1), 20–29.

Hitchings, S.P. and Beebee, T.J. (1997). Genetic substructuring as a result of barriers to gene flow in urban *Rana temporaria* (common frog) populations: implications for biodiversity conservation. *Heredity*, 79(2), 117.

Jablonka, E. and Raz, G. (2009). Transgenerational epigenetic inheritance: prevalence, mechanisms, and implications for the study of heredity and evolution. *Quarterly Review of Biology*, 84, 131–76.

Jeremias, G., Barbosa, J., Marques, S.M., et al. (2018). Synthesizing the role of epigenetics in the response and adaptation of species to climate change in freshwater ecosystems. *Molecular Ecology*, 27, 2790–806.

Johnson, M.T.J. and Munshi-South, J. (2017). Evolution of life in urban environments. *Science*, 358, eaam8327.

Jones, P.A. (2012). Functions of DNA methylation: islands, start sites, gene bodies and beyond. *Nature Publishing Group*, 13, 484–92.

Laine, V.N., Gossmann, T.I., Schachtschneider, K.M., et al. (2016). Evolutionary signals of selection on cognition from the great tit genome and methylome. *Nature Communications*, 7, 10474.

Lande, R. and Arnold, S.J. (1983). The measurement of selection on correlated characters. *Evolution*, 37(6), 1210–26.

Lenormand, T. (2002). Gene flow and the limits to natural selection. *Trends in Ecology & Evolution*, 17, 183–9.

Lowry, D.B., Hoban, S., Kelley, J.L., et al. (2016). Breaking RAD: an evaluation of the utility of restriction site-associated DNA sequencing for genome scans of adaptation. *Molecular Ecology Resources*, 17, 142–52.

Lowry, D.B., Hoban, S., Kelley, J.L., et al. (2017). Responsible RAD: striving for best practices in population genomic studies of adaptation. *Molecular Ecology Resources*, 17(3), 366–9.

McGeary, J. (2009). The DRD4 exon 3 VNTR polymorphism and addiction-related phenotypes: a review. *Pharmacology, Biochemistry and Behavior*, 93, 222–9.

McKinney, G.J., Larson, W.A., Seeb, L.W., and Seeb, J.E. (2017). RADseq provides unprecedented insights into molecular ecology and evolutionary genetics: comment on Breaking RAD by Lowry et al. (2016). *Molecular Ecology Resources*, 17, 356–61.

McNew, S.M., Beck, D., Sadler-Riggleman, I., et al. (2017). Epigenetic variation between urban and rural populations of Darwin's finches. *BMC Evolutionary Biology*, 17(183), 1–14.

Merilä, J. (2012). Evolution in response to climate change: in pursuit of the missing evidence. *BioEssays*, 34(9), 811–18.

Micheletti, S.J. and Narum, S.R. (2018). Utility of pooled sequencing for association mapping in nonmodel organisms. *Molecular Ecology Resources*, 18, 825–37.

Miles, L.S., Dyer, R.J., and Verrelli, B.C. (2018). Urban hubs of connectivity: contrasting patterns of gene flow within and among cities in the western black widow spider. *Proceedings of the Royal Society B: Biological Sciences*, 285, 20181224.

Morrissey, M.B. and Sakrejda, K. (2013). Unification of regression-based methods for the analysis of natural selection. *Evolution*, 67(7), 2094–100.

Mueller, J.C., Partecke, J., Hatchwell, B.J., Gaston, K.J., and Evans, K.L. (2013). Candidate gene polymorphisms for behavioural adaptations during urbanization in blackbirds. *Molecular Ecology*, 22, 3629–37.

Mueller, J.C., Kuhl, H., Boerno, S., et al. (2018). Evolution of genomic variation in the burrowing owl in response to recent colonization of urban areas. *Proceedings of the Royal Society B: Biological Sciences*, 285, 20180206–9.

Mullen, L.M. and Hoekstra, H.E. (2008). Natural selection along an environmental gradient: a classic cline in mouse pigmentation. *Evolution: International Journal of Organic Evolution*, 62(7), 1555–70.

Munshi-South, J., Zak, Y., and Pehek, E. (2013). Conservation genetics of extremely isolated urban populations of the northern dusky salamander (*Desmognathus fuscus*) in New York City. *PeerJ*, 1, e64.

Narum, S.R., Buerkle, C.A., Davey, J.W., Miller, M.R., and Hohenlohe, P.A. (2013). Genotyping-by-sequencing in ecological and conservation genomics. *Molecular Ecology*, 22, 2841–7.

Oziolor, E.M., Reid, N.M., Yair, S., et al. (2019). Adaptive introgression enables evolutionary rescue from extreme environmental pollution. *Science*, 364, 455–7.

Pelletier, F. and Coltman, D.W. (2018). Will human influences on evolutionary dynamics in the wild pervade the Anthropocene? *BMC Biology*, 16(1), 7.

Perrier, C., Lozano del Campo, A., Szulkin, M., et al. (2018). Great tits and the city: distribution of genomic diversity and gene–environment associations along an urbanization gradient. *Evolutionary Applications*, 30, 114.

Postma, E. (2014). Four decades of estimating heritabilities in wild vertebrate populations: improved methods, more data, better estimates. In: Charmantier, A., Garant, D., and Kruuk, L.E.B. (eds) *Quantitative Genetics in the Wild*, pp. 16–33. Oxford University Press, Oxford.

Reid, N.M., Proestou, D.A., Clark, B.W., et al. (2016). The genomic landscape of rapid repeated evolutionary adaptation to toxic pollution in wild fish. *Science*, 354, 1305–8.

Riyahi, S., Sánchez-Delgado, M., Calafell, F., Monk, D., and Senar, J.C. (2015). Combined epigenetic and intraspecific variation of the DRD4 and SERT genes influence novelty seeking behavior in great tit *Parus major*. *Epigenetics*, 10, 516–25.

Riyahi, S., Björklund, M., Mateos-Gonzalez, F., and Senar, J.C. (2016). Personality and urbanization: behavioural traits and DRD4 SNP830 polymorphisms in great tits in Barcelona city. *Journal of Ethology*, 1–11. Doi: 10.1007/s10164-016-0496-2.

Saxena, G., Mitra, S., Marzinelli, E.M., et al. (2018). Metagenomics reveals the influence of land use and rain on the benthic microbial communities in a tropical urban waterway. *mSystems*, 3, 1–14.

Schell, C.J. (2018). urban evolutionary ecology and the potential benefits of implementing genomics. *Journal of Heredity*, 109, 138–51.

Schield, D.R., Walsh, M.R., Card, D.C., et al. (2015). EpiRADseq: scalable analysis of genomewide patterns of methylation using next-generation sequencing. *Methods in Ecology and Evolution*, 7, 60–69.

Schlötterer, C., Tobler, R., Kofler, R., and Nolte, V. (2014). Sequencing pools of individuals—mining genome-wide polymorphism data without big funding. *Nature Publishing Group*, 15, 749–63.

Schuttler, S.G., Sorensen, A.E., Jordan, R.C., Cooper, C., and Shwartz, A. (2018). Bridging the nature gap: can citizen science reverse the extinction of experience? *Frontiers in Ecology and the Environment*, 16(7), 405–11.

Slate, J. (2015). Why I'm wary of candidated gene studies. http://jon-slate.staff.shef.ac.uk/why-im-wary-of-candidate-gene-studies/.

Somers, C.M. and Cooper, D.N. (2009). Air pollution and mutations in the germline: are humans at risk? *Human Genetics*, 125, 119–30.

Somers, C.M., Yauk, C.L., White, P.A., Parfett, C.L., and Quinn, J.S. (2002). Air pollution induces heritable DNA mutations. *Proceedings of the National Academy of Sciences of the United States of America*, 99(25), 15904–7.

Stapley, J., Reger, J., Feulner, P.G., et al. (2010). Adaptation genomics: the next generation. *Trends in Ecology & Evolution*, 25(12), 705–12.

Su, J.-Q., An, X.-L., Li, B., et al. (2017). Metagenomics of urban sewage identifies an extensively shared antibiotic resistome in China. *Microbiome*, 5(1), 84.

Taddei, F., Radman, M., Maynard-Smith, J., et al. (1997). Role of mutator alleles in adaptive evolution. *Nature*, 387(6634), 700.

Teyssier, A., Rouffaer, L.O., Hudin, N.S., et al. (2018). Inside the guts of the city: urban-induced alterations of the gut microbiota in a wild passerine. *Science of the Total Environment*, 612, 1276–86.

Theodorou, P., Radzevičiūtė, R., Kahnt, B., et al. (2018). Genome-wide single nucleotide polymorphism scan suggests adaptation to urbanization in an important pollinator, the red-tailed bumblebee (*Bombus lapidaries* L.). *Proceedings of the Royal Society B: Biological Sciences*, 285, 20172806.

Thompson, K.A., Rieseberg, L.H., and Schluter, D. (2018). Speciation and the city. *Trends in Ecology & Evolution*, 33(11), 815–26.

van't Hof, A.E. and Saccheri, I.J. (2010). Industrial melanism in the peppered moth is not associated with genetic variation in canonical melanisation gene candidates. *PLOS ONE*, 5, e10889.

van't Hof, A.E., Edmonds, N., Dalíková, M., Marec, F., and Saccheri, I.J. (2011). Industrial melanism in British peppered moths has a singular and recent mutational origin. *Science*, 332(6032), 958–60.

van't Hof, A.E., Campagne, P., Rigden, D.J., et al. (2016). The industrial melanism mutation in British peppered moths is a transposable element. *Nature*, 534, 102–5.

Verhoeven, K.J.F., von Holdt, B.M., and Sork, V.L. (2016). Epigenetics in ecology and evolution: what we know and what we need to know. *Molecular Ecology*, 25, 1631–8.

Wandeler, P., Funk, S.M., Largiader, C.R., Gloor, S., and Breitenmoser, U. (2003). The city-fox phenomenon: genetic consequences of a recent colonization of urban habitat. *Molecular Ecology*, 12, 647–56.

Watson, H., Videvall, E., Andersson, M.N., and Isaksson, C. (2017). Transcriptome analysis of a wild bird reveals physiological responses to the urban environment. *Scientific Reports*, 7, 44180.

Whitehead, A., Clark, B.W., Reid, N.M., Hahn, M.E., and Nacci, D. (2017). When evolution is the solution to pollution: key principles, and lessons from rapid repeated adaptation of killifish (*Fundulus heteroclitus*) populations. *Evolutionary Applications*, 10(8), 762–83.

Yauk, C., Polyzos, A., Rowan-Carroll, A., et al. (2008). Germline mutations, DNA damage, and global hypermethylation in mice exposed to particulate air pollution in an urban/industrial location. *Proceedings of the National Academy of Sciences of the United States of America*, 105, 605–10.

Zhu, M. and Zhao, S. (2007). Candidate gene identification approach: progress and challenges. *International Journal of Biological Sciences*, 3, 420–7.

CHAPTER 6

Evolutionary Consequences of the Urban Heat Island

Sarah E. Diamond and Ryan A. Martin

Diamond, S.E. and Martin, R.A., *Evolutionary Consequences of the Urban Heat Island* In: *Urban Evolutionary Biology*. Edited by Marta Szulkin, Jason Munshi-South and Anne Charmantier, Oxford University Press (2020). © Oxford University Press.
DOI: 10.1093/oso/9780198836841.003.0006

6.1 Introduction

The expansion of human settlements leads to the proliferation of surfaces impermeable to water, including roads, pavements, and buildings. Because these surfaces have high heat retention capacity and distinct albedo properties, they elevate local environmental temperature, in what is commonly referred to as the urban heat island (UHI) effect (Figure 6.1; Grimm et al. 2008). With some exceptions, much of the elevated temperature in cities is driven by warmer night-time temperatures relative to non-urban areas; daytime temperatures are generally not much warmer in urban versus non-urban areas (Heisler and Brazel 2010). Because temperature drives biochemical rates that manifest from molecules to ecosystems (Angilletta 2009), there is enormous potential for UHI effects to radically alter biological systems.

Although warming is potentially one of the most consistent features of urbanization, the magnitude of warming varies considerably within and among cities. The level and nature of urban development each contribute to the amount of urban warming in a particular location. The magnitude of warming compared with nearby non-urban areas is generally greater in large cities than in small cities and in dense city centres rather than more peripheral areas, although the precise nature of urban development can alter these broad patterns (Imhoff et al. 2010). For example, large UHI effects can be found relatively far away from city centres with recent shifts in construction practices towards large houses on small lots that contain few trees (Grimmond 2007). Furthermore, although most habitats in cities contribute to UHIs (Heisler and Brazel 2010; Brans et al. 2018a), the magnitude of warming differs greatly

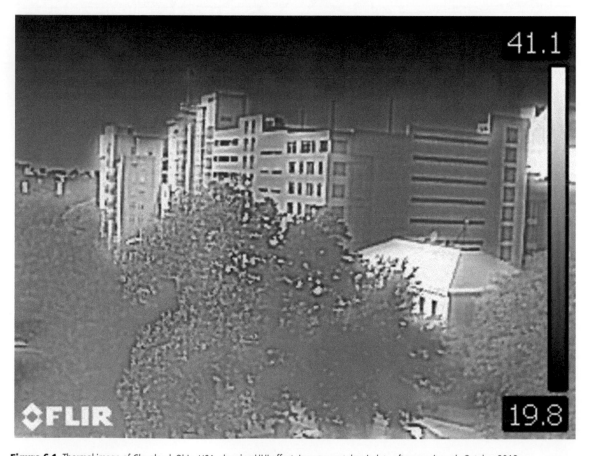

Figure 6.1 Thermal image of Cleveland, Ohio, USA, showing UHI effect. Image was taken in late afternoon in early October 2018.

among microsites. Ground temperatures are generally much higher compared with air temperatures, though localized buffering of these effects may occur through shading by vegetation or even other man-made structures (Grimmond 2007). We do not treat this complexity of the UHI effect here, but rather explore the effects of mean temperature rise on urban organisms. There are two reasons for this: (1) given that the field of urban evolution is nascent in many respects, there are few empirical examples of how populations might evolve in response to the complexity of UHI effects; and (2) there are excellent studies and reviews of heterogeneity in UHIs from an ecological perspective (Grimmond 2007; Heisler and Brazel 2010; Imhoff et al. 2010), which we prefer not to duplicate.

The magnitude of urban warming (averaging 1 to 5+ °C) approximates the global warming projected by the end of the century (Youngsteadt et al. 2015). In this manner, the localized warming that presently occurs within a city's footprint can serve as an approximation of future warming at the global scale. Although such an approximation increases the scope of the inferences that can be drawn from studying UHI effects, it is important to also recognize that cities are expanding rapidly in many areas, with over half the world's population now living in urbanized environments (Seto et al. 2012). As a consequence, understanding the capacity of populations

to evolve in response to rapid temperature rise in cities is an important research area in and of itself.

We have established that UHIs affect organisms greatly and create putatively strong selective pressures on populations, but a key question is how we should quantify the magnitude of these effects, from the perspectives of both the environment and the organism. Ideally, we should quantify urban temperature in a way that captures how organisms actually experience the urban environment. In other words, we should account for how organisms use habitats in urban environments, and how they experience these environments over diurnal and seasonal scales. However, quantifying relevant thermal environments remains a challenge (Gilman et al. 2006). Often, we must rely on proxies of the thermal environment that organisms experience. Ground-based sensors positioned in or near the habitats of organisms are generally good proxies. In the absence of such data, remote sensing can be used to estimate environmental temperature. The percentage of developed impervious surface within a given area (ISA) is a common metric, and is available at a global scale (at a resolution of 1 km (Figure 6.2; Chapter 2; Elvidge et al. 2007). More localized data, for example, within the coterminous USA, document ISA at a finer scale (a resolution of 30 m; Homer et al. 2015). In many places, ISA is highly correlated with environmental temperature

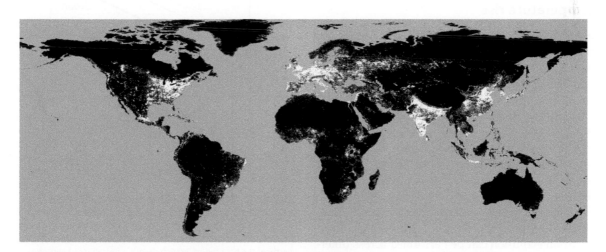

Figure 6.2 Global extent and magnitude of urbanization, expressed as a percentage of developed ISA using 1-km grid data from Elvidge et al. 2007. On landmasses, greyscale shading indicates amount of urbanization, from 0 percent impervious surface area in black to100 percent impervious surface area in white.

(Figure 2.1; Imhoff et al. 2010), and can therefore be used as a proxy of urban warming. A notable exception concerns desert cities, which can exhibit urban cooling due to the effects of irrigation (Imhoff et al. 2010). Land-cover types can also provide relative rankings of urban temperature rise, with the expectation that industrial centres will be warmer than the periphery of city centres, though again structure density and vegetation cover can disrupt these patterns (Yan et al. 2015).

In many cases, researchers refer to 'urban versus rural' environments or 'urban versus non-urban' environments, though these are less rigidly defined groups and are more understood to be shorthand for habitat differences of interest. The important part here is simply for researchers to be diligent about reporting their criteria for assigning particular habitats to urban or rural (or non-urban) classifications. In our case, for UHI effects, this means reporting the relative differences in environmental temperature (or relevant proxy) across an urbanization gradient. Despite variation in how we quantify UHI effects and the expression of urban warming across different cities and different locations within cities, urbanization has yielded replicated warming experiments that provide us with an unprecedented opportunity to explore contemporary adaptation to temperature.

6.2 Evolution in response to urban temperature rise

Studies that show strong evidence of evolutionary change in response to UHIs are surprisingly few in number. A recent review of urban evolution (Johnson and Munshi-South 2017) documented only 4 out of 192 studies that were explicitly focused on the UHI effect. In part, this scarcity might reflect the fact that the field of urban evolution is relatively new and more focused on broad-scale effects of urbanization rather than specific drivers of evolution. Indeed, it is almost certainly easier to assess whether or not urban evolution has occurred than it is to determine which specific aspects of urbanization are responsible for urban evolution. Although the assessment of urban evolution can be performed across different levels of biological organization, here we focus on evidence of evolved (genetic) changes at the level of the individual and population. For assess-

ment of urban evolution at the molecular level, we refer readers to Chapters 4 and 5.

The core issue for determining whether evolution has occurred in response to urbanization in general, and UHIs in particular, is disentangling the contribution of genetic change from phenotypic plasticity (Figure 6.3; Donihue and Lambert 2015; Diamond and Martin 2016). Plasticity enables organisms to enhance their fitness in an urban environment without genetic change. As a consequence, shifts in phenotypes across urban and non-urban areas could reflect evolutionary change or phenotypic plasticity. To illustrate, consider a recent synthesis of body size changes in response to urbanization (Merckx et al. 2018). We should expect urban organisms to be smaller because of the plastic effects of elevated urban temperature on body size. This result of smaller adult body size under warmer conditions is a well-established pattern of phenotypic plasticity. For ectothermic species, over 80 per cent of species follow this temperature–size rule, wherein a genotype exhibits smaller body size when reared under warmer conditions (Atkinson 1994). In cities, we might therefore develop a reasonable expectation for plastic shifts to smaller body size under UHIs, whereas our predictions for evolved shifts are likely to be more

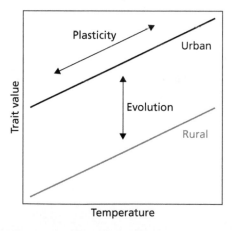

Figure 6.3 Plastic and evolved trait responses to UHIs. Plasticity describes the within-individual (and thus within-generation) effect of increasing temperature on trait values; here, urban and rural populations have similar plastic responses to temperature. Evolution describes the between-generation genetic differentiation between urban and rural population trait values; here, urban populations always have higher trait values compared with rural populations regardless of environmental temperature conditions.

specific to the taxa under consideration. Indeed, in a recent synthesis of body size responses to urbanization, Merckx and colleagues (2018) find that while many species become smaller in cities, putatively as a consequence of urban warming, several become larger (though, in part, this unexpected result might reflect an interaction with dispersal ability; see section 6.3). But the question remains: Which of these phenotypic shifts in body size are the result of evolution and which represent plasticity?

To distinguish among the contributions of these mechanisms, evolution, and plasticity, we need to separate environmental effects from genetic effects. Fortunately, standard procedures exist for accomplishing this task, and which share a long history of application in the field of evolutionary biology (De Witt and Scheiner 2004; for more recent applications in the context of environmental change see Chapter 5; Merilä and Hendry 2014; Donihue and Lambert 2015). First, using common garden experiments, we can mitigate environmental effects by raising organisms from adjacent urban and rural populations in the same environment for at least a generation. If the differences in traits persist, the differences are likely genetically based and indicate evolutionary divergence between populations. Similarly, we can perform reciprocal transplant experiments in the field, swapping urban organisms and rural organisms between environments and evaluating phenotypes in each environment. Importantly, reciprocal transplants allow us to assess evidence for local adaptation, wherein fitness of each population is highest in its 'home' environment. Although these methods allow us to disentangle plastic from evolved responses, they require greater investment than measuring phenotypes along an urban-to-rural gradient. This extra step might explain why so few studies have disentangled plastic from evolved responses to anthropogenic environmental change.

In the remainder of this chapter, we review the current findings from studies that explore evolutionary change in response to UHIs. We divide our chapter into subsections that focus on urban thermal adaptation of morphology, physiology, and life history. To cover this broad array of traits, we develop predictions for the effects of UHIs on the magnitude and direction of evolutionary change based largely on empirical patterns from the rich literature on thermal ecology and evolution (see Angilletta 2009 for relevant predictions from theory). This structure also enables us to examine whether these predictions, often derived from long-term evolutionary responses to temperature, agree with patterns of contemporary evolution in response to urban warming.

To begin, we might consider which traits are likely to respond quickly to natural selection caused by urban warming. Heritability, or the resemblance between parents and offspring, provides an estimate of additive genetic variation within a population and can thus, at least over short evolutionary timescales, provide information about the evolutionary potential for traits to respond to natural selection (note that using heritability to predict long-term evolutionary responses has been criticized (Pemberton 2010)). Greater heritability should lead to a greater evolutionary response to urban warming. In general, heritabilities of morphological and physiological traits tend to be quite high, whereas heritabilities of many life-history traits tend to be comparatively low (Mousseau and Roff 1987; Geber and Griffen 2003). This pattern would suggest a greater likelihood of finding evolutionary differentiation in response to urban warming for morphological or physiological traits compared with other traits. However, this prediction must be tempered by the fact that the measure of heritability can be environmentally dependent, and so the heritability of a trait at high temperatures likely differs from heritability measured in most studies (e.g., Chown et al. 2009). In contrast, comparative analyses of trait values among species in the context of their shared evolutionary history (phylogenetic signal) suggest the opposite prediction might be made: morphological and physiological traits show high phylogenetic signal (Blomberg et al. 2003), which can (but does not necessarily) suggest shared underlying genetic or developmental constraints that might limit their response to selection. Of course, both of these metrics are imperfect, and given their contrasting predictions for how urban thermal traits might evolve, further reinforces the need to develop a diverse database of evolutionary responses to urban warming. We now synthesize the empirical data available at present.

6.3 Morphology

Morphology comprises a diverse category of traits, from body size and shape, to colouration, limb length, and flower number. Although only a handful of studies have evaluated evolutionary divergence in morphology in response to UHIs, we can pull from other research areas to make predictions about how morphological traits may evolve in response to UHIs. The temperature variation associated with latitudinal clines is a natural analogue for the UHI effect, and there is a wealth of both theory and empirical data exploring how morphological traits may evolve along these natural biogeographic clines.

Perhaps the most studied morphological cline is that associated with Bergmann's rule, a trend of increasing body size within taxa (populations or related species) with increasing latitude from the equator to the poles (Bergmann 1848). While this biogeographic pattern is not universal—it is widespread in mammals and birds, but species within other taxa sometimes follow the reverse cline or exhibit no cline at all (Kingsolver and Huey 2008)—interest in this pattern has led to the formulation of several hypotheses for how temperature may influence body size through both evolution and plasticity. The original explanation for Bergmann's rule, formulated for endotherms, posits that in colder climates larger individuals should benefit from reduced heat loss (and therefore reduced energetic costs) due to a lower surface area to volume ratio (Bergmann 1848). The heat loss mechanism likely only applies to some ectotherms (e.g., those that actively thermoregulate), and there may be no single explanation for latitudinal clines in the body size of ectotherms (Vinarski 2014).

Temperature can also directly affect size in ectothermic species via development, with warmer temperatures often leading to smaller body sizes and more rapid development (the temperature–size rule) (Sibly and Atkinson 1994). In recent syntheses, Horne et al. (2015, 2017) found that across arthropods, taxa closely following the temperature–size rule also tended to follow Bergmann clines. This pattern was especially strong and consistent for aquatic organisms, where costs to large body sizes are imposed by oxygen limitations in warmer water.

Latitudinal clines for terrestrial arthropods were more variable, but also predictable. Univoltine arthropods (producing one brood of offspring per year) tend to follow the reverse cline, with larger body sizes at warmer temperatures and lower latitudes, while multivoltine species (producing multiple broods of offspring per year) better match the patterns of aquatic taxa. Life-history tradeoffs provide a potential evolutionary explanation for contrasting patterns like these across ectothermic species, and suggest that how temperature affects growth and body size can evolve (Angilletta et al. 2004).

Notwithstanding the diverse plastic and evolutionary responses of body size to temperature, we can assess how responses of body size to UHIs align with the patterns predicted from latitudinal clines in body size, and see if these responses have an evolutionary component. For example, we might predict endotherms and thermoregulating, aquatic, and multivoltine ectotherms to generally evolve smaller body sizes in cities if the UHI is an important selective pressure. In support of these predictions, urban birds tend to be smaller than their rural counterparts (Liker et al. 2008; Meillère et al. 2015; Caizergues et al. 2018), although whether temperature is responsible for this pattern and if it represents evolutionary divergence is unknown. Urban phenotypic clines in body size have been fairly widely examined in invertebrates, at either the species or community level. Results of these studies are mixed, with some urban species and populations smaller than their rural counterparts and others larger (Merckx et al. 2018). However, whether these patterns are the result of plasticity or evolution is generally unknown.

The few studies evaluating evolutionary divergence in body size between urban and rural populations have done so in a diverse group of organisms, including water fleas, grasshoppers, *Anolis* lizards, and plants in the Asteraceae and Brassicaceae families. Of these, common garden experiments in water fleas from urban and rural ponds provide the clearest link between the UHI as a selective agent acting on body size (Brans et al. 2017). Brans et al. (2018a) found that urban ponds were more than 3 °C hotter than rural ponds. In response, urban water flea populations have evolved smaller body sizes than nearby rural populations, matching the biogeographic patterns found in aquatic arthropods

(Brans et al. 2017). Moreover, because smaller water fleas also have higher thermal tolerances, these evolved changes in body size at least partially explain the evolution of increased heat tolerance in these urban populations (Brans et al. 2017). While aquatic invertebrates are predicted to evolve smaller body sizes in urban environments, univoltine insects are predicted to evolve larger sizes (Horne et al. 2015). Rearing urban and rural grasshoppers (*Chorthippus brunneus*) from six paired populations in a common garden experiment, San Martin y Gomez and Van Dyck (2012) found that, as predicted, urban females grew faster and were larger (i.e., higher mass) than females from rural populations.

Evidence for evolved differences in body size between urban and rural populations has also been assessed in both native and invasive populations of the Puerto Rican crested anole (*Anolis cristatellus*). Intriguingly, urban *A. cristatellus* grew to a larger size under common garden rearing conditions in an invasive Miami population (Hall and Warner 2017), but did not differ in body size between urban and rural populations in their native range (Winchell et al. 2016). Lizards often follow reverse Bergmann's clines, reaching larger body sizes in warmer climates (Ashton and Feldman 2003), presumably because warmer daily and seasonal temperatures allow for longer periods of activity and growth for these thermoregulating species (Sears and Angilletta 2004). Exceptions to this pattern may occur when other selective pressures, such as predation, covary with temperature, altering age-specific mortality schedules, and resulting in selection for faster times to maturity and smaller adult body size at warmer temperatures (Sears and Angilletta 2004). Predation risk was greater for native urban *Anolis* populations (Tyler et al. 2016), and so differences in the relative predation pressures among urban populations could potentially explain the variable responses of body size evolution in *A. cristatellus*.

Finally, several studies have used either greenhouse common garden designs or reciprocal transplants to investigate evolved differences in plant size (among other traits) between urban and rural populations. Two of three species evolved larger sizes in urban environments (Yakub and Tiffin 2017, height; Lambrecht et al. 2016, biomass), while the third generally displayed little difference in plant

size (Gorton et al. 2018). Because temperature was not directly manipulated in these studies, the selective agent for these responses is not clear. In plants, size seems to increase with decreasing latitude within species (e.g., Li et al. 1998), suggesting that the evolution of larger sizes in urban environments may be an adaptation to longer growing seasons in cities. However, other potential explanations have been proposed, such as larger leaf areas in cities as an adaptation to compensate for increased respiration due to night-time-biased warming (Searle et al. 2012).

Beyond overall size, temperature has the potential to influence the evolution of many other morphological traits. For example, appendage size may increase in warmed environments either as an adaptation to dissipate heat (i.e., Allen's rule) or to minimize thermal radiation from the ground, as in a desert-adapted ant (*Cataglyphis bombycina*). Miller et al. (2018) tested if the pattern of Allen's rule held for northern cardinal (*Cardinalis cardinalis cardinalis*) bill size, measured from museum specimens, in relation to housing density (a measure of urbanization) and changes in average minimum temperature over time in three North American cities. While bill size increased with increased minimum temperatures in two cities, housing density was associated with larger bills in only one city, and only for females. Finally, the Puerto Rican crested anole (*A. cristatellus*) has evolved longer fore limbs and hind limbs in urban habitats, although these traits have been linked to adaptive locomotor performance, and any potential role in thermal performance is yet unknown (Chapter 12; Winchell et al. 2016).

As a final trait of interest, the evolution of melanism, while historically a very important topic in urban evolution, has not been examined in the context of UHIs. The evolution of increased melanism has been linked with the effects of pollution in urban areas, both as a mechanism of camouflage, such as in the peppered moth (*Biston betularia*) (Cook 2003), and to aspects of immunity and antioxidant capacity (Jacquin et al. 2011; Chatelain et al. 2016). However, melanin also has important thermal properties. All else being equal, darker individuals will heat up faster than lighter individuals (Clusella-Trullas et al. 2007). As a consequence, we might predict that the UHI would select for less melanistic populations. However, this benefit could depend upon background climate (Diamond et al. 2015), and perhaps disappear in cities in colder climates. Moreover, whether this selective pressure would be greater than that caused by the effects of pollution or parasitism favouring more melanistic individuals is unclear. Nevertheless, evaluating if melanism plays a role in the evolutionary response to UHI is certainly an interesting topic for future studies.

How the UHI will influence morphological evolution is only beginning to be explored. Among the studies discussed above, only the research with water fleas (Brans et al. 2017) functionally links morphological evolution to the UHI. It is clear, however, that temperature can mediate selection on body size and other morphological traits in myriad direct and indirect ways. But it is also clear that our understanding of temperature's role in morphological size evolution is incomplete. The idea of cities as replicated experiments is a recurring one in this chapter and throughout the book; through careful design, future studies should be able to use cities to tease apart the selective mechanisms by which temperature, versus other correlated environmental selective pressures, drives broad patterns of morphological evolution.

6.4 Physiology

Much in the same way that the evolution of morphology under different climatic regimes has a rich body of work supporting it, so too does the evolution of thermal physiology. In this section, we draw on this literature when we consider how both the lethal limits of thermal performance and performance at intermediate temperatures evolve in response to contemporary urban warming. Relative to the other trait classes under examination, the evolution of thermal physiology in cities is currently the most well-studied trait class with respect to number of cities and taxa examined (see section 6.7).

The thermal limits to performance exhibit strong relationships with climate across ectothermic taxa. Thermal limits specify the lowest and highest temperatures an organism can tolerate either before death or before loss of muscular coordination where an organism could not remove itself to a thermal refuge (Figure 6.4). Along biogeographic gradients in environmental temperature including increasing latitude

and elevation, the ability to tolerate colder temperatures increases and the ability to tolerate warmer temperatures decreases (though more modestly than the magnitude of change in cold tolerance, typically leading to smaller tolerance breadths in warmer climates) (Sunday et al. 2012). These patterns apply among species and among populations within a given species. The interspecific biogeographic patterns suggest a role for evolution in shaping thermal tolerances. A smaller subset of studies show, using laboratory common garden experiments, that climate also generates evolutionary divergence in thermal tolerance within species across climatic gradients; however, it is important to bear in mind that warmer environments also induce plastic responses such that heat tolerance increases while cold tolerance decreases (Sørensen et al. 2016). Interestingly, early work on leaf cutter ant (*Atta sexdens rubropilosa*) responses to UHIs revealed increases in heat tolerance, but no change in cold tolerance in urban populations compared with rural populations (Angilletta et al. 2007). However, these results reflect field-caught ants, not ants reared under common garden conditions, so the relative contributions of evolution versus plasticity are unclear.

There are only a few studies that disentangle plastic from evolved variation in phenotypic shifts of thermal tolerance across urban warming gradients, but these studies generally find results consistent with biogeographic patterns. Water fleas collected from multiple urban and rural populations in Flanders and reared under common garden conditions in the laboratory show evolved increases in heat tolerance (Brans et al. 2017). Similarly, acorn

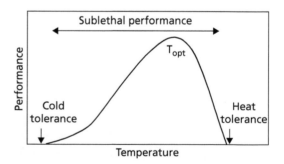

Figure 6.4 Thermal performance curve describing lethal limits of performance (cold tolerance and heat tolerance for lower and upper endpoints, respectively) and sublethal performance. Optimum temperature for performance (T_{opt}) is also indicated.

ants (*Temnothorax curvispinosus*) collected from multiple urban and rural populations across the eastern USA and reared under common garden conditions in the laboratory show evolved increases in heat tolerance, losses in cold tolerance, and decreases in thermal tolerance breadth (Diamond et al. 2017, 2018a). Notably, however, in the acorn ant system, one of three urbanization gradients exhibited contrasting patterns. Along this urbanization gradient, there was no evolutionary differentiation in heat tolerance among urban and rural populations and, surprisingly, evolved increases in cold tolerance of urban populations. Despite this outlier, in two out of three cities, acorn ants exhibited the expected pattern of evolved response. Although it is certainly a noteworthy result that UHIs have largely predictable effects on the evolution of thermal tolerance traits, the truly impressive result is that these changes—on the order of what is seen for long-term evolution in response to major differences in climate across biogeographic regions—are seen over very short timescales in response to UHIs (Brans et al. 2017; Diamond et al. 2018a). For example, in acorn ants, evolutionary divergence in thermal tolerance has occurred over a mere century in absolute time, and over as few as 20 generations of acorn ant colonies (Diamond et al. 2018a).

In the absence of directly assessed thermal tolerances, proxies of thermal tolerance can provide insight into urban thermal evolution. Cyanogenesis in white clover (*Trifolium repens*) is one such example: the production of hydrogen cyanide in clover confers protection against herbivory, but also makes the plants more vulnerable to cold stress. Reduced snow cover in cities exposes urban clover to cooler air temperatures, and thus selects for plants with less hydrogen cyanide. Indeed, this pattern is evident in clover among dozens of cities (Chapter 3; Thompson et al. 2016; Johnson et al. 2018). Exceptions to this pattern further support a relationship between environmental temperature and cyanogenesis in clover. For example, in cities positioned in cool climates with high snowfall, this cline in cyanogenesis is disrupted owing to the insulating effects of snow in these particular urban environments. Indeed, these 'exceptions' suggest the importance of measuring local thermal environments as organisms experience them in cities.

Acorn ant evolutionary responses to the *rate* of temperature change in cities further demonstrate why microsite variation is important. Acorn ants can experience temperature change in two settings, including within the acorn nest environment and as ants move across the landscape to forage. In cities, acorn ants experience faster rates of temperature change over time from early morning to midday, and across space, as urban environments are more thermally heterogeneous due to patchier canopy cover. Remarkably, urban acorn ants are better able to translate these faster rates of temperature change into greater heat tolerance compared with rural acorn ants, and these differences persist under common garden conditions in the lab, indicating evolutionary differentiation among urban and rural populations in response to the rate of temperature change in cities (Diamond et al. 2018b). This study emphasizes that there are more than simple mean UHI effects on the evolution of thermal traits. From a biological perspective, it also demonstrates a case of how plasticity itself (i.e., how heat tolerance responds to faster versus slower rates of temperature change) can evolve under UHIs (Figure 6.5).

Of course, the evolution of thermal physiological plasticity is not restricted to variation in the rate of temperature change over fine-grain spatial and diurnal timescales. Temperature variation over seasonal and annual cycles could also lead to evolved

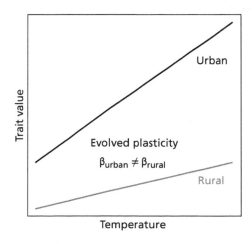

Figure 6.5 Evolution of plasticity in response to UHIs. In this example, relative increase in trait values with increasing temperature is greater for the urban population compared with the rural population.

differences in thermal physiological plasticity among populations. The broad expectation for biogeographic studies is that organisms from more thermally variable environments (at high latitude and elevation) will exhibit greater thermal plasticity. That is, such organisms will show a greater magnitude of change in their thermal tolerance across a range of environmental temperatures. However, comparative biogeographic studies have found weak, contrary, or no support (dependent upon the specific thermal physiological trait) for greater thermal physiological plasticity in more thermally variable climates (Gunderson and Stillman 2015; Seebacher et al. 2015). Interestingly, early findings from UHIs support this pattern. Cities are often less thermally variable over time compared with nearby non-urban areas (though urban aquatic habitats may be an exception (Brans et al. 2018a)). In particular, night-time-biased warming (higher night-time temperatures in urban areas relative to non-urban areas, with little change in daytime temperatures) compresses the thermal variance experienced during the diurnal cycle, and these changes to thermal variance persist throughout the year. We might then expect reduced thermal physiological plasticity in urban populations compared with non-urban populations. Yet despite this reduction in urban thermal variance, neither acorn ant thermal tolerances nor water flea stress physiology show strong evidence of evolved plasticity across their respective urbanization gradients (Brans et al. 2017; Diamond et al. 2018a).

Although the endpoints of thermal performance are clearly important for understanding the evolution of physiology in response to UHIs, performance at intermediate temperatures is also likely to respond (Figure 6.4). Examples of such traits include growth rate, stress physiology, and metabolic rate. Water flea stress physiology and oxidative damage, which capture responses to heat stress and other types of urban stressors such as pollution, show evolved differences among urban and rural populations. Specifically, urban populations exhibit either lower or comparable stress physiology to rural populations, suggesting the evolution of greater efficiency of the stress response (Brans et al. 2018b). In some cases, the shape of performance across a range of intermediate temperatures is of interest. The ther-

mal performance curve describes the rise of a given thermal performance trait from its minimum temperature threshold up to the optimum temperature and its decline to the maximum temperature threshold. Evolutionary differentiation among populations in their thermal performance curves can take a number of forms—for example, through shifts in the height, breadth, and location of optimal thermal performance. Growth rate is presently the only trait for which evolved shifts in these components of thermal performance curves have been explored across multiple species. Although there are few data available, it is notable that a diverse range of evolved shifts in thermal performance curves have been documented. In two species of chitinolytic fungus, increases in the thermal optima for growth rate were observed; however, in another two species of chitinolytic fungus, urban populations grew as fast or faster across all test temperatures (McLean et al. 2005). A damselfly (*Coenagrion puella*) showed yet another type of response wherein a vertical shift in the urban thermal performance curve was observed, consistent with a pattern of countergradient variation (Tüzün et al. 2017).

6.5 Life history

A fundamental question in evolutionary biology is how the suite of selective pressures organisms face shape when and how energy is allocated towards the basic functions of life, i.e., growth, development, maintenance, and reproduction. Taken together, the evolution of these traits represents a continuum of life-history strategies. As temperature affects many aspects of energy allocation across an organism's life, we expect that the UHI is an important driver of life-history evolution in cities. Phenology, the timing of lifecycle events, has been the focus of many urban studies, and so we first discuss the theory and evidence for its evolution in cities before discussing the evolution of integrated urban life-history strategies.

For phenological traits that are at least partly cued by temperature, there are many well-documented examples, some extending decades to centuries, of shifts in the timing and length of lifecycle events, such as plant flowering (Primack et al. 2009), butterfly development (Roy and Sparks 2000), and bird migration (Cotton 2003), in response to global change. These patterns lead to the prediction that phenological events should generally advance with urbanization due to the UHI effect. In support of this prediction, plants leaf out and flower earlier in cities (Jochner and Menzel 2015) and urban birds begin to sing and to lay eggs earlier than rural birds (Møller et al. 2015). However, phenological advances are not always so straightforward. For instance, while many butterfly species advance their flight periods in urban areas, some species show delays in their peak abundance and first appearance in a season, likely due to the effects of thermal stress on development (Diamond et al. 2014). Similarly, scale insect development has shifted earlier in urban habitats, but the development of these insects' parasitoid natural enemies has not (Meineke et al. 2014). As is true for our other trait classes, however, whether phenological change in cities is due to plasticity or evolution is often unknown.

Three studies of urban plants have used either common garden or reciprocal transplant designs to separate the plastic from evolved responses of phenological change in cities, each finding a different evolutionary response of flowering time to urbanization. Using a reciprocal transplant, Gorton et al. (2018) found that common ragweed (*Ambrosia artemisiifolia*) originating from both rural and urban source populations in the Minneapolis, MN area flowered later when transplanted to urban sites compared to their phenology in rural sites. However, the evolved differences between the two populations showed a counter-gradient response, with plants from urban source populations flowering earlier than those from rural source populations regardless of the transplant environment. The direction of this evolved response differs from those of Virginia pepperweed (*Lepidium virginicum*) collected from five urban–rural gradients in the northeastern USA. Yakub and Tiffin (2017) found that when grown in a common garden experiment, urban-sourced Virginia pepperweed bolted earlier than plants from rural populations but also had delayed time to flowering after bolting, resulting in similar overall flowering phenology across the populations. In contrast, *Crepis sancta* grown in a common garden

using seeds collected from urban and rural habitats near Montpellier, France showed phenological delays for urban plants, flowering and senescing later than plants from the rural source population (Chapter 8; Lambrecht et al. 2016). While these studies were unable to isolate temperature as the selective agent, the results suggest that plant phenology might not always evolve earlier within cities.

Beyond looking at one life-history trait in isolation, how might urban warming affect the evolution of life-history strategies? Phenotypic comparisons of life-history traits in birds suggest that urban environments may favour 'fast' life-history strategies (fast development, early maturation, and earlier senescence (Charmantier et al. 2017)), although our ignorance of the general patterns of selection in cities for these traits makes forming theoretical predictions difficult (Møller 2009; Caizergues et al. 2018). Moreover, predicting the evolutionary responses of life-history strategies to temperature per se are not obvious, but may often be related to the UHI effects on seasonal length and activity. As we have discussed, temperature strongly impacts physiology, growth, and metabolism in ectotherms, suggesting that warmer environments may generally select for fast life histories in these taxa as well.

Brans and De Meester (2018) were able to disentangle and quantify the plastic and evolved components of life-history divergence using a common garden experiment, directly manipulating rearing temperatures in water fleas. Matching predictions, urban water fleas and water fleas reared at higher temperatures had faster maturation, earlier release of progeny, smaller size at maturity, increased fecundity, and higher maximal population growth rate (r) compared to genotypes isolated from rural ponds and animals reared at lower temperatures. Evolution in response to urbanization accounted for 30 per cent of the total trait change in life history, but plastic responses to experimental warming were similar across urban and rural populations (Brans and De Meester 2018). Two important open questions are whether temperature itself is a major driver of life-history evolution in cities across taxa, and how temperature-driven selection may interact with other potentially important selective agents that vary with urbanization (e.g., predation or food abundance).

6.6 Fitness

Lurking behind each consideration of a trait's evolutionary response is whether the changes are beneficial or detrimental, that is, whether the changes represent adaptive evolution (Figure 6.6). The classic test of adaptive divergence is in documenting the existence of fitness tradeoffs between contrasting environments using reciprocal transplant experiments (Donihue and Lambert 2015). A handful of studies to date have evaluated whether urban populations do indeed achieve higher fitness than rural populations in urban environments, and hence are locally adapted. Among these studies, we can perhaps make the clearest link to the UHI as a selective agent behind adaptive divergence in acorn ants (*T. curvispinosus*). When reared under temperature regimes designed to match urban and rural habitats in the lab, colonies achieved higher fitness when reared at the temperature profile matching their native habitat (e.g., Figure 6.6), and the direction of selection on thermal tolerance was also divergent between the two environments. Interestingly, this pattern was repeated across the two urbanization

gradients where *T. curvisponosis* have diverged in thermal tolerance, but not in a third city where divergence was not found (Diamond et al. 2018a).

Brans et al. (2018b) used a similar lab-manipulated temperature design to test if urban and rural water fleas have diverged in their pace-of-life (see section 6.5), and measured several aspects of fitness across both urban and rural temperature environments. If viewed in isolation, urban populations would appear to achieve higher fitness, as estimated by fecundity and population growth rate, than rural populations in both temperatures. However, urban populations also mature earlier, and at a smaller size than rural populations, consistent with the hypothesis that rural and urban populations have evolved slow and fast paces of life, respectively, leading to the prediction that urban and rural populations each have higher lifetime fitness in their native environments, consistent with adaptive evolutionary divergence. This highlights the potential complexities of measuring and interpreting fitness data where tradeoffs among fitness components exist.

While these lab-based experiments have the benefit of measuring fitness while manipulating a single

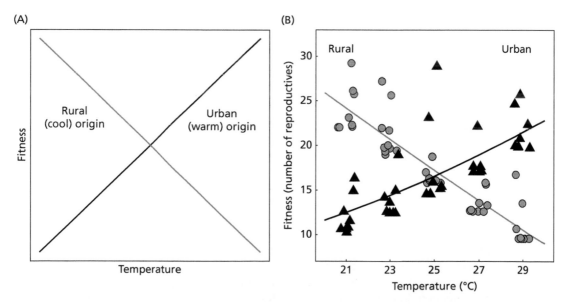

Figure 6.6 Fitness tradeoffs across warm, urban environments and cool, rural environments of urban and rural source populations. Urban source populations have high fitness in warm, urban environments but low fitness in cool, rural environments. Rural populations have high fitness in cool, rural environments but low fitness in warm, urban environments, demonstrating local adaptation. (A) is the expectation; (B) is actual data on fitness of acorn ant colonies (in terms of number of alates, or winged, reproductive individuals produced) from urban (triangles) and rural (circles) populations in Knoxville, TN, reared under common garden conditions across five laboratory temperature regimes (Diamond et al. 2018a).

selective agent, field-based reciprocal transplant experiments measure each population's fitness under more natural conditions. Using a reciprocal transplant experiment with urban and rural populations of common ragweed (*Ambrosia artemisiifolia*), Gorton et al. (2018) found that rural populations had higher lifetime fitness in both urban and rural habitats, a result not consistent with local adaptation to the urban environment. However, other lines of evidence provided support for local adaptation. Specifically, urban and rural populations showed evolved differences in flowering phenology, consistent with adaptation to the UHI (see also Lambrecht et al. 2016). Furthermore, selection on several traits was more intense on populations transplanted into the foreign environment. The authors suggest these seemingly conflicting results may be the result of greater environmental heterogeneity in the urban environment. Thompson et al. (2016) likewise found complex fitness responses in white clover, with selection favouring greater cyanogenesis in urban populations during the summer (in contrast with the decrease in cyanogenesis in urban populations) when measured with final biomass as a fitness proxy, but this relationship was not apparent when fitness was measured as seed mass. Together, these case studies highlight the importance and challenges of combining measures of fitness and selection over multiple life-history stages in the field along with manipulative experiments to evaluate the selective agents responsible for adaptive divergence between urban and rural populations (see also Chapter 9).

6.7 Synthesis: vote-counting meta-analysis

Although we are clearly in the early stages of summarizing evolutionary responses to UHIs, we still have a sufficient number of studies to perform a cursory vote-counting meta-analysis of the major findings to date. Our goals with this analysis are two-fold: first, we aimed to summarize the current patterns (or lack of pattern) among different trait classes and taxonomic groups in their evolutionary responses to UHIs; and second, we aimed to highlight considerations for future endeavours that would best allow extraction and usage of

relevant data for ongoing quantitative syntheses of the field.

We collated all studies with either directly or likely putative effects of UHIs on the evolution of morphological, physiological and life-history traits; because this area is a relatively new area of study, we identified studies from recent special issues and syntheses (Santangelo et al. 2018) and checked Google Scholar by searching for terms that included 'urban', 'temperature', 'heat', and 'evolution'. Using these criteria, we identified 16 studies that we were able to include in our synthesis (McLean et al. 2005; San Martin y Gomez and Van Dyck 2012; Lambrecht et al. 2016; Thompson et al. 2016; Winchell et al. 2016; Brans et al. 2017, 2018b; Diamond et al. 2017, 2018a; Hall and Warner 2017, 2018; Tüzün et al. 2017; Yakub and Tiffin 2017; Brans and De Meester 2018; Gorton et al. 2018; Johnson et al. 2018). Notably, a number of these studies did not explicitly state whether UHIs were the major agent driving evolutionary change; instead, studies typically reported urbanization as the agent of natural selection. Of course, it can be challenging to disentangle various drivers of evolution in cities. However, providing these data, even if there are multiple potential drivers of evolution, will be useful to develop generalizations regarding how evolution unfolds in cities, including in response to UHIs.

We used a simple vote-counting meta-analysis to determine whether patterns found in the study matched or deviated from expectations for two reasons: (1) the traits we considered ranged from morphology to life histories, and (2) relatively few trait values and few studies were available. The assessment of whether expectations were satisfied only became complex for body size. Indeed, expectations for climatic effects on body size qualitatively differ across terrestrial and aquatic systems and even across different taxonomic groups (Horne et al. 2015; see section 6.3 for taxon-specific expectations). Other traits, including heat tolerance, cold tolerance, tolerance breadth, shifts in performance at intermediate temperatures involving changes in breadth, height, and location of the thermal optimum, and life history (which was made to encompass phenology to improve cross-study replication), were comparable across taxa. For example, we expected the evolution of greater heat tolerance in

Figure 6.7 Species represented in UHI evolution meta-analysis. Note that only one exemplar species for the four chitinolytic fungi is shown. (Photo credits from top, left to right: Anolis cristatellus by Bjoertvedt (CC BY-SA 4.0 (https://creativecommons.org/licenses/by-sa/4.0), from Wikimedia Commons); Ascomycota fungus by Ulitca (CC BY-SA 4.0 (https://creativecommons.org/licenses/by-sa/4.0), from Wikimedia Commons); Coenagrion puella by Jörg Hempel (CC BY-SA 3.0 de (https://creativecommons.org/licenses/by-sa/3.0/de/deed.en), from Wikimedia Commons); Trifolium repens (public domain, Wikimedia Commons); Temnothorax curvispinosus by Lauren Nichols; Chorthippus brunneus by Jiří Berkovec (CC BY-SA 2.5 (https://creativecommons.org/licenses/by-sa/2.5), from Wikimedia Commons); Lepidium virginicum by Forest and Kim Starr (CC BY 3.0 (https://creativecommons.org/licenses/by/3.0), via Wikimedia Commons); Ambrosia artemisiifolia by Forest and Kim Starr (CC BY 3.0 (https://creativecommons.org/licenses/by/3.0), via Wikimedia Commons); Crepis sancta by ymm at Hebrew Wikipedia (GFDL (http://www.gnu.org/copyleft/fdl.html) or CC BY-SA 3.0 (https://creativecommons.org/licenses/by-sa/3.0), via Wikimedia Commons); Daphnia magna by Hajime Watanabe (PLoS Genetics, March 2011) (CC BY 2.5 (https://creativecommons.org/licenses/by/2.5), via Wikimedia Commons).

cities compared with non-urban areas. We did not analyse specialized traits that were unreplicated among studies and taxa, such as photosynthetic capacity and haemoglobin levels.

We used the level of analysis according to the presentation of results in the original study. For example, if a study measured multiple urbanization gradients, either within or among different cities, but performed a single analysis of urbanization effects with all studies, we interpreted the result pooled across all replicates. The taxa we considered were exclusively restricted to ectothermic species,

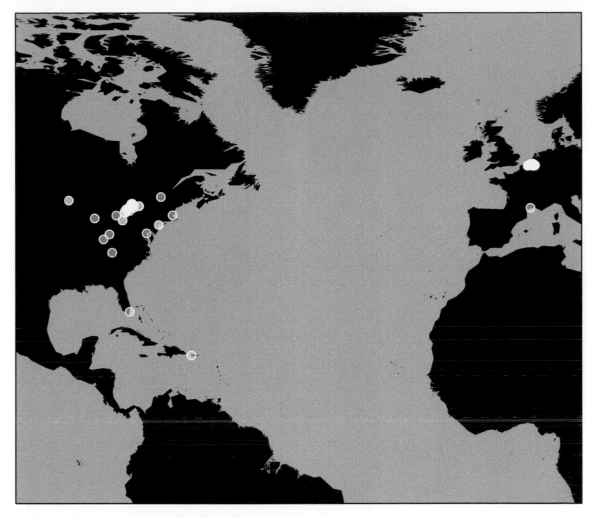

Figure 6.8 Geographic locations of studies that explore the evolution of responses to UHI effects; note that these data include within-study replication among different urbanization gradients. Points are semi-transparently shaded white. As a consequence, lighter areas indicate a greater number of urban evolution studies performed in that geographic location.

and were mostly invertebrates. We had one verte-brate species (a lizard, *Anolis cristatellus*) present in the dataset, represented across three individual studies (Figure 6.7). We detected a high level of geo-graphic clustering of studies that explored evolution in response to UHIs, with most studies occurring in temperate regions in eastern North America and central Europe (Figure 6.8).

Several patterns emerge from the meta-analysis, but perhaps the most striking is that the effect of UHIs varied across all trait classes (Figure 6.9). Our original assertions regarding the relative degree of parallelism in urban evolution across different trait

classes were supported in a broad sense. Specifically, we expected that the evolution of morphology would be fairly consistent across different UHI gradients. Although expectations for temperature effects on body size vary among taxa and habitat types, this trait appears to exhibit fairly predictable (the results go in the expected direction) and parallel (multiple taxa and cities show the predicted responses for body size) responses to urban warming. Heat tolerance and cold tolerance responded consistently to UHIs, a finding that we expected for physio-logical traits. By contrast, performance at intermedi-ate temperatures, expressed as evolved differences

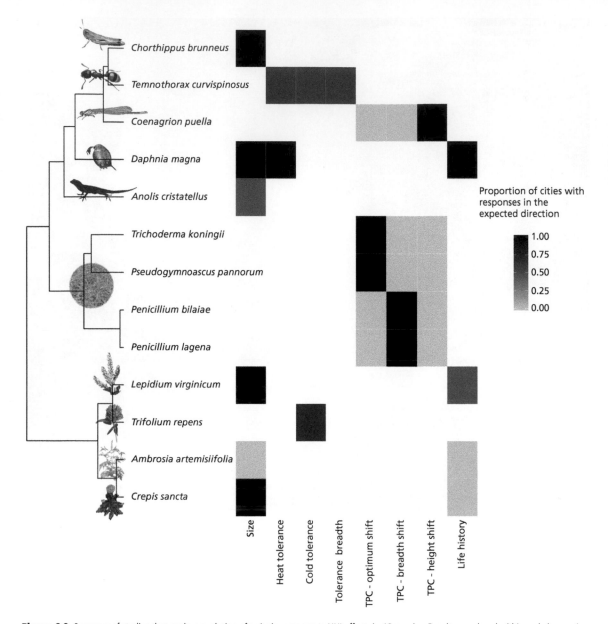

Figure 6.9 Summary of studies that explore evolution of traits in response to UHI effects in 13 species. Results are placed within a phylogenetic context. Heatmap describes proportion of cities (including responses from different studies and different taxa) that exhibit the predicted response to UHI effects based on theory and empirical data syntheses. Traits under consideration include body size, heat tolerance, cold tolerance, tolerance breadth, location of thermal optimum of thermal performance curve (TPC), breadth of TPC, height of TPC, and life history which includes phenology (in order, from left to right, on x-axis). For example, in half of the cities where body size of Anolis cristatellus was examined, the expected evolved decrease in body size in urban environments was found.

in thermal performance curves, was much more variable. The breadth, height, and optimum of thermal performance curves responded differently among studies. Life-history traits, including pace-of-life syndromes and phenology, were more variable among species and urbanization gradients as to whether the expectations for evolved faster pace-of-life and earlier phenological events were met. In summary, the evolution of traits in response to UHIs, while limited in terms of number of cities and taxa studied, is remarkably repeatable for body size and thermal limits, and less so for performance at intermediate temperatures and life history.

6.8 Future directions: beyond standard evolutionary biology in a warmer environment

As the field of urban evolution in response to warming is nascent, an obvious future direction is to continue to build the empirical literature, in particular regarding traits that might be important for adaptation to city life but have been underinvestigated, such as behavioural traits. One important aspect highlighted by our synthesis of urban thermal evolution was the critical role of null results. Indeed, many of the studies that deviated from our expectations were those with null results, rather than studies with significant effects in the unexpected direction. Publishing null results is, and will continue to be, important for avoiding publication bias to adequately quantify the magnitude and direction of evolutionary change in cities (Møller and Jennions 2001). Relatedly, reporting of standard errors of estimates will likewise be important for future summary efforts. Here we have performed a simple vote-counting analysis, but ideally numerical comparisons of evolved responses will be performed in the future. By making point estimates and their errors available (or providing access to the raw data via online repositories including Dryad and FigShare), formal meta-analytical techniques can be applied which take into account not only the magnitude and direction of phenotypic change but also the variability and precision of these estimates (Nakagawa and Santos 2012). In this way, more variable, less well-replicated estimates will be down-weighted relative to more precise estimates of evolutionary change.

Although further development of urban thermal evolution datasets should remain a priority, it is also worth considering what aspects of cities are unique, beyond their use as replicated warming experiments. Within cities, rates of temperature change and temperature variability are two elements that have received little attention (Speights et al.

2017; Diamond et al. 2018b), but which represent unique signatures of UHI effects as compared with other sources of warming, including biogeographic clines in environmental temperature. Further, because cities occur across a broad suite of different background climates, cities present an opportunity to examine the context-dependence of warming effects. A major open question is whether urban warming at already warm locations is the same or different from warming at cool locations (Diamond et al. 2015).

Finally, although we focused on thermal adaptation in this chapter, mostly as a consequence of the types of studies available to synthesize, non-adaptive evolution in response to urban warming should be considered in future efforts. Some of these broad themes are developed elsewhere (Chapters 4, 5, and 11). In the context of UHIs specifically, such explorations could involve the effects of warming on faster mutation rates (Gillooly et al. 2007). Thermal barriers to dispersal could lead to patchier metapopulations and potentially greater local extinction. And temperature-driven population declines could lead to a more prominent role for genetic drift in cities.

Acknowledgements

We thank Michael Angilletta, Anne Charmantier, Marc Johnson, Jason Munshi-South, and Marta Szulkin for helpful comments on a previous version of this manuscript.

References

Angilletta, M.J. (2009). *Thermal Adaptation: A Theoretical and Empirical Synthesis*. Oxford University Press, Oxford.

Angilletta, M.J., Steury, T.D., and Sears, M.W. (2004). Temperature, growth rate, and body size in ectotherms: fitting pieces of a life-history puzzle. *Integrative and Comparative Biology*, 44, 498–509.

Angilletta, M.J. Jr, Wilson, R.S., Niehaus, A.C., et al. (2007). Urban physiology: city ants possess high heat tolerance. *PLOS ONE*, 2, e258.

Ashton, K.G. and Feldman, C.R. (2003). Bergmann's rule in nonavian reptiles: turtles follow it, lizards and snakes reverse it. *Evolution*, 57, 1151–63.

Atkinson, D. (1994). *Temperature and Organism Size: A Biological Law for Ectotherms?* Academic Press, Cambridge, MA.

Bergmann, C. (1848). *Über die Verhältnisse der Wärmeökonomie der Thiere zu ihrer Größe*. Vandenhoek und Ruprecht, Göttingen.

Blomberg, S.P., Garland, T., Ives, A.R., and Crespi, B. (2003). Testing for phylogenetic signal in comparative data: behavioral traits are more labile. *Evolution*, 57, 717–45.

Brans, K.I. and De Meester, L. (2018). City life on fast lanes: urbanization induces an evolutionary shift towards a faster life style in the water flea *Daphnia*. *Functional Ecology*, 32, 2225–40.

Brans, K.I., Jansen, M., Vanoverbeke, J., Tüzün, N., Stoks, R., and De Meester, L. (2017). The heat is on: genetic adaptation to urbanization mediated by thermal tolerance and body size. *Global Change Biology*, 23(12), 5218–27.

Brans, K.I., Engelen, J.M.T., Souffreau, C., and De Meester, L. (2018a). Urban hot-tubs: local urbanization has profound effects on average and extreme temperatures in ponds. *Landscape and Urban Planning*, 176, 22–9.

Brans, K.I., Stoks, R., and De Meester, L. (2018b). Urbanization drives genetic differentiation in physiology and structures the evolution of pace-of-life syndromes in the water flea *Daphnia magna*. *Proceedings of the Royal Society B: Biological Sciences*, 285, 20180169.

Caizergues, A.E., Grégoire, A., and Charmantier, A. (2018). Urban versus forest ecotypes are not explained by divergent reproductive selection. *Proceedings of the Royal Society B: Biological Sciences*, 285, 1882.

Charmantier, A., Demeyrier, V., Lambrechts, M., Perret, S., and Grégoire, A. (2017). Urbanization is associated with divergence in pace-of-life in great tits. *Frontiers in Ecology and Evolution*, 5, 53.

Chatelain, M., Gasparini, J., and Frantz, A. (2016). Do trace metals select for darker birds in urban areas? An experimental exposure to lead and zinc. *Global Change Biology*, 22, 2380–91.

Chown, S.L., Jumbam, K.R., Sørensen, J.G., and Terblanche, J.S. (2009). Phenotypic variance, plasticity and heritability estimates of critical thermal limits depend on methodological context. *Functional Ecology*, 23, 133–40.

Clusella Trullas, S., van Wyk, J.H., and Spotila, J.R. (2007). Thermal melanism in ectotherms. *Journal of Thermal Biology*, 32, 235–45.

Cook, L.M. (2003). The rise and fall of the *Carbonaria* form of the peppered moth. *Quarterly Review of Biology*, 78, 399–417.

Cotton, P.A. (2003). Avian migration phenology and global climate change. *Proceedings of the National Academy of Sciences of the United States of America*, 100, 12219–22.

DeWitt, T.J. and Scheiner, S.M. (2004). *Phenotypic Plasticity: Functional and Conceptual Approaches*. Oxford University Press, Oxford.

Diamond, S.E. and Martin, R.A. (2016). The interplay between plasticity and evolution in response to human-induced environmental change. *F1000Research*, 5, 2835.

Diamond, S.E., Cayton, H., Wepprich, T., et al. (2014). Unexpected phenological responses of butterflies to the interaction of urbanization and geographic temperature. *Ecology*, 95, 2613–21.

Diamond, S.E., Dunn, R.R., Frank, S.D., Haddad, N.M., and Martin, R.A. (2015). Shared and unique responses of insects to the interaction of urbanization and background climate. *Current Opinion in Insect Science*, 11, 71–7.

Diamond, S.E., Chick, L., Perez, A., Strickler, S.A., and Martin, R.A. (2017). Rapid evolution of ant thermal tolerance across an urban-rural temperature cline. *Biological Journal of the Linnean Society*, 121, 248–57.

Diamond, S.E., Chick, L.D., Perez, A., Strickler, S.A., and Martin, R.A. (2018a). Evolution of thermal tolerance and its fitness consequences: parallel and non-parallel responses to urban heat islands across three cities. *Proceedings of the Royal Society B: Biological Sciences*, 285, 20180036.

Diamond, S.E., Chick, L.D., Perez, A., Strickler, S.A., and Zhao, C. (2018b). Evolution of plasticity in the city: urban acorn ants can better tolerate more rapid increases in environmental temperature. *Conservation Physiology*, 6, coy030.

Donihue, C.M. and Lambert, M.R. (2015). Adaptive evolution in urban ecosystems. *Ambio*, 44, 194–203.

Elvidge, C.D., Tuttle, B.T., Sutton, P.C., et al. (2007). Global distribution and density of constructed impervious surfaces. *Sensors*, 7, 1962–79.

Geber, M.A. and Griffen, L.R. (2003). Inheritance and natural selection on functional traits. *International Journal of Plant Sciences*, 164, S21–42.

Gillooly, J.F., McCoy, M.W., and Allen, A.P. (2007). Effects of metabolic rate on protein evolution. *Biology Letters*, 3, 655–60.

Gilman, S.E., Wethey, D.S., and Helmuth, B. (2006). Variation in the sensitivity of organismal body temperature to climate change over local and geographic scales. *Proceedings of the National Academy of Sciences of the United States of America*, 103, 9560–65.

Gorton, A.J., Moeller, D.A., and Tiffin, P. (2018). Little plant, big city: a test of adaptation to urban environments in common ragweed (*Ambrosia artemisiifolia*). *Proceedings of the Royal Society B: Biological Sciences*, 285(1881), 20180968.

Grimm, N.B., Faeth, S.H., Golubiewski, N.E., et al. (2008). Global change and the ecology of cities. *Science*, 319, 756–60.

Grimmond, S. (2007). Urbanization and global environmental change: local effects of urban warming. *The Geographical Journal*, 173, 83–8.

Gunderson, A.R. and Stillman, J.H. (2015). Plasticity in thermal tolerance has limited potential to buffer ectotherms from global warming. *Proceedings of the Royal Society B: Biological Sciences*, 282, 20150401.

Hall, J.M. and Warner, D.A. (2017). Body size and reproduction of a non-native lizard are enhanced in an urban environment. *Biological Journal of the Linnean Society*, 122, 860–71.

Hall, J.M. and Warner, D.A. (2018). Thermal spikes from the urban heat island increase mortality and alter physiology of lizard embryos. *Journal of Experimental Biology*, 221, 181552.

Heisler, G.M. and Brazel, A.J. (2010). The urban physical environment: temperature and urban heat islands. In: Aitkenhead-Peterson, J. and Volder, A. (eds) *Urban Ecosystem Ecology*, pp. 29–56. American Society of Agronomy, Crop Science Society of America, Soil Science Society of America, Madison, WI.

Homer, C., Dewitz, J., Yang, L., et al. (2015). Completion of the 2011 national land cover database for the conterminous United States–representing a decade of land cover change information. *Photogrammetric Engineering & Remote Sensing*, 81, 345–54.

Horne, C.R., Hirst, A.G., and Atkinson, D. (2015). Temperature-size responses match latitudinal-size clines in arthropods, revealing critical differences between aquatic and terrestrial species. *Ecology Letters*, 18, 327–35.

Horne, C.R., Hirst, A.G., and Atkinson, D. (2017). Seasonal body size reductions with warming covary with major body size gradients in arthropod species. *Proceedings of the Royal Society B: Biological Sciences*, 284, 20170238.

Imhoff, M.L., Zhang, P., Wolfe, R.E., and Bounoua, L. (2010). Remote sensing of the urban heat island effect across biomes in the continental USA. *Remote Sensing of Environment*, 114, 504–13.

Jacquin, L., Lenouvel, P., Haussy, C., Ducatez, S., and Gasparini, J. (2011). Melanin-based coloration is related to parasite intensity and cellular immune response in an urban free living bird: the feral pigeon *Columba livia*. *Journal of Avian Biology*, 42, 11–15.

Jochner, S. and Menzel, A. (2015). Urban phenological studies—past, present, future. *Environmental Pollution*, 203, 250–61.

Johnson, M.T.J. and Munshi-South, J. (2017). Evolution of life in urban environments. *Science*, 358, eaam8327.

Johnson, M.T.J., Prashad, C.M., Lavoignat, M., and Saini, H.S. (2018). Contrasting the effects of natural selection, genetic drift and gene flow on urban evolution in white clover (*Trifolium repens*). *Proceedings of the Royal Society B: Biological Sciences*, 285, 20181019.

Kingsolver, J.G. and Huey, R.B. (2008). Size, temperature, and fitness: three rules. *Evolutionary Ecology Research*, 10, 251–68.

Lambrecht, S.C., Mahieu, S., and Cheptou, P.-O. (2016). Natural selection on plant physiological traits in an urban environment. *Acta Oecologica*, 77, 67–74.

Li, B., Suzuki, J.-I., and Hara, T. (1998). Latitudinal variation in plant size and relative growth rate in *Arabidopsis thaliana*. *Oecologia*, 115, 293–301.

Liker, A., Papp, Z., Bókony, V., and Lendvai, A.Z. (2008). Lean birds in the city: body size and condition of house sparrows along the urbanization gradient. *Journal of Animal Ecology*, 77, 789–95.

McLean, M.A., Angilletta, M.J., and Williams, K.S. (2005). If you can't stand the heat, stay out of the city: thermal reaction norms of chitinolytic fungi in an urban heat island. *Journal of Thermal Biology*, 30, 384–91.

Meillère, A., Brischoux, F., Parenteau, C., and Angelier, F. (2015). Influence of urbanization on body size, condition, and physiology in an urban exploiter: a multi-component approach. *PLOS ONE*, 10, e0135685.

??Meineke, E.K., Dunn, R.R., and Frank, S.D. (2014). Early pest development and loss of biological control are associated with urban warming. *Biology Letters*, 10.

Merckx, T., Souffreau, C., Kaiser, A., et al. (2018). Body-size shifts in aquatic and terrestrial urban communities. *Nature*, 558, 113.

Merilä, J. and Hendry, A.P. (2014). Climate change, adaptation, and phenotypic plasticity: the problem and the evidence. *Evolutionary Applications*, 7, 1–14.

Miller, C.R., Latimer, C.E., and Zuckerberg, B. (2018). Bill size variation in northern cardinals associated with anthropogenic drivers across North America. *Ecology and Evolution*, 8, 4841–51.

Møller, A.P. (2009). Successful city dwellers: a comparative study of the ecological characteristics of urban birds in the Western Palearctic. *Oecologia*, 159, 849–58.

Møller, A.P. and Jennions, M.D. (2001). Testing and adjusting for publication bias. *Trends in Ecology & Evolution*, 16, 580–86.

Møller, A.P., Díaz, M., Grim, T., et al. (2015). Effects of urbanization on bird phenology: a continental study of paired urban and rural populations. *Climate Research*, 66, 185–99.

Mousseau, T.A. and Roff, D.A. (1987). Natural selection and the heritability of fitness components. *Heredity*, 59(2), 181–97.

Nakagawa, S. and Santos, E.S.A. (2012). Methodological issues and advances in biological meta-analysis. *Evolutionary Ecology*, 26, 1253–74.

Pemberton, J.M. (2010). Evolution of quantitative traits in the wild: mind the ecology. *Philosophical Transactions of the Royal Society B: Biological Sciences*, 365, 2431–8.

Primack, R.B., Higuchi, H., and Miller-Rushing, A.J. (2009). The impact of climate change on cherry trees and other species in Japan. *Biological Conservation*, 142, 1943–9.

Roy, D.B. and Sparks, T.H. (2000). Phenology of British butterflies and climate change. *Global Change Biology*, 6, 407–16.

San Martin y Gomez, G. and Van Dyck, H. (2012). Ecotypic differentiation between urban and rural populations of the grasshopper Chorthippus brunneus relative to climate and habitat fragmentation. *Oecologia*, 169, 125–33.

Santangelo, J.S., Rivkin, L.R., and Johnson, M.T.J. (2018). The evolution of city life. *Proceedings of the Royal Society B: Biological Sciences*, 285, 20181529.

Searle, S.Y., Turnbull, M.H., Boelman, N.T., et al. (2012). Urban environment of New York City promotes growth in northern red oak seedlings. *Tree Physiology*, 32, 389–400.

Sears, M.W. and Angilletta, M.J. (2004). Body Size Clines in Sceloporus Lizards: Proximate Mechanisms and Demographic Constraints. *Integrative and Comparative Biology*, 44, 433–42.

Seebacher, F., White, C.R., and Franklin, C.E. (2015). Physiological plasticity increases resilience of ectothermic animals to climate change. *Nature Climate Change*, 5, 61–6.

Seto, K.C., Güneralp, B., and Hutyra, L.R. (2012). Global forecasts of urban expansion to 2030 and direct impacts on biodiversity and carbon pools. *Proceedings of the National Academy of Sciences of the United States of America*, 109, 16083–8.

Sibly, R.M. and Atkinson, D. (1994). How Rearing Temperature Affects Optimal Adult Size in Ectotherms. *Functional Ecology*, 8, 486–93.

Sørensen, J.G., Kristensen, T.N., and Overgaard, J. (2016). Evolutionary and ecological patterns of thermal acclimation capacity in Drosophila: is it important for keeping up with climate change? *Current Opinion in Insect Science*, 17, 98–104.

Speights, C.J., Harmon, J.P., and Barton, B.T. (2017). Contrasting the potential effects of daytime versus night-time warming on insects. *Current Opinion in Insect Science*, 23, 1–6.

Sunday, J.M., Bates, A.E., and Dulvy, N.K. (2012). Thermal tolerance and the global redistribution of animals. *Nature Climate Change*, 2, 686–90.

Thompson, K.A., Renaudin, M., and Johnson, M.T.J. (2016). Urbanization drives the evolution of parallel clines in plant populations. *Proceedings of the Royal Society B: Biological Sciences*, 283, 20162180.

Tüzün, N., Op de Beeck, L., Brans, K.I., Janssens, L., and Stoks, R. (2017). Microgeographic differentiation in thermal performance curves between rural and urban populations of an aquatic insect. *Evolutionary Applications*, 10, 1067–75.

Tyler, R.K., Winchell, K.M., and Revell, L.J. (2016). Tails of the City: Caudal Autotomy in the Tropical Lizard, Anolis cristatellus, in Urban and Natural Areas of Puerto Rico. *Journal of Herpetology*, 50, 435–41.

Vinarski, M.V. (2014). On the applicability of Bergmann's rule to ectotherms: the state of the art. *Biology Bulletin Reviews*, 4, 232–42.

Winchell, K.M., Reynolds, R.G., Prado-Irwin, S.R., Puente-Rolón, A.R., and Revell, L.J. (2016). Phenotypic shifts in urban areas in the tropical lizard Anolis cristatellus. *Evolution*, 70, 1009–22.

Yakub, M. and Tiffin, P. (2017). Living in the city: urban environments shape the evolution of a native annual plant. *Global Change Biology*, 23, 2082–9.

Yan, W.Y., Shaker, A., and El-Ashmawy, N. (2015). Urban land cover classification using airborne LiDAR data: A review. *Remote Sensing of Environment*, 158, 295–310.

Youngsteadt, E., Dale, A.G., Terando, A.J., Dunn, R.R., and Frank, S.D. (2015). Do cities simulate climate change? A comparison of herbivore response to urban and global warming. *Global Change Biology*, 21:97–105.

CHAPTER 7

The Evolutionary Ecology of Mutualisms in Urban Landscapes

Rebecca E. Irwin, Elsa Youngsteadt, Paige S. Warren, and Judith L. Bronstein

Irwin, R.E., Youngsteadt, E., Warren, P.S. and Bronstein, J.L., *The Evolutionary Ecology of Mutualisms in Urban Landscapes* In: *Urban Evolutionary Biology*.
Edited by Marta Szulkin, Jason Munshi-South and Anne Charmantier, Oxford University Press (2020). © Oxford University Press.
DOI: 10.1093/oso/9780198836841.003.0007

7.1 Introduction

Mutualisms—reciprocal positive interactions—are ubiquitous in natural communities, conferring net benefits to the survival, growth, and/or reproduction of the interacting species (Bronstein 2015). Nearly all species are involved in one or more mutualisms. Mutualisms provide ecosystem services (e.g., Klein et al. 2006; Clemmensen et al. 2015), and also supported or sparked major evolutionary events, such as the evolution of the eukaryotic cell. Thus, they are critically important in the sustainability and diversity of life we see today.

Mutualisms are not static components of ecosystems. Rather, differences in the biotic and abiotic environment among sites can affect the strength of most mutualisms, and even shift them away from being mutually beneficial to commensal or even parasitic. There is widespread appreciation that natural spatial and temporal variation in biotic and abiotic factors can drive this variation. For example, whether *Greya* moths act as mutualists (pollinators) or antagonists (floral parasites) of the host plant *Lithophragma parviflorum* depends on spatial variation in the presence and abundance of effective co-pollinator species (Thompson and Cunningham 2002). *Greya* and *L. parviflorum* interact as mutualists in sites with few co-pollinators, with both species being co-dependent on one another for reproduction. However, abundant co-pollinators in sites change the outcome of the interaction between *Greya* and *L. parviflorum* to commensalism and antagonism. There is also growing recognition that human activities are altering Earth's natural systems, with the potential for global environmental change to drive alterations in the evolutionary ecology of mutualisms (Kiers et al. 2010). Drivers of change, such as CO_2 enrichment, nitrogen deposition, habitat loss and fragmentation, biological invasions, and temperature increases, have the potential to affect forces of evolution important for mutualism outcome and persistence.

These drivers of environmental change occur not only at global scales but also at local scales, for example in the urban landscape. Over half the global human population currently lives in cities, and although urban areas only make up ca. 3 per cent of the land surface on Earth, urbanization and impacts of the urban footprint are expanding (Grimm et al.

2008). Relative to adjacent non-urban areas, urban areas are associated with abiotic changes—including increased temperature, nutrient loading, habitat loss, and fragmentation—as well as biotic changes, including reduced biodiversity, increased density of invasive species and human commensals, and altered species interactions (reviewed in Niemela and Breuste 2011). These physical and biological changes in the urban landscape can alter forces of evolution and the evolutionary trajectories of species. There are striking examples of non-adaptive and adaptive evolutionary differences in urban vs non-urban areas (Johnson and Munshi-South 2017). Species interactions have the potential to magnify, dampen, or alter the direction of evolution. However, implications for mutualism in particular have received minimal attention, in spite of their importance to the persistence and health of entire communities.

Here, we build from an ecological perspective on mutualisms in urban landscapes to consider how urbanization affects their evolutionary ecology. First, we review the adaptive mechanisms by which urbanization may affect selection pressures important for mutualisms. As much as possible, we draw from concepts and theories describing natural systems to make predictions about urban ones, although we note that urban ecosystems may be so unusual that some predictions relevant to natural systems may not apply. Second, we survey three main classes of mutualism: transportation, protection, and nutrition. Within each class, we consider how urbanization affects ecological patterns and processes involved in the evolution of mutualist hosts and partners. Our survey includes terrestrial examples only; while mutualisms are important to the functioning of aquatic ecosystems, no studies of urban aquatic mutualisms have been conducted from an evolutionary perspective to our knowledge, suggesting an important gap in knowledge (see also Chapter 10). Even in terrestrial mutualisms, we know much more about their ecology than about their evolutionary trajectories, both in and out of urban settings. However, we note that urbanization often encapsulates multiple global change drivers simultaneously (e.g., increased temperature, light pollution, and altered species interactions), and when different drivers act in opposite ways, it is not always obvious how mutualisms will respond to combined effects in an urban context. This complexity suggests that the study of interactions among factors is an important future direction. We end by outlining research directions to further the study of the evolutionary ecology of mutualisms in urban landscapes. Given that many mutualisms confer critical ecosystem functions and services, understanding the evolutionary ecology of mutualisms in urban landscapes has important implications for human and ecosystem health and well-being.

7.2 A mechanistic perspective on the evolutionary ecology of urban mutualisms

In this chapter, we emphasize the potential for adaptive evolutionary change in urban mutualisms as a function of changing selection pressures. We note, however, that mutualisms such as pollination and seed dispersal likely alter gene flow of host plants with increasing urbanization (see section 7.3), and that the potential for adaptive evolution may be constrained by gene flow and genetic drift, especially in small, fragmented urban landscapes (Chapter 4; Rivkin et al. 2019).

Evolutionary ecologists are fundamentally interested in spatial variation in natural selection and selective mechanisms. Studies have shown that mutualisms can experience geographic mosaics of adaptive evolution, with variation in the cost:benefit ratio between mutualist partners affecting natural selection and evolution of traits involved in the mutualism (Thompson and Cunningham 2002). Given that the urban landscape modifies a variety of abiotic and biotic factors important to mutualisms, it is reasonable to suspect that the urban landscape may alter the selection mosaic and evolutionary trajectory of mutualisms beyond that already occurring across varied natural systems. We review five key ways in which this can occur: (1) shifts from mutualism to antagonism, (2) changes in trait–fitness relationships, (3) partner switching, (4) plastic changes in partner behaviour, and (5) partner loss (Figure 7.1). We note that these five mechanisms are neither mutually exclusive nor all-encompassing. Nor are they unique to mutualisms in urban settings; they have been noted, for example, in other anthropogenically modified environments, including agricultural ones (Kiers et al. 2010).

7.2.1 Shifts from mutualism to antagonism

Partner species can vary greatly in the degree to which they benefit one another, ranging from mutually beneficial (benefits exceed costs of interaction) to antagonistic (costs exceed benefits for one or both partners) depending on the ecological context. Mutualisms are commonly defined as interactions in which benefits exceed costs, but in fact cost:benefit ratios can vary spatially, and in some circumstances costs exceed benefits (Hoeksema and Bruna 2015). The urban landscape can affect these cost:benefit ratios, potentially facilitating an evolutionary shift towards antagonism if selection is intense, its direction is consistent, and gene flow is relatively low. This might be expected to result because selection should favour individuals that maximize their own fitness at the expense of their partners, leading to the evolution of decreased partner quality or selection for antagonistic traits (Sachs et al. 2004; Werner et al. 2018). Aspects of the urban environment (biotic or abiotic) that weaken partner fitness feedbacks, such as high soil nutrient availability that reduces beneficial feedbacks between plants and their root symbionts, have been argued to select for less-cooperative mutualists (West et al. 2002), especially if there are tradeoffs among traits that confer cooperation and traits that confer partner fitness in the absence of the host. However, it is important to note that a variety of stabilizing factors are known to counteract selection for decreased cooperation. In cases where hosts cannot live without their partners, sanctions against poor cooperators might be expected to select for increased partner cooperation (Kiers et al. 2010).

7.2.2 Changes in trait–fitness relationships

Selection on mutualism traits in a host should be strong when there is a strong positive relationship between these traits and host fitness (Ashman and Morgan 2004). Patterns of selection should thus change if urban landscapes weaken trait–fitness relationships. A weakening of the trait–fitness relationship could occur if the mutualistic interaction no longer limits host fitness. For example, if mutualist partners are abundant in the urban landscape, any variation in traits involved in partner attraction or reward may have little effect on host fitness.

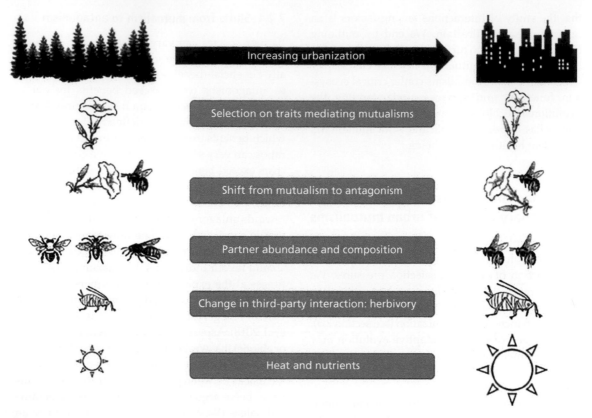

Figure 7.1 Some mechanisms driving differential selection on traits mediating mutualisms in urban compared to non-urban sites. These differences can include, but are not limited to, shifts from mutualism to antagonism (such as bees shifting from pollinating to robbing nectar from flowers; see section 7.2.1), changes in partner abundance and composition (and at the extreme, partner loss; see sections 7.2.3 and 7.2.5), and differences in third-party interactions (such as herbivory in protection mutualisms; see section 7.4). Changes in mutualism interactions can also be driven and shaped by altered abiotic factors in urban vs non-urban sites, such as differences in heat and nutrient availability (see section 7.5).

Alternatively, reductions in mutualist availability could intensify trait–fitness relationships for traits involved in attracting them. For example, reductions in pollinator populations should increase pollen limitation, which is expected to intensify selection on floral traits that are important to pollinator attraction or that confer reproductive assurance (Harder and Aizen 2010).

7.2.3 Partner switching

The urban landscape has well-documented effects on species richness and composition. These effects are particularly evident in comparisons of different urban habitat types (e.g., urban parks vs residential lands vs city centres), urban habitat fragments of different sizes or ages, and along gradients from urban to non-urban areas (Mckinney 2008; Aronson et al. 2014). Some of the affected species are critical mutualists for resident species. As a consequence, changes in richness and composition could shift mutualistic interactions towards increasing generalization, especially with the replacement of native species by, or the addition of, exotics and human commensals in urban habitats. These partner switches could affect selection on traits that hosts use to attract and/or reward mutualist partners, or traits partners use to access those rewards. Mutualist partners vary strikingly in the quality and quantity of benefits they provide (Schupp et al. 2017), and so partner switching also has the potential to affect variation in host fitness and the opportunity for selection.

7.2.4 Changes in partner behaviour

Urban landscapes could influence the behaviour of mutualistic partners in at least two ways, with potential effects on patterns of natural selection and evolution. First, the urban environment may affect the preferences of partners, and the traits that partners use to discriminate among hosts. Partner preferences may change if the urban environment masks some traits in hosts or affects the ability of partners to detect variation in the traits, requiring partners to use other traits to discriminate among potential hosts. For example, air pollution can degrade floral volatiles and modify the chemical composition of those volatiles, affecting the ability of pollinators to locate their host plants from scent plumes (Fuentes et al. 2016). Air pollution also affects background odours, which can impede pollinator signal detection and behaviour (Riffell et al. 2014; Jürgens and Bischoff 2017), which could affect plant–pollinator interactions and patterns of selection. Second, the human-built environment may alter the movement behaviour of mutualist partners, such as impervious surface cover altering the foraging behaviour and nesting density of bee pollinators (Jha and Kremen 2013a), which could have subsequent effects on the evolution of host-plant traits (see section 7.3).

7.2.5 Partner loss

Given declines in species richness with urbanization, native species in urban landscapes may find themselves without any mutualist partners at all. The complete loss of mutualists is known in some cases to select for traits that confer the service in the absence of partners, such as shifts from biotic pollination to the ability to self-fertilize (Roels and Kelly 2011). In facultative mutualisms, loss of partners could also result in selection against traits that confer the mutualism, especially if the traits are costly. Loss of mutualist partners can affect the pace of morphological change in mutualism-related traits in hosts over evolutionary time (Chomicki and Renner 2017). Whether loss of mutualist partners in urban ecosystems affects trait change over shorter timescales warrants investigation, especially in urban systems that experience limited gene flow with non-urban ones.

In sections 7.3, 7.4, and 7.5, we draw upon these mechanisms to highlight how urban landscapes affect the evolutionary ecology of different types of mutualisms. The usual way to group mutualisms is according to the nature of the benefits that partners exchange (Bronstein 2015). In the most common scheme, almost all benefits fall into one of three categories. *Transportation* includes movement of partners themselves, or movement of partners' gametes. *Protection* involves the provision of defence against the biotic and abiotic environment. It includes aggression toward or consumption of the partner's enemies, as well as the conferring of resistance to those enemies. Finally, *nutrition* involves the provision of one or more limiting nutrients to the partner. Mutualisms involve the reciprocal exchange of these benefits. For example, plant–pollinator mutualists exchange transportation for nutrition, plant–ant defender mutualists exchange protection for nutrition, and plant–rhizobial mutualisms involve a nutrition (carbon) for nutrition (nitrogen) exchange. As we show below, urbanization likely has different evolutionary effects depending on the natural history of this exchange (Table 7.1).

7.3 Transportation mutualisms

Classic examples of transportation mutualisms include pollination and seed dispersal. Most information about the evolutionary ecology of mutualisms in urban environments comes from the study of plant–pollinator mutualisms (see section 7.3.1 and references therein). This stems in part from concerns over pollinator declines and the conservation of pollination services in human-modified habitats.

7.3.1 Pollination mutualisms

The spectacular diversification of angiosperm flowers has been driven, in part, by the activity of animal pollinators. In situations where pollen receipt or export limits a plant's reproductive success, selection should favour floral traits that increase the frequency and quality of pollinator visits. Indeed, research has detected pollinator-mediated selection on numerous floral traits, including flower number, height, size, shape, spur length, phenology,

Table 7.1 Predictions of how urban drivers will affect phenotypic selection and potentially adaptive evolution of hosts and/or partners in transportation (pollination and dispersal), protection, and nutritional mutualisms.

Mutualism	Urban driver	Prediction
Pollination	Increased pollen limitation (some systems)	Selection for floral traits that promote more or more effective visits, or reproductive assurance
	Reduced pollen limitation (some systems)	Relaxed selection and increased variability in floral traits
	Altered pollinator community	Partner switching, or selection for traits that promote frequent or efficient visits by new partner
	Urban warming shifts flower or pollinator phenology	Partner switching, weakened mutualism, or selection to match partner phenology
	Altered antagonist community (florivores, nectar robbers)	Selection for floral traits that reduce antagonism
Dispersal	Impervious surface/inhospitable germination sites	High cost of dispersal, or loss of traits that attract dispersers
	Reduced density or quality of dispersing partners	Partner switching, loss of traits that attract dispersers, or selection for traits that attract new partner
Protection	Reduced density or quality of protective partners	Partner switching, loss of traits that attract partners, or selection for alternative defences
	Altered herbivore density	Increased or decreased selection for traits that attract protective partners
Nutritional	Nutrient enrichment	Shift from mutualism to antagonism with selection for antagonistic traits in partner or increased sanctions by the host, or altered partner choice
	Reduced diversity of mutualist partners	Weakened mutualisms and partner loss, or reduced selection on traits that generate positive partner feedbacks
	Pesticides, heat, altered predator/parasite communities	Selection for partner strains that confer protective traits

colour, nectar-guide pattern, pollen production, and scent (Lloyd and Barrett 1996).

Plant–pollinator mutualisms are highly variable through space and time, and changes in the abundance, identity, phenology, or behaviour of pollinators can alter the strength or direction of natural selection on floral traits (Lloyd and Barrett 1996). However, flowers and pollinators do not interact in isolation. In some times and places, plant reproduction is limited by non-pollination phenomena, such as seed predation, growing season length, or other limiting resources, and these may relax or counteract selection mediated by pollinators. These varying pressures, and their micro- and macroevolutionary consequences for floral form and diversity, are well demonstrated in non-urban systems. There is growing recognition that these selection pressures can differ in urban systems (Harrison and Winfree 2015), with implications for patterns of selection and floral evolution.

A key question is whether plant reproduction is pollen-limited in urban areas. Pollen limitation of seed set is widely documented in non-urban systems, detected in up to three-quarters of studies that test for it (Ashman et al. 2004). Pollen limitation is expected to increase in fragmented habitats because small, isolated plant populations include fewer pollen donors and attract or support fewer pollinators (Eckert et al. 2010). Urbanization, however, goes beyond fragmentation alone. It is not obvious whether urbanization consistently intensifies or weakens pollen limitation. It could intensify pollen limitation by exacerbating pollen scarcity or by releasing plants from other resource limitations that previously masked pollen scarcity; the reverse processes are equally plausible. Indeed, the few studies that have compared pollen limitation of seed set in the same plant species in urban and non-urban areas have reported different patterns. Most examples detected some degree of pollen limitation at both ends of an urban gradient, with urban populations experiencing less (e.g., Parker 1997), more (e.g., Wee et al. 2015), or equally (e.g., Verboven et al. 2012) intense pollen limitation, while one study (Minnefors 2016) detected no pollen limitation regardless of urbanization. In most, but not all, cases, changes in pollen limitation corresponded to changes in pollinator visitation rates. Urbanization thus acts in system-specific ways to increase or reduce the potential for pollinators to act as agents of selection on floral traits, but we note that few studies have assessed the degree to which urban–rural differences in pollen limitation are temporally stable within a given system.

Theory predicts that in the presence of pollen limitation, selection could favour traits that increase the frequency or quality of pollinator visits, or it could favour reproductive assurance mechanisms such as self-pollination (Table 7.1; section 7.2.2; Eckert et al. 2010; Harder and Aizen 2010). These possibilities have been examined in an urban context for a limited number of plant species (e.g., Bode and Tong 2017). We highlight two examples—a native hawksbeard (*Crepis sancta*) growing in and around the city of Montpellier, France (see also Chapter 8), and a native dayflower (*Commelina communis*) (Figure 7.2) in the Osaka-Kobe metropolitan area of Japan. Urban hawksbeard plants

Figure 7.2 Reduced availability of mates and pollinators, such as the syrphid fly *Episyrphus balteatus*, in urban populations favours the evolution of traits associated with reproductive assurance in *Commelina communis*. (Photo source: Koki R. Katsuhara.)

grow in small, sparse populations and receive greatly reduced pollinator visitation compared to larger and denser rural populations (Cheptou et al. 2006; Andrieu et al. 2009). Despite heritable variation in the extent of self-compatibility, a common garden experiment detected no increase in self-pollination ability in urban populations (Cheptou et al. 2006). Furthermore, despite pollinator preference for larger-diameter inflorescences, *in situ* urban inflorescences were smaller than rural ones—contrary to the expected pattern if pollen-limited urban plants were under stronger selection for attractive traits (Andrieu et al. 2009). Absence of genetic variation seems an unlikely explanation; urban populations of the same species in the same city evolved adaptive changes in seed morphology in an estimated five to twelve generations (section 7.3.2; Cheptou et al. 2008).

The dayflower example shares some general ecological similarities with the hawksbeard example, in that urban populations produced fewer flowers and experienced reduced pollinator visitation (Ushimaru et al. 2014). Dayflowers in sites with lower visitation exported less pollen and set fewer seeds (Ushimaru et al. 2014). They also had shorter stigmas, less anther–stigma separation, and a higher ratio of hermaphroditic flowers to male flowers, all traits thought to promote self-pollination. Although this study did not confirm heritability of these traits, nor their contribution to realized self-pollination rates, the observed pattern was consistent with the prediction that reduced availability of both mates and pollinators in urban habitats could favour the evolution of reproductive assurance (section 7.2.2).

Although plants in these two studies experienced reduced visitation rates in urban areas, this is not universal. Studies of focal plant species along urbanization gradients have found that pollinator visitation rates may increase (e.g., Leong et al. 2014), decrease (e.g., Wee et al. 2015), or stay the same (e.g., Williams and Winfree 2013). The causes of this variation are unclear, but likely include differences in the urbanization gradient and the type of non-urban land use investigated. In one striking example, urban starthistle (*Centaurea solstitialis*) flowers in San Francisco, CA, USA received approximately twice the visitation rate of non-urban flowers, but had significantly lower seed set (Leong et al. 2014). Urban flowers may have been limited by resources rather than pollen, or urban pollinators may have been less efficient or more generalized (section 7.2.3), resulting in heterospecific (i.e., cross-species) pollen transfer. The starthistle study did not distinguish between these possibilities, but other work corroborates the potential for greater transfer of heterospecific pollen in urban areas. For example, three studies have compared plant–pollinator network structure between urban and non-urban sites with high taxonomic resolution (Gotlieb et al. 2011; Baldock et al. 2015; Martins et al. 2017). Each found that urban gardens had more species of insect-visited flowering plants than did natural or semi-natural habitats. In turn, urban insect species visited more plant species on average, but a smaller proportion of the total available species, than did non-urban insects. These changes in network structure could reduce pollinator efficiency and promote heterospecific pollen transfer. Plants that experience high visitation rates while remaining pollen limited may be subject to relatively strong selection for any trait that favours more effective pollinator visits (Table 7.1).

Although effects on visitation rate vary across studies, urbanization reliably produces changes in pollinator community composition—documented in observations of focal plant species (e.g., Cane et al. 2006; Geerts and Pauw 2012) and in trap-based surveys (e.g., Fortel et al. 2014). Changes may be pronounced, with urban and non-urban communities differing in pollinating bird bill length (Geerts and Pauw 2012), bee body size or phenology (Harrison et al. 2018), or dominant insect order (Ushimaru et al. 2014). The evolutionary consequences of these pollinator partner switches (section 7.2.3) have not been investigated in urban sites, but shifts in pollinator identity are a key driver of floral evolution in non-urban systems. Even in generalist-pollinated plants, changes in pollinator community composition can create selective mosaics across fine spatial scales (Table 7.1). In one (non-urban) example, an alpine mustard (*Erysimum mediohispanicum*) with at least 130 species of pollinators experienced differential selection on—and corresponding variation in—corolla shape over distances as little as 250 m, linked to consistent spatial differences in pollinator species composition (Gómez et al. 2009). Similar patterns may await detection along urban–rural gradients or among habitat types within urban areas.

On the other hand, pollinator-mediated selection may be dampened or opposed by other forces.

These include abiotic factors such as temperature or resource availability, as well as biotic antagonists, including herbivores, florivores, and nectar robbers (Table 7.1). For example, in resource-limited environments, selection may favour reduced floral display size, which could oppose selection by pollinators for increased display size. Studies are needed that manipulate both resources and pollination in a factorial design and measure subsequent patterns of selection in urban and non-urban sites. Moreover, if selection is temporally unstable, directional evolution will be less likely.

Some of these factors were investigated in the vine *Gelsemium sempervirens* in urban and non-urban sites in the Raleigh-Durham, NC, USA metropolitan area over multiple years. In the first two sampling periods (2005 and 2007), urban plants experienced more florivory, nectar robbing, and heterospecific pollen deposition than those in non-urban sites. These antagonistic interactions were more common in plants that had longer, narrower corollas, narrower petals, and larger display sizes, suggesting that these traits should be more strongly selected against in urban sites. Partly supporting this prediction, urban populations did have significantly wider corollas, and the larger size of urban flowers was partly retained in a common garden experiment (Irwin et al. 2014). However, in a follow-up study measuring similar traits along with seed set in 2009, different patterns emerged (Irwin et al. 2018). For example, the correlation between floral display size and florivory reversed, as did differences in conspecific pollen receipt between urban and non-urban sites. The only trait under differential selection between urban and non-urban sites in 2009 was floral display size, with larger displays more strongly associated with higher pollen receipt and plant fitness in urban sites. Taken together, these studies provide evidence that urbanization can drive significant differences in phenotypic selection on floral traits relevant to plant–pollinator mutualisms. However, the work also highlights the possibility that the strength and direction of these selection gradients may fluctuate extensively between years, such that the consequences of evolution in this system remain difficult to interpret. This could be a common pattern in other mutualisms as well, and selection studies should ideally compare urban with non-urban sites over multiple years (Bode and Tong 2017).

Finally, urban evolution may also be constrained by genetic drift or gene flow, if adaptive variation is lost in small populations or if genetic exchange with non-urban populations swamps local adaptation. This point is equally true for all types of mutualisms, but it should be particularly important for transportation mutualisms, in which the benefits specifically involve the movement of propagules. Although cities may act as geographic barriers to pollinator dispersal on a regional scale (e.g., Jha and Kremen 2013b), several population genetic studies of plants (e.g., Culley et al. 2007) and pollinators (e.g., López-Uribe et al. 2015) detected weak to no differentiation between urban and adjacent rural populations (but see Bartlewicz et al. 2015; Yakub and Tiffin 2017). Nevertheless, recent studies of white clover (Johnson et al. 2018) and a European bumble bee (Theodorou et al. 2018) each detected signatures of adaptive evolution in urban areas, despite extensive gene flow with non-urban populations. Although these studies were unrelated to pollination, their results suggest that the genetic potential for evolution of plant–pollinator mutualisms is likely to exist in cities. The European bumble bee example (Theodorou et al. 2018) points to the need for more studies that measure the potential for evolution not only in plants but also in pollinators in urban compared to non-urban landscapes.

7.3.2 Seed dispersal mutualisms

Dispersal is a crucial process in the ecology and evolution of organisms. Dispersal influences individual fitness, population persistence, and biodiversity across scales, and is key to a population's ability to track changing habitats and adapt to novel environments. There is a well-developed literature documenting how habitat loss and fragmentation affect seed dispersal, often limiting seed dispersal and gene flow and resulting in strong selection on dispersal traits (reviewed in Cheptou et al. 2017). However, there is a lack of studies on the effects of urbanization on dispersal mutualisms in an evolutionary framework, and a similar dearth of studies considering how dispersal mutualisms affect gene flow in urban vs non-urban landscapes. The most striking example documenting differential evolution of dispersal with urbanization does not involve

mutualism. Hawksbeard (*C. sancta*; section 7.3.1; Chapter 8) grows in small patches in urban areas and produces wind-dispersed and non-dispersed seeds; Cheptou et al. (2008) documented rapid evolution of non-dispersing seeds of hawksbeard in urban patches due to a high cost of dispersal (falling onto impervious surface not suitable for germination).

Evolutionary theory provides testable predictions for how urbanization may affect dispersal mutualisms. For example, high costs of dispersal measured as loss or death of propagules should select against traits that confer the mutualism (Table 7.1). In urban areas, the costs of dispersal mutualisms (and the cost:benefit ratio) could change in urban relative to non-urban areas due to partner switching (loss of higher-quality dispersers in urban areas and reliance on lower-quality partners; section 7.2.3), changes in disperser behaviour (due to the human-built environment; section 7.2.4), and complete partner loss and breakdown of the mutualism (section 7.2.5). Partner loss and switching have resulted in the evolution of seed traits in the Atlantic Forest in Brazil, where palms in forest fragments lacking toucans and cotingas (due to fragmentation and hunting) quickly evolved smaller seeds that could be dispersed by the remaining small birds (Galetti et al. 2013). Urban seed characteristics might be expected to change in similar ways as a result of partner loss or switching and warrant further investigation.

Although there are no studies to our knowledge that link changes in cost:benefit ratios of dispersal mutualisms in urban landscapes to evolutionary response, ecological studies have documented urban-mediated changes in the disperser community and behaviour. For example, invasive ants have become numerically dominant in many urban areas (reviewed in Philpott et al. 2009), and these tend to be relatively poor seed dispersers compared to native ants (Ness and Bronstein 2004). The well-studied Argentine ant (*Linepithema humile*), for example, displaces native seed dispersers, resulting in reduced rates of seed removal and seedling establishment (Rodriguez-Cabal et al. 2009). Urbanization can change ant communities even in the absence of invasive ants, which could affect seed removal rates and dispersal distances. One study that measured seed removal rates in urban vs non-urban forests of Manitoba, Canada documented higher seed removal rates of the plant *Viola pubescens* by ants in urban relative to non-urban forests, accompanied by a simplification of the ant community in urban areas; this result may have been driven by a community shift that favoured species that were inherently efficient seed dispersers, or by changes in ant foraging behaviour associated with urbanization (Thompson and McLachlan 2007). In comparison, in some areas, highly disturbed habitats along roads are characterized by reduced ant abundance and distinctly different species composition, with lower seed removal rates and dispersal distances along roads relative to intact sites (Zhu and Wang 2018). Vertebrate-mediated seed dispersal can also be altered in urban habitats. Some bird species experience large increases in population size in urban areas, including corvids and parrots, which can act as effective seed dispersers (Czarnecka et al. 2013; Luna et al. 2018). However, the human-built environment can limit vertebrate movement, and subsequent seed removal rates and seed movement distances (Niu et al. 2018). Given the effects of the urban landscape on dispersal partner switching, behaviour, and loss, this creates significant potential for differential adaptive evolution of dispersal traits in urban vs non-urban sites (Table 7.1), assuming available standing genetic variation on which selection can act.

7.4 Protection mutualisms

Ant defence is the most thoroughly studied protection mutualism from the ecological perspective. There are at least two common types of ant protection. (1) Ant-rewarding plants provide food and, in some cases, nesting space for ants in return for defence against herbivores and competing plants. (2) Ants and other insects provide protection to phloem-feeding insects (trophobionts) in return for honeydew, a sugary substance secreted from the anus or glands. Theory predicts that when ant protectors are lost (section 7.2.5), plant or insect hosts should experience selection for reduced expression of rewards such as food, assuming a cost to reward production (Table 7.1). A few studies in natural systems have suggested that when ants are naturally absent, specialized ant-plants do not produce food rewards (Keeler 1985). Unlike other forms of mutualism,

protection mutualisms require a third party, specifically an enemy (i.e., a predator or herbivore) that can be deterred by the host's mutualists, for their benefit to be realized by the host. Thus, there are additional mechanisms not outlined in section 7.2 by which urbanization could affect protection mutualism evolution via changes in enemy identity or abundance, which appear to be common in urban habitats (Table 7.1, section 7.6.1). For example, if enemy pressure is also low in urban areas, then the absence of ants may select for plants that express no or highly reduced traits to attract and/or reward ants, or ones in which the mutualism itself is inducible (e.g., Ness et al. 2013). Alternatively, if enemy pressure is high or variable in urban areas, the absence of ants may select for other or additional forms of defence, such as secondary chemistry, which may also be inducible (Heil and McKey 2003).

Studies suggest that urban ecosystems differ in ant abundance and composition (Philpott et al. 2009) as well as enemy damage to hosts (Raupp et al. 2010), both of which can drive potential selection on host traits important in protection mutualisms. For example, a number of studies have assessed the effects of the urban landscape on ant abundance, and species richness and composition, and most find differences between urban and nearby non-urban habitats; however, the direction of the effect is not consistent across urban areas. There are studies that find that ant species richness declines with increasing levels of urbanization and decreasing size and increasing age of habitat fragments (e.g., Lessard and Buddle 2005; Buczkowski and Richmond 2012), although these patterns of negative effects of urbanized landscapes on ants are not universal (e.g., Ives et al. 2013; Melliger et al. 2018). Studies of ants in urban habitats typically include all ants, but not all ant species function as protectors. In studies that have focused specifically on ants or ant activity associated with protection mutualisms, some find evidence suggesting negative effects of the urban landscape on plant-defending ants (Rios et al. 2008; but see Rocha and Fellowes 2018). Ant species can vary in their efficacy as protectors of plants and trophobionts (Rico-Gray and Oliveira 2007), and in some cases invasive and urban ants can be equally or more effective protectors than are native species (Ness and Bronstein 2004). Given that urbanization is associated with changes in ant community composition, studies that assess the degree to which the urban landscape is associated with partner switching and impacts on the mutualism may yield important insights. Taken together, these variable effects of urbanization on ant protectors suggest that selection on traits associated with protection mutualisms may also vary (Table 7.1).

Only one study to our knowledge has measured population variation in plant traits associated with a protection mutualism in an urban landscape. Rios et al. (2008) reported lower insect abundance (ant and herbivore) in urban compared to non-urban sites. They used a common garden experiment to show that plants of the annual *Chamaecrista fasciculata* (Figure 7.3) grown from seed collected from

Figure 7.3 Reduced abundance of ants and herbivores in urban populations of *Chamaecrista fasciculata* promotes differential evolution of extrafloral nectar traits associated with protection mutualism. (A) *Chamaecrista fasciculata* (photo source: Wikimedia Commons) and (B) an extrafloral nectary of *C. fasciculata* (photo source: Mary Anne Borge, the-natural-world.org).

urban sites had smaller extrafloral nectaries (the organs that secrete rewards for ant mutualists) and lower extrafloral nectar production than plants grown from seed collected from non-urban sites. For this plant species, decreasing population-level herbivore density and leaf damage was associated with decreases in the volume of nectar and amount of sugar per extrafloral nectary. Moreover, populations with less leaf damage had traits associated with less ant attraction and variation in these traits was high, suggesting relaxation of selection on plant traits that confer the mutualism. In combination with a lack of maternal effects noted in the common garden experiment, this study provides strong evidence of the potential for differential evolution of traits associated with ant rewards in urban vs non-urban sites. This study was conducted in multiple urban and rural populations in the St Louis metropolitan area of Missouri and Illinois, USA. However, given the wide geographic distribution of *C. fasciculata* across eastern and midwestern North America, it provides an exciting opportunity to conduct a parallel evolution study across multiple urban–rural pairs (see Chapter 3). In addition, this study supports the theoretical prediction that reductions in ants and herbivores should reduce selection for traits that confer protection mutualisms. How common this evolutionary response is in other urban landscapes remains to be investigated.

7.5 Nutritional mutualisms

Nutritional mutualisms are essential to ecosystem functioning. Most importantly, plants provide nutrients (nitrogen and phosphorus) to rhizobial bacteria and mycorrhizal fungi, respectively, in exchange for a photosynthate reward. As a rule, mutualists exchange nutrients they have in abundance for those that are unavailable or in short supply (Hoeksema and Schwartz 2003). Nutrients are often highly enriched in urban landscapes, as a consequence of atmospheric deposition, combustion, importation of food, and fertilizer application (Kaye et al. 2006). In soils, macronutrients, such as nitrogen (N) and phosphorus (P), accumulate in some urban ecosystems, depending on soil depth (reviewed in Pickett et al. 2001). For example, in the Central Arizona–Phoenix (CAP) metropolitan ecosystem, USA, total N inputs exceed outputs, with N accumulation estimated at up to 21 000 metric tons per year, from a variety of inputs including food importation, fertilizer application, deposition, pet excretion, and wastewater effluent and groundwater used for irrigation (Baker et al. 2001).

How does this nutrient enrichment in urban ecosystems affect the evolutionary ecology of nutritional mutualisms (Table 7.1)? Theory predicts that nutrient enrichment should ameliorate nutrient limitation. At that point, the cost of maintaining nutrient-providing microorganisms should exceed the benefits they provide, leading in some cases to plants severing connections to their erstwhile mutualists. Alternatively or additionally, selection may favour microorganisms able to retain connection to the plant in a more antagonistic relationship (section 7.2.1). For example, selection may favour more antagonistic microbial genotypes (Thrall et al. 2007) that allocate more resources to internal storage for the microbe compared to providing benefits to the host (Johnson 2010). In the extreme scenario, nutrient enrichment could be disadvantageous to the evolutionary persistence of some nutritional mutualisms (Johnson 2010; Kiers et al. 2010).

Studies of nutritional mutualisms in urban ecosystems have mainly focused on ecological patterns and processes. Given the potential for high nutrient enrichment, heavy-metal accumulation, and limited dispersal characteristics of urban environments, most studies predict (1) a reduced abundance and diversity of partners (and the potential for microbial partner loss), (2) differences in partner composition, and potentially (3) partner switching, and higher cost:benefit ratios for the host (Table 7.1, section 7.2), especially for host–plant interactions with their root associates. In accordance with these predictions, for plant–mycorrhizal fungi mutualisms, studies have documented lower mycorrhizal diversity in urban compared to non-urban sites (Bainard et al. 2011) and differences in fungal community composition (Karpati et al. 2011). Some mycorrhizal fungi fail to sporulate or colonize roots under high nutrient or heavy-metal conditions (Treseder and Allen 2000; Yang et al. 2015). These negative effects of the urban landscape on the mycorrhizal community have the potential to influence the strength of the mutualism. For example, plant response (i.e., growth) to

mycorrhizal fungi is often more positive when the soil community is more diverse, and when nutrients are limiting, especially phosphorus (Hoeksema et al. 2010). However, these effects of the urban landscape on mycorrhizal fungi and their mutualism with plants are not always found (e.g., Tonn and Ibáñez 2017). Such variation in findings among studies is likely related to the magnitude of difference in nutrient availability, soil moisture, light, and plant species composition in the urban vs non-urban sites and the scale of urbanization studied (i.e., large metropolitan areas vs smaller cities and suburban areas).

Despite a growing body of urban ecological studies showing variation in the strength of mutualism, and partner switching and loss, little attention has been paid to the question of whether urban nutrient enrichment influences the evolution of microbial partners. However, insights can be gleaned from the global change and agricultural literature. For example, in an exemplary study using a 22-year N fertilization experiment in an old field successional habitat in Michigan, USA, N addition caused the evolution of less-cooperative microbial mutualists (Figure 7.4; Weese et al. 2015). Rhizobial bacteria housed in the root nodules of leguminous plants convert atmospheric N into biologically active N (in the form of ammonium) in exchange for plant photosynthates. Because N is an important traded nutrient in this mutualism, amelioration of N limitation is expected to select for rhizobia with fewer growth or fitness benefits to the host. Weese et al. (2015) documented that *Trifolium* plants inoculated with rhizobial strains taken from the N-addition treatments produced up to 30 per cent less biomass, 28 per cent fewer leaves, and 21 per cent fewer stolons, and had 17 per cent lower leaf chlorophyll content compared to *Trifolium* inoculated with strains from unfertilized (control) treatments. Moreover, a phylogenetic analysis revealed that the differences observed were due to microevolutionary genetic changes in the rhizobial bacteria and not due to differences in species composition (Weese et al. 2015). A population genomic analysis revealed evolutionary differentiation at a symbiosis gene region (symbiotic plasmid (pSym)) that contributes to partner quality decline (Klinger et al. 2016), suggesting an adaptive response by the rhizobia. The

Figure 7.4 Nitrogen fertilization causes the evolution of less-cooperative rhizobial bacteria in *Trifolium*. (A) *Trifolium repens* (photo source: Wikimedia Commons) and (B) representative root nodules harbouring rhizobial bacteria from a legume (photo source: Wikimedia Commons).

evolutionary responses of rhizobia may be system-specific, however. In a 24-year fertilizer and tillage experiment in a corn–soybean–wheat row crop system, Schmidt et al. (2017) found little evidence of evolutionary change in the net growth benefits that rhizobia provided to soybean. Nonetheless, given that urban landscapes, especially lawns and landscaped vegetation, receive significant amounts of nutrient input, it is reasonable to predict similar adaptive evolutionary responses in urban relative to non-urban rhizobia and other nutritional mutualists. However, no studies to our knowledge have explored such patterns in urban landscapes.

The potential for evolution of partners and hosts in nutritional mutualisms is not limited to plants and their microbial associates. There is growing evidence that, in a wide diversity of animals, nutritional endosymbionts can alter host evolution and potentially host speciation, with effects dependent on the environment (Table 7.1; Shapira 2016). Striking examples include microbial-dependent agrochemical detoxification, enhanced heat tolerance, and parasite resistance in host insects, allowing for host adaptation. For example, aphids harbour the obligate symbiont *Buchnera*, which supplies essential nutrients to aphids. Dunbar et al. (2007) documented that a single-nucleotide mutation in a heat-shock transcriptional promoter of *Buchnera* results in differential aphid heat tolerance and aphid fitness. These heat-tolerant and intolerant *Buchnera* strains also occur in field populations, suggesting the potential for temperature-mediated selection on the holobiont (aphid and *Buchnera*) in the field. Such temperature-dependent symbiont effects could have important implications for host adaptation to urban heat islands. In an additional example, stinkbugs harbour the gut symbiont *Burkholderia* in their midguts that they acquire from soil in their second instar. These bacteria can confer pesticide resistance in hosts, increasing host fitness, and amplification of the bacteria in specialized host organs, allowing the potential for host adaptation to the agrochemical environment (Kikuchi et al. 2012). At least 78 million households in the USA use home and garden pesticides (Kiely et al. 2004), and suburban lawns and gardens can receive more pesticide applications per acre than some agricultural fields. This creates significant potential for microbial-dependent host adaptation in urban environments and warrants further investigation.

7.6 Future directions

Because few studies have compared the evolutionary ecology and evolution of mutualisms in urban vs non-urban sites, or along urban gradients, many basic questions remain unanswered. This also means that exciting challenges remain. At a basic level, there is a need for studies that test whether and why differential genetic changes occur in urban vs non-urban mutualisms, and for studies that focus on heritable phenotypic traits using common gardens. Moreover, studies are needed that examine mutualisms in urban landscapes from both partners' perspectives. Beyond this basic level, we identify three areas of research to move the field forward.

7.6.1 Do mutualisms respond differently (ecologically and evolutionarily) to urbanization than do other species interactions?

There are many other types of species interactions beyond mutualisms that play critical roles in natural communities, including herbivory, parasitism, predation, and competition. Research is beginning to show that they too may function and evolve differently in urban settings (Alberti et al. 2017). Herbivory often increases with urbanization, driven by changes in factors such as temperature and habitat complexity (Raupp et al. 2010). By contrast, predation risk is often reduced in urban landscapes, at least for some vertebrate taxa (Vincze et al. 2017), although the importance of urban pets as predators cannot be overlooked (Loss and Marra 2017). In one study, impacts of predators and parasitoids on a herbivorous insect were overwhelmed by bottom-up factors, such as increased temperature (Dale and Frank 2014). For songbirds, some authors have suggested that predation rates may be decoupled from predator densities by elevated levels of anthropogenic food resources (Rodewald et al. 2011). Parasitic interactions show many contrasting findings. Transmission of parasites on wildlife has been observed to increase in urban areas (Bradley and Altizer 2007). However, parasite loads in birds are sometimes higher and sometimes lower in urban areas relative to non-urban areas (Geue and Partecke 2008; Giraudeau et al. 2014).

An important direction for future research is to explore whether responses to urbanization are predictably different across different species interactions. In particular, are mutually beneficial interactions more at risk than are ones that benefit only one of the two partner species (commensalism, antagonism) or that involve reciprocal antagonism (competition)? Of course, other features might be more important predictors of the trajectories and fates of interactions in an urban landscape. For example, the degree of specialization might be more predictive of interaction evolution than is the beneficial or antagonistic nature of the interaction (section 7.6.2); similarly, highly context-dependent interactions might be more resilient evolutionarily than ones with more fixed outcomes, regardless of the interaction's natural history. These are critical questions for the future, if we

hope to build a broader perspective of the evolution of interactions as a whole in urban environments.

7.6.2 What forms of mutualism will be most affected evolutionarily by urbanization?

Beyond asking whether mutualisms will evolve differently in urban settings, we should also explore whether certain mutualisms will respond differently from others. In this chapter, we structured our discussion around three broad categories of mutualism: transportation (with a particular focus on pollination and seed dispersal), protection (especially focusing on ant protection of plants), and nutrition (mostly focusing on plant symbioses with root bacteria and fungi). As we have highlighted, there is a growing literature on the effects of urbanization on each of these. However, different research groups have studied each of these forms of mutualism, often with different questions in mind. For example, the rarity of mutualisms in urban systems and the evolutionary consequences of this rarity have attracted attention with regard to pollination (section 7.3.1) and, to a lesser extent, ant protection (section 7.4), but has barely been studied in seed dispersal mutualisms. Similarly, evolutionary consequences of nutrient enrichment have been explored in nutrition mutualisms, albeit not specifically in urban ecosystems, but minimally in other mutualisms. The resulting patchwork of knowledge across different combinations of urban drivers and interaction types makes it difficult to align our understanding across all of the forms of mutualism. We simply do not yet know which mutualisms are most at risk, nor which of them shows more evidence for each of the mechanisms of change introduced in section 7.2.

Simultaneous, comparative studies across different mutualisms, ideally conducted along the same urbanization gradient (or even better: different mutualisms followed across replicated urbanization gradients), would be a major step forward. A particularly interesting route would be to focus on individual plant species that interact with different guilds of mutualists simultaneously, e.g., pollinators, seed dispersers, and root symbionts. An interesting complication is that different mutualisms with a shared partner can themselves interact and influence each other's evolution. Relatively few such integrative, comparative studies of mutualism evolution have yet been conducted in either urbanized or natural settings (but see, for example, Dutton et al. 2016). To our knowledge, no study has yet asked whether different mutualisms involving the same plant species evolve differently in response to any form of anthropogenic change, including urbanization.

For the purposes of our discussion, we chose to categorize mutualisms based on the nature of the benefits exchanged, but other partitions might be equally or even more informative. Rafferty et al. (2015) argued that the risk of a temporal mismatch between mutualists—one pervasive threat to persistence of mutualism (section 7.3.1)—is better predicted by the obligacy, specificity, seasonality, and intimacy of the interaction than by the nature of the benefit exchanged. In particular, they predicted and then offered early evidence that a risk of phenological mismatch should be particularly high in four kinds of mutualisms: those that (1) are free-living rather than symbiotic; (2) do not co-disperse as a unit; (3) are brief and seasonal; and (4) are facultative and generalized, rather than obligate and species-specific. Do these patterns—which, as Rafferty et al. (2015) point out, are biased towards pollination systems—hold true for risks associated with other threats posed by anthropogenic change, including urbanization? We suspect that the answer is no. For example, highly specialized mutualisms commonly exhibit finely tuned features that reduce the risk of disassociation in unpredictable environments. Flowers of some plants with highly specialized pollination systems, for instance, release species-specific volatiles detectable by obligate pollinators over extraordinarily long distances (Hossaert-McKey et al. 2010). Yet, these same mutualisms are likely to be those that will suffer the most from the regional absence of a partner, and are likely to experience the strongest trait selection in response to such loss.

7.6.3 Is urbanization a unique evolutionary threat for mutualisms?

Building from the last point, an essential open question is whether mutualisms that are particularly fragile in urban settings are also those most susceptible to other anthropogenic risks. Many of the abiotic changes associated with urbanization, including hotter temperatures and habitat fragmentation, are becoming increasingly evident in non-urbanized settings as well. We suggest that the fine-grained spatial structure of cities, as well as their many co-varying disturbances such as warming and air pollution, likely produce unique selective landscapes that are qualitatively different from those in 'natural' systems—even as they, too, experience some of the same drivers of global change. Future work should compare the spatial and temporal scales over which selection fluctuates in replicated urban and non-urban environments (see also Chapters 2 and 3). It should also examine the consistency of realized trait evolution between cities of different ages and sizes to determine whether urbanization produces consistent ecotypes. In addition, human activities, such as bee-keeping, gardening for pollinators, and use of pesticides, could promote or dampen evolutionary responses in mutualisms. Studies are needed that explicitly address the direct and indirect effects of human activities as part of the evolutionary ecology of mutualisms in urban landscapes. Taken together, these future directions should spur progress toward understanding what is unique, general, and predictable about the evolution of mutualism in urban environments.

Acknowledgements

The authors thank the National Science Foundation (DEB-1354061/1641243 and DEB-0743535) and the USDA NIFA (Hatch project 1018689) Any opinions, findings, and conclusions or recommendations expressed in this material are those of the authors and do not necessarily reflect the views of the funding agency.

References

Alberti, M., Correa, C., Marzluff, J.M., et al. (2017). Global urban signatures of phenotypic change in animal and plant populations. *Proceedings of the National Academy of Sciences of the United States of America*, 114, 8951–6.

Andrieu, E., Dornier, A., Rouifed, S., Schatz, B., and Cheptou, P.-O. (2009). The town *Crepis* and the country *Crepis*: how does fragmentation affect a plant–pollinator interaction? *Acta Oecologica*, 35, 1–7.

Aronson, M.F. J., La Sorte, F.A., Nilon, C.H., et al. (2014). A global analysis of the impacts of urbanization on bird and plant diversity reveals key anthropogenic drivers. *Proceedings of the Royal Society B: Biological Sciences*, 281, 20133330.

Ashman, T.-L. and Morgan, M.T. (2004). Explaining phenotypic selection on plant attractive characters: male function, gender balance or ecological context? *Proceedings of the Royal Society B: Biological Sciences*, 271, 553–9.

Ashman, T.-L., Knight, T.M., Steets, J.A., et al. (2004). Pollen limitation of plant reproduction: ecological and evolutionary causes and consequences. *Ecology*, 85, 2408–21.

Bainard, L.D., Klironomos, J.N., and Gordon, A.M. (2011). The mycorrhizal status and colonization of 26 tree species growing in urban and rural environments. *Mycorrhiza*, 21, 91–6.

Baker, L.A., Hope, D., Xu, Y., Edmonds, J., and Lauver, L. (2001). Nitrogen balance for the Central Arizona-Phoenix (CAP) ecosystem. *Ecosystems*, 4, 582–602.

Baldock, K.C., Goddard, M.A., Hicks, D.M., et al. (2015). Where is the UK's pollinator biodiversity? The importance of urban areas for flower-visiting insects. *Proceedings of the Royal Society B: Biological Sciences*, 282, 20142849.

Bartlewicz, J., Vandepitte, K., Jacquemyn, H., and Honnay, O. (2015). Population genetic diversity of the clonal self-incompatible herbaceous plant *Linaria vulgaris* along an urbanization gradient. *Biological Journal of the Linnean Society*, 116, 603–13.

Bode, R.F. and Tong, R. (2017). Pollinators exert positive selection on flower size on urban, but not rural Scotch broom (*Cytisus scoparius* L. Link). *Journal of Plant Ecology*, 11, 493–501.

Bradley, C.A. and Altizer, S. (2007). Urbanization and the ecology of wildlife diseases. *Trends in Ecology and Evolution*, 22, 95–102.

Bronstein, J.L. (2015). *Mutualism*. Oxford University Press, Oxford.

Buczkowski, G. and Richmond, D.S. (2012). The effect of urbanization on ant abundance and diversity: a temporal examination of factors affecting biodiversity. *PLOS ONE*, 7, e41729.

Cane, J.H., Minckley, R.L., Kervin, L.J., and Williams, N.M. (2006). Complex responses within a desert bee guild (Hymenoptera: Apiformes) to urban habitat fragmentation. *Ecological Applications*, 16, 632–44.

Cheptou, P.-O., Avendano, V., and Lyz, G. (2006). Pollination processes and the Allee effect in highly fragmented populations: consequences for the mating system in urban environments. *New Phytologist*, 172, 774–83.

Cheptou, P.-O., Carrue, O., Rouifed, S., and Cantarel, A. (2008). Rapid evolution of seed dispersal in an urban environment in the weed *Crepis sancta*. *Proceedings of the National Academy of Sciences of the United States of America*, 105, 3796–9.

Cheptou, P.-O., Hargreaves, A. L., Bonte, D., and Jacquemyn, H. (2017). Adaptation to fragmentation: evolutionary dynamics driven by human influences. *Philosophical Transactions of the Royal Society of London B*, 372, 20160037.

Chomicki, G. and Renner, S.S. (2017). Partner abundance controls mutualism stability and the pace of morphological change over geological time. *Proceedings of the National Academy of Sciences of the United States of America*,114, 3951–6.

Clemmensen, K.E., Finlay, R.D., Dahlberg, A., et al. (2015). Carbon sequestration is related to mycorrhizal fungal community shifts during long-term succession in boreal forests. *New Phytologist*, 205, 1525–36.

Culley, T.M., Sbita, S.J., and Wick, A. (2007). Population genetic effects of urban habitat fragmentation in the perennial herb *Viola pubescens* (Violaceae) using ISSR markers. *Annals of Botany*, 100, 91–100.

Czarnecka, J., Kitowski, I., Sugier, P., et al. (2013). Seed dispersal in urban green space—does the rook *Corvus frugilegus* L. contribute to urban floral homogenization. *Urban Forestry & Urban Greening*, 12, 359–66.

Dale, A.G. and Frank, S.D. (2014). Urban warming trumps natural enemy regulation of herbivorous pests. *Ecological Applications*, 24, 1596–607.

Dunbar, H.E., Wilson, A.C.C., Ferguson, N.R., and Moran, N.A. (2007). Aphid thermal tolerance is governed by a point mutation in bacterial symbionts. *PLOS Biology*, 5, e96.

Dutton, E.M., Luo, E.Y., Cembrowski, A.R., Shore, J.S., and Frederickson, M.E. (2016). Three's a crowd: tradeoffs between attracting pollinators and ant bodyguards with nectar rewards in *Turnera*. *The American Naturalist*, 188, 38–51.

Eckert, C.G., Kalisz, S., Geber, M.A., et al. (2010). Plant mating systems in a changing world. *Trends in Ecology & Evolution*, 25, 35–43.

Fortel, L., Henry, M., Guilbaud, L., et al. (2014). Decreasing abundance, increasing diversity and changing structure of the wild bee community (Hymenoptera: Anthophila) along an urbanization gradient. *PLOS ONE*, 9, e104679.

Fuentes, J.D., Chamecki, M., Roulston, T., Chen, B., and Pratt, K.R. (2016). Air pollutants degrade floral scents and increase insect foraging times. *Atmospheric Environment*, 141, 361–74.

Galetti, M., Guevara, R., Côrtes, M.C., et al. (2013). Functional extinction of birds drives rapid evolutionary changes in seed size. *Science*, 340, 1086–90.

Geerts, S. and Pauw, A. (2012). The cost of being specialized: pollinator limitation in the endangered geophyte *Brunsvigia litoralis* (Amaryllidaceae) in the Cape Floristic region of South Africa. *South African Journal of Botany*, 78, 159–64.

Geue, D. and Partecke, J. (2008). Reduced parasite infestation in urban Eurasian blackbirds (*Turdus merula*): a factor favoring urbanization? *Canadian Journal of Zoology/Revue Canadienne de Zoologie*, 86, 1419–25.

Giraudeau, M., Mousel, M., Earl, S., and McGraw, K. (2014). Parasites in the city: degree of urbanization predicts poxvirus and coccidian infections in house finches (*Haemorhous mexicanus*). *PLOS One*, 9, e86747.

Gómez, J., Perfectti, F., Bosch, J., and Camacho, J. (2009). A geographic selection mosaic in a generalized plant–pollinator–herbivore system. *Ecological Monographs*, 79, 245–63.

Gotlieb, A., Hollender, Y., and Mandelik, Y. (2011). Gardening in the desert changes bee communities and pollination network characteristics. *Basic and Applied Ecology*, 12, 310–20.

Grimm, N.B., Faeth, S.H., Golubiewski, N.E., et al. (2008). Global change and the ecology of cities. *Science*, 319, 756–60.

Harder, L.D. and Aizen, M.A. (2010). Floral adaptation and diversification under pollen limitation. *Philosophical Transactions of the Royal Society of London B*, 365, 529–43.

Harrison, T. and Winfree, R. (2015). Urban drivers of plant–pollinator interactions. *Functional Ecology*, 29, 879–88.

Harrison, T., Gibbs, J., and Winfree, R. (2018). Forest bees are replaced in agricultural and urban landscapes by native species with different phenologies and life-history traits. *Global Change Biology*, 24, 287–96.

Heil, M. and McKey, D. (2003). Protective ant–plant interactions as model systems in ecological and evolutionary research. *Annual Review of Ecology and Systematics*, 34, 425–53.

Hoeksema, J.D. and Bruna, E.M. (2015). Context-dependent outcomes of mutualistic interactions. In: Bronstein, J.L. (ed.) *Mutualism*. Oxford University Press, Oxford.

Hoeksema, J.D. and Schwartz, M.W. (2003). Expanding comparative-advantage biological market models: contingency of mutualism of partners' resource requirements and acquisition trade-offs. *Proceedings of the Royal Society B: Biological Sciences*, 270, 913–19.

Hoeksema, J.D., Chaudhary, V.B., Gehring, C.A., et al. (2010). A meta-analysis of context-dependency in plant response to inoculation with mycorrhizal fungi. *Ecology Letters*, 13, 394–407.

Hossaert-McKey, M., Soler, C., Schatz, B., and Profitt, M. (2010). Floral scents: their role in nursery pollination systems. *Chemoecology*, 20, 75–88.

Irwin, R.E., Warren, P.S., Carper, A.L., and Adler, L.S. (2014). Plant–animal interactions in suburban environments: implications for floral evolution. *Oecologia*, 174, 803–15.

Irwin, R.E., Warren, P.S., and Adler, L.S. (2018). Phenotypic selection on floral traits in an urban landscape. *Proceedings of the Royal Society B: Biological Sciences*, 285, 20181239.

Ives, C.D., Taylor, M.P., Nipperess, D.A., and Hose, G.C. (2013). Effect of catchment urbanization on ant diversity in remnant riparian corridors. *Landscape and Urban Planning*, 110, 155–63.

Jha, S. and Kremen, C. (2013a). Resource diversity and landscape-level homogeneity drive native bee foraging. *Proceedings of the National Academy of Sciences of the United States of America*, 110, 555–8.

Jha, S. and Kremen, C. (2013b). Urban land use limits regional bumble bee gene flow. *Molecular Ecology*, 22, 2483–95.

Johnson, M.T.J. and Munshi-South, J. (2017). Evolution of life in urban environments. *Science*, 358, eaam8327.

Johnson, M.T., Prashad, C.M., Lavoignat, M., and Saini, H.S. (2018). Contrasting the effects of natural selection, genetic drift and gene flow on urban evolution in white clover (*Trifolium repens*). *Proceedings of the Royal Society B: Biological Sciences*, 285, 20181019.

Johnson, N.C. (2010). Resource stoichiometry elucidates the structure and function of arbuscular mycorrhizas across scales. *New Phytologist*, 185, 631–47.

Jürgens, A. and Bischoff, M. (2017). Changing odour landscapes: the effect of anthropogenic volatile pollutants on plant–pollinator olfactory communication. *Functional Ecology*, 31, 56–64.

Karpati, A.S., Handel, S.N., Dighton, J., and Horton, T.R. (2011). *Quercus rubra*-associated ectomycorrhizal fungal communities of disturbed urban sites and mature forests. *Mycorrhiza*, 21, 537–47.

Kaye, J.P., Groffman, P.M., Grimm, N.B., Baker, L.A., and Pouyat, R.V. (2006). A distinct urban biogeochemistry. *Trends in Ecology and Evolution*, 21, 192–9.

Keeler, K.H. (1985). Extrafloral nectaries on plants in communities without ants: Hawaii. *Oikos*, 44, 407–14.

Kiely, T., Donaldson, D., and Grube, A. (2004). Pesticides industry sales and usage: 2000 and 2001 market estimates. EPA-733-R-04-001. US-EPA, Washington, DC.

Kiers, E.T., Palmer, T.M., Ives, A.R., Bruno, J.F., and Bronstein, J.L. (2010). Mutualisms in a changing world: an evolutionary perspective. *Ecology Letters*, 13, 1459–74.

Kikuchi, Y., Hayatsu, M., Hosokawa, T., et al. (2012). Symbiont-mediated insecticide resistance. *Proceedings of the National Academy of Sciences of the United States of America*, 109, 8618–22.

Klein, A.-M., Vaissière, B.E., Cane, J.H., et al. (2006). Importance of pollinators in changing landscapes for world crops. *Proceedings of the Royal Society B: Biological Sciences*, 274, 303–13.

Klinger, C.R., Lau, J.A., and Heath, K.D. (2016). Ecological genomics of mutualism decline in nitrogen-fixing bacteria. *Proceedings of the Royal Society B: Biological Sciences*, 283, 20152563.

Leong, M., Kremen, C., and Roderick, G.K. (2014). Pollinator interactions with yellow starthistle (*Centaurea solstitialis*) across urban, agricultural, and natural landscapes. *PLOS ONE*, 9, e86357.

Lessard, J.P. and Buddle, C.M. (2005). The effects of urbanization on ant assemblages (Hymenoptera: Formicidae) associated with the Molson Nature Reserve, Quebec. *Canadian Entomologist*, 137, 215–25.

Lloyd, D.G. and Barrett, S.C.H. (1996). *Floral Biology: Studies on Floral Evolution in Animal-Pollinated Plants*. Chapman and Hall, New York.

López-Uribe, M.M., Morreale, S.J., Santiago, C.K., and Danforth, B.N. (2015). Nest suitability, fine-scale population structure and male-mediated dispersal of a solitary ground nesting bee in an urban landscape. *PLOS ONE*, 10, e0125719.

Loss, S.R. and Marra, P.P. (2017). Population impacts of free-ranging domestic cats on mainland vertebrates. *Frontiers in Ecology and the Environment*, 15, 502–9.

Luna, A., Romero-Vidal, P., Hiraldo, F., and Tella, J.L. (2018). Cities may save some threatened species but not their ecological functions. *PeerJ*, 6, e4908.

Martins, K.T., Gonzalez, A., and Lechowicz, M.J. (2017). Patterns of pollinator turnover and increasing diversity associated with urban habitats. *Urban Ecosystems*, 20, 1359–71.

Mckinney, M.L. (2008). Effects of urbanization on species richness: a review of plants and animals. *Urban Ecosystems*, 11, 161–76.

Melliger, R.L., Braschler, B., Rusterholz, H.-P., and Baur, B. (2018). Diverse effects of degree of urbanisation and forest size on species richness and functional diversity of plants, and ground surface-active ants and spiders. *PLOS ONE*, 13, e0199245.

Minnefors, A. (2016). *Pollen Limitation in the City: Measuring Floral Traits and Pollination Services in Urban* Chamerion angustifolium. Uppsala University, Sweden.

Ness, J.H. and Bronstein, J.L. (2004). The effect of invasive ants on prospective ant mutualists. *Biological Invasions*, 6, 445–61.

Ness, J.H., Morales, M.A., Kenison, E., et al. (2013). Reciprocally beneficial interactions between introduced plants and ants are induced by the presence of a third introduced species. *Oikos*, 122, 695–704.

Niemela, J. and Breuste, J.H. (2011). *Urban Ecology: Patterns, Processes, and Applications*. Oxford University Press, Oxford.

Niu, H.-Y., Xing, J.-J., Zhang, H.-M., Wang, D., and Wang, X.-R. (2018). Roads limit of seed dispersal and seedling recruitment of *Quercus chenii* in an urban hillside forest. *Urban Forestry & Urban Greening*, 30, 307–14.

Parker, I. M. (1997). Pollinator limitation of *Cytisus scoparius* (Scotch broom), an invasive exotic shrub. *Ecology*, 78, 1457–70.

Philpott, S.M., Perfecto, I., Armbrecht, I., and Parr, C.L. (2009). Ant diversity and function in disturbed and changing habitats. In: Lach, L., Parr, C.L., and Abbott, K.L. (eds) *Ant Ecology*. Oxford University Press, Oxford.

Pickett, S.T.A., Cadenasso, M.L., Grove, J.M., et al. (2001). Urban ecological systems: linking terrestrial ecological,

physical, and socioeconomic components of metropolitan areas. *Annual Review of Ecology and Systematics*, 32, 127–57.

Rafferty, N.E., Caradonna, P.J., and Bronstein, J.L. (2015). Phenological shifts and the fate of mutualisms. *Oikos*, 124, 14–21.

Raupp, M.J., Shrewsbury, P.M., and Herms, D.A. (2010). Ecology of herbivorous arthropods in urban landscapes. *Annual Review of Entomology*, 55, 19–38.

Rico-Gray, V. and Oliveira, P.S. (2007). *The Ecology and Evolution of Ant–Plant Interactions*. University of Chicago Press, Chicago.

Riffell, J.A., Schlizerman, E., Sanders, E., et al. (2014). Flower discrimination by pollinators in a dynamic chemical environment. *Science*, 344, 1515–18.

Rios, R.S., Marquis, R.J., and Flunker, J.C. (2008). Population variation in plant traits associated with ant attraction and herbivory in *Chamaecrista fasciculata* (Fabaceae). *Oecologia*, 156, 577–88.

Rivkin, L.R., Santangelo, J.S., Alberti, M., et al. (2019). A roadmap for urban evolutionary ecology. *Evolutionary Applications*, 12, 384–98.

Rocha, E.A. and Fellowes, M.D.E. (2018). Does urbanization explain differences in interactions between an insect herbivore and its natural enemies and mutualists? *Urban Ecosystems*, 21, 405–17.

Rodewald, A.D., Kearns, L.J., and Shustack, D.P. (2011). Anthropogenic resource subsidies decouple predator–prey relationships. *Ecological Applications*, 21, 936–43.

Rodriguez-Cabal, M.A., Stuble, K.L., Nunez, M.A., and Sanders, N.J. (2009). Quantitative analysis of the effects of the exotic Argentine ant on seed-dispersal mutualisms. *Biology Letters*, 5, 499–502.

Roels, S.A. and Kelly, J.K. (2011). Rapid evolution caused by pollinator loss in *Mimulus guttatus*. *Evolution*, 65, 2541–52.

Sachs, J.L., Mueller, U.G., Wilcox, T.P., and Bull, J.J. (2004). The evolution of cooperation. *Quarterly Review of Biology*, 79, 135–60.

Schmidt, J.E., Weese, D.J., and Lau, J.A. (2017). Long-term agricultural management does not alter the evolution of a soybean–rhizobium mutualism. *Ecological Applications*, 27, 2487–96.

Schupp, E.W., Jordano, P., and Gómez, J.M. (2017). A general framework for effectiveness concepts in mutualisms. *Ecology Letters*, 20, 577–90.

Shapira, M. (2016). Gut microbiotas and host evolution: scale up symbiosis. *Trends in Ecology & Evolution*, 31, 539–49.

Theodorou, P., Radzeviciute, R., Kahnt, B., Soro, A., Grosse, I., and Paxton, R.J. (2018). Genome-wide single nucleotide polymorphism scan suggests adaptation to urbanization in an important pollinator, the red-tailed bumblebee (*Bombus lapidarius* L.). *Proceedings of the Royal Society B: Biological Sciences*, 285, 20172806.

Thompson, B. and McLachlan, S. (2007). The effects of urbanization on ant communities and myrmecochory in Manitoba, Canada. *Urban Ecosystems*, 10, 43–52.

Thompson, J.N. and Cunningham, B.M. (2002). Geographic structure and dynamics of coevolutionary selection. *Nature*, 417, 735–8.

Thrall, P.H., Hochberg, M.E., Burdon, J.J., and Bever, J.D. (2007). Coevolution of symbiotic mutualists and parasites in a community context. *Trends in Ecology & Evolution*, 22, 120–6.

Tonn, N. and Ibáñez, I. (2017). Plant–mycorrhizal fungi associations along an urbanization gradient: implications for tree seedling survival. *Urban Ecosystems*, 20, 823–37.

Treseder, K.K. and Allen, M.F. (2000). Mycorrhizal fungi have a potential role in soil carbon storage under elevated CO_2 and nitrogen deposition. *New Phytologist*, 147, 189–200.

Ushimaru, A., Kobayashi, A., and Dohzono, I. (2014). Does urbanization promote floral diversification? Implications from changes in herkogamy with pollinator availability in an urban–rural area. *The American Naturalist*, 184, 258–67.

Verboven, H.A., Brys, R., and Hermy, M. (2012). Sex in the city: reproductive success of *Digitalis purpurea* in a gradient from urban to rural sites. *Landscape and Urban Planning*, 106, 58–164.

Vincze, E., Seress, G., Lagisz, M., et al. (2017). Does urbanization affect predation of bird nests? A meta-analysis. *Frontiers in Ecology and Evolution*, 5, 29.

Wee, A.K., Low, S.Y., and Webb, E.L. (2015). Pollen limitation affects reproductive outcome in the bird-pollinated mangrove *Bruguiera gymnorrhiza* (Lam.) in a highly urbanized environment. *Aquatic Botany*, 120, 240–3.

Weese, D.J., Heath, K.D., Dentinger, B.T.M., and Lau, J.A. (2015). Long-term nitrogen addition causes the evolution of less-cooperative mutualists. *Evolution*, 69, 631–2.

Werner, G.D.A., Cornelissen, J.H.C., Cornwell, W.K., et al. (2018). Symbiont switching and alternate resource acquisition strategies drive mutualism breakdown. *Proceedings of the National Academy of Sciences of the United States of America*, 115, 5229–34.

West, S.A., Kiers, E.T., Simms, E.L., and Denison, R.F. (2002). Sanctions and mutualism stability: why do rhizobia fix nitrogen? *Proceedings of the Royal Society B: Biological Sciences*, 269, 685–94.

Williams, N.M. and Winfree, R. (2013). Local habitat characteristics but not landscape urbanization drive pollinator visitation and native plant pollination in forest remnants. *Biological Conservation*, 160, 10–18.

Yakub, M. and Tiffin, P. (2017). Living in the city: urban environments shape the evolution of a native annual plant. *Global Change Biology*, 23, 2082–9.

Yang, Y., Song, Y., Scheller, H.V., et al. (2015). Community structure of arbuscular mycorrhizal fungi associated with *Robinia pseudoacacia* in uncontaminated and heavy metal contaminated soils. *Soil Biology and Biochemistry*, 86, 146–58.

Zhu, Y. and Wang, D. (2018). Response of ants to human-altered habitats with reference to seed dispersal of the myrmecochore *Corydalis giraldii* Fedde (Papaveraceae). *Nordic Journal of Botany*, 36, e01882.

CHAPTER 8

Sidewalk Plants as a Model for Studying Adaptation to Urban Environments

Pierre Olivier Cheptou and Susan C. Lambrecht

Cheptou, P.O. and Lambrecht, S.C., *Sidewalk Plants as a Model for Studying Adaptation to Urban Environments* In: *Urban Evolutionary Biology*. Edited by Marta Szulkin, Jason Munshi-South and Anne Charmantier, Oxford University Press (2020). © Oxford University Press.
DOI: 10.1093/oso/9780198836841.003.0008

8.1 Introduction

Urban plants are as old as cities themselves. While city parks with trees and ornamental plants have been introduced into urban settings, the idea that ecological or evolutionary processes can occur in urban environments is, by nature, outside the concept of urban planning. Yet, it is possible to find wild plants in urban habitats growing independently of human management. Such plants are subject to urban ecological constraints. Though rarely investigated, naturalists have documented lists of urban species in the past, for instance in Paris, France, in the nineteenth century (Lizet 1997). Cities are also the crossroads of commercial exchange, which can accidentally supply a source of exotic species, as was the case in the nineteenth century in Montpellier, France (Godron 1854). However, these early botanical investigations remain descriptive, with little concern for ecological dynamics (population viability) or evolutionary processes (population adaptation). In an era of conservation, ecologists have realized that urban flora potentially provide a reservoir of biodiversity, whose richness is sometimes higher than that of surrounding agricultural fields. Ecological dynamics of urban plants have been investigated only in a few cases (Dornier et al. 2011) and we have little evidence of whether urban populations are viable or just sinks supplied by countryside sources. Ironically, we have mounting evidence that contemporary evolution is acting in cities in response to ecological specificities such as fragmentation or reduced pollinator diversity and abundance (see also Chapter 7). For instance, in Belgium, Brys and Jacquemyn (2012) have demonstrated that urban populations of the common centaury (*Centaurium erythraea*) have recently evolved the ability to self-fertilize as a response to pollinator impoverishment. Urban noise has also been shown to cause plastic changes in mating songs in urban populations of great tits (*Parus major*) (Slabbekoorn and Peet 2003).

The ability of species to adapt rapidly to the urban environment, as shown in several empirical studies (see Johnson and Munshi-South 2017 for review), may appear somewhat paradoxical with the classical theory of adaptation. Indeed, evolutionary changes require either standing variation in populations or *de novo* mutations. The maintenance of standing variation is favoured by large effective population sizes minimizing genetic drift. Also, the appearance of new beneficial mutations is more likely when population sizes are large. The small population sizes and supposedly reduced gene flow in urban populations (Johnson and Munshi-South 2017) are thus expected to make adaptation more difficult in urban environments. The proven examples of rapid evolution in urban environments (Johnson and Munshi-South 2017) show, however, that both selection and genetic variance of traits can be present in urban populations.

Moreover, we think that the urban environment has the potential to inspire new evolutionary studies for several reasons. The urban environment is an atypical environment for plants; for instance, high fragmentation, with little equivalent in natural systems, contrasts with the surrounding countryside and potentially generates divergent selection at a small spatial scale. Because many cities around the world share a number of characteristics, this opens the possibility to analyse if there is convergent evolution as plants face supposedly similar selection pressures (Chapter 3; Alberti et al. 2017). Interestingly, Grimm et al. (2008) have proposed that the urban environment encompasses five major characteristics of global change (land-use and land-cover change, altered biogeochemical cycles, climate change, human modifications of hydrologic systems, and biodiversity changes), which suggests that cities could provide a 'laboratory' to analyse the impact of global change on biodiversity and adaptation. In line with this, urban studies may also help us understand ecological processes, such as dynamics of small populations (extinction, role of stochasticity (Caughley 1994)), that are identified as important drivers of biodiversity dynamics and adaptation. Providing that genetic variance of traits involved in adaptation is still present in urban populations, we expect rapid adaptation in a few generations. In their review, Johnson and Munshi-South (2017) have reported examples of evolution in various types of organisms such as viruses, plants, insects, fish, amphibians and reptiles, birds, and mammals. In particular, a recent study in the plant *Lepidium virginicum* across five North American cities has revealed convergent evolution towards faster growth, larger size, and earlier flowering (Yakub and Tiffin 2017).

In this chapter, we report the results of an empirical research programme studying plants growing in small patches around street trees on pavements (sidewalks), a typical plant habitat in urban environments. More specifically, we studied life-history traits, physiological change, and adaptation in urban patchy populations in the annual plant *Crepis sancta* (hawksbeard, Asteraceae), a widespread weed in the south of France. While the focal habitat of our study is the urban patch, our study highlights that the diversity of urban habitats may lead to heterogeneous selection in the city.

8.2 The sidewalk plants model

8.2.1 Taking advantage of the urban geometry

As human constructs, cities have a rather regular geometry, with straight lines and ordered patterns. As a consequence, the spatial patterns of urban ecosystems differ somewhat from natural ecosystems per se, thus providing an interesting feature for ecological experiments. As a typical urban habitat, we studied plants living in small patches around trees on sidewalks (Figure 8.1). These habitats provide several interesting aspects for plant ecologists. First, they constitute highly fragmented habitats surrounded by a concrete matrix. Sidewalks are thus a binary habitat where only patches are suitable for plant establishment. Second, they are akin to a formal experimental design, since they are of constant surface, regularly spaced, and numerous in cities. Third, these patches are drier and hotter than rural populations, and thus represent an interesting model for investigating the effect of drought stress and elevated temperature (Lambrecht et al. 2016). Fourth, they are not under intentional human pressure, as plants that have colonized such habitats have escaped urban planning. In Montpellier, more than 100 plant species can be found in this habitat (P.O. Cheptou, unpublished data). Such species are often annual and ruderal plants. Among the most frequent are annual meadow grass (*Poa annua*), chickweed (*Stellaria media*), and common groundsel (*Senecio vulgaris*). The study species, *Crepis sancta*, is among the five most frequent species in the city of Montpellier.

(A)

(B)

Figure 8.1 Crepis sancta in urban patches. (A) In the city centre of Montpellier, southern France (Antigone district) (photos courtesy of G. Przetak), the species grows in small patches around trees. (B) The species produces two types of fruits (achenes): (left) heavy achenes deprived of pappus at the periphery of the inflorescence surrounded by floral tissue, and (right) light achenes with pappus at the centre of the inflorescence (drawing courtesy of R. Ferris).

8.2.2 *Crepis sancta* along the rural–urban gradient

Crepis sancta is an annual Mediterranean species. Seeds germinate in the autumn with rainfall. Plants grow during winter and start to flower in early spring, and disperse achenes in early April–May. It is an allogamous species (Cheptou et al. 2002), partially self-incompatible. Of interest, *C. sancta* produces large, light-coloured, non-dispersing fruits (achenes) lacking a pappus located at the periphery of the inflorescence, as well as small, brown-coloured, dispersing fruits with a pappus at the centre of the infloresence (Imbert et al. 1996). The relative proportion of pappus-bearing fruits provides an estimate of an individual's dispersal ability (Cheptou et al. 2008). Dornier et al. (2011) have shown that no seeds survive in the soil after germination in autumn, i.e. the species has no seed

bank and both achene types do not exhibit interannual dormancy.

In the French Mediterranean region, this species forms very large populations in the countryside of up to several hundred thousand plants per hectare in early successional stages. In contrast, urban patches contain no more than 50 plants per patch (Dornier and Cheptou 2012), but no evidence of density-dependent mortality or individual sizes was found. In the city, extinction and colonization dynamics have been found to be high (Dornier et al. 2011), and population size is the major determinant of pollination success and seed production (the Allee effect) (Cheptou and Avendano 2006) and of patch extinction (Dornier and Cheptou 2012). In addition, urban patches in Montpellier have been shown to be 2–3 °C hotter and about 30 per cent drier than countryside populations (Figure 8.2; Lambrecht et al. 2016).

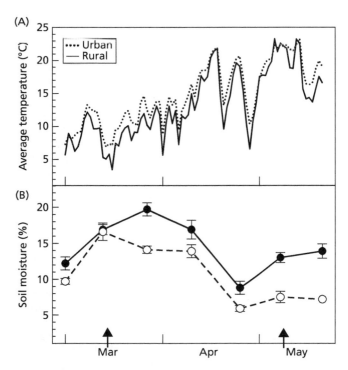

Figure 8.2 Dryness and temperature in rural populations and urban patches in Montpellier in 2013. (A) Average daily temperatures measured near the soil surface in each of the field populations. Loggers were placed at the height of *C. sancta* leaves. Each line is the mean (± SEM) of four measurements. (B) Average soil moisture in urban and rural field sites. Each line is the mean (± SEM) of five measurements. Vertical arrows along the x-axis indicate approximate dates of flowering and seed set. (Redrawn from Lambrecht et al. (2016).)

The high fragmentation of urban patches contrasts with the continuous habitat of rural populations. However, a few large populations located in vineyards and orchards in private gardens can also be found in the city. This means that fragmentation and urbanity, per se, are not confounded. This specificity allows us to disentangle the role of urbanity and the role of fragmentation in trait variation.

Neutral genetic differentiation using a set of polymorphic microsatellites has been studied by estimating hierarchical $F_{ST}:F_{ST}$ among habitats and F_{ST} among populations within habitats (Table 8.1; Dubois and Cheptou 2017). First, this study reveals very low F_{ST} values (low genetic structure, with $F_{ST} < 0.001$) among habitats for both urban versus rural and fragmented versus unfragmented habitats. Second, within habitats, population differentiation is also low. This shows that, in spite of the low census population size in patches, effective population sizes are large and the role of genetic drift low, probably because of substantial gene flow. Such gene flow may potentially impede local adaptation. The important consequence of such low neutral population differentiation is that shifts in quantitative traits among habitats can be assigned to adaptation; i.e., drift is likely to have a minor effect on trait variation.

Table 8.1 Neutral genetic differentiation estimated from eight microsatellites markers. Using a set of six populations—two urban fragmented, two urban unfragmented, and two rural unfragmented—hierarchical F_{ST} was estimated using two hierarchical partitions. The first partition (fragmented vs unfragmented) and second partition (urban vs rural) were considered. F_{ST} among populations within habitat was also estimated. Second column provides *p-values* for significance of F_{ST} ($F_{ST} \neq 0$) obtained from the randomization function implemented in HierFstat (Dubois and Cheptou 2017).

	F_{ST}	*p-values*
Fragmentation		
Fragmented vs unfragmented	0.00032	0.511
Among populations within habitat	0.01254	0.001
Urbanization		
Urban vs rural	0.00087	0.616
Among populations within habitat	0.01226	0.001

8.3 Natural selection on dispersal traits in response to urban fragmentation

We used a rural–urban gradient in Montpellier to test for the possibility of reduced dispersal in urban patches. Like in oceanic islands (Carlquist 1974), we hypothesized that dispersal of seeds in the urban fragmented patches is costly because of the low probability for dispersing seeds to reach a suitable habitat. In contrast, non-dispersing seeds may be more successful by germinating in the mother plant's patch. The heterocarpy of *C. sancta* allowed us to define an individual plant's ability for seed dispersal, which can be captured by the dispersal index (Rd)—the ratio of non-dispersing seeds to total seeds (Cheptou et al. 2008).

8.3.1 Is dispersal costly in urban patches?

While fragmentation has been hypothesized to reduce dispersal (Carlquist 1974; Cody and Overton 1996; Riba et al. 2009), the strength of selection on dispersal exerted by fragmentation is often difficult to estimate in natural systems. To estimate seed loss in urban patches, we used artificial patches with sticky surfaces. By placing plants (before dispersal) on the artificial patches, a comparison of Rd before dispersal and Rd estimated from seeds stuck on the patches (after dispersal) could be made, and it was possible to estimate the probability of a dispersing achene falling out of the patch relative to non-dispersing seeds. Using several thousand seeds, Cheptou et al. (2008) found that dispersing achenes had a 55 per cent greater chance of falling out of the patch than non-dispersing achenes, which indicates that dispersal is costly in urban patches. Note that our artificial patches potentially catch dispersing seeds from other patches, and thus provide an unbiased measure of seed loss in urban patches. Among-patch dispersal is likely to be very small, given that patch surfaces represent less than 1 per cent of the total surface.

8.3.2 Shift of the seed dispersal ratio

In one experiment, we analysed the pattern of dispersal in seven urban fragmented populations and four large non-fragmented populations (three rural populations and one urban population). Seeds were collected in the spring and were used to grow about thirty plants per population in a greenhouse. At flowering, cross pollination within each population was performed using commercial bumblebees. When seeds were mature, two capitula per plant were harvested and seeds were counted to estimate Rd. The results show that urban patchy populations have a significantly higher Rd than continuous populations (Figure 8.3). Interestingly, the urban continuous population exhibited a ratio as low as the rural populations. This points to the role of fragmentation in the shift of Rd more than urbanity, per se. To disentangle the role of fragmentation and urbanity, we performed a second experiment by sampling six populations: two rural (non-fragmented) populations, two urban (non-fragmented) populations, and two urban patchy populations. Thus, the popu-

lations could be classified as urban versus rural populations or fragmented versus non-fragmented populations. As in the previous experiment, plants were grown in the greenhouse until fructification to measure Rd. We analysed Rd in a linear model using either fragmentation or urbanity as explanatory variables (Dubois and Cheptou 2017). Dubois and Cheptou (2017) showed that the linear model including fragmentation was the best model, which identifies the role of fragmentation but not urbanity, per se, in the evolution of Rd in urban patches. In addition, Dubois and Cheptou (2017) analysed the quantitative differentiation (Q_{ST}) on Rd. They used the same partition as used for the F_{ST} neutral differentiation (Table 8.1). Q_{ST} among fragmented versus unfragmented habitats was much larger than Q_{ST} among urban versus rural habitats. Because the corresponding F_{ST} for neutral markers was close

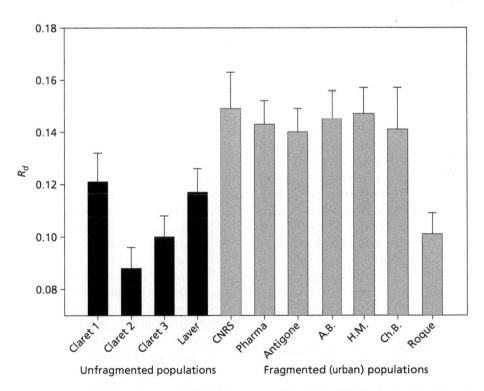

Figure 8.3 Mean (SE) for proportion of non-dispersing achenes (Rd) in the rural–urban gradient. (Left) Rural unfragmented populations; (right) urban fragmented 'patchy' populations, except the urban unfragmented population 'Roque'. General linear model (GLM) was used to test the effect of fragmentation on Rd. Population type (fragmented vs unfragmented) was considered as a fixed factor and was tested against population factor (nested in population type factor) considered as a random factor. A significant effect of fragmentation [F(1, 9) = 6.72, p = 0.05] was found. (Redrawn from Cheptou et al. (2008).)

to zero, this shows a strong diverging selection for dispersal between fragmented and non-fragmented populations.

8.3.3 An evolutionary scenario for reduced dispersal in urban patches

To estimate the pace of adaptation in urban patches, we built a quantitative genetic model to account for the shift of Rd observed in urban patches. We estimated the fitness of plants in urban patches as the sum of non-dispersing and dispersing seeds weighted by the probability of staying in the patch, P_{stay}, which equates to $w(Rd) = Rd + (1 - Rd) \times P_{stay}$. We used the experimental value of $P_{stay} = 0.45$. From this equation, we were able to estimate the selection differential, S, on Rd caused by the loss of dispersing seeds in the urban fragmented habitats. The response to selection in one generation was estimated using the breeder's equation ($R = h^2 S$). For the calculations, we used the narrow sense heritability of Rd = 0.25 estimated on a rural population by Imbert (2001). We found that the shift observed in Figure 8.3 (from Rd = 10 per cent to Rd = 15 per cent) is consistent with fewer than 15 years of selection in urban patches. The number of generations estimated is dependent on the constancy of heritability, which, admittedly, could have changed during selection in urban patches. The fact that the phenotypic variance of Rd estimated in the greenhouse for both urban and rural populations is nearly equivalent in both habitats suggests that heritability has not been substantially reduced in urban populations. Moreover, our estimate is always lower than the maximum age of the populations deduced from the date of building of the sidewalks (20–30 years old) (Cheptou et al. 2008).

Overall, our results show that adaptation has occurred on a short timescale and the shift observed is not due to urbanity but to fragmentation in the city.

8.4 Natural selection on physiological traits in the urban environment

8.4.1 Plant physiological traits related to the urban heat island

Because urban environments experience warmer temperatures, higher CO_2 levels, and drier soils due to enhanced runoff and evaporation as compared with rural environments (Gregg et al. 2003; Carreiro and Tripler 2005; Pataki et al. 2006; Grimm et al. 2008; Searle et al. 2012; Lahr et al. 2018), they provide a pertinent ecological model to investigate the adaptation of physiological traits to climate change. In spite of the importance of physiological traits to survival, the physiological adaptation of plants to an urban climate has been little investigated. The plasticity of physiological traits in response to the urban environment has been studied in trees (Gregg et al. 2003; Searle et al. 2012), but their generation time precludes the evaluation of evolutionary processes. In 2013, we compared physiological traits of urban and rural *C. sancta* plants grown together in a greenhouse. We then led a study of selection on these traits on plants of both origins planted together in urban patches. While Lambrecht et al. (2016) focused on a suite of traits, we will concentrate here on phenology and water-use efficiency, two traits that are widely hypothesized to respond to drought associated with global change.

Water-use efficiency and phenology are genetically correlated, and both traits are involved in plant adaptation to drought (McKay et al. 2003; Lovell et al. 2013; Kenney et al. 2014; Monroe et al. 2018). Water-use efficiency (WUE) is the ratio of the amount of carbon gained via photosynthesis relative to the amount of water transpired, and may be assessed by instantaneous measures of photosynthesis and transpiration, or by the use of stable carbon isotope ratios ($\delta^{13}C$) to estimate intrinsic WUE. Several studies have revealed that most plants exhibit one of two primary drought-coping strategies that involve phenology and WUE (McKay et al. 2003; Sherrard and Maherali 2006; Franks et al. 2007; Lovell et al. 2013; Kenney et al. 2014; Kooyers 2015). One strategy is 'drought escape', whereby plants grow and reproduce rapidly while water is available, but then become dormant during periods of low water availability. This rapid growth relies on high physiological rates, leading to low WUE in order to support an earlier phenology (see also Chapter 9 for further discussion on urban-driven changes in phenology). An alternative strategy is 'drought or dehydration avoidance', whereby plants grow slowly and exhibit traits that enable them to conserve water, including a high WUE, so they can survive and reproduce longer into the drought period. In Mediterranean regions, many annual species follow the drought escape strategy to some degree, by growing during winter and senescing during summer.

Due to the warmer, drier conditions found in urban patches as compared with rural environments (Figure 8.2), we expected to observe a shift in phenology and WUE. To test for differences between *C. sancta* plants from urban and rural populations, we grew sixty plants from field-collected seeds of each population under common conditions in a greenhouse. We found that urban plants delayed flowering by ~ 2.7 days and delayed senescence by ~ 9 days compared with rural plants (Table 8.2). In conjunction with those results, urban plants exhibited a higher WUE, as shown by both instantaneous and intrinsic measures, than rural plants (Table 8.2). These results suggest more a move toward a dehydration avoidance strategy in this annual Mediterranean species, than a drought escape strategy.

Most studies of plant phenology in urban environments have identified accelerated phenology, consistent with a drought escape strategy, when compared with rural areas via experiments (Yakub and Tiffin 2017; Gorton et al. 2018), direct observations (Roetzer et al. 2000; Lu et al. 2006), and remote sensing (White et al. 2002; Zhou et al. 2016). These results contrast with our own of delayed flowering. However, these other studies were conducted in climates where warmer temperatures and associated factors (e.g., more rapid snowmelt and greater

snow clearing in urban environments) indicate the onset of the growing season in spring. In Mediterranean climates, where snow cover is generally not present, plants are typically physiologically active during the rainy season in winter and spring. Therefore, phenological cues may be different for our study species. Although uncommon, selection for later flowering times in response to warming and drought associated with climate change has been identified in a few studies. For example, warmer temperatures are expected to favour later flowering in both native and introduced ranges of purple loosestrife (*Lythrum salicaria*), perhaps because during warmer springs, plants may grow larger and accumulate more resources before the onset of flowering (Colautti et al. 2017). An investigation of wild *Arabidopsis* populations in Spain identified a population that flowered later after a 10-year warming period and moderate drought, even while other populations flowered earlier, indicating variation in phenological response to drought (Gomez et al. 2018). In fact, a study of *Arabidopsis* ecotypes found selection toward later flowering under mild drought conditions, but selection toward earlier flowering under extreme drought, demonstrating that severity of drought may lead to divergent responses (Schmalenbach et al. 2014). Such contrasting results highlight the need for broader studies of the physiological adaptation of plants to drought and to urban environments located in different climatic zones.

Table 8.2 Phenological and physiological differences between urban and rural plants. Mean values (SE) were measured from plants from an urban and a rural population grown together in a greenhouse under common conditions ($N = 120$). Selection differentials (S) and selection gradients (β) were estimated for urban and rural plants growing in urban patches ($N = 120$) (Lambrecht et al. 2016).

	Greenhouse study			Field study			
	Rural	Urban	*p-values*	S	*p-values*	β	*p-values*
Phenology							
Days to first flower	32.0 (0.4)	34.7 (0.6)	< 0.001	–	–	–	–
Days to senescence	81.6 (0.8)	90.6 (1.2)	< 0.001	0.31 (0.06)	0.001	0.17 (0.05)	0.001
Water-use efficiency							
Instantaneous (mmol CO_2/mmol H_2O)	49.5 (1.7)	60.6 (2.3)	< 0.001	−0.07 (0.1)	0.62	–	–
Intrinsic ($\delta^{13}C$, ‰)	−30.5 (0.1)	−29.8 (0.2)	< 0.001	0.15 (0.07)	0.10	0.04 (0.06)	0.45

8.4.2 Are selection gradients in urban patches consistent with physiological traits?

We tested for selection on phenology and WUE in urban environments to determine whether the observed differences between rural and urban plants was a result of adaptation. We grew both urban and rural plants in separate patches around trees along sidewalks in Montpellier, France (Figure 8.1). We measured the same traits as in our greenhouse study, with the exception of date to first flowering. We used measurements of fitness (capitula number) to estimate two forms of selection on these traits. First, direct selection was quantified with selection gradients (β), which were calculated as the slope of the regression between standardized trait values and relative fitness (Lande and Arnold 1983). Additionally, selection differentials (S) were used to quantify total selection, including both direct and indirect selection. These, which indicate the difference between population means before and after selection, were calculated as the covariance between standardized trait values and relative fitness (Lande and Arnold 1983). We included only a subset of measured traits in our estimates of selection gradients, in order to avoid collinearity that arises when measured traits are highly correlated or a function of one another (Lande and Arnold 1983). While we found evidence of selection for delayed phenology in the urban environment (as indicated by later senescence), selection for increased WUE was not significant (Table 8.2; Lambrecht et al. 2016). Evidence for selection on physiological traits is typically difficult to detect (Donovan et al. 2009), particularly when traits are correlated, as are phenology and WUE. Furthermore, variability among the patches may have influenced the strength of selection. For example, although patches may be homogeneous in size, shadows from nearby buildings and differential run-off from paved areas may lead to heterogeneity in temperature and moisture among the patches.

8.5 Contemporary evolution: what can we learn from urban systems?

8.5.1 Compelling evidence for rapid evolution in an urban environment

Overall, our research programme studying the adaptation of plants to urban patches has revealed compelling evidence for trait evolution, including dispersal and some physiological traits. Interestingly, we are confident that the driver of adaptation for seed dispersal is fragmentation, while it is likely drought and elevated temperature for physiological traits. This difference points out an interesting feature of the urban system: because urban ecosystems provide diverse habitats for plants, from very fragmented to continuous populations, it is, to a certain extent, possible to identify the drivers of adaptation to the urban setting. In our study of physiological traits, our protocol of contrasting only rural populations and patchy populations does not allow us to identify if such adaptation is general to urban settings or only to fragmented habitats. While patches were found to be drier and hotter than rural populations, probably because of the concrete matrix, temperature and soil humidity are likely heterogeneous among urban habitats, as reported in several cities (see, for instance, Quénol et al. 2010). As a consequence, the shift in physiological traits may not be extrapolated to the urban environment as a whole.

The important consequence is that the urban ecosystem is not homogeneous in terms of selection on plants and evolutionary responses are likely to be idiosyncratic. On this point, evolutionary studies have revealed mixed results (see Chapter 9). In Virginia pepperweed (*Lepidium virginicum*), Yakub and Tiffin (2017) found rather a consistent shift in phenology in five North American cities, suggesting convergent evolution. However, Thompson et al. (2016) found contrasting patterns of clines in cyanogenesis in white clover (*Trifolium repens*). The authors hypothesized that reduced herbivory or higher temperature should have selected for reduced cyanogenesis in urban populations. This trend was found in three cities and the reverse trend was found in one city. While this result may seem inconsistent with regard to urban adaptation, the response to temperature was, however, homogeneous. Indeed, the temperature in the city exhibiting the 'opposite pattern' was actually lower in winter because of snow removal management in this city. This points to the need to see beyond the rural/urban pattern and to identify drivers of selection (see also Chapter 2).

Nevertheless, studying adaptation along the rural–urban gradient may be facilitated by the low genetic divergence between rural and urban populations together with their highly contrasting ecological features. This allows us to study small trait shifts. However, we have to keep in mind that some traits for which we expected a shift may have not evolved. For instance, the lower pollination services compared to rural populations have not led to higher ability of autonomous self-fertilization in urban populations of *C. sancta* (Cheptou and Avendano 2006). Yet this shift has been found in the species *Centaurium erythraea* in an urban population in Belgium (Brys and Jacquemyn 2012). Such contrasting results across species can result from the fact that the genetic architecture of a trait in a given species makes a trait more or less likely to evolve in the short term. The reduction in anther stigma distance in urban populations of *Centaurium erythraea* is a quantitative trait that can easily respond to selection (see Roels and Kelly 2011). In contrast, self-incompatibility in the species *Crepis sancta* associated with substantial inbreeding depression is probably the reason for the absence of mating system shift; the dissolution of self-incompatibility is not expected to evolve in a few generations in plants, but rather over the long term (Cheptou 2019).

8.5.2 Adaptation to global change

As mentioned in section 8.1, the urban environment exhibits some of the major components of global change. We would like to stress that urban ecosystems may provide a powerful system to study adaptation to global change. In our study, the 2 to 3 °C increase in temperature associated with 20–30 per cent drier soil in urban patches fits well with the climate change scenarios expected in Mediterranean France (Giorgi and Lionello 2008). In the same vein, lower pollination may be a surrogate for pollinator decline, which is being experienced in many parts of the world. However, the low pollination services observed in Montpellier compared with the surrounding countryside may not be generalizable to all cities. A comparison of plant communities in Paris and Montpellier revealed that, contrary to Montpellier, insect-pollinated species were found more frequently in Paris than in the surrounding countryside, suggesting higher pollinator activity in Paris than out of Paris. This is consistent with the intensive cereals monoculture unsuitable for

pollinators in the Parisian region (P.O. Cheptou, unpublished results). It is thus important to identify the drivers of selection in urban environments, since the ecological heterogeneity of cities is likely to generate heterogeneous responses (Thompson et al. 2016). Urban adaptation may not be unequivocal.

8.5.3 Modes and tempo of evolutionary processes

Our study has revealed that adaptation is likely and sometimes rapid (in a dozen generations). This suggests that evolvability of urban populations is substantial. This contrasts with the classical view of small urban populations that should have deprived genetic variance of traits. In the model *C. sancta*, while the demographic populations of urban patches are small, we found no evidence of reduced genetic diversity in either neutral markers or quantitative traits, which suggests large effective population sizes and substantial gene flow. The rapid evolution of major life-history traits is, however, puzzling given that life-history traits are expected to be under strong stabilizing selection (Houle 1992). The rapid evolution in spite of a supposedly reduced genetic variance of traits is definitely an interesting perspective in evolutionary ecology that could be pursued in analysing adaptation in urban environment and suggests that epigenetics may be at play.

8.6 Conclusions

Our research programme has revealed that not only adaptation to the ecological features of urban patches (temperature, drought, fragmentation, etc.) is possible in the short term, but also adaptation is probably not unequivocal in cities because of their ecological complexity. Sidewalk patches represent a typical urban habitat that can be found in most cities in the world, with different local flora. The urban patches provide a unique opportunity to study parallel evolution in the face of fragmentation and global change in many parts of the world.

According to us, urban evolution should not be considered as separate from evolution in natural systems. Indeed, urban ecosystems and natural ecosystems often share the same ecological characteristics but at varying degrees. We think that studying urban adaptation may help us understand evolutionary processes in natural systems. As an example, the evolution of dispersal in fragmented areas has been widely studied in oceanic islands. Such studies hypothesized the same selection pressure as hypothesized in urban patches (the cost of dispersal), but the facilities offered by urban patches (regular geometry, low genetic divergence between urban and rural populations) have allowed us to formally demonstrate the role of fragmentation in dispersal reduction. Lastly, urban patches may provide a suitable system in which to involve citizens in data collection. These habitats are easily geolocalized and tools such as smartphone applications will allow us to generate massive data on those systems that can be useful for evolutionary ecologists.

Acknowledgements

We thank Stephanie Mahieu for assistance during field and greenhouse studies. S.C.L. acknowledges support from the Department of Biological Sciences and the Office of Faculty Affairs, San José State University, CA, USA.

References

Alberti, M., Marzluff, J., and Hunt, V.M. (2017). Urban driven phenotypic changes: empirical observations and theoretical implications for eco-evolutionary feedback. *Philosophical Transactions of the Royal Society B: Biological Sciences*, 372, 1712.

Brys, R. and Jacquemyn, H. (2012). Effects of human-mediated pollinator impoverishment on floral traits and mating patterns in a short-lived herb: an experimental approach. *Functional Ecology*, 26, 189–97.

Carlquist, S. (1974). *Island Biology*. Colombia University Press, New York.

Carreiro, M.M. and Tripler, C.E. (2005). Forest remnants along rural-urban gradients: examining their potential for global change research. *Ecosystems*, 8, 568–82.

Caughley, G. (1994). Directions in conservation biology. *Journal of Animal Ecology*, 63, 215–44.

Cheptou, P.O. (2019). Does the evolution of self-fertilization rescue populations or increase the risk of extinction? *Annals of Botany*, 123, 337–45.

Cheptou, P.O. and Avendano, L.G. (2006). Pollination processes and the Allee effect in highly fragmented populations: consequences for the mating system in urban environments. *New Phytologist*, 172, 774–83.

Cheptou, P.O., Lepart, J., and Escarre, J. (2002). Mating system variation along a successional gradient in the allogamous and colonizing plant *Crepis sancta* (Asteraceae). *Journal of Evolutionary Biology*, 15, 753–62.

Cheptou, P.O., Carrue, O., Rouifed, S., and Cantarel, A. (2008). Rapid evolution of seed dispersal in an urban environment in the weed *Crepis sancta*. *Proceedings of the National Academy of Sciences of the United States of America*, 105, 3796–9.

Cody, M.L. and Overton, J.M. (1996). Short-term evolution of reduced dispersal in island plant populations. *Journal of Ecology*, 84, 53–61.

Colautti, R.I., Agren, J., and Anderson, J.T. (2017). Phenological shifts of native and invasive species under climate change: insights from the *Boechera–Lythrum* model. *Philosophical Transactions of the Royal Society B: Biological Sciences*, 372, 20160032.

Donovan, L.A., Ludwig, F., Rosenthal, D.M., Rieseberg, L.H., and Dudley, S.A. (2009). Phenotypic selection on leaf ecophysiological traits in *Helianthus*. *New Phytologist*, 183, 868e879.

Dornier, A. and Cheptou, P.O. (2012). Determinants of extinction in fragmented plant populations: *Crepis sancta* (asteraceae) in urban environments. *Oecologia*, 169, 703–12.

Dornier, A., Pons, V., and Cheptou, P.O. (2011). Colonization and extinction dynamics of an annual plant metapopulation in an urban environment. *Oikos*, 120, 1240–6.

Dubois, J. and Cheptou, P.O. (2017). Effects of fragmentation on plant adaptation to urban environments. *Philosophical Transactions of the Royal Society B: Biological Sciences*, 372, 1712.

Franks, S.J., Sim, S., and Weiss, A.E. (2007). Rapid evolution of flowering time by an annual plant in response to a climate fluctuation. *Proceedings of the National Academy of Sciences of the United States of America*, 104, 1278–82.

Giorgi, F. and Lionello, P. (2008). Climate change projections for the Mediterranean region. *Global and Planetary Change*, 63, 90–104.

Godron, D.A. (1854). *Florula juvenilis ou Énumération des Plantes Étrangères qui Croissent Naturellement au Port Juvénal, près de Montpellier*. Grumblot et veuve Raybois, Nancy.

Gomez, R., Mendez-Vigo, B., Marcer, A., Alonso-Blanco, C., and Pico, F.X. (2018). Quantifying temporal change in plant population attributes: insights from a resurrection approach. *AoB PLANTS*, 10, ply063.

Gorton, A.J., Moeller, D.A., and Tiffin, P. (2018). Little plant, big city: a test of adaptation to urban environments in common ragweed (*Ambrosia artemisiifolia*). *Proceedings of the Royal Society B: Biological Sciences*, 285, 20180968.

Gregg, J.W., Jones, C.G., and Dawson, T.E. (2003). Urbanization effects on tree growth in the vicinity of New York City. *Nature*, 424, 183–7.

Grimm, N.B., Faeth, S.H., Golubiewski, N.E., et al. (2008). Global change and the ecology of cities. *Science*, 319, 756–60.

Houle, D. (1992). Comparing evolvability and variability of quantitative traits. *Genetics*, 130, 195–204.

Imbert, E. (2001). Capitulum characters in a seed heteromorphic plant, *Crepis sancta* (Asteraceae): variance partitioning and inference for the evolution of dispersal rate. *Heredity*, 86, 78–86.

Imbert, E., Escarre, J., and Lepart, J. (1996). Achene dimorphism and among-population variation in *Crepis sancta* (Asteraceae). *International Journal of Plant Sciences*, 157, 309–15.

Johnson, M.T.J. and Munshi-South, J. (2017). Evolution of life in urban environments. *Science*, 358, eaam8327.

Kenney, A.M., McKay, J.K., Richards, J.H., and Juenger, T.E. (2014) Direct and indirect selection on flowering time, water-use efficiency (WUE, d13C), and WUE plasticity to drought in *Arabidopsis thaliana*. *Ecology and Evolution*, 4, 4505–21.

Kooyers, N.J. (2015). The evolution of drought escape and avoidance in natural herbaceous populations. *Plant Science*, 234, 155–62.

Lahr, E.C., Dunn, R.R., and Frank, S.D. (2018). Getting ahead of the curve: cities as surrogates for global change. *Proceedings of the Royal Society B: Biological Sciences*, 285, 20180643.

Lambrecht, S.C., Mahieu, S., and Cheptou, P.O. (2016). Natural selection on plant physiological traits in an urban environment. *Acta Oecologica*, 77, 67–74.

Lande, R. and Arnold, S.J. (1983). The measurement of selection on correlated characters. *Evolution*, 37, 1210–26.

Lizet, B. (1997). Au jardin d'Athis—portrait de Paul Jovet. *Journal d'Agriculture Traditionnelle et de Botanique Appliquée*, 39, 131–55.

Lovell, J.T., Juenger, T.E., Michaels, S.D., et al. (2013). Pleiotropy of *FRIGIDA* enhances the potential for multivariate adaptation. *Proceedings of the Royal Society B: Biological Sciences*, 280, 20131043.

Lu, P., Yu, Q., Liu, J., and Lee, X. (2006). Advance of tree flowering dates in response to urban climate change. *Agricultural and Forest Meteorology*, 138, 120–31.

McKay, J.K., Richards, J.H., and Mitchell-Olds, T. (2003). Genetics of drought adaptation in *Arabidopsis thaliana*: I. Pleiotropy contributes to genetic correlations among ecological traits. *Molecular Ecology*, 12, 1137–51.

Monroe, J.G., Powell, T., Price, N., et al. (2018). Drought adaptation in *Arabidopsis thaliana* by extensive genetic loss-of-function. *eLife*, 7, e41038.

Pataki, D.E., Alig, R.J., Fung, A.S., et al. (2006). Urban ecosystems and the North American carbon cycle. *Global Change Biology*, 12, 2092–102.

Quénol, H., Dubreuil, V., Mimet, A., et al. (2010). Climat urbain et impact sur la phénologie végétale printanière. *La Météorologie*, 68, 50–57.

Riba, M., Mayol, M., Giles, B.E., et al. (2009). Darwin's wind hypothesis: does it work for plant dispersal in fragmented habitats? *New Phytologist*, 183, 667–77.

Roels, S.A.B. and Kelly, J.K. (2011). Rapid evolution caused by pollinator loss in *Mimulus guttatus*. *Evolution*, 65, 2541–52.

Roetzer, T., Wittenzeller, M., Haeckel, H., and Nekavar, J. (2000). Phenology in central Europe—differences in spring phenophases in urban and rural areas. *International Journal of Biometeorology*, 44, 60–66.

Schmalenbach, I., Zhang, L., Reymond, M., and Jiménez-Gómez, J.M. (2014). The relationship between flowering time and growth responses to drought in the *Arabidopsis* Landsberg erecta × Antwerp-1 population. *Frontiers in Plant Science*, 5, 609.

Searle, S.Y., Turnbull, M.H., Boelman, N.T., et al. (2012). Urban environment in New York City promotes growth in northern red oak seedlings. *Tree Physiology*, 32, 389–400.

Sherrard, M.E. and Maherali, H. (2006). The adaptive significance of drought escape in *Avena barbata*, an annual grass. *Evolution*, 60, 2478–89.

Slabbekoorn, H. and Peet, M. (2003). Ecology: birds sing at a higher pitch in urban noise—great tits hit the high notes to ensure that their mating calls are heard above the city's din. *Nature*, 424, 267–7.

Thompson, K.A., Renaudin, M., and Johnson, M.T.J. (2016). Urbanization drives the evolution of parallel clines in plant populations. *Proceedings of the Royal Society B: Biological Sciences*, 283, 1845.

White, M.K., Nemani, R.R., Thornton, P.E., and Running, S.W. (2002). Satellite evidence of phenological differences between urbanized and rural areas of the Eastern United States deciduous broadleaf forest. *Ecosystems*, 5, 260–73.

Yakub, M. and Tiffin, P. (2017). Living in the city: urban environments shape the evolution of a native annual plant. *Global Change Biology*, 23, 2082–9.

Zhou D., Zhao, S., Zhang, L., and Liu, S. (2016). Remotely sensed assessment of urbanization effects on vegetation phenology in China's 32 major cities. *Remote Sensing of Environment*, 176, 272–81.

CHAPTER 9

Adaptive Evolution of Plant Life History in Urban Environments

Amanda J. Gorton, Liana T. Burghardt, and Peter Tiffin

Gorton, A.J., Burghardt, L.T. and Tiffin, P., *Adaptive Evolution of Plant Life History in Urban Environments* In: *Urban Evolutionary Biology*. Edited by Marta Szulkin, Jason Munshi-South and Anne Charmantier, Oxford University Press (2020). © Oxford University Press.
DOI: 10.1093/oso/9780198836841.003.0009

9.1 Introduction

In recent years, biologists have become increasingly interested in how urban environments affect the evolution of plant and animal populations. This interest follows the growth of the field of urban ecology, which originally developed from the recognition that cities represent important and complex ecosystems, as well as from the desire to describe urban biodiversity (McDonnell 2011).

The increasing interest in evolution in urban environments is driven by multiple factors. One driver is that biologists work to understand their environments and many of us live in urban areas—it is only natural that we try to understand the biology of these environments. Also contributing to interest in evolution in urban environments is a growing acknowledgement of the inadvertent effects human activities have on evolutionary processes (Fugère and Hendry 2018). In fact, some biologists have been motivated to study urban evolution because cities can be used as partial analogues of future environmental conditions (Ziska et al. 2003; Harrison and Winfree 2015; Johnson et al. 2015). In particular, urban environments differ from surrounding rural environments in ways that are similar to those projected under future climate change (e.g., increased temperatures, higher concentrations of atmospheric CO_2, and altered patterns of water and nutrient availability) (Parmesan 2006; Sih et al. 2011; Franks et al. 2013; Lau et al. 2014; Epps and Keyghobadi 2015). Urban environments also provide fruitful ground for investigating basic evolutionary processes, including: adaptation to novel environments, adaptation in the face of gene flow, adaptive convergence (see Chapter 3), and the consequences of selection at temporal scales that are longer than what can be imposed under artificial selection experiments but shorter than what can be investigated in most natural environments. We note that we limit our use of the term adaptation to refer to evolutionary adaptation, or a change in the genetic composition of a population that occurs in response to selection. Plastic responses to environmental conditions do not result in a change in the genetic composition of a population and therefore we do not consider a plastic response as adaptation, even though plasticity can itself result from adaptive evolution.

Plant life-history traits are of central importance to organismal fitness. Life-history traits capture many aspects of phenology and provisioning to offspring. For plants, these traits include growth rates, lifespan, size at the time of reproduction, investment in survival versus reproduction, time of flowering, flower number, the number of seeds plants produce, the size of those seeds, seed dormancy, and sex ratio of offspring (Stearns 1976; Roff 2002; Liu et al. 2017). Life-history traits are a natural place to start investigating urban adaptation because they often experience strong, environmentally-dependent selection (Mazer and LeBuhn 1999; Kingsolver et al. 2001). This selection often leads to local adaptation, as demonstrated repeatedly in studies examining adaptation to local climates (e.g., Wilczek et al. 2014; Peterson et al. 2016; Postma and Ågren 2016).

Many of the environmental factors that differentiate urban from rural environments, and therefore drive the urban–rural differences in selection that influence urban adaptation, also vary within urban environments and within rural environments. However, the spatial and temporal scale at which these environmental factors vary might differ (see also Chapter 2). In fact, one of the attractions of researching life-history adaptation to urban environments as well as within urban environments is that we can find stark environmental contrasts over very small spatial scales. For example, temperature within a given city can vary at a scale of 1–2 km (Smoliak et al. 2015) and mowed and unmowed habitats, which can alter selection acting on plant growth form (Lennartsson et al. 1998), may be found side by side. Furthermore, habitats vary from railways to riverbanks to pavement cracks, resulting in changes in water availability as well as temperature. In addition to spatial variation, the temporal dynamics of urban environments could differ from that in nearby rural environments. With few exceptions, urban environments, at least in their current size and distinctness, were created relatively recently from an evolutionary perspective, and thus populations have had less time to adapt to these environments.

While the evolution of animal life histories to urban environments has been relatively well studied in several species (McDonnell and Hahs 2015), the evolution of life histories of urban plant populations is in a nascent stage. As sessile organisms that are comparatively easy to manipulate, plants are excellent empirical study systems to advance our understanding of the effects of urban environments on evolution and adaptation. Moreover, given the close relationship between life-history traits and fitness, life-history traits provide an obvious starting point for understanding evolution in urban environments. Indeed, although the studies of plant evolution in urban environments are still relatively few, much of the work that has been conducted has focused on urban–rural divergence in key life-history traits, e.g., flowering time, plant size at the time of reproduction, and seed dispersal. These studies provide a foundation for understanding adaptation to urban environments. In addition, these studies highlight the potential power of using urban environments to understand the evolution of plant life histories. In this chapter, we review what empirical studies have revealed about the adaptive divergence of plant life histories among urban and rural populations, provide an overview of the strengths and weaknesses of the empirical approaches used to study evolutionary processes in urban plant populations, and discuss ideas for future work. Although we focus on plants, many of the conceptual issues we discuss are relevant to animals and microbes.

9.2 Potential effects of urban environments on plant life-history adaptation

Decades of research on life-history adaptation in non-urban settings has shown that selection acting on life-history traits can be strongly dependent on the environment (e.g., Hall and Willis 2006; Savolainen et al. 2007). Moreover, many of the environmental factors that drive adaptive differentiation of life histories in non-urban settings also differ between urban and rural environments. As such, studies in non-urban systems provide guidance for identifying traits that are likely to be subject to differential selection in urban versus rural environments, as well as specific environmental factors that may be driving that selection. Urban environments differ from rural environments in a myriad of ways, only

some of which are well characterized. In the next paragraphs, we provide a brief overview of how some important urban–rural differences might be expected to alter evolution of plant life histories. The list of environmental factors we discuss is by no means exhaustive; rather, we focus on specific environmental factors that might be important for driving adaptation of plant life histories: the patchiness of urban environments, higher temperatures and longer growing seasons, altered water availability, and changes to the biotic community.

One of the most evident differences between urban and rural environments is the patchiness of urban environments. Areas suitable for plant growth are interspersed among buildings, roads, pavements, etc. This patchiness can directly alter the selection acting on dispersal-related traits (as shown in Cheptou et al. 2008; see also Chapter 8), which we discuss below. Habitat fragmentation is likely to affect non-adaptive evolution, also discussed below, by breaking populations into subpopulations (see Chapter 4). The evolution of these subpopulations will be affected by gene flow from other populations as well as genetic drift, which is greater in small than large populations and can reduce the efficacy of selection (Cote et al. 2017).

While fragmentation caused by buildings, roads, and pavements can directly alter selection, the concentration of asphalt and concrete in urban areas affects the temperature and water availability. Thus, fragmentation has important indirect effects on plant growth, fitness, and selection. The effects of urbanization on temperature are two-fold: daily temperatures are expected to be higher in cities than surrounding non-urban areas (Kalnay and Cai 2003), and, at least in northern latitudes, urban areas are likely to have longer potential growing seasons than surrounding rural areas. By contrast, in lower latitudes and warmer climates, the higher temperatures in urban areas might shorten the length of the growing season due to combined effects of heat and water shortage. On the other hand, lawn watering might increase water availability in urban areas, or at least parts of urban areas, relative to surrounding rural areas. Changes in the length of the growing season have the potential to greatly alter tradeoffs between plant growth and reproduction. Given that plants need to germinate, grow, flower, and produce seeds before the end of the growing season, a longer growing season may allow plants to grow much larger before they initiate reproduction. Similarly, longer growing seasons in urban areas may make it selectively advantageous for perennials to accelerate the transition from dormancy to growth and/or delay the transition from growth to dormancy.

Warmer daily temperatures along with high concentrations of impervious surfaces have the potential to strongly affect plant water demands in urban areas (e.g., Zipper et al. 2017). Ample evidence from non-urban systems indicates that changes in the timing or amount of water availability can alter growing season length and selection on germination timing (Donohue et al. 2010; Anderson 2016) and flowering time (Franks et al. 2007; Haggerty and Galloway 2010) and/or cause changes in seedling or adult survival that shift selection from favouring annuals to perennials (Galloway and Burgess 2009; Kim and Donohue 2013). In colder climates, in which water is not limiting, warmer daily temperatures may be advantageous for genotypes that are able to extend their growing periods before transitioning to reproduction. However, when water is limiting, hotter day and night temperatures, particularly during the peak of the summer, are likely to produce stressful conditions that favour genotypes that are able to tolerate or avoid stress, possibly through phenological shifts. These shifts could take the form of favouring individuals that are able to complete their cycles—allowing plants to reproduce before the hottest times of the year, as has been shown in populations of plants that experience seasonal droughts (Kooyers 2015). Alternatively, when coupled with the longer growing seasons in urban areas, selection might favour delays in reproduction, especially if plants are able to avoid or tolerate drought stress during the midst of the growing season.

Warmer temperatures in urban environments might alter selection acting on plant populations in unpredictable ways that are not directly related to life history (see also Chapter 6). For example, Thompson et al. (2016) found that the frequency of cyanogenesis (the ability to produce hydrogen cyanide (HCN)) producing genotypes in the

perennial plant white clover (*Trifolium repens*) increased with the distance from the urban centre in three of four cities. HCN production has been extensively studied for its role in protecting plants against herbivores, and thus the higher frequency of HCN in rural than urban areas seemed likely due to greater herbivore pressure in rural areas. A field experiment, however, revealed no evidence for HCN concentrations having differential consequences for herbivore damage in rural compared to urban environments (Thompson et al. 2016). A more likely driver of the putative adaptive clines in the frequency of HCN producing genotypes is that cyanogenesis can have negative fitness consequences if freezing temperatures occur when there is no snow cover. Despite the urban heat island, minimum ground temperature during the winter, and the location where plants are overwintering, it is actually colder in some urban areas than rural areas because snow acts as an insulator during the cold winter months. Since snow cover can be less persistent in urban environments, selection could be acting on winter survival, with adaptation occurring through a reduction in HCN production. The lesson here is that although we might expect life histories to evolve in response to shifts in temperature or water availability, populations might adapt to these changes through less obvious mechanisms.

Of course, it is not only the abiotic, but also the biotic environment that shapes selection, and biotic communities can be profoundly different in urban and rural areas. From the perspective of life histories, one of the most important shifts in biotic community is likely to be shifts in the herbivore and pollinator communities (Denys and Schmidt 1998; Martins et al. 2017). Changes in the population sizes or identity of herbivores and pollinators can strongly shape selection on life histories (e.g., Moeller 2006; Williams 2009). Although studies have investigated how pollinator communities change along urban–rural gradients and how these compositional changes may affect phenotypic selection acting on floral traits (Chapter 7; Irwin et al. 2018), the selective effects of these changes on life-history strategies have not been thoroughly investigated. As we discuss in section 9.7, we think this is a particularly promising area for future study.

9.3 Life-history syndromes and tradeoffs

Just as reciprocal transplant and common garden experiments can provide guidance on which traits or environmental variables may be most important in a population's adaptation to the urban environment, lessons we have learned about selection on life-history traits in natural populations can help guide studies of urban adaptation. Studies of urban adaptation of life-history traits have primarily focused on examining single traits. However, life-history traits do not evolve independently and the field of life-history theory is premised on the notion of correlated suites of traits and tradeoffs among traits. Covariances among traits can both facilitate and constrain adaptation (Burgess et al. 2007; Kimball et al. 2013; Lovell et al. 2013). A clear lesson from studies in non-urban environments is that covariance among traits as well as covariance among environmental factors can make it difficult to identify the specific traits experiencing selection, as well as the causative agents of selection (Hoban et al., 2016; Wadgymar et al., 2017; Mitchell and Whitney, 2018). This means that we need to be cautious when claiming that selection is acting on a specific trait—it may be acting on an unmeasured correlated trait. Relatively little work in urban environments has focused on measuring tradeoffs and trait correlations that constrain adaptation or lead to co-selected combinations of traits, which has been a key theme of life-history research (Stearns, 1976). Nevertheless, urban environments provide a valuable setting for such studies given that there is potential for selection to be strong and favour novel combinations of traits.

The complications of trait covariances are mirrored in the covariance of the agents of selection, which are likely to often interact, possibly in complex non-additive ways. A meta-analysis by Anderson (Anderson, 2016) concluded that life-history components such as germination, survival, and fecundity were differentially sensitive to snow removal, heating, drought and heating, and drought. For instance, shorter growing seasons due to an earlier onset of drought can select against late flowering (Franks, Sim and Weis, 2007) and perenniality (Rohde and Bhalerao 2007) and potentially many other correlated traits (Lovell et al., 2013; Schmalenbach et al., 2014). The lessons from non-urban systems are three-fold: (1) agents of selection and targeted phenotypes can be difficult to pin down (Wadgymar et al., 2017; Mitchell and Whitney, 2018); (2) interactions between climate variables can lead to complex and nonlinear effects on traits and fitness (Anderson, 2016; Wadgymar et al., 2017); and (3) life-history traits can evolve in response to population characteristics such as population size and genetic diversity (reviewed in Mitchell and Whitney 2018).

9.4 Empirical approaches to studying urban evolution

Empirical studies of urban adaptation, just like other studies of local adaptation, have employed a variety of approaches and sampling schemes (Table 9.1). Some studies have focused on a single trait, allowing for relatively in-depth examination of the specific environmental factors responsible for selection. Other studies have examined multiple traits and identified which trait(s) are under the strongest selection and/or are most diverged. Experiments have also been conducted in a variety of environments including growth chambers, greenhouses, single common gardens, and reciprocal common gardens in field conditions, the latter of which approach the natural conditions of urban–rural environments. There are also differences in the definition of urban and rural environments and the collection of samples—some have pooled individuals sampled from multiple spatially defined subpopulations from urban areas and from rural areas and then treated them as either urban or rural, respectively (e.g., Yakub and Tiffin 2017). By contrast, others have retained individual sampling identities in order to track the environmental and spatial heterogeneity found within urban and rural environments (e.g., Gorton et al. 2018). Of course, these approaches differ in their strengths and weaknesses (Table 9.1), which are important to consider when both designing and interpreting results from urban adaptation studies.

Table 9.1 Strengths and weaknesses of empirical approaches for investigating urban adaptation.

	Strengths/insights	Weaknesses/complications
Experimental approaches		
Observational data (Ziska et al. 2003; Neil and Wu 2006)	Identify environmental factors that covary with specific traits, survival, and fecundity. Inform future manipulative experiments. Easy to collect data on multiple species	Environmental (i.e., plastic) and genetic effects are confounded. No estimates of selection or past adaptation
Common garden, single site (Cheptou et al. 2008; Yakub and Tiffin 2017)	Characterize genetic variation within and between/among populations. If in a semi-natural environment, can characterize strength of selection acting on specific traits	Challenging for long-lived, large species. Site preparation can result in unnatural environments. Genotype × environment interactions may limit generality of results. Inferences of local adaptation limited
Environmental manipulations in common gardens (Thompson et al. 2016)	Characterize genetic variation within and between populations. Identify and estimate relative importance of specific selective agents	Challenging for long-lived, large species. Site preparation can result in unnatural environments. Some selective agents are difficult to manipulate (e.g., temperature, soil properties, atmospheric conditions). Difficult to know if manipulation is biologically relevant
Reciprocal common garden (Gorton et al. 2018)	Strongest evidence for local adaptation. Estimates of plasticity as well as genetic divergence among environments. Estimates of environmental/site-specific selection. Ability to estimate fitness in a biologically relevant environment	Challenging for long-lived, large species. Identifying agent of selection is difficult without manipulating potential selective agents. Site preparation can result in unnatural environments. Labour intensive
Sampling schemes		
Pooled individuals from multiple spatially defined subpopulations (Yakub and Tiffin 2017)	Efficient for characterizing overall urban–rural differences	No insight into within-habitat variance
Single population from each of two environments (Lambrecht et al. 2016)	Enables in-depth characterization of population	Lack of replication of urban/rural environments. Divergence could be non-adaptive or not due to urban/rural selection
Multiple populations from each habitat (Cheptou et al. 2008; Gorton et al. 2018)	Characterize variation within rural/urban environments. Urban/rural replication	Labour intensive
Along urban–rural gradients (Cheptou et al. 2008; Thompson et al. 2016)	Can use correlation/regression approaches to test associations between environment and phenotype	Can miss considerable within-habitat heterogeneity when single population samples are collected

9.5 Empirical evidence

The first study to show divergence and putative adaptation of plant populations to an urban environment focused on seed dispersal in *Crepis sancta* populations in Montpellier, France (Chapter 8; Cheptou et al. 2008). This study revealed that plants growing in small soil patches under urban trees surrounded by pavements, roads, and buildings showed an increase in the proportion of non-dispersing seeds relative to plants growing in large populations in non-urban areas. The high concentration of asphalt and concrete in urban areas results in patchier distributions of suitable habitat than in rural environments. Thus, these results are consistent with dispersal being costly in patchy environments.

However, the authors were careful to attribute changes in seed morphs specifically to fragmentation that accompanies urbanization, rather than other urban–rural differences (Cheptou et al. 2008). The importance of fragmentation, rather than other urban–rural differences, as the selective force driving adaptive difference in dispersal was confirmed in follow-up work by Dubois and Cheptou (2017). While there may be costs to seeds landing on inhospitable surfaces, as illustrated in this example, for other species there could be benefits to dispersal, such as avoidance of intraspecific competition, avoidance of inbreeding, and the colonization of new habitats that outweigh the costs (reviewed in Levin et al. 2003; Lowe and McPeek 2014). The value of Cheptou et al.'s study is twofold. First, it established that urban environments can drive adaptation of plant life history. Second, it provide an example of how urban environments can be used to test predictions of evolutionary theory—Cheptou was motivated by dispersal theory and this theory provided clear a priori predictions of what to expect in urban environments.

Cheptou et al. (2008) focused only on seed dispersal. A limit of focusing on a single life-history trait is that we gain no insight into the relative importance of that versus other traits in contributing to urban–rural divergence. An alternative approach is to collect data on multiple traits, with the goal of identifying which life-history traits show the greatest adaptive divergence between urban and rural environments. It is likely that these are the traits that are most affected by, and perhaps most responsible for, urban adaptation. Moreover, by measuring and analysing multiple traits, researchers have the opportunity to untangle the relative importance of selection acting directly on a trait of interest relative to selection acting through traits that are genetically correlated with the trait of interest (Lande and Arnold 1983). Examining multiple traits also could lead to the identification of an 'urban life-history strategy', i.e., a suite of traits that lead to higher fitness in urban environments.

Yakub and Tiffin (2017) used a common garden in a greenhouse and conducted the first experiment examining multiple traits in urban and rural plant populations. They were interested in determining if urban–rural populations of Virginia pepperweed

(*Lepidium virginicum*) from multiple metropolitan areas show similar phenotypic differences. Similar urban–rural differences across multiple cities would provide strong evidence for urban adaptation. A greenhouse experiment using seeds collected from multiple urban and rural locations from each of five metropolitan areas in the northern USA (Minneapolis–St Paul, Chicago, Detroit, Baltimore, and New York City) revealed several phenotypic differences that were consistent in each of the five urban–rural comparisons. Plants grown from seeds collected from urban areas bolted earlier, grew larger, had fewer leaves, had an extended time between bolting and flowering, and produced more seeds than plants grown from seeds collected from rural areas. For two of these traits, time of bolting and time between bolting and flowering, the proportion of phenotypic variation due to urban–rural differences was greater than the proportion of variance due to the metropolitan area—suggesting that shifts in developmental timing might be an important life-history adaptation to urban settings in general. For the remaining traits, differences between proximate urban–rural environments were smaller than differences between metropolitan areas. Thus, this sampling scheme and experimental structure (Table 9.1) can provide insight into whether urban adaptation is consistent across metropolitan areas or if urban environments vary widely in their effects on a resident population.

Gorton et al. (2018) also used a common garden approach to investigate patterns of adaptation of common ragweed (*Ambrosia artemiisifolia*) populations to urban and rural environments. Similar to Yakub and Tiffin (2017), Gorton et al. (2018) found evidence that adaptation to urban environments may result from a shift in developmental timing; urban populations flowered earlier than rural populations. However, unlike Yakub and Tiffin (2017) who grew plants in a greenhouse, Gorton et al. (2018) used a reciprocal transplant experiment, which provided additional insights about local adaptation to urban environments. First, although genotypes from urban populations had earlier average time to flowering, plants growing in urban experimental sites flowered later than plants in rural experimental sites. This counter-gradient variation (Levins 1968, 1969; Conover and Schultz 1995), in which genetic differences lead to earlier flowering but environmental differences lead to later flowering, may reflect adaptation to urban environments. In particular, when urban populations were first established by rural migrants, selection may have favoured earlier-flowering plants because the environmental shift from rural to urban locations resulted in delayed flowering (Figure 9.1). Thompson et al. (2016) reported a similar pattern of counter-gradient variation in HCN concentration in *T. repens*.

Taken together, the evidence for counter-gradient variation suggests caution in using observational approaches for making inferences about how selection may have driven the evolution of urban populations. The counter-gradient variation also draws attention to another complicating issue: life-history traits are highly plastic, i.e., their expression is often environmentally sensitive. For this reason, differentiating adaptive divergence among populations requires common garden experiments to disentangle the influence of genetic versus environmental variation (e.g., Haggerty and Galloway 2011; Wilczek et al. 2014). It is important to remember that observational studies that reveal life-history differences between plants growing in urban and rural environments do not provide direct evidence for adaptation. Moreover, the counter-gradient variation indicates that selection might drive adaptive divergence in the opposite direction of plastic responses.

Consistent with adaptive divergence of urban and rural populations, Gorton et al. (2018) found that net selection on flowering time tended to be stronger on foreign seed sources in both urban and rural common gardens, presumably because foreign populations are further from a selective optimum (Figure 9.2). However, despite the evidence for adaptive divergence of urban and rural populations, Gorton et al. (2018) found that rural populations had higher lifetime fitness in both rural and urban environments. We can speculate that the consistently higher fitness is indicative of selection being more effective in rural sites, either because it is more

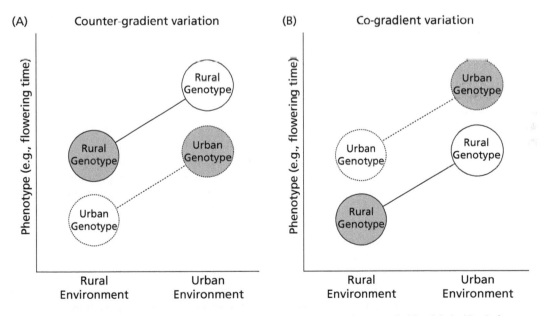

Figure 9.1 Counter-gradient variation (A) occurs when the plastic (i.e., between environment) response (solid and dashed lines) of genotypes is in an opposite direction of the genetically determined differences (i.e., urban–rural genotypic differences); by contrast, co-gradient variation (B) describes a pattern in which the plastic response is in the same direction as genetically determined differences. In the counter-gradient example shown in (A), a study relying on observing genotypes only in urban environments and rural genotypes only in rural environments, would reveal no evidence for urban–rural divergence. This mirrors results from Gorton et al. (2018), who found that flowering time was earlier in rural sites but that urban plant genotypes flower earlier. Thus, when rural seeds are grown in rural environments and when urban seeds are grown in urban environments, plants flower at approximately the same time (i.e., grown in home environments, grey shaded circles in (A)). Counter-gradient variation highlights the risk of relying on observational data of phenotypic shifts alone to make inferences about urban adaptation.

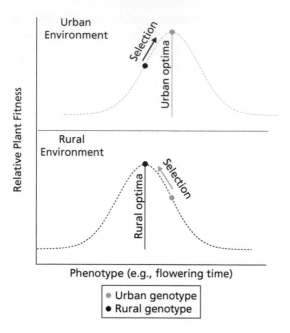

Figure 9.2 Stronger selection in foreign than home environments can be an indication of past local adaptation if there are different trait optima in each environment (lines) and trait variation is largely due to genetic differences between populations. In the scenario shown here, when plants are grown in the 'foreign' environment they will be further from trait optima than 'home' plants, resulting in stronger directional selection on the recently arrived migrants than on the resident population.

consistent across rural areas or because population sizes may be larger. However, we note that the reason for the consistently higher fitness remains an open question. Yakub and Tiffin (2017) found consistent fitness differences between urban and rural populations, but the pattern they found was in the opposite direction of that of Gorton; urban populations had consistently higher fecundity than nearby rural populations. While these differences could be indicative of more effective selection, in this case on urban populations, the use of a single common garden by Yakub and Tiffin (2017) makes it difficult to determine the cause of the higher fitness of urban populations. It is possible that the higher fitness in urban populations is due to efficacy of selection, but it is also possible that the greenhouse environment they used is more similar to urban than rural environments.

Many of the insights of Gorton et al. (2018) and Thompson et al. (2016) were only possible because they used reciprocal transplant experiments in which plants of both urban and rural origin were grown together in both urban and rural common gardens. Unlike experiments conducted in only a single location, reciprocal common gardens can provide estimates of selection and local adaptation that are not obtainable from greenhouse-grown plants (Table 9.1). Furthermore, the plastic responses and countergradient variation results of Thompson et al. (2016) and Gorton et al. (2018) both suggest plastic responses may not always reflect underlying genetic differences. The complicating effects of phenotypic plasticity underscore the importance of manipulative experiments for properly disentangling environmental and genetic effects to determine the role of selection in driving divergence between urban and rural populations.

9.6 Non-adaptive evolution

While identifying adaptive life-history divergence between urban and rural environments is interesting, it is important to remember that urbanization might also affect non-adaptive evolution. For instance, urban–rural differences in life-history traits might result from genetic drift associated with genetic bottlenecks that occur during the founding of urban populations, or through fragmentation within urban landscapes (Cheptou et al. 2008, 2017; Dubois and Cheptou 2016). In particular, fragmented landscapes may reduce the effective population sizes of local populations due to bottlenecks during population establishment or restriction of gene-flow between patches. Smaller local population sizes, which might be found in more fragmented landscapes, will increase the effects of drift on populations and also can reduce the efficacy of selection to drive adaptation. Moreover, non-selective processes can generate phenotypic clines across urban–rural gradients due to decreasing gene flow among populations with increases in geographic distance (i.e., isolation by distance) or barriers to gene flow.

A few studies have characterized the effect of urbanization on patterns of genetic differentiation and drift in urban plant populations. The limited data available from these studies indicate that in some species, urban environments will not have strong effects on genetic diversity (Culley 2007; Johnson et al. 2018) or gene flow (Culley et al. 2007), whereas in other species, urban populations may harbour less diversity than rural populations (Bartlewicz et al. 2015; Yakub and Tiffin 2017). Those systems that do not show reduced genetic diversity in urban populations may be systems that have not experienced strong genetic bottlenecks or perhaps even strong selection, at least relative to that experienced by rural populations. Clines in genetic diversity might be particularly important to study when one is trying to infer adaptation of phenotypic traits with fairly simple genetic basis, such as HCN production in *T. repens*. For these traits, phenotypic clines along environmental gradients might be due to genetic drift (Santangelo et al. 2018), although the analyses of Johnson et al. (2018) suggest this is not the case with HCN production in *T. repens*. Genetic data provide only limited evidence for fragmentation of urban plant populations, and some of the results from phenotypic studies are consistent with population fragmentation. For example, Gorton et al. (2018) found that genetically determined phenotypic variation in male and female flowering time was greater among urban ragweed populations than among rural ragweed populations. The greater variation might reflect adaptation to habitats found within urban environments or it might result from genetic drift among fragmented populations found in urban areas.

9.7 Opportunities for the future

Although the study of plant life-history evolution in urban environments is in early stages, it is clear that cities have the potential to drive adaptive evolution of plant life history. It should not be surprising that some traits experience different selection in urban than rural environments; we know from decades of work on selection and local adaptation that environmental differences can drive the evolutionary divergence of populations (Leimu and Fischer 2008; Savolainen et al. 2013). Nevertheless, establishing evidence of adaptation of life-history traits was an important first step in the study of evolution in urban environments. Moreover, given that the number of empirical studies characterizing how urban environments shape the evolution of plant populations is still too few to make broad generalizations, more studies that address questions and use approaches similar to those of the studies described in this

chapter are still valuable. At the same time, now that we know that urban environments can shape the evolution of plant populations, we should consider what potential aspects of urban adaptation have yet to be explored. This research can both advance our understanding of adaptation to urban environments and be used to broaden our understanding of life-history evolution. In addition, the study of plant life history in urban environments offers the opportunity to advance our understanding of plant adaptation in response to global climate change (Box 9.1).

Most studies of urban adaptation have focused on selection acting during later stages of plant development. However, early stages in the life cycle may also be important: seed dormancy, germination cues, and germination speed are all subject to strong selection (Donohue et al. 2010; Walck et al. 2011). Particularly intriguing in this light is that the ecological and evolutionary consequences of seed dormancy and germination can depend on the stability of the seasonal environment (Liu et al. 2017). In addition, there is observational evidence that plants surrounded by artificial lighting may display altered phenological and physiological traits (Bennie et al. 2016). For plant species with strong photoperiodic responses, the light pollution present in urban areas may disrupt patterns of germination and dormancy. Furthermore, adaptation in seed traits can have cascading effects on later life stages (Donohue 2002; Galloway and Etterson 2007; Burghardt et al. 2015). Not only have studies of urban adaptation not focused on these traits, but also some have eliminated any effect they may have on results by planting

> **Box 9.1 Promising areas for future research into plant urban adaptation**
>
> - Early life-history traits such as dormancy, germination, and below-ground traits such as allocation to roots.
> - Role of phenotypic plasticity in facilitating or undermining colonization and adaptation to urban environments.
> - Community-level consequences of urban evolution.
> - Determine if gene flow out of urban environments facilitates or constrains adaptation of non-urban populations.
> - Investigate cities in tropical and southern latitudes.

multiple seeds and then thinning to obtain only a single experimental plant, transplanting young seedlings, or using clones. Seed traits tend to respond plastically, and environmental factors, including temperature during seed development, can drive changes in seed dormancy and subsequent germination timing (Walck et al. 2011; Burghardt et al. 2016). These shifts might be important given that climatic changes in cities, i.e., hotter drier summers and extended growing seasons, may alter the timing of seed production and also the temperature environment plants experience during seed development. These shifts could result in immediate, although not necessarily adaptive, phenotypic changes when plants encounter urban environments that could shift the balance of selection on survival vs reproduction, i.e., semelparity vs iteroparity. That said, the plasticity of germination traits may facilitate establishment of plant populations in urban areas also allow populations to persist long enough to adapt to those environments (Price et al. 2003; Ghalambor et al. 2015).

Changes in timing of plant development, as well as other life-history components, might have consequences beyond the species in which that evolution occurs. In particular, evolution of life history in one species can feedback to affect its own ecology and the fitness of the organisms that interact with it (Lancaster et al. 2017). Given that abundance, composition, and selection imposed by herbivore and pollinator communities (Denys and Schmidt 1998; Meineke et al. 2013; Martins et al. 2017) can change along urban–rural gradients, it seems likely that biotic interactions play an important role in the evolution of urban plant populations. These interactions need not be negative—mutualist interactions as well as the traits that mediate mutualisms might also change in urban environments (Chapter 7; Harrison and Winfree 2015). Consistent with this, pollinators have been shown to selectively favour larger flowers in urban, but not rural populations of some species (Bode and Tong 2018; Irwin et al. 2018). Belowground processes, including root competition, herbivory, and interactions with both mutualistic and pathogenic microbes, may be important (Chapter 7). Although there are few data characterizing plant-associated microbes in urban environments, a survey comparing microbial communities in road medians vs parks *within* Manhattan found differences in both bacteria and fungi communities associated with differences in soil chemical properties (Reese et al. 2016).

Another potentially promising research direction is to investigate the extent to which gene flow out of cities affects the evolution of rural populations. Gene flow has the potential to inhibit or accelerate the rate of adaptation, depending on whether adaptive alleles are introduced or whether non-adaptive alleles overwhelm adaptation (Slatkin 1987). If urban–rural divergence is due to adaptation to environmental differences, we might expect gene flow from cities to rural areas to be maladaptive. The reciprocal common garden experiments from Gorton et al. (2018) appear consistent with this; urban genotypes had lower fitness than rural genotypes in rural common gardens. However, it is also possible that gene flow out of urban areas will be beneficial to rural populations. Cities differ from rural areas in ways that reflect recent and predicted climate change (e.g., warmer temperatures, longer growing season, altered precipitation patterns, higher concentrations of atmospheric CO_2), and each of these environmental changes could directly or indirectly alter selection acting on plant populations (Lau et al. 2014). As such, gene flow from populations adapted to urban environments might provide genetic material that increases the adaptive potential of rural populations to current or future environmental conditions. Evaluating the fitness consequences of gene flow out of urban areas will require estimates of directional gene flow, something that is possible with molecular data (Sousa and Hey 2013), as well as common garden experiments that compare the fitness of urban, rural, and hybrid genotypes in rural areas.

A final, perhaps glaring, bias in current studies is that they have all been conducted at higher latitudes of the northern hemisphere, areas with strong seasonal and temperature climates. This is not unique to plant urban evolution: in an analysis of 110 cities on the impacts of urbanization on plant biodiversity, Aronson et al. (2014) found that fewer than ten studies were conducted in tropical climates or at southern latitudes. However, the effect of urbanization and the urban heat island on plants in temperate climates might have very different biological consequences than in warmer climates and in the southern hemisphere. For example, in seasonal climates, the urban heat island might extend the growing season. By contrast, in tropical environments, urban environments might create stressful growing conditions via higher temperatures and possibly reduced water availability, due to changes in either precipitation or runoff. Thus, differences in the selective outcomes of the same environmental factors in temperate vs tropical urban environments may cause plant life histories to evolve in different ways. Differences between temperate and tropical climates are not limited to direct effects of temperature and precipitation. For example, the heavy use of road salt during the winter in some northern US cities can increase sodium concentrations in roadside soils and the plants that grow in those soils (Bryson and Barker 2002). Parallel studies, using similar methodologies, conducted along urban–rural gradients across broad geographic distributions would be particularly valuable for determining how the effects of urbanization on plant evolution may vary across climate zones.

9.8 Conclusions

Although the study of adaptive evolution of plant life histories is in its early stages, current knowledge provides a baseline of empirical support for plant adaptation to urban environments. These studies clearly show that urban environments can alter selection in ways that drive the adaptive divergence of plant life histories, although more work is needed before we can make robust predictions about how urbanization will affect life-history evolution. In addition to establishing that life histories are subject to adaptive evolution in urban environments, existing studies provide a foundation for the field of urban evolutionary ecology to move forward. We think it will be particularly interesting to focus on specific aspects of urban adaptation that will broaden our understanding of both urban biology and evolutionary processes. For example, identifying the traits that have been most (and least) affected by urban adaptation may not only provide insight into urban environments but also tell us about the evolutionary lability of plant phenotypes. Similarly, more rigorous characterization

of the selective differences between urban and rural environments and identification of the differences that have driven adaptation can provide insight into the efficacy of selection in urban environments, in rapidly changing climates, and likely also in natural environments. Finally, given the proximity of many biologists to urban areas, highly fragmented and dynamic urban landscapes provide the potential to investigate the complex interplay between genetic drift, gene flow, and selection in shaping evolution. Taken together, studies that focus on these areas promise to not only increase our understanding of urban biology—contributing to effective management of plant populations in the face of increasing urbanization and changing climate—but also advance our understanding of plant evolution.

Acknowledgements

We thank Anne Charmantier and two anonymous reviewers whose thoughtful comments on an earlier version of this chapter prompted revisions that greatly improved the manuscript. We also thank the University of Minnesota for a Doctoral Dissertation Fellowship that funded A.J.G. during the writing of this manuscript.

References

Anderson, J.T. (2016). Plant fitness in a rapidly changing world. *New Phytologist*, 210(1), 81–7.

Aronson, M.F.J., La Sorte, F.A., Nilon, C.H., et al. (2014). A global analysis of the impacts of urbanization on bird and plant diversity reveals key anthropogenic drivers. *Proceedings of the Royal Society B: Biological Sciences*, 281(1780), 20133330.

Bartlewicz, J., Vandepitte, K., and Jacquemyn, H. (2015). Population genetic diversity of the clonal self-incompatible herbaceous plant *Linaria vulgaris* along an urbanization gradient. *Biological Journal of the Linnean Society*, 116(3), 603–13.

Bennie, J., Davies, T.W., Cruse, D., and Gaston, K.J. (2016). Ecological effects of artificial light at night on wild plants. *Journal of Ecology*, 104(3), 611–20.

Bode, R.F. and Tong, R. (2018). Pollinators exert positive selection on flower size on urban, but not on rural Scotch broom (*Cytisus scoparius* L. Link). *Journal of Plant Ecology*, 11(3), 493–501.

Bryson, G.M. and Barker, A.V. (2002). Sodium accumulation in soils and plants along Massachusetts roadsides. *Communications in Soil Science and Plant Analyses*, 33(1–2), 67–78.

Burgess, K.S., Etterson, J.R., and Galloway, L.F. (2007). Artificial selection shifts flowering phenology and other correlated traits in an autotetraploid herb. *Heredity*, 99(6), 641–8.

Burghardt, L.T., Metcalf, C.J., Wilczek, A.M., Schmitt, J., and Donohue, K. (2015). Modeling the influence of genetic and environmental variation on the expression of plant life cycles across landscapes. *The American Naturalist*, 185(2), 212–27.

Burghardt, L.T., Edwards, B.R., and Donohue, K. (2016). Multiple paths to similar germination behavior in *Arabidopsis thaliana*. *New Phytologist*, 209(3), 1301–12.

Cheptou, P.-O., Carrue, O., Rouifed, S. and Cantarel, A. (2008). Rapid evolution of seed dispersal in an urban environment in the weed *Crepis sancta*. *Proceedings of the National Academy of Sciences of the United States of America*, 105(10), 3796–9.

Cheptou, P.-O., Hargreaves, A.L., Bonte, D., and Jacquemyn, H. (2017). Adaptation to fragmentation: evolutionary dynamics driven by human influences. *Philosophical Transactions of the Royal Society B*, 372(1712), 20160037.

Conover, D. and Schultz, E.T. (1995). Phenotypic similarity and the evolutionary significance of countergradient variation. *Trends in Ecology & Evolution*, 10(6), 248–52.

Cote, J., Bestion, E., Jacob, S., et al. (2017). Evolution of dispersal strategies and dispersal syndromes in fragmented landscapes. *Ecography*, 40(1), 56–73.

Culley, T.M., Sbita, S.J., and Wick, A. (2007). Population genetic effects of urban habitat fragmentation in the perennial herb *Viola pubescens* (Violaceae) using ISSR markers. *Annals of Botany*, 100(1), 91–100.

Denys, C. and Schmidt, H. (1998). Insect communities on experimental mugwort (*Artemisia vulgaris* L.) plots along an urban gradient. *Oecologia*, 113(2), 269–77.

Donohue, K. (2002). Germination timing influences natural selection on life-history characters in *Arabidopsis thaliana*. *Ecology*, 83(4), 1006–16.

Donohue, K., Rubio de Casas, R., Burghardt, L., Kovach, K., and Willis, C.G.. (2010). Germination, postgermination adaptation, and species ecological ranges. *Annual Review of Ecology, Evolution, and Systematics*, 41, 293–319.

Dubois, J. and Cheptou, P.-O. (2016). Effects of fragmentation on plant adaptation to urban environments. *Philosophical Transactions of the Royal Society B*, 372(1712), 20160038.

Epps, C.W. and Keyghobadi, N. (2015). Landscape genetics in a changing world: disentangling historical and contemporary influences and inferring change. *Molecular Ecology*, 24(24), 6021–40.

Franks, S.J., Sim, S., and Weis, A.E. (2007). Rapid evolution of flowering time by an annual plant in response to a climate fluctuation. *Proceedings of the National Academy of Sciences of the United States of America*, 104(4), 1278–82.

Franks, S.J., Weber, J.J., and Aitken, S.N. (2013). Evolutionary and plastic responses to climate change in terrestrial plant populations. *Evolutionary Applications*, 7(1), 123–39.

Fugère, V. and Hendry, A.P. (2018). Human influences on the strength of phenotypic selection, *Proceedings of the National Academy of Sciences of the United States of America*, 115(40), 10070–75.

Galloway, L.F. and Burgess, K.S. (2009). Manipulation of flowering time: phenological integration and maternal effects. *Ecology*, 90(8), 2139–48.

Galloway, L.F. and Etterson, J.R. (2007). Transgenerational plasticity is adaptive in the wild. *Science*, 318(5853), 1134–6.

Ghalambor, C.K., Hoke, K.L., Ruell, E.W., et al. (2015). Non-adaptive plasticity potentiates rapid adaptive evolution of gene expression in nature. *Nature*, 525(7569), 372–5.

Gorton, A.J., Moeller, D.A., and Tiffin, P. (2018). Little plant, big city: a test of adaptation to urban environments in common ragweed (*Ambrosia artemisiifolia*). *Proceedings of the Royal Society B: Biological Sciences*, 285(1881), 20181579.

Haggerty, B.P. and Galloway, L.F. (2011). Response of individual components of reproductive phenology to growing season length in a monocarpic herb. *Journal of Ecology*, 99(1), 242–53.

Hall, M.C. and Willis, J.H. (2006). Divergent selection on flowering time contributes to local adaptation in *Mimulus guttatus* populations. *Evolution*, 60(12), 2466–77.

Harrison, T. and Winfree, R. (2015). Urban drivers of plant–pollinator interactions. *Functional Ecology*, 29(7), 879–88.

Hoban, S., Kelley, J.L., Lotterhos, K.E., et al. (2016). Finding the genomic basis of local adaptation: pitfalls, practical solutions, and future directions. *The American Naturalist*, 188(4), 379–97.

Irwin, R.E., Warren, P.S., and Adler, L.S. (2018). Phenotypic selection on floral traits in an urban landscape. *Proceedings of the Royal Society B: Biological Sciences*, 285(1884).

Johnson, M.T.J., Thompson, K.A., and Saini, H.S. (2015). Plant evolution in the urban jungle. *American Journal of Botany*, 102(12), 1951–3.

Johnson, M.T.J., Prashad, C.M., Lavoignet, M., and Saini, H.S. (2018). Contrasting the effects of natural selection, genetic drift and gene flow on urban evolution in white clover (*Trifolium repens*). *Proceedings of the Royal Society B: Biological Sciences*, 285(1883).

Kalnay, E. and Cai, M. (2003). Impact of urbanization and land-use change on climate. *Nature*, 423(6939), 528–31.

Kim, E. and Donohue, K. (2013). Local adaptation and plasticity of *Erysimum capitatum* to altitude: its implications for responses to climate change. *Journal of Ecology*, 101(3), 796–805.

Kimball, S., Gremer, J.R., Huxman, T.E., et al. (2013). Phenotypic selection favors missing trait combinations in coexisting annual plants. *The American Naturalist*, 182(2), 191–207.

Kingsolver, J.G., Gomulkiewicz, R., and Carter, P.A. (2001). Variation, selection and evolution of function-valued traits. *Genetica*, 2001(112–113), 87–104.

Kooyers, N.J. (2015). The evolution of drought escape and avoidance in natural herbaceous populations. *Plant Science*, 234, 155–62.

Lambrecht, S.C., Mahieu, S., and Cheptou, P.-O. (2016). Natural selection on plant physiological traits in an urban environment. *Acta Oecologica*, 77, 67–74.

Lancaster, L.T., Morrison, G., and Fitt, R.N. (2017). Life history trade-offs, the intensity of competition, and coexistence in novel and evolving communities under climate change. *Philosophical Transactions of the Royal Society B*, 372, 20160046.

Lande, R. and Arnold, S.J. (1983). The measurement of selection on correlated characters. *Evolution*, 37(6), 1210–26.

Lau, J.A., Shaw, R.G., Reich, P.B., and Tiffin, P. (2014). Indirect effects drive evolutionary responses to global change. *New Phytologist*, 201(1), 35–343.

Leimu, R. and Fischer, M. (2008). A meta-analysis of local adaptation in plants. *PLOS ONE*, 3(12), 1–8.

Lennartsson, T., Nilsson, P., and Tuomi, J. (1998). Induction of overcompensation in the field gentian, *Gentianella campestris*. *Ecology*, 79(3), 1061–72.

Levin, S.A., Muller-Landau, H.C., Nathan, R., and Chaveet, J. (2003). The ecology and evolution of seed dispersal: a theoretical perspective. *Annual Review of Ecology, Evolution, and Systematics*, 34(2003), 575–604.

Levins, R. (1968). *Evolution in Changing Environments: Some Theoretical Explorations*. Princeton University Press, New Jersey.

Levins, R. (1969). Thermal acclimation and heat resistance in *Drosophila* species. *The American Naturalist*, 103(933), 483–99.

Liu, Y., Walck, J.L., and El-Kassaby, Y.A. (2017). Roles of the environment in plant life-history trade-offs. In: Jimenez-Lopez, J.C. (ed.) *Advances in Seed Biology*. InTech Open, London.

Lovell, J.T., Juenger, T.E., Michaels, S.D., et al. (2013). Pleiotropy of FRIGIDA enhances the potential for multivariate adaptation. *Proceedings of the Royal Society B: Biological Sciences*, 280(1763), 20131043.

Lowe, W.H. and McPeek, M.A. (2014). Is dispersal neutral? *Trends in Ecology & Evolution*, 29(8), 444–50.

Martins, K.T., Gonzalez, A., and Lechowicz, M.J. (2017). Patterns of pollinator turnover and increasing diversity associated with urban habitats. *Urban Ecosystems*, 20(6), 1359–71.

Mazer, S.J. and LeBuhn, G. (1999). Genetic variation in life-history traits: heritability estimates within and genetic differentiation among populations. In: Vuorisalo, T.O. and Mutikainen, P.K. (eds) *Life History Evolution in Plants*, pp. 85–172. Kluwer Academic, Dordrecht.

McDonnell, M.J. (2011). The history of urban ecology: an ecologist's perspective. In: Niemelä, J. (ed.) *Urban Ecology: Patterns, Processes and Applications*, pp. 5–13. Oxford University Press, Oxford.

McDonnell, M.J. and Hahs, A.K. (2015). Adaptation and adaptedness of organisms to urban environments. *Annual Review of Ecology, Evolution and Systematics*, 46, 261–80.

Meineke, E.K., Dunn, R.R., Sexton, J.O., and Frank, S.D. (2013). Urban warming drives insect pest abundance on street trees. *PLOS ONE*, 8(3), e59687.

Mitchell, N. and Whitney, K.D. (2018). Can plants evolve to meet a changing climate? The potential of field experimental evolution studies. *American Journal of Botany*, 105(10), 1–4.

Moeller, D.A. (2006). Geographic structure of pollinator communities, reproductive assurance, and the evolution of self-pollination. *Ecology*, 87(6), 1510–22.

Neil, K. and Wu, J. (2006). Effects of urbanization on plant flowering phenology: a review. *Urban Ecosystems*, 9(3), 243–57.

Parmesan, C. (2006). Ecological and evolutionary responses to recent climate change. *Annual Review of Ecology and Systematics*, 37(1), 637–69.

Peterson, M.L., Kay, K.M., and Angert, A.L. (2016). The scale of local adaptation in *Mimulus guttatus*: comparing life history races, ecotypes, and populations. *New Phytologist*, 211(1), 345–56.

Postma, F.M. and Ågren, J. (2016). Early life stages contribute strongly to local adaptation in *Arabidopsis thaliana*. *Proceedings of the National Academy of Sciences of the United States of America*, 113(27), 201606303.

Price, T.D., Qvarnström, A., and Irwin, D.E. (2003). The role of phenotypic plasticity in driving genetic evolution. *Proceedings of the Royal Society B: Biological Sciences*, 270(1523), 1433–40.

Reese, A.T., Savage, A., Youngsteadt, E., et al. (2016). Urban stress is associated with variation in microbial species composition—but not richness—in Manhattan. *ISME Journal*, 10(3), 751–60.

Roff, D.A. (2002). *Life History Evolution*. Sinauer Associates, Sunderland, MA.

Santangelo, J.S., Johnson, M.T., and Ness, R.W. (2018). Urban spandrels: the roles of genetic drift, gene flow and natural selection in the formation of parallel clines. *Proceedings of the Royal Society B: Biological Sciences*, 285, 20180230.

Savolainen, O., Pyhajarvi, T., and Knurr, T. (2007). Gene flow and local adaptation in trees. *Annual Review of Ecology and Systematics*, 38, 595–619.

Savolainen, O., Lascoux, M., and Merila, J. (2013). Ecological genomics of local adaptation. *Nature Reviews Genetics*, 14(11), 807–20.

Schmalenbach, I., Zhang, L., Reymond, M., and Jiménez-Gómez, J.M. (2014). The relationship between flowering time and growth responses to drought in the *Arabidopsis* Landsberg *erecta* × Antwerp-1 population. *Frontiers in Plant Science*, 5(609).

Sih, A., Ferrari, M.C.O., and Harris, D.J. (2011). Evolution and behavioural responses to human-induced rapid environmental change. *Evolutionary Applications*, 4(2), 367–87.

Slatkin, M. (1987). Gene flow and the geographic structure of natural populations. *Science*, 236(4803), 787–92.

Smoliak, B.V., Snyder, P.K., Twine, T.E., Mykleby, P.M., and Hertel, W.F. (2015). Dense network observations of the twin cities canopy-layer urban heat island. *Journal of Applied Meteorology and Climatology*, 54(9), 1899–917.

Sousa, V. and Hey, J. (2013). Understanding the origin of species with genome-scale data: modelling gene flow. *Nature Review Genetics*, 14(6), 404–14.

Stearns, S.C. (1976). Life-history tactics: a review of the ideas. *Quarterly Review of Biology*, 51(1), 3–47.

Thompson, K.A., Renaudin, M., and Johnson, M.T.J. (2016). Urbanization drives parallel adaptive clines in plant populations. *Proceedings of the Royal Society B: Biological Sciences*, 8(11), 1–51.

Wadgymar, S.M., Lowry, D.B., Gould, B.A., et al. (2017). Identifying targets and agents of selection: innovative methods to evaluate the processes that contribute to local adaptation. *Methods in Ecology and Evolution*, 8(6), 738–49.

Walck, J. L., Dixon, K., Thompson, K., and Hidayatiet, S. (2011). Climate change and plant regeneration from seed. *Global Change Biology*, 17(6), 2145–61.

Wilczek, A.M., Cooper, M.D., Korves, T.M., and Schmit, J. (2014). Lagging adaptation to warming climate in *Arabidopsis thaliana*. *Proceedings of the National Academy of Sciences of the United States of America*, 111(22), 7906–13.

Williams, J.L. (2009). Flowering life-history strategies differ between the native and introduced ranges of a monocarpic perennial. *The American Naturalist*, 174(5), 660–72.

Yakub, M. and Tiffin, P. (2017). Living in the city: urban environments shape the evolution of a native annual plant. *Global Change Biology*, 23(5), 2082–9.

Zipper, S.C, Schatz, J., Kucharik, C.J., and Loheide, S.P. (2017). Urban heat island-induced increases in evapotranspirative demand. *Geophysical Research Letters*, 44(2), 873–81.

Ziska, L.H., Gebhard, D.E., Frenz, D.A., et al. (2003). Cities as harbingers of climate change: common ragweed, urbanization, and public health. *Journal of Allergy and Clinical Immunology*, 111(2), 290–95.

CHAPTER 10

Urbanization and Evolution in Aquatic Environments

R. Brian Langerhans and Elizabeth M.A. Kern

Langerhans, R.B. and Kern, E.M.A., *Urbanization and Evolution in Aquatic Environments* In: *Urban Evolutionary Biology*. Edited by Marta Szulkin, Jason Munshi-South and Anne Charmantier, Oxford University Press (2020). © Oxford University Press.
DOI: 10.1093/oso/9780198836841.003.0010

10.1 Introduction

Throughout the history of our species, we have typically built our settlements, and especially city centres, near water—a pattern that persists to this day (Kummu et al. 2011). This is because of not only the necessity of freshwater for life, but also the utility of water in travel, commerce, food production, security, power generation, and other human uses. Unfortunately, this means that urbanization has had a heavy impact on aquatic species (Vörösmarty et al. 2010). This impact has been well studied from an ecological perspective (Paul and Meyer 2001; Wenger et al. 2009), but the effect of urbanization on aquatic species' evolution is just beginning to be recognized.

Understanding evolutionary responses to urbanization has many practical applications, from human health to pest control to designing sustainable cities (Alberti et al. 2017; Johnson and Munshi-South 2017). Contemporary evolution may affect species persistence, inform conservation decisions, interact with ecological processes and ecosystem services, and even alter how human impacts are gauged. For example, measuring the toxicity of a specific pollutant will lead to inaccurate guidelines if the assayed species is from a population with evolved tolerance (Brady et al. 2017). Owing to the ubiquity of cities, urbanization also presents opportunities to explore evolutionary questions using many replicates and a 'natural experiment' design (e.g., see Chapter 3).

Here we discuss how evolution in aquatic systems is influenced by four major categories of changes that cities can cause: (1) biotic interactions, (2) physical environment, (3) temperature, and (4) pollution. In reviewing evolutionary impacts, we focus on selection, but also discuss genetic drift and gene flow. Along the way, we take into account as many aquatic taxa as possible: plants, algae, invertebrates, fish, and other water-associated organisms (e.g., amphibians, reptiles, and insects with an aquatic larval stage), and we consider all types of aquatic habitats, from backyard puddles to sweeping coastal waters. Despite casting this broad net, we find that research on evolution in urban aquatic habitats is still nascent, with a modest number of truly well-documented cases. We highlight such cases where possible, but in areas where no research has been done, we use theory and empirical studies from

other systems to outline testable hypotheses about how evolution might be proceeding (Table 10.1).

The timescale of urban evolution is typically viewed on the scale of years to decades—indeed, the burgeoning research on the topic has largely emerged from studies of contemporary evolution. Consequently, most examples comprise such recent evolutionary change. However, cities first began altering aquatic habitats thousands of years ago. For example, cities of Mesopotamia included large-scale systems of dikes, dams, canals, levees, and gated ditches, causing major alterations to river environments—modifications occurring as early as 5000 years ago. More recently, the Aztecs built their huge city Tenochtitlán on an island in Lake Texcoco, which flourished from 1325 to 1521. The Aztecs heavily modified the aquatic environment, with a large, complex system of canals, causeways, and aqueducts. Thus, city-induced changes to aquatic environments have likely been causing evolution in aquatic organisms for longer timescales than the past several decades. Nevertheless, we inevitably focus on more recent impacts owing to the paucity of studies on longer timescales.

10.2 Biotic interactions

We begin by briefly reviewing the kinds of evolutionary changes that can be caused by disruptions in predator–prey dynamics, competition, and resource–consumer interactions (diet). Urbanization affects these biotic interactions in a myriad of ways: for instance, artificial light can increase night-time predation and alter foraging strategies (Dwyer et al. 2013); pharmaceuticals in urban wastewater affect species' feeding rates (Brodin et al. 2013); and modified structural habitat can alter prey availability and foraging efficiency (Smokorowski and Pratt 2007; Bulleri and Chapman 2010). More directly, urbanization alters biotic interactions by causing the decline or disappearance of sensitive species (especially benthic macroinvertebrates; Brown et al. 2009), introducing non-natives, and increasing the abundance of species that prefer the altered conditions (e.g., when slow-water species colonize a reservoir) (Paul and Meyer 2001).

How species in aquatic habitats respond evolutionarily to urban changes in biotic interactions is

Table 10.1 Major impacts of urbanization on evolution in aquatic species reviewed in this chapter.

Type of urban impact	Types of changes	Likely selection targets
BIOTIC INTERACTIONS		
Predation	Novel predators Loss of predators Changes in predator abundance or behaviour Higher/lower predation pressure	Prey life history, morphology, chemical defenses, anti-predator behaviour
Competition	Novel competitors Loss of competitors Changes in competitor abundance or behaviour Higher/lower intraspecific competition	Foraging and feeding behaviour, body size, trophic traits, diet specialization
Diet	Changes in abundance and diversity of food, consumer behaviour, or foraging habitat Less terrestrial inputs to streams	Broader/narrower niche width, gut length, eye position and morphology, locomotor traits, trophic traits
PHYSICAL ENVIRONMENT		
Habitat fragmentation	Smaller populations More isolated populations Reduced movement distances Altered habitat	Life-history strategies, dispersal traits, body size, locomotor traits, mating system, sexual signals, anti-predator behaviour, trophic traits
Urban stream flow	Increased stream flashiness Higher maximum velocity and variance	Body morphology, fin morphology, body size, swimming performance
TEMPERATURE		
Urban heat island effect	Warmer habitat Longer growing seasons	Timing of spring and autumn events, morphology, body size, pace-of-life syndrome traits, growth rate, thermal tolerance, sex determination
POLLUTION		
Metals/inorganics	Higher concentrations of lead, zinc, copper, cadmium, chromium, arsenic, nickel, and salt	Resistance/tolerance mechanisms
Synthetic organic compounds	Presence of PHCs, PAHs, PCBs, endocrine disruptors, antibiotics, and pesticides Altered ecosystem productivity Altered chemical signalling	Resistance/tolerance mechanisms
Artificial light at night	Altered light intensity and spectra at night Altered community composition, predator size, and behaviour Disruption of hormone expression	Diel behavioural patterns, movement/migration traits, sexual signals, phenology, endocrine systems, foraging traits, schooling behaviour
Anthropogenic sound	Elevated noise levels Altered frequencies of background sounds	Acoustic signal and receiver traits, endocrine systems, social behaviours, foraging traits, stress response, schooling behaviour
Nutrients and suspended particles	Increased presence of sewage and nitrogen Increased turbidity Lower dissolved oxygen levels Eutrophication Phytoplankton blooms	Toxin resistance, body size, oxygen uptake mechanisms, blood pigments, metabolic rate, visual signal traits

PHCs, petroleum hydrocarbons; PAHs, polycyclic aromatic hydrocarbons; PCBs, polychlorinated biphenyls.

mostly unknown. However, prior work in urban terrestrial systems and the large body of research on evolution in aquatic species allow us to predict how urban aquatic species should evolve in response to altered predation, competition, and diet. We will refer to mosquitofish (*Gambusia* spp.) several times here and throughout the chapter because it has become a model genus for studying rapid and human-induced evolutionary changes.

10.2.1 Predation

Predation strongly influences evolution, with many well-documented examples from aquatic species. To name a few, threespine stickleback, poeciliid live-bearing fishes, and crustacean zooplankton have all served as models for how predation drives evolutionary divergence. Although there is not yet enough research for us to generalize how predation regimes affect evolution in urban aquatic systems, it seems reasonable to predict that patterns will probably be similar to those already documented in natural and experimental settings, where predation commonly shapes life histories, morphologies, chemical defences, behaviours, and locomotor abilities.

We offer two pieces of advice to future researchers investigating such predictions. First, an important initial step will be to determine whether the urban population of interest is under higher, lower, or otherwise altered predation pressure. A general pattern of lower predation pressure (and, paradoxically, higher predator density) has been identified among urban vertebrates on land, but hasn't been investigated in the water (Fischer et al. 2012). Importantly, prey often evolve trait changes in response to increases *or* decreases in predation risk from a given source, as well as in response to novel predators and release from predation entirely.

Second, in urban settings, careful study design is necessary to demonstrate that predation, and not some other covarying factor(s), is the driving agent behind the evolution of prey traits. This is because other aspects of urbanization that also influence mortality rates, morphology, behaviour, and life history could be confounded with altered predation regimes. To give just two examples, roadside salt reduces embryonic survival in salamanders (Brady 2012) and both predation and urban light pollution

independently affect vertical migration in zooplankton (Gliwicz 1986; Moore et al. 2001). In such situations, model-selection approaches are useful. Research on Bahamian mosquitofish (*Gambusia* spp.) in tidal creeks—where road construction across the creeks has caused dramatic ecological changes that include reduction of predatory fish—has employed model-selection approaches to identify the putative agents underlying phenotypic changes. Changes in predation by piscivorous fishes was found to drive trait changes over ~ 35–50 years in male genital morphology, muscle mass, and fat content, but not male colouration (Heinen-Kay et al. 2014; Giery et al. 2015; Riesch et al. 2015).

10.2.2 Competition

Competition is another strong evolutionary force that can be disrupted by urbanization. A meta-analysis of mostly North American studies indicated that for fish, birds, and amphibians, urban habitats are less densely populated, while macroinvertebrate population densities are higher in urban sites (Sievers et al. 2018). Whether these density changes result in higher or lower competition depends of course on habitat quality and resource abundance, since low densities could just as well correlate with more or scarcer prey. Additionally, competition in urban waters can be affected by altered primary productivity, and for parasites, competition might be affected by increased susceptibility of a host population that is stressed by pollution and warm urban temperatures.

Changes in competition could have several evolutionary impacts. For one thing, competing with novel species can alter selection for traits involved in reducing interference or exploitative competition. Character displacement, where species evolve phenotypic differences to reduce competition for the same resources, should occur when urban conditions increase competition among species. On the other hand, when urban conditions decrease interspecific competition (e.g., extirpation of competitors), this can also lead to changes in traits to better exploit the newly available resources. For instance, as road construction that restricts connectivity between tidal creeks and the ocean has led to reductions of interspecific competitors of Bahamian mosquitofish, their populations have subsequently shown increases in body size (Riesch et al. 2015). Urban environments can affect intraspecific competition as well. Elevated intraspecific competition can lead to the evolution of prey specialization. Bahamian mosquitofish increase individual diet specialization subsequent to anthropogenically induced changes in competition (Araujo et al. 2014).

10.2.3 Diet

In addition to changes in competition, there are many reasons why urban aquatic species might consume different food resources. Urbanization alters the abundance, distribution, and diversity of food, affects consumer behaviour, and modifies the structural habitat consumers navigate while foraging (Smokorowski and Pratt 2007; El-Sabaawi 2018). For instance, urban coastal infrastructure such as breakwaters, jetties, and seawalls support altered epibiota and fish assemblages (Bulleri and Chapman 2010). As another example, urban streams and lakes receive less allochthonous inputs (e.g., terrestrial insects and vegetation fragments) from their immediate surroundings. Such changes can cause dietary shifts. Stable isotope analysis suggests that turtles have different diets in urban settings (Ferronato et al. 2016), while urban freshwater fish consistently eat fewer terrestrial insects and fewer sensitive aquatic insects, relying more heavily on detritus and pollution-tolerant fly larvae (i.e., Diptera) (Francis and Schindler 2009). Not surprisingly, urban diet changes may be taxa-specific. Road construction across tidal creeks has led to altered diets in Bahamian mosquitofish and grey snapper (*Lutjanus griseus*), but in very different manners owing to their different trophic positions (Layman et al. 2007; Araujo et al. 2014).

Dietary changes are notorious for driving evolutionary change, even over short timescales, but specific types of evolutionary changes in urban aquatic species have yet to be described. Some logical areas for hypothesis testing would be evolutionary changes in body size, gut length, trophic morphology, eye position and morphology, and locomotor traits, or changes in the plasticity of these traits. As with other biotic interactions discussed in this section, evolutionary changes caused by dietary shifts in urban aquatic species are a 'frontier topic', an understudied

area with rich potential for illustrating major ideas in evolutionary biology.

10.3 Physical environment

Urbanization transforms the physical landscape in and around aquatic habitats. While urbanization can greatly modify the structural habitat within which an organism resides (potentially altering its diet, foraging and anti-predator behaviours, and locomotor demands; see section 10.2 and Chapter 12), we focus here on two of the major consequences of urban-induced physical alterations: aquatic habitat fragmentation and disrupted hydrologic regimes. In this section we briefly address how urbanized watersheds impede organism movement and dispersal, and how altered flow regimes are driving evolution in urban streams.

10.3.1 Habitat fragmentation

Barriers that commonly fragment aquatic habitat and interfere with organism movement (reviewed in Fuller et al. 2015) can be quite conspicuous, such as roads, culverts, dams, buried channels, and stream dewatering, while other barriers like thermal pollution (e.g., heated water discharge from manufacturing plants) and wastewater plumes (e.g., treated or untreated sewage fluid moving through waterways) are less visible to the naked eye. Some types of barriers may be surprising, such as artificial night lighting that attracts flying aquatic insects and prevents them from migrating (Perkin et al. 2014) or reduces insect dispersal by affecting larval drift (Henn et al. 2014). Aquatic species that disperse overland, like turtles, amphibians, and some aquatic insects, are also often, but not always, affected by urban habitat fragmentation (e.g., flying aquatic insects can be fatally attracted to polarized reflected light on urban surfaces (reviewed in Smith et al. 2009)). Consequences of a given barrier might be difficult to predict, since closely related species can respond differently to the same barrier: for example, when it comes to travelling upstream through culverts, three crayfish species within a single genus significantly varied in their impedance velocity, meaning the same water speed can be passable to one species but not to a congener (Foster and Keller 2011). To further complicate matters, fragmentation (like other urban changes) can come with a plethora of additional environmental impacts. For instance, fragmentation by roads alone can involve multifarious shifts in selection via changes to runoff patterns, chemical pollution, and light levels, among others (Trombulak and Frissell 2000).

Fragmentation tends to lead to smaller, more isolated populations (although a small proportion of species actually increase in population size or in mobility in urban settings) (Sievers et al. 2018). In theory, since fragmentation decreases gene flow and reduces effective population size, it should increase inbreeding, genetic homogeneity, and genetic drift, while potentially facilitating local adaptation if reduced diversity and drift are not severe (see Chapter 4). To date, many studies have investigated the genetic effects of urban-induced aquatic fragmentation, especially for semi-aquatic organisms, and have overall yielded mixed results: many cases of reduced genetic diversity and increased population genetic differentiation have been uncovered, but this finding is not ubiquitous (e.g., Smith et al. 2009; Mather et al. 2015; Benjamin et al. 2016; Lourenco et al. 2017). It is further important to consider outside factors (such as life history, dispersal ability, small amounts of gene flow, or fragmentation timescale) in such studies of fragmentation effects, as they can have an equal or greater influence on how fragmented populations evolve (Ewers and Didham 2006).

In contrast to the attention on genetic impacts, research on the evolutionary effects of urban aquatic habitat fragmentation on phenotypes is scarce. Theoretically, targets of selection from fragmentation (reviewed in Cheptou et al. 2017) should include life-history strategies (such as those affecting dispersal between fragments), niche shifts (towards more generalist niches or more edge-habitat usage), mating systems (favouring strategies like self-fertilization that are advantageous in small populations), and dispersal traits (although there are arguments for the advantages of both higher and lower dispersal rates). Another possible consequence of fragmentation is altered sexual selection regimes (see Chapter 14). Altered and strengthened sexual selection subsequent to fragmentation has led to enhanced flight performance in damselflies (*Coenagrion puella*) (Tüzün

et al. 2017a) and altered male genital shape and the allometry of male genital size in Bahamian mosquitofish (Heinen-Kay et al. 2014).

With the diverse array of possible adaptations and few empirical studies, we are a long way from being able to describe overall trends. However, some existing evidence and hypotheses do point toward future directions in fragmented aquatic systems. For amphibians, it has been proposed that selection in fragmented urban environments should favour 'philopatry, relaxed anti-predator behaviour, and larger body size', since there are fewer predators in isolated urban wetlands and since dispersal into the hostile urban matrix is not likely to result in higher fitness (Munshi-South et al. 2013). For fish, predictions for responses to fragmentation appear to depend on the focal species. In fragmented tidal creeks throughout the Bahamian archipelago, a top fish predator shows narrower trophic niche width (due to decreased prey diversity), while a small, livebearing fish shows a broader food niche (due to higher resource competition) (Layman et al. 2007; Araujo et al. 2014). Such heterogeneity of responses can even occur among close relatives: three mosquitofish species show different kinds of changes in dorsal-fin colouration as a result of habitat fragmentation (Giery et al. 2015). Clearly, expectations for altered selection, and consequently trait evolution, will depend on taxa-specific natural history.

10.3.2 Urban stream flow

Urbanization strongly affects the hydrological regime of many types of aquatic habitat, especially wetlands, stormwater retention ponds, and lotic environments (i.e., rivers and streams) (Jacobson 2011). Owing to their well-documented impacts and relatively clear predictions for altered selection regimes, we centre on urban-induced changes in flow regimes of lotic habitats. One of the most widespread impacts of urbanization on lotic environments occurs via the increased imperviousness of urban land cover. This causes precipitation to reach streams much more quickly, causing water levels to rise and fall sharply. These 'flashy' urban streams have increased maximum water velocity and variance. Flashy streams consequently increase scouring of algal and plant communities, transport nutrients downstream more rapidly, and physically move fauna at a higher rate than less flashy rural streams (Wenger et al. 2009; Jacobson 2011). Notably, not all streams respond by becoming flashy in urban environments: streams in arid areas (surrounded by hardpan) and tropical areas (with already saturated soil from frequent, heavy rainstorms) are flashy by nature, and don't show much difference in flow regime when urbanized (Brown et al. 2009). Very small streams and those in steep drainage areas are also naturally somewhat flashy. Additionally, sometimes human activities reduce flow variability instead of increasing it, as when streamflow is artificially regulated for human purposes. Water levels can also become severely depleted in urban streams and rivers due to diversion for commercial and residential uses in cities.

To date, very few studies have looked at the evolutionary impacts of these dramatic changes in flow regime, but there is evidence that fish respond to flashy urban streamflow regimes by evolving changes in body morphology that affect manoeuvrability or endurance (Kern and Langerhans 2018; Pease et al. 2018) and changes in swimming abilities (Nelson et al. 2003; Nelson et al. 2008; Kern and Langerhans 2019) or swimming plasticity (Nelson et al. 2015) that are related to locomotor efficiency. Both morphology and swimming ability in fishes vary with water velocity in undisturbed systems (Langerhans 2008), so it is likely that these changes in urban fish are caused by altered flow regimes selecting for higher swimming efficiency in fast-moving water. Notably, different fish species exhibit different morphological responses to urbanization—even within the same drainage—apparently reflecting multiple morphological solutions for enhancing steady-swimming performance (Kern and Langerhans 2019). Future work could consider investigating the generality of these findings, the genes underlying fish body shape and locomotor change in urban settings, and hydrological impacts on urban aquatic invertebrates.

10.4 Temperature

Aquatic habitats in cities are often warmer due to the urban heat island effect (see Chapter 6), removal of riparian vegetation, stream widening, water removal, and heated water discharge from treatment plants, power plants, or factories (e.g., Brans et al. 2018; also see Chapter 11). Urban streams and ponds are also characterized by dramatic temporary temperature spikes, brought by stormwater running off hot pavements and other urban surfaces (e.g., Somers et al. 2013). The effect of urbanization on water temperature does, however, depend on local climate—the largest temperature jumps occur when temperate forests are urbanized, while in desert biomes cities may actually be cooler during the day than their natural surroundings (Imhoff et al. 2010). Urban temperatures can strongly influence which aquatic species persist in these settings, and species that can take the heat should experience subsequent evolutionary changes.

There is a sizeable research gap on the evolutionary impact of urban heating on aquatic species, with the notable exception of careful work in the water flea (*Daphnia magna*) by K.I. Brans and colleagues (Brans et al. 2017; Brans and De Meester 2018; see also Chapter 11). However, extrapolating both from these studies and from climate-change research, it seems probable that in urban-tolerant species, traits related to phenology, morphology, size, and sex determination could often evolve in response to urban heating.

10.4.1 Phenology

Few studies have addressed aquatic species' phenological responses to urban heat, although it is known, for example, that the seasonal larval abundance of *Culex* mosquitoes peaks earlier in urban areas (Townroe and Callaghan 2014), and some (but not all) odonates show urban shifts in flight dates (Villalobos-Jimenez and Hassall 2017). Taking a hint from the vast literature on climate change and phenology, we might make a number of predictions: (1) urban heat islands will select for earlier timing of spring biological events and later timing of autumn events; (2) species in middle latitudes might have stronger responses to urban warming than species in tropical zones, due to the former's evolutionary history of reliance on temperature as a phenological cue; (3) small invertebrates, amphibians, and larval bony fishes might have the strongest phenological responses to temperature change; and (4) responses will likely be highly heterogeneous among taxa (Cohen et al. 2018). Each of these pre-dictions is based on phenotypic data: as yet, we know little about how these trends will extend to genetic changes.

10.4.2 Morphology

Increased urban water temperature might have evo-lutionary consequences for morphological traits. One pathway to this outcome might be through phenology evolution, which has downstream con-sequences for morphology, size, growth, and preda-tor defences. Another pathway might be through ontogeny: fish morphology can be plastically altered by rearing temperature, and this plasticity has a genetic basis and can evolve (Ramler et al. 2014). There might be an adaptive advantage to certain shape changes associated with temperature: in a latitudinal survey of eastern mosquitofish (*Gambusia holbrooki*), fish with shallower bodies and smaller heads were associated with colder climates (Riesch et al. 2018). Still, there are substantial gaps in research about how widespread such morphological responses are, their plastic vs genetic basis, and whether responses to temperature are parallel across different species.

10.4.3 Body size and pace-of-life

General ecological rules predict that in hotter habi-tats, smaller species should typically be more suc-cessful. In controlled experiments, ectotherms—and especially aquatic species—mature at smaller sizes at higher temperatures (Forster et al. 2012), and in the wild the trend towards smaller sizes with increas-ing temperatures often holds true on the pheno-typic, population, or community level in aquatic bacteria, fish, plankton, and diatoms (Daufresne et al. 2009; Winder et al. 2009; Riesch et al. 2018). This suggests that we might expect selection in hot urban habitats to favour smaller body sizes. Experimental studies have shown that urban *Daphnia* populations do evolve smaller body sizes, which in turn contributes to thermal tolerance (Brans et al. 2017). On the other hand, rotifer and ostracod assem-blages in urban ponds do not show a strong effect of temperature on mean community body size (in con-trast to cladocerans). This could mean that different

taxa will experience different temperature-induced selection, or that for some species, selection for smaller body size might conflict with selection for larger sizes to facilitate adequate dispersal across fragmented urban landscapes (Merckx et al. 2018).

Body size change has received considerable atten-tion in amphibians in an effort to understand the impact of long-term global warming trends. It appears that for this group, size response to tem-perature varies from one species to another and can sometimes depend more highly on other variables like precipitation, density, or prey abundance, rather than temperature per se (Sheridan et al. 2018). Past studies have mostly focused on decades of frog data under a slowly warming climate, but future research conducted in urban heat islands, where temperat-ures have risen much higher and faster than overall global warming, might be useful (together with common garden or reciprocal transplant experi-ments) in predicting future responses of amphibians to climate change.

Predicted smaller body sizes in warmer urban environments are also part of a broader expect-ation for many organisms experiencing increased temperatures: selection for a faster 'pace-of-life'. The pace-of-life concept is an extension of life-history theory, describing a life-history syndrome, or a suite of covarying life-history traits, where fast-living organisms exhibit rapid growth and maturation, small body size, high fecundity, and potentially reduced lifespan (Debecker et al. 2016). For instance, eastern mosquitofish exhibit a smaller body size, larger reproductive investment, and smaller offspring size in colder environments (Riesch et al. 2018). This fish species is ubiquitous in urban waters, and while temperature can affect phenotypes in both its native and introduced ranges (Ouyang et al. 2018b; Riesch et al. 2018), specific impacts of urban warm-ing have not yet been examined. In *D. magna*, urban populations have evolved faster maturation, earlier release of progeny, and a smaller body size at matur-ity (Brans and De Meester 2018). These phenotypic changes could prove relatively common in some aquatic invertebrates.

Aquatic insect larvae may often exhibit higher growth rates in warmer urban settings. The larvae of various mosquito species develop faster at higher temperatures and fare better in cities, and this rapid larval development is part of the reason why differences in the incidence of dengue fever (spread by mosquitoes) among neighbourhoods of a major city are explained more by temperature differences than by income level or even human population density (Araujo et al. 2015). This might in turn have implications for virulence evolution. In contrast to mosquitos, damselfly larvae from urban habitats show a genetically based tendency to mature more slowly, possibly because longer growing seasons in warm urban habitats have relaxed selection for fast growth rates (Tüzün et al. 2017b). Considering the well-documented urban heat island effect and the clear-cut predictions in many cases for selection on thermal tolerance, thermal adaptation in aquatic species merits much more attention.

10.4.4 Sex determination

Many egg-laying reptiles and some fish have temperature-dependent sex determination, and at least two studies have shown that sex ratios in snapping turtles and sea turtles can be altered by nest temperature differences caused by shading in urban sites. In these cases, residential vegetation or coastal development actually lowered nest temperatures, resulting in biased hatchling sex ratios (Hanson et al. 1998; Kolbe and Janzen 2002). Temperature-dependent sex determination could often vary between populations with different thermal regimes (for instance, different leatherback turtle populations have different temperature ranges at which males and females are produced) (Chevalier et al. 1999), so it would be interesting to investigate whether urban populations show signs of evolving changes in sex determination. Unfortunately, although the trait of temperature-dependent sex determination can have high heritability, the rate of its adaptive evolution may be too slow in long-lived reptiles to keep pace with urban warming (or even global warming) (Mitchell and Janzen 2010).

The possible evolutionary consequences of anthropogenically altered sex ratios are not clear-cut, partly because the adaptive value of environmental sex determination is rather enigmatic itself. However, this topic is especially important to urban evolutionary biology because multiple aspects of urbanization—temperature, nitrates, pH, and

endocrine disrupting compounds—can affect sex determination in aquatic species (Wedekind 2017). We discuss these latter impacts further in the next section.

10.5 Pollution

The effects of pollution on ecology and human health have long been topics of intense research. Perhaps for this reason, evolutionary responses to pollution are fairly well documented compared to other human impacts. This has allowed researchers to delve deeper into this topic, beyond the simple question of 'Do species evolve pollution tolerance?' and into questions like whether tolerance involves shared mechanisms (Whitehead et al. 2017), how evolution affects persistence (Veprauskas et al. 2018), and the fitness costs that might accompany adaptation to pollutants (Pedrosa et al. 2017). Since urban pollution takes many forms, we divide this section into inorganic pollutants, synthetic compounds, light, sound, and nutrients (thermal pollution is treated separately, under section 10.4).

10.5.1 Metals and other inorganic pollutants

Common pollutants from urban stormwater runoff and atmospheric deposition include a number of metals or metalloids such as lead, zinc, copper, cadmium, chromium, arsenic, and nickel. In regions where road salting is common in winter, salt is another major urban inorganic pollutant. Metal tolerance has been documented in exposed aquatic species too many times to mention here (reviewed in Klerks and Weis 1987), in taxa ranging from fish to algae to isopods to coral to oligochaetes. Since the genetic basis for observed tolerance is not often tested, many questions about the evolution of tolerance remain unanswered, but research in specific areas, like the metallothionein gene and its regulatory regions, has already revealed some common patterns in how metal tolerance evolves (Janssens et al. 2008).

Despite widespread documentation of metal tolerance, sometimes resistance has surprisingly not been found or has even proceeded in the opposite direction. Exposed *Daphnia* actually evolved reduced tolerance to copper and cadmium (Rogalski 2017),

possibly due to genotoxic effects of these metals. It
is clear that metal tolerance evolution can be fairly
complex: duckweed more easily acquires tolerance
to some metals over others (Van Steveninck et al.
1992), while in algae, resistance to one metal often
confers resistance to another (Hall 1980). Temperature
stress interacts with pollution stress (Sokolova and
Lannig 2008), as do infectious disease and parasites
(Morley 2010), and multiple antibiotic resistances
(Sabry et al. 1997). One factor in determining pollution resistance may be genetic diversity, which in
turn is itself affected by pollutants (Maes et al. 2005).

The widespread salinization of rivers is generally
thought to primarily lead to a shift to salt-tolerant
fauna, especially eliminating salt-sensitive insects
(Buchwalter et al. 2008). Some fish are capable of
adapting to increased salinity (Brennan et al. 2016),
and genetic variation in salt tolerance suggests that
newts possess at least the raw material needed for
evolutionary responses (Hopkins et al. 2013), but
the ability of some insects to rapidly adapt to altered
salt conditions requires further study (Kefford et al.
2016). There is strong suggestive evidence, supported by a reciprocal transplant experiment, that
salamanders in roadside habitats have adaptively
evolved to handle saltier conditions than salamanders in neighbouring, unaffected habitats (Brady
2012). Surprisingly, however, amphibians from salt-
contaminated pools showed reduced chloride resistance in lab experiments, despite apparent adaptation
in the field (Brady 2012; Brady et al. 2017). Additional
taxa will need to be studied to decipher the trends
and ecological impacts of adaptive evolution to salt
in aquatic environments.

10.5.2 Synthetic organic compounds, endocrine disruptors, and antibiotics

An unpleasant assortment of synthetic pollutants
are common in urban waters. Runoff from paved
surfaces brings gasoline-related compounds (e.g.,
petroleum hydrocarbons and polycyclic aromatic
hydrocarbons), while a broad array of endocrine
disruptors, pharmaceuticals, and personal care products easily pass through wastewater treatment plants
and enter surface waters around the world (Ebele
et al. 2017). Other synthetic organic compounds like
pesticides regularly occur in urban water bodies,

and an emerging concern is microplastics, which
allow persistent organic pollutants to enter aquatic
food webs (do Sul and Costa 2014).

What are the evolutionary consequences of these
pollutants? Loss of genetic variation is one response
(Fasola et al. 2015), since pollution can greatly reduce
population sizes or generate a selective sweep.
Although other evolutionary responses are not
well studied, the immediate biological or ecological
impacts of synthetic organic compounds have been
documented in a large number of species and systems, and include changes in ecosystem productivity, fish behaviour (aggression, boldness, predation,
migration, activeness, social behaviours), insect
emergence, and chemical signalling (Brodin et al.
2013; Van Donk et al. 2016; Richmond et al. 2017).
The diversity of impacts makes it difficult to generalize or enumerate all the possible changes in
selection they could induce. On the simplest level,
however, we know that the evolution of resistance
can occur: mosquitos (*Anopheles gambiae*) to insecticide (Kamdem et al. 2017); amphipods (*Hyalella
azteca*) to pyrethroid insecticide (Major et al. 2018);
golden shiners (*Notemigonus crysoleucas*) to the
piscicide rotenone (Orciari 1979); and tomcod
(*Microgadus tomcod*) and killifish (*Fundulus heteroclitus*)
to polychlorinated biphenyls (PCBs) (Wirgin et al.
2011; Whitehead et al. 2017).

Endocrine disrupting compounds (oestrogenic,
androgenic, and thyroidal) from human wastewater have been a topic of particularly intense focus.
These compounds mimic hormones and have reproductive toxicity in aquatic life, causing vitellogenesis,
feminization, and deformities. Models of the evolutionary consequences of disturbed sex determination
indicate various potential consequences: the extinction of a sex chromosome, switching to a different
sex-determination system (reviewed in Wedekind
2017), or altered sensitivity and reproductive behaviour (Mitchell and Janzen 2010). However, evidence
for any of these changes is scant, and researchers
have predicted that extinction may be more likely
than evolutionary rescue in this situation (Mizoguchi
and Valenzuela 2016).

Urban waters often contain a range of antibiotics,
as well as bacterial groups carrying and disseminat-
ing antibiotic resistance genes (Rizzo et al. 2013).
Urban wastewater treatment plants represent the

primary source of antibiotics in aquatic environments, and thus regions closest to these point sources often harbour the greatest concentration of both antibiotics and antibiotic resistance genes. The occurrence of antibiotics can drive strong selection for antibiotic resistance genes and antibiotic-resistant bacteria. Urban waters could thus serve as reservoirs for rapid bacterial evolution that can pose serious health risks to humans and animals. Future research is needed to understand the extent and nature of such evolution, its impacts, and ways to mitigate these concerns.

10.5.3 Light pollution

The effect of artificial light at night is an active area of research in ecology and public health. Already several reviews have covered its impact on aquatic and marine ecology (e.g., Rich and Longcore 2006; Gaston et al. 2014). Evolutionary impacts, on the other hand, have largely gone untested (Swaddle et al. 2015; Hopkins et al. 2018), but known ecological impacts point toward where evolutionary forces may be at work.

First, artificial light has well-documented effects on freshwater and marine community composition, predator size and presence, diel activity patterns (i.e., behaviours over a 24-hour period), predation behaviour, and prey assemblages (e.g., Dwyer et al. 2013; Bolton et al. 2017), all of which can have downstream evolutionary consequences (see section 10.2). Altered diel behavioural patterns can particularly influence urban evolutionary trajectories, as these can obviously alter interactions with predators, competitors, resources, and structural habitat, and could even affect hybridization risk, with sympatric heterospecifics typically isolated from one another due to different timings of mating behaviours.

A second mechanism by which light may influence evolution is through sexual selection or changes in reproductive biology. Night lighting disrupts circadian rhythms and hormone production. For example, light can decrease gene expression of gonadotropin, an important reproductive hormone, in perch (*Perca fluviatilis*) (Bruning et al. 2016); change melatonin (which regulates reproduction) levels in carp (*Catla catla*) (Maitra et al. 2013); and affect reproductive biology by causing gene expression changes that lead to ovarian tumours in zebrafish (*Danio rerio*) (Khan et al. 2018). Pheromones are important to aquatic animals and anurans, and artificial light could shape evolution by affecting pheromone-mediated sexual selection (Henneken and Jones 2017), or mate choice and mating efficiency (Botha et al. 2017). How reproductive biology is altered by artificial light may vary by species; for instance, grey treefrogs (*Hyla versicolor*) showed no mating response to light (Underhill and Höbel 2018), while seven other frog species respond to artificial light by decreasing calls (Hall 2016), perhaps a previously adaptive response to increased predation risk on moonlit nights.

Third, light might generate selection for changes in species movement (Gaston et al. 2014), including migration and drift in aquatic invertebrates. Cities could select for non-migrating and non-drifting invertebrates (Rich and Longcore 2006); animals like zooplankton usually migrate at night to avoid fish predation, but if those formerly safe drift/migrate times are now fraught with predators, migration could be selected against. Predation does in fact affect vertical migration in zooplankton (Gliwicz 1986), and urban light pollution decreases vertical migration in *Daphnia*, although whether that is due to predation or is a plastic response to illumination is unknown (Moore et al. 2001).

Besides predation, sexual selection, and movement, there are a host of other studied impacts that might have evolutionary consequences. One such target of selection might be phenology, because artificial light appears to alter biological timings (Gaston et al. 2017) and timings can evolve in some cases; for example, Urbanski et al. (2012) showed rapid photoperiod evolution in an invasive mosquito. Other traits known to be affected by light are foraging, phototaxis, schooling, activity, gene expression, physiology, body condition, and salmon smoltification. Light should probably induce strong selection on endocrine system evolution (Ouyang et al. 2018a) to compensate for negative effects. In some cases, it may not be possible for organisms to adapt to altered light conditions: for instance, if attraction to night lights historically offered fitness advantages, but now yields drastic fitness costs by inducing movement toward artificial lights, and the organism has no sensory means of reliably distinguishing between natural and artificial lights. Future research is clearly needed on this topic.

Finally, it is worth cautioning that even with an increase in data, it will be nearly impossible to formulate general rules about the evolutionary impact of light on aquatic species. Responses to light are highly species-specific, differing dramatically even between closely related species (Bruning et al. 2011), and are highly dependent on ontogenetic stage even within the same species. Depending on ontogenetic stage, fish either avoid light or are attracted to it, and either start or cease various behaviours. The type and colour of the light and duration of its flashes (if any) is also important; for example, the fact that strobe lights deter fish, while mercury vapour lights attract them has been used to direct fish traffic at dams, away from hazardous areas and towards safe bypassages (Rich and Longcore 2006). Type of light matters to amphipod behaviour (Navarro-Barranco and Hughes 2015), and different light wavelengths have different impacts on melatonin production in fish (Bruning et al. 2016).

10.5.4 Anthropogenic sound

Noise from human activities has recently been recognized as a major pollutant and potential evolutionary force (Swaddle et al. 2015). Underwater noise produced by marine resource extraction and seismic surveys (high-decibel blasts in search of oil and gas reserves) causes traumatic injuries and mortality in many species, but we will limit our discussion here to urban noise, such as car and boat traffic and construction. The impacts of noise on animal behaviour and physiology are well documented: immediate effects on fish and frogs have been measured in endocrinological stress responses, metabolic rate change, and behavioural responses such as foraging efficiency, startle response, schooling, and activity level (Kunc et al. 2016); reactions in crustaceans include altered social behaviour, foraging, and predator response (Tidau and Briffa 2016). Noise can have strong negative impacts on fitness, which should influence evolutionary trajectories by diminishing population sizes and/or by selecting for resilient or resistant phenotypes. Since anthropogenic noise infringes on animal communication space, it also has the potential to disrupt acoustic courtship signals, with potential consequences for sexual selection (Amorim et al. 2015). It could even conceivably create a breakdown in reproductive isolation maintained by acoustic mate discrimination. However, to our knowledge, evolution in response to anthropogenic noise has not yet been tested in any aquatic species. Two promising areas of investigation might be (1) stress responses, which have already been shown to be genetically determined and respond to selection in fish, and (2) sexual signalling in anurans, which are known to be affected by traffic noise.

10.5.5 Nutrients and suspended particles

In many cities and especially in developing countries, untreated sewage is a major source of urban water pollution. Worldwide, large cities with inadequate wastewater treatment systems offload prodigious amounts of nitrogen into rivers, sometimes bringing dissolved oxygen levels to nearly zero. Nitrogenous compounds have immediate toxic effects on aquatic life, as well as indirect effects via phytoplankton blooms, eutrophication, and hypoxia (oxygen deficiency) (Camargo and Alonso 2006). Populations can undergo mass mortality before evolutionary responses to nitrogen pollution can arise (there is a good reason why eutrophied coastal waters are termed 'dead zones'). However, in species that are robust enough to persist, subsequent adaptations to toxins, hypoxia, and turbidity have been observed.

Harmful algal blooms (formerly called 'red tides') that follow nutrient enrichment produce high levels of toxins. Evolved resistance to phytoplankton toxins has been documented numerous times in zooplankton (Hairston et al. 2001; Jiang et al. 2011), suggesting it might be widespread. However, resistance in taxa besides zooplankton appears to be either nonexistent or understudied. Interestingly, such toxin resistance seems to have a fitness cost (Dam 2013), which means that in places with seasonal or periodic algae blooms, oscillating loss and gain of resistance could occur regularly.

Adaptations to hypoxia in urban-impacted waters are difficult to predict with certainty. In non-urban scenarios, many species have adapted to low-oxygen environments such as intertidal waters, hypoxic sulfidic waters, swamps, and oceanic benthic zones (e.g., Hoback and Stanley 2001; Bickler and Buck 2007). This gives us some ideas of what kinds of changes might be expected to evolve in anthropogenically impacted waters for a wide variety of taxa (e.g., smaller, thinner bodies and modified gills to increase oxygen uptake; altered blood pigments; lowered metabolic rates). On the other hand, urban environments might complicate matters because organisms must deal with multiple stressors besides hypoxia: for example, in killifish, the evolution of resistance to industrial pollutants comes at the cost of reduced resistance to hypoxia (Meyer and Di Giuliuo 2003).

Another consequence of nutrient pollution, and of the input of fine particles that do not settle to the bottom, can be increased turbidity, which alters the visual communication environment used during mate choice and other visually mediated behaviours. Several examples exist of sexual selection in fish being weakened by turbidity (e.g., Seehausen et al. 1997; Candolin 2009). However, sexual selection can also be enhanced by turbidity in some situations (Sundin et al. 2017), and can lead to increased sexual ornamentation (Dugas and Franssen 2011). Giery et al. (2015) showed that increased turbidity and changes in water colour in Bahamian tidal creeks altered by road construction resulted in changes in male colouration of Bahamian mosquitofish. Turbidity can have other evolutionary effects too, such as morphological divergence driven by altered predator behaviour in turbid environments (Bartels et al. 2012). In light of these findings, it appears that urban-induced changes in turbidity deserve increased attention with respect to rapid evolution related to altered sexual selection.

10.6 Conclusions

To date, the most clear-cut cases of rapid evolution in urban aquatic species come from responses to metals and certain organic compound pollutants, altered temperature, and hydrologic shifts. These anthropogenic factors may commonly drive phenotypic evolution in urban aquatic taxa. Yet, even in these areas, we have gaping holes in our understanding of the predictability, repeatability, magnitude, and frequency of urban-driven phenotypic evolution, and its role in eco-evolutionary dynamics. Building

on the existing evidence, additional research in these areas can rapidly begin to address such questions.

A common theme that emerges from our survey is that evolutionary changes are often taxa specific, such that generalized predictions about responses to urbanization may not uniformly apply to disparate species. Even in areas that have received considerable attention (such as evolved metal tolerance or the genetic impacts of urban fragmentation), results have varied across taxa. Predictability of responses may also commonly depend on the 'scale' of inquiry—e.g., predictability and parallelism may be greater for whole-organism performance than for morphology than for genes; e.g., a wide range of genetic changes could result in similar performance and fitness values. It is also worth noting that the urban impacts and selective agents we have discussed are not mutually exclusive—they may often influence and interact with one another, as when temperature stress interacts with pollution toxicity.

The clearest take-home message of this chapter is that we direly need more research into the evolutionary consequences of urbanization on aquatic organisms. The urban factors that seem most likely to be affecting aquatic evolution but have so far received very little attention are changes in biotic interactions, fragmentation, artificial light at night, and sound pollution. In each of these areas there is strong evidence for altered selection regimes, and sometimes altered phenotypes, but little-to-no research on evolutionary changes. Since disparate urban factors can influence the same phenotypic traits (e.g., as when both predation and night lighting affect vertical migration in zooplankton), future work may sometimes require sophisticated study design to disentangle the effects of particular selective agents.

References

Alberti, M., Correa, C., Marzluff, J.M., et al. (2017). Global urban signatures of phenotypic change in animal and plant populations. *Proceedings of the National Academy of Sciences of the United States of America*, 114, 8951–6.

Amorim, M.C.P., Vasconcelos, R.O., and Fonseca, P.J. (2015). Fish sounds and mate choice. In: Ladich, F. (ed.) *Sound Communication in Fishes*, Volume 4, 1–33. Animal Signals and Communication. Springer Nature, Basel.

Araujo, M.S., Langerhans, R.B., Giery, S.T., and Layman, C.A. (2014). Ecosystem fragmentation drives increased diet variation in an endemic livebearing fish of the Bahamas. *Ecology and Evolution*, 4, 3298–308.

Araujo, R.V., Albertini, M.R., Costa-da-Silva, A.L., et al. (2015). São Paulo urban heat islands have a higher incidence of dengue than other urban areas. *Brazilian Journal of Infectious Diseases*, 19, 146–55.

Bartels, P., Hirsch, P.E., Svanback, R., and Eklov, P. (2012). Water transparency drives intra-population divergence in Eurasian perch (*Perca fluviatilis*). *PLOS ONE*, 7(8), e43641.

Benjamin, A., May, B., O'Brien, J., and Finger, A.J. (2016). Conservation genetics of an urban desert fish, the Arroyo chub. *Transactions of the American Fisheries Society*, 145, 277–86.

Bickler, P.E. and Buck, L.T. (2007). Hypoxia tolerance in reptiles, amphibians, and fishes: life with variable oxygen availability. *Annual Review of Physiology*, 69, 145–70.

Bolton, D., Mayer-Pinto, M., Clark, G.F., et al. (2017). Coastal urban lighting has ecological consequences for multiple trophic levels under the sea. *Science of the Total Environment*, 576, 1–9.

Botha, L.M., Jones, T.M., and Hopkins, G.R. (2017). Effects of lifetime exposure to artificial light at night on cricket (*Teleogryllus commodus*) courtship and mating behaviour. *Animal Behaviour*, 129, 181–8.

Brady, S.P. (2012). Road to evolution? Local adaptation to road adjacency in an amphibian (*Ambystoma maculatum*). *Scientific Reports*, 2(235).

Brady, S.P., Richardson, J.L., and Kunz, B.K. (2017). Incorporating evolutionary insights to improve ecotoxicology for freshwater species. *Evolutionary Applications*, 10, 829–38.

Brans, K.I. and De Meester, L. (2018). City life on fast lanes: urbanization induces an evolutionary shift towards a faster lifestyle in the water flea *Daphnia*. *Functional Ecology*, 32, 2225–40.

Brans, K.I., Jansen, M., Vanoverbeke, J., et al. (2017). The heat is on: genetic adaptation to urbanization mediated by thermal tolerance and body size. *Global Change Biology*, 23, 5218–27.

Brans, K.I., Engelen, J.M.T., Souffreau, C., and De Meester, L. (2018). Urban hot-tubs: local urbanization has profound effects on average and extreme temperatures in ponds. *Landscape and Urban Planning*, 176, 22–9.

Brennan, R.S., Hwang, R., Tse, M., Fangue, N.A., and Whitehead, A. (2016). Local adaptation to osmotic environment in killifish, *Fundulus heteroclitus*, is supported by divergence in swimming performance but not by differences in excess post-exercise oxygen consumption or aerobic scope. *Comparative Biochemistry and Physiology Part A: Molecular & Integrative Physiology*, 196, 11–19.

Brodin, T., Fick, J., Jonsson, M., and Klaminder, J. (2013). Dilute concentrations of a psychiatric drug alter behavior of fish from natural populations. *Science*, 339, 814–15.

Brown, L.R., Cuffney, T.F., Coles, J.F., et al. (2009). Urban streams across the USA: lessons learned from studies in 9 metropolitan areas. *Journal of the North American Benthological Society*, 28, 1051–69.

Bruning, A., Holker, F., and Wolter, C. (2011). Artificial light at night: implications for early life stages development in four temperate freshwater fish species. *Aquatic Sciences*, 73, 143–52.

Bruning, A., Holker, F., Franke, S., Kleiner, W., and Kloas, W. (2016). Impact of different colours of artificial light at night on melatonin rhythm and gene expression of gonadotropins in European perch. *Science of the Total Environment*, 543, 214–22.

Buchwalter, D.B., Cain, D.J., Martin, C.A., et al. (2008). Aquatic insect ecophysiological traits reveal phylogenetically based differences in dissolved cadmium susceptibility. *Proceedings of the National Academy of Sciences of the United States of America*, 105, 8321–6.

Bulleri, F. and Chapman, M.G. (2010). The introduction of coastal infrastructure as a driver of change in marine environments. *Journal of Applied Ecology*, 47, 26–35.

Camargo, J.A. and Alonso, A. (2006). Ecological and toxicological effects of inorganic nitrogen pollution in aquatic ecosystems: a global assessment. *Environment International*, 32, 831–49.

Candolin, U. (2009). Population responses to anthropogenic disturbance: lessons from three-spined sticklebacks *Gasterosteus aculeatus* in eutrophic habitats. *Journal of Fish Biology*, 75, 2108–21.

Cheptou, P.O., Hargreaves, A.L., Bonte, D., and Jacquemyn, H. (2017). Adaptation to fragmentation: evolutionary dynamics driven by human influences. *Philosophical Transactions of the Royal Society B*, 372(1712), 20160037.

Chevalier, J., Godfrey, M.H., and Girondot, M. (1999). Significant difference of temperature-dependent sex determination between French Guiana (Atlantic) and Playa Grande (Costa-Rica, Pacific) leatherbacks (*Dermochelys coriacea*). *Annales des Sciences Naturelles-Zoologie et Biologie Animale*, 20, 147–52.

Cohen, J.M., Lajeunesse, M.J., and Rohr, J.R. (2018). A global synthesis of animal phenological responses to climate change. *Nature Climate Change*, 8, 224–8.

Dam, H.G. (2013). Evolutionary adaptation of marine zooplankton to global change. *Annual Review of Marine Science*, 5, 349–70.

Daufresne, M., Lengfellner, K., and Sommer, U. (2009). Global warming benefits the small in aquatic ecosystems. *Proceedings of the National Academy of Sciences of the United States of America*, 106, 12788–93.

Debecker, S., Sanmartin-Villar, I., de Guinea-Luengo, M., Cordero-Rivera, A., and Stoks, R. (2016). Integrating the pace-of-life syndrome across species, sexes and individuals: covariation of life history and personality under pesticide exposure. *Journal of Animal Ecology*, 85, 726–38.

do Sul, J.A.I. and Costa, M.F. (2014). The present and future of microplastic pollution in the marine environment. *Environmental Pollution*, 185, 352–64.

Dugas, M.B. and Franssen, N.R. (2011). Nuptial coloration of red shiners (*Cyprinella lutrensis*) is more intense in turbid habitats. *Naturwissenschaften*, 98, 247–51.

Dwyer, R.G., Bearhop, S., Campbell, H.A., and Bryant, D.M. (2013). Shedding light on light: benefits of anthropogenic illumination to a nocturnally foraging shorebird. *Journal of Animal Ecology*, 82, 478–85.

Ebele, A.J., Abou-Elwafa Abdallah, M., and Harrad, S. (2017). Pharmaceuticals and personal care products (PPCPs) in the freshwater aquatic environment. *Emerging Contaminants*, 3, 1–16.

El-Sabaawi, R. (2018). Trophic structure in a rapidly urbanizing planet. *Functional Ecology*, 32, 1718–28.

Ewers, R.M. and Didham, R.K. (2006). Confounding factors in the detection of species responses to habitat fragmentation. *Biological Reviews*, 81, 117–42.

Fasola, E., Ribeiro, R., and Lopes, I. (2015). Microevolution due to pollution in amphibians: a review on the genetic erosion hypothesis. *Environmental Pollution*, 204, 181–90.

Ferronato, B.O., Marques, T.S., Lara, N.R.F., et al. (2016). Isotopic niche in the eastern long-necked turtle, *Chelodina longicollis* (Testudines: Chelidae), along a natural–urban gradient in southeastern Australia. *Herpetological Journal*, 26, 297–304.

Fischer, J.D., Cleeton, S.H., Lyons, T.P., and Miller, J.R. (2012). Urbanization and the predation paradox: the role of trophic dynamics in structuring vertebrate communities. *Bioscience*, 62, 809–18.

Forster, J., Hirst, A.G., and Atkinson, D. (2012). Warming-induced reductions in body size are greater in aquatic than terrestrial species. *Proceedings of the National Academy of Sciences of the United States of America*, 109, 19310–14.

Foster, H.R. and Keller, T.A. (2011). Flow in culverts as a potential mechanism of stream fragmentation for native and nonindigenous crayfish species. *Journal of the North American Benthological Society*, 30, 1129–37.

Francis, T.B. and Schindler, D.E. (2009). Shoreline urbanization reduces terrestrial insect subsidies to fishes in North American lakes. *Oikos*, 118, 1872–82.

Fuller, M.R., Doyle, M.W., and Strayer, D.L. (2015). Causes and consequences of habitat fragmentation in river networks. *Annals of the New York Academy of Sciences*, 1355, 31–51.

Gaston, K.J., Duffy, J.P., Gaston, S., Bennie, J., and Davies, T.W. (2014). Human alteration of natural light cycles: causes and ecological consequences. *Oecologia*, 176, 917–31.

Gaston, K.J., Davies, T.W., Nedelec, S.L., and Holt, L.A. (2017). Impacts of artificial light at night on biological timings. *Annual Review of Ecology, Evolution, and Systematics*, 48, 49–68.

Giery, S.T., Layman, C.A., and Langerhans, R.B. (2015). Anthropogenic ecosystem fragmentation drives shared and unique patterns of sexual signal divergence among three species of Bahamian mosquitofish. *Evolutionary Applications*, 8, 679–91.

Gliwicz, M.Z. (1986). Predation and the evolution of vertical migration in zooplankton. *Nature*, 320, 746–8.

Hairston, N.G., Holtmeier, C.L., Lampert, W., et al. (2001). Natural selection for grazer resistance to toxic cyanobacteria: evolution of phenotypic plasticity? *Evolution*, 55, 2203–14.

Hall, A. (1980). Heavy-metal co-tolerance in a copper-tolerant population of the marine fouling alga *Ectocarpus siliculosus* (Dillw.) Lyngbye. *New Phytologist*, 85, 73–8.

Hall, A.S. (2016). Acute artificial light diminishes central Texas anuran calling behavior. *American Midland Naturalist*, 175, 183–93.

Hanson, J., Wibbels, T., and Martin, R.E. (1998). Predicted female bias in sex ratios of hatchling loggerhead sea turtles from a Florida nesting beach. *Canadian Journal of Zoology*, 76, 1850–61.

Heinen-Kay, J.L., Noel, H.G., Layman, C.A., and Langerhans, R.B. (2014). Human-caused habitat fragmentation can drive rapid divergence of male genitalia. *Evolutionary Applications*, 7, 1252–67.

Henn, M., Nichols, H., Zhang, Y.X., and Bonner, T.H. (2014). Effect of artificial light on the drift of aquatic insects in urban central Texas streams. *Journal of Freshwater Ecology*, 29, 307–18.

Henneken, J. and Jones, T.M. (2017). Pheromones-based sexual selection in a rapidly changing world. *Current Opinion in Insect Science*, 24, 84–8.

Hoback, W.W. and Stanley, D.W. (2001). Insects in hypoxia. *Journal of Insect Physiology*, 47, 533–42.

Hopkins, G.R., French, S.S., and Brodie, E.D. (2013). Potential for local adaptation in response to an anthropogenic agent of selection: effects of road deicing salts on amphibian embryonic survival and development. *Evolutionary Applications*, 6, 384–92.

Hopkins, G.R., Gaston, K.J., Visser, M.E., Elgar, M.A., and Jones, T.M. (2018). Artificial light at night as a driver of evolution across urban–rural landscapes. *Frontiers in Ecology and the Environment*, 16, 472–9.

Imhoff, M.L., Zhang, P., Wolfe, R.E., and Bounoua, L. (2010). Remote sensing of the urban heat island effect across biomes in the continental USA. *Remote Sensing of Environment*, 114, 504–13.

Jacobson, C.R. (2011). Identification and quantification of the hydrological impacts of imperviousness in urban catchments: a review. *Journal of Environmental Management*, 92, 1438–48.

Janssens, T.K.S., Lopez, R.D.R., Marien, J., et al. (2008). Comparative population analysis of metallothionein promoter alleles suggests stress-induced microevolution in the field. *Environmental Science & Technology*, 42, 3873–8.

Jiang, X.D., Lonsdale, D.J., and Gobler, C.J. (2011). Rapid gain and loss of evolutionary resistance to the harmful dinoflagellate *Cochlodinium polykrikoides* in the copepod *Acartia tonsa*. *Limnology and Oceanography*, 56, 947–54.

Johnson, M.T.J. and Munshi-South, J. (2017). Evolution of life in urban environments. *Science*, 358(6363), eaam8327.

Kamdem, C., Fouet, C., Gamez, S., and White, B.J. (2017). Pollutants and insecticides drive local adaptation in African malaria mosquitoes. *Molecular Biology and Evolution*, 34, 1261–75.

Kefford, B.J., Buchwalter, D., Canedo-Argulles, M., et al. (2016). Salinized rivers: degraded systems or new habitats for salt-tolerant faunas? *Biology Letters*, 12(3), 20151072.

Kern, E.M.A. and Langerhans, R.B. (2018). Urbanization drives contemporary evolution in stream fish. *Global Change Biology*, 24, 3791–803.

Kern, E.M.A. and Langerhans, R.B. (2019). Urbanization alters swimming performance of a stream fish. *Frontiers in Ecology and Evolution*, 6, 229.

Khan, Z.A., Labala, R.K., Yumnamcha, T., et al. (2018). Artificial light at night (ALAN), an alarm to ovarian physiology: a study of possible chronodisruption on zebrafish (*Danio rerio*). *Science of the Total Environment*, 628–629, 1407–21.

Klerks, P.L. and Weis, J.S. (1987). Genetic adaptation to heavy metals in aquatic organisms: a review. *Environmental Pollution*, 45, 173–205.

Kolbe, J.J. and Janzen, F.J. (2002). Impact of nest-site selection on nest success and nest temperature in natural and disturbed habitats. *Ecology*, 83, 269–81.

Kummu, M., de Moel, H., Ward, P.J., and Varis, O. (2011). How close do we live to water? A global analysis of population distance to freshwater bodies. *PLOS ONE*, 6(6), e20578.

Kunc, H.P., McLaughlin, K.E., and Schmidt, R. (2016). Aquatic noise pollution: implications for individuals, populations, and ecosystems. *Proceedings of the Royal Society B: Biological Sciences*, 283(1836).

Langerhans, R.B. (2008). Predictability of phenotypic differentiation across flow regimes in fishes. *Integrative and Comparative Biology*, 48, 750–68.

Layman, C.A., Quattrochi, J.P., Peyer, C.M., and Allgeier, J.E. (2007). Niche width collapse in a resilient top predator following ecosystem fragmentation. *Ecology Letters*, 10, 937–44.

Lourenco, A., Alvarez, D., Wang, I.J., and Velo-Anton, G. (2017). Trapped within the city: integrating demography, time since isolation and population-specific traits to assess the genetic effects of urbanization. *Molecular Ecology*, 26, 1498–514.

Maes, G.E., Raeymaekers, J.A.M., Pampoulie, C., et al. (2005). The catadromous European eel *Anguilla anguilla* (L.) as a model for freshwater evolutionary ecotoxicology: relationship between heavy metal bioaccumulation, condition and genetic variability. *Aquatic Toxicology*, 73, 99–114.

Maitra, S.K., Chattoraj, A., Mukherjee, S., and Moniruzzaman, M. (2013). Melatonin: a potent candidate in the regulation of fish oocyte growth and maturation. *General and Comparative Endocrinology*, 181, 215–22.

Major, K.M., Weston, D.P., Lydy, M.J., Wellborn, G.A., and Poynton, H.C. (2018). Unintentional exposure to terrestrial pesticides drives widespread and predictable evolution of resistance in freshwater crustaceans. *Evolutionary Applications*, 11, 748–61.

Mather, A., Hancox, D., and Riginos, C. (2015). Urban development explains reduced genetic diversity in a narrow range endemic freshwater fish. *Conservation Genetics*, 16, 625–34.

Merckx, T., Souffreau, C., Kaiser, A., et al. (2018). Body-size shifts in aquatic and terrestrial urban communities. *Nature*, 558, 113–16.

Meyer, J.N. and Di Giuliuo, R.T. (2003). Heritable adaptation and fitness costs in killifish (*Fundulus beteroclitus*) inhabiting a polluted estuary. *Ecological Applications*, 13, 490–503.

Mitchell, N.J. and Janzen, F.J. (2010). Temperature-dependent sex determination and contemporary climate change. *Sexual Development*, 4, 129–40.

Mizoguchi, B.A. and Valenzuela, N. (2016). Ecotoxicological perspectives of sex determination. *Sexual Development*, 10, 45–57.

Moore, M.V., Pierce, S.M., Walsh, H.M., Kvalvik, S.K., and Lim, J.D. (2001). Urban light pollution alters the diel vertical migration of *Daphnia*. *SIL Proceedings*, 27, 779–82.

Morley, N.J. (2010). Interactive effects of infectious diseases and pollution in aquatic molluscs. *Aquatic Toxicology*, 96, 27–36.

Munshi-South, J., Zak, Y., and Pehek, E. (2013). Conservation genetics of extremely isolated urban populations of the northern dusky salamander (*Desmognathus fuscus*) in New York City. *PeerJ*, 1, e64.

Navarro-Barranco, C. and Hughes, L.E. (2015). Effects of light pollution on the emergent fauna of shallow marine ecosystems: amphipods as a case study. *Marine Pollution Bulletin*, 94, 235–40.

Nelson, J.A., Gotwalt, P.S., and Snodgrass, J.W. (2003). Swimming performance of blacknose dace (*Rhinichthys atratulus*) mirrors home-stream current velocity. *Canadian Journal of Fisheries and Aquatic Sciences*, 60, 301–308.

Nelson, J.A., Gotwalt, P.S., Simonetti, C.A., and Snodgrass, J.W. (2008). Environmental correlates, plasticity, and repeatability of differences in performance among blacknose dace (*Rhinichthys atratulus*) populations across a gradient of urbanization. *Physiological and Biochemical Zoology*, 81, 25–42.

Nelson, J.A., Atzori, F., and Gastrich, K.R. (2015). Repeatability and phenotypic plasticity of fish swimming performance across a gradient of urbanization. *Environmental Biology of Fishes*, 98, 1431–47.

Orciari, R.D. (1979). Rotenone resistance of golden shiners from a periodically reclaimed pond. *Transactions of the American Fisheries Society*, 108, 641–5.

Ouyang, J.Q., Davies, S., and Dominoni, D. (2018a). Hormonally mediated effects of artificial light at night on behavior and fitness: linking endocrine mechanisms with function. *Journal of Experimental Biology*, 221(Pt 6), jeb156893.

Ouyang, X., Gao, J.C., Xie, M.F., et al. (2018b). Natural and sexual selection drive multivariate phenotypic divergence along climatic gradients in an invasive fish. *Scientific Reports*, 8(11164).

Paul, M.J. and Meyer, J.L. (2001). Streams in the urban landscape. *Annual Review of Ecology and Systematics*, 32, 333–65.

Pease, J.E., Grabowski, T.B., Pease, A.A., and Bean, P.T. (2018). Changing environmental gradients over forty years alter ecomorphological variation in Guadalupe bass *Micropterus treculii* throughout a river basin. *Ecology and Evolution*, 8, 8508–22.

Pedrosa, J., Campos, D., Cocchiararo, B., et al. (2017). Evolutionary consequences of historical metal contamination for natural populations of *Chironomus riparius* (Diptera: Chironomidae). *Ecotoxicology*, 26, 534–46.

Perkin, E.K., Holker, F., and Tockner, K. (2014). The effects of artificial lighting on adult aquatic and terrestrial insects. *Freshwater Biology*, 59, 368–77.

Ramler, D., Mitteroecker, P., Shama, L.N.S., Wegner, K.M., and Ahnelt, H. (2014). Nonlinear effects of temperature on body form and developmental canalization in the threespine stickleback. *Journal of Evolutionary Biology*, 27, 497–507.

Rich, C. and Longcore, T. (eds). (2006). *Ecological Consequences of Artificial Night Lighting*. Island Press, Washington, DC.

Richmond, E.K., Grace, M.R., Kelly, J.J., Reisinger, A.J., Rosi, E.J., and Walters, D.M. (2017). Pharmaceuticals and personal care products (PPCPs) are ecological disrupting compounds (EcoDC). *Elementa—Science of the Anthropocene*, 5, 66.

Riesch, R., Easter, T., Layman, C.A., and Langerhans, R.B. (2015). Rapid human-induced divergence of lifehistory strategies in Bahamian livebearing fishes (family Poeciliidae). *Journal of Animal Ecology*, 84, 1732–43.

Riesch, R., Martin, R.A., Diamond, S.E., et al. (2018). Thermal regime drives a latitudinal gradient in morphology and life history in a livebearing fish. *Biological Journal of the Linnean Society*, 125, 126–41.

Rizzo, L., Manaia, C., Merlin, C., et al. (2013). Urban wastewater treatment plants as hotspots for antibiotic resistant bacteria and genes spread into the environment: a review. *Science of the Total Environment*, 447, 345–60.

Rogalski, M.A. (2017). Maladaptation to acute metal exposure in resurrected *Daphnia ambigua* clones after decades of increasing contamination. *The American Naturalist*, 189, 443–52.

Sabry, S.A., Ghozlan, H.A., and AbouZeid, D.M. (1997). Metal tolerance and antibiotic resistance patterns of a bacterial population isolated from sea water. *Journal of Applied Microbiology*, 82, 245–52.

Seehausen, O., Alphen, J.J.M.v., and Witte, F. (1997). Cichlid fish diversity threatened by eutrophication that curbs sexual selection. *Science*, 277, 1808–11.

Sheridan, J.A., Caruso, N.M., Apodaca, J.J., and Rissler, L.J. (2018). Shifts in frog size and phenology: testing predictions of climate change on a widespread anuran using data from prior to rapid climate warming. *Ecology and Evolution*, 8, 1316–27.

Sievers, M., Hale, R., Parris, K.M., and Swearer, S.E. (2018). Impacts of human-induced environmental change in wetlands on aquatic animals. *Biological Reviews*, 93, 529–54.

Smith, R.F., Alexander, L.C., and Lamp, W.O. (2009). Dispersal by terrestrial stages of stream insects in urban watersheds: a synthesis of current knowledge. *Journal of the North American Benthological Society*, 28, 1022–37.

Smokorowski, K.E. and Pratt, T.C. (2007). Effect of a change in physical structure and cover on fish and fish habitat in freshwater ecosystems: a review and meta-analysis. *Environmental Reviews*, 15, 15–41.

Sokolova, I.M. and Lannig, G. (2008). Interactive effects of metal pollution and temperature on metabolism in aquatic ectotherms: implications of global climate change. *Climate Research*, 37, 181–201.

Somers, K.A., Bernhardt, E.S., Grace, J.B., et al. (2013). Streams in the urban heat island: spatial and temporal variability in temperature. *Freshwater Science*, 32, 309–26.

Sundin, J., Aronsen, T., Rosenqvist, G., and Berglund, A. (2017). Sex in murky waters: algal-induced turbidity increases sexual selection in pipefish. *Behavioral Ecology and Sociobiology*, 71(5), 78.

Swaddle, J.P., Francis, C.D., Barber, J.R., et al. (2015). A framework to assess evolutionary responses to anthropogenic light and sound. *Trends in Ecology & Evolution*, 30, 550–60.

Tidau, S. and Briffa, M. (2016). Review on behavioral impacts of aquatic noise on crustaceans. *Proceedings of Meetings on Acoustics*, 27, 1–14.

Townroe, S. and Callaghan, A. (2014). British container breeding mosquitoes: the impact of urbanisation and climate change on community composition and phenology. *PLOS ONE*, 9(4), e95325.

Trombulak, S.C. and Frissell, C.A. (2000). Review of ecological effects of roads on terrestrial and aquatic communities. *Conservation Biology*, 14, 18–30.

Tüzün, N., de Beeck, L.O., and Stoks, R. (2017a). Sexual
selection reinforces a higher flight endurance in urban
damselflies. *Evolutionary Applications*, 10, 694–703.

Tüzün, N., Op de Beeck, L., Brans, K.I., Janssens, L., and
Stoks, R. (2017b). Microgeographic differentiation in thermal performance curves between rural and urban populations of an aquatic insect. *Evolutionary Applications*, 10,
1067–75.

Underhill, V.A. and Höbel, G. (2018). Mate choice behavior of female eastern gray treefrogs (*Hyla versicolor*) is
robust to anthropogenic light pollution. *Ethology*, 124,
537–48.

Urbanski, J., Mogi, M., O'Donnell, D., DeCotiis, M., Toma,
T., and Armbruster, P. (2012). Rapid adaptive evolution
of photoperiodic response during invasion and range
expansion across a climatic gradient. *The American
Naturalist*, 179, 490–500.

Van Donk, E., Peacor, S., Grosser, K., Domis, L.N.D., and
Lurling, M. (2016). Pharmaceuticals may disrupt natural
chemical information flows and species interactions in
aquatic systems: ideas and perspectives on a hidden
global change. *Reviews of Environmental Contamination
and Toxicology*, 238, 91–105.

Van Steveninck, R.F.M., Van Steveninck, M.E., and Fernando,
D.R. (1992). Heavy-metal (Zn, Cd) tolerance in selected
clones of duck weed (*Lemna minor*). *Plant and Soil*, 146,
271–80.

Veprauskas, A., Ackleh, A.S., Banks, J.E., and Stark, J.D.
(2018). The evolution of toxicant resistance in daphniids
and its role on surrogate species. *Theoretical Population
Biology*, 119, 15–25.

Villalobos-Jimenez, G. and Hassall, C. (2017). Effects of the
urban heat island on the phenology of Odonata in London,
UK. *International Journal of Biometeorology*, 61, 1337–46.

Vörösmarty, C.J., McIntyre, P.B., Gessner, M.O., et al.
(2010). Global threats to human water security and river
biodiversity. *Nature*, 467, 555–61.

Wedekind, C. (2017). Demographic and genetic consequences of disturbed sex determination. *Philosophical
Transactions of the Royal Society B*, 372(1729), 20160326.

Wenger, S.J., Roy, A.H., Jackson, C.R., et al. (2009). Twenty-
six key research questions in urban stream ecology: an
assessment of the state of the science. *Journal of the North
American Benthological Society*, 28, 1080–98.

Whitehead, A., Clark, B.W., Reid, N.M., Hahn, M.E., and
Nacci, D. (2017). When evolution is the solution to pollution: key principles, and lessons from rapid repeated
adaptation of killifish (*Fundulus heteroclitus*) populations.
Evolutionary Applications, 10, 762–83.

Winder, M., Reuter, J.E., and Schladow, S.G. (2009). Lake
warming favours small-sized planktonic diatom species. *Proceedings of the Royal Society B: Biological Sciences*,
276, 427–35.

Wirgin, I., Roy, N.K., Loftus, M., Chambers, R.C., Franks,
D.G., and Hahn, M.E. (2011). Mechanistic basis of resist-
ance to PCBs in Atlantic tomcod from the Hudson River.
Science, 331, 1322–5.

CHAPTER 11

Evolutionary Dynamics of Metacommunities in Urbanized Landscapes

Kristien I. Brans, Lynn Govaert, and Luc De Meester

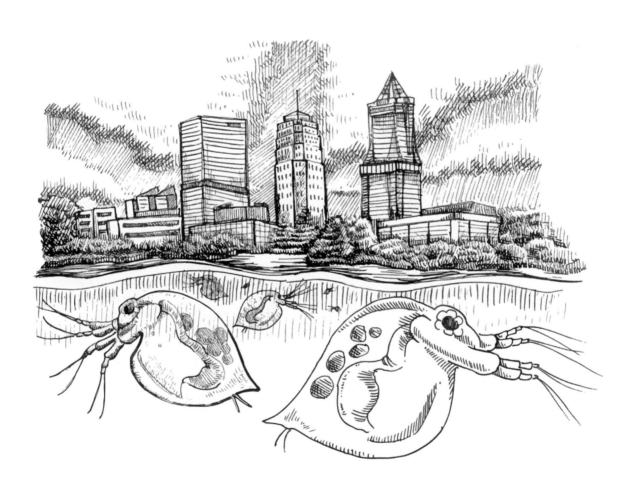

Brans, K.I., Govaert, L. and De Meester, L., *Evolutionary Dynamics of Metacommunities in Urbanized Landscapes* In: *Urban Evolutionary Biology*. Edited by Marta Szulkin, Jason Munshi-South and Anne Charmantier, Oxford University Press (2020). © Oxford University Press. DOI: 10.1093/oso/9780198836841.003.0011

11.1 Introduction

Urbanization is a rapidly growing aspect of global change, with already 50 percent of people worldwide living in cities, and this percentage is predicted to increase to 66 percent by 2050 (Seto et al. 2012). Currently, 3 percent of the terrestrial area globally is urbanized (Liu et al. 2014), and urban cores are rapidly expanding, with a predicted increase in global urban land cover between 2000 and 2030 estimated to reach 185 percent (1.2 million km^2) (Seto et al. 2012). Given the high density of humans and the high level of traffic and industrial activity linked to cities, urbanization represents a very strong human-induced selection pressure (Donihue and Lambert 2015; Parris 2016; Alberti et al. 2017; Rivkin et al. 2018). Compared to nearby natural ecosystems, urban environments are characterized by higher temperatures (the so-called urban heat island effect) (Chapter 6; Oke 1973; Zhao et al. 2018), and relatively high levels of overall disturbance, including light pollution, noise pollution, chemical pollution, and nutrient enrichment (Parris 2016). Furthermore, cities in general harbour more non-native species than natural areas via accidental or deliberate introductions (McKinney 2008). As a result, plant, animal, and microbial communities in urbanized areas often show strong changes in species composition compared to rural areas, with communities often characterized by a lower species diversity, a higher abundance of generalist, opportunistic species, and a higher level of homogenization across sites than in undisturbed areas (McKinney 2008; Aronson et al. 2014; Parris 2016; Barnum et al. 2017; Piano et al. 2017).

In addition to changes in species composition, urbanization leads to shifts in community trait value distributions (e.g., Barnum et al. 2017; Brans et al. 2017a). In a recent study covering nine different organism groups (including rotifers, spiders, cladocerans, ostracods, butterflies, and orthopterans) sampled along shared urbanization gradients, Merckx et al. (2018) demonstrated that most taxa show a decrease in body size with increasing urbanization. These observations are in line with expectations based on the presence of urban heat islands (Brans et al. 2017a, 2017b; Brans and De Meester 2018) and the temperature–size rule in

ectotherms (Atkinson 1994). Merckx et al. (2018) also revealed, however, that three taxa (orthopterans, macromoths, and butterflies), for which body size was positively correlated with dispersal capacity, showed an increase in body size with increasing urbanization, likely linked to habitat fragmentation. Dahirel et al. (2017) reported that urban spider communities show changes in spider-web structure (i.e., web height, mesh size, and surface area) in response to smaller prey size in urban environments. Such trait-based patterns provide insight into how abiotic and biotic urbanization-associated environmental filters and selection pressures shape communities, and do so differently for different taxonomic groups (Concepción et al. 2015; Piano et al. 2017; Merckx et al. 2018).

Given the strong environmental gradients imposed by urbanization, it can be expected that selective regimes in urban areas are altered compared to those in natural areas (Parris 2016; Alberti et al. 2017; Rivkin et al. 2018). As a result, even those species that persist in cities might have experienced changes in their trait values compared to undisturbed populations. In some cases, these changes might be driven by phenotypic plasticity (Parris 2016). In other cases, one might suspect that species can persist in cities because they genetically responded to altered environmental conditions in urban centres via adaptive evolution (Reid et al. 2016; Winchell et al. 2018). The relative importance of phenotypic plasticity versus evolution in explaining *intraspecific trait variation* along urbanization gradients has not been systematically determined, and requires an analysis of common gardening transplant experiments across environmental gradients (e.g., Chapter 2; Donihue and Lambert 2015; Tüzün et al. 2017b; Brans and De Meester 2018). An increasing number of studies have documented genetic responses to urbanization (see also Chapter 5). These studies report changes in allele frequencies as detected using molecular data (Reid et al. 2016; Harris and Munshi-South 2017; Theodorou et al. 2018), as well as genetic trait change quantified via common garden or transplant experiments (e.g., Brans et al. 2017a, 2017b, 2018; Brans and De Meester 2018; Gorton et al. 2018; Winchell et al. 2018), or change in traits for which the Mendelian inheritance is well documented (e.g., Thompson et al. 2016). For example, in plants, transplant experiments using common ragweed (*Ambrosia artemisiifolia*) revealed local adaptation to an earlier onset of the growing season via changes in flowering time in urban compared to rural populations (Chapter 9; Gorton et al. 2018). Work on the water flea (*Daphnia magna*) involving common garden experiments (Brans et al. 2017a, 2017b, 2018; Brans and De Meester 2018) showed genetic adaptation to urban heat islands for not less than 14 traits linked to heat tolerance, life history, and physiology (e.g., energy storage).

Urbanization is characterized by strong and often quite repeatable selection gradients linked to, among others, the urban heat island effect, light pollution, and increased levels of disturbance. Increased urbanization also results in a change in overall landscape connectivity both for natural areas and for urban centres (Beninde et al. 2016, 2018; Miles et al. 2018). Cities can in this context be viewed as expanding patches that are connected by transportation corridors enabling increased within- and among-city dispersal (Crispo et al. 2011; Miles et al. 2018). Given their abundance, their dynamic nature and the strong and repeatable environmental gradients they generate, cities are uniquely suited as a 'natural experiment', unintendedly repeated on a global scale (Chapter 2; Johnson and Munshi-South 2017). Urbanizing areas allow us to study how a key aspect of human-induced environmental change impacts populations of organisms, communities, and ecosystems. Given that there is an explicit spatial context to urbanization, it offers excellent opportunities for studying metacommunity implications of human-induced stressors (Baldissera et al. 2012; Johnson et al. 2013; Gianuca et al. 2018), the repeatability of evolution (e.g., Chapter 3; Johnson et al. 2018; Winchell et al. 2018), landscape genetic implications of urbanization (e.g., Chapter 4; Munshi-South and Kharchenko 2010; Beninde et al. 2018; Combs et al. 2018; Miles et al. 2018), and the *evolving metacommunity* structure in the face of urbanization and climate change (Urban and Skelly 2006; Urban et al. 2016). In section 11.2, we focus on this interaction between ecology, evolution, and geographical distance ('space') and introduce the building blocks of the urban evolving metacommunity framework.

11.2 The urban evolving metacommunity framework

11.2.1 Metacommunity ecology and landscape genetics

Given that urban areas develop as patches in heterogeneous landscapes and are linked via corridors and urban sprawling, they require a landscape perspective. Metacommunity ecology examines how community composition and traits among and within communities vary across space (Leibold and Chase 2017). It integrates local community assembly dynamics that are driven by local environmental filters (abiotic conditions and biotic interactions) with regional processes such as dispersal, driven by the regional species pool, landscape connectivity, and the dispersal ability of the taxa (Vellend 2015; Leibold and Chase 2017). In the case of *species sorting*, dispersal rates are sufficiently high so that species reach their preferred habitats, and community assembly is strongly determined by niche-based processes (i.e., by responses to local environmental conditions) (Leibold and Chase 2017). In contrast, very high dispersal rates can result in source–sink dynamics (i.e., mass effects). Mass effects and dispersal limitation result in a reduced match between species occurrences and environmental gradients (Leibold and Chase 2017). A limited number of studies explicitly address metacommunity structure in heterogeneous, urbanized landscapes (Niemelä and Klotze 2009; Johnson et al. 2013; Turrini and Knop 2015; Bourassa et al. 2017; Piano et al. 2017; Gianuca et al. 2018). In one example, metacommunity analyses of web-spider assemblages in the Atlantic Forest area (Brazil) undergoing urbanization revealed environmental (bush density, urban land use in the surroundings of patches, and shape of patches) rather than spatial factors structuring local community composition (Baldissera et al. 2012).

At another level of biological organization, the field of landscape genetics aims at clarifying the impact of selection, gene flow, and genetic drift on the distribution of intraspecific genetic variation in heterogeneous landscapes (Manel et al. 2010). In landscape genetics, similar to metacommunity analyses, one quantifies the relative importance of environment and space on the distribution of genetic variation across the landscape (Orsini et al. 2013). Patterns of genetic differentiation between urban populations across multiple cities are increasingly studied (e.g., Chapters 4 and 5; Balbi et al. 2018; Beninde et al. 2018; Combs et al. 2018; Miles et al. 2018; Mueller et al. 2018). Yet, most studies do not involve parallel assessments of adaptive phenotypic trait divergence for traits along the same urbanization gradients (but see, for example, Reid et al. 2016 for pollution tolerance in killifish and Johnson et al. 2018 for cyanogenesis in white clover). Vice versa, most studies focusing on parallel evolutionary trait change do not include genomic analyses (Brans and De Meester 2018; Winchell et al. 2018). In the context of landscape genetic structure in urbanized landscapes, the degree to which genetic trait change in response to urbanization is the result of parallel evolution is important (see Chapter 3). For systems with polygenic inheritance, an analysis of the repeatability of trait responses combined with population genomics can elucidate whether parallel trait evolution is driven by gene flow (of pre-adapted genotypes) combined with selective sweeps (Mueller et al. 2013) or by localized, independent evolution of similar trait changes across urban–rural gradients (Evans et al. 2009; Reid et al. 2016; Mueller et al. 2018).

11.2.2 Evolving metacommunities in urbanized landscapes

From the previous, it is clear that there are striking parallels between how communities and genetic variation can be structured along heterogeneous landscapes. Just as the landscape genetic structure of species in heterogeneous, urbanized landscapes can be a function of selection by the environment, genetic drift, and gene flow, the metacommunity structure in these same landscapes can be a function of selection by the environment, ecological drift, and dispersal (Vellend 2015). Comparative studies quantifying the relative importance of selection, dispersal, and drift in structuring both populations and communities in the same landscapes are needed to understand how the patterns at both levels of biological organization might influence each other. Trait values are key to the impact of selection, and in some cases, it has been shown that environmental selection at the population (i.e., within-species) and community (i.e., among-species) level can involve the same traits. One example is provided by zooplankton, where body size is an important trait structuring zooplankton communities along urbanization gradients (Gianuca et al. 2018) and a key trait showing genetic changes along the same urbanization

---

Box 11.1 Zooplankton body size, metacommunity structure, and urban evolution along urbanization gradients

In zooplankton, metacommunity structure (Gianuca et al. 2018) and urban evolution of the key large-bodied zooplankton species *Daphnia magna* (Brans et al. 2017a, 2017b, 2018; Brans and De Meester 2018) have been studied along the same urbanization gradients in Flanders (Belgium). For community-weighted mean trait values as well as for population-specific genetic trait values in *D. magna*, body size showed a clear response to urbanization (Figure 11.1). In both cases, the overarching pattern is that body size declines with increasing urbanization. Given that body size varies at the population as well as the community level along the same urbanization gradients, both may influence each other.

Brans et al. (2017a, 2017b) quantified *D. magna* body size under common garden conditions at two temperatures (20 and 24 °C), reflecting summer temperatures in rural and urban ponds in Flanders (Brans et al. 2017a; Brans and De Meester 2018). They also quantified the body size of ten

individuals of each species and each sample in the community samples. Variation in body size of individuals within species can be the result of genetic differences, plasticity in response to among-site variation in environmental conditions, and/or ontogenetic change. Brans et al. (2017a) then used the method of Lepš et al. (2012) to partition the total variation in community body size along the urbanization gradient in contributions of species sorting (species composition), intraspecific trait variation (across all species, measured directly on samples taken from the field and thus integrating all sources of intraspecific variation), *D. magna* genotypic trait variation (common garden experiment), *D. magna* intraspecific trait variation linked to plasticity to temperature (as a key variable linked to urbanization and tested for in the common garden experiment), and *D. magna* intraspecific trait variation linked to other environmental differences among sites or ontogenetic changes.

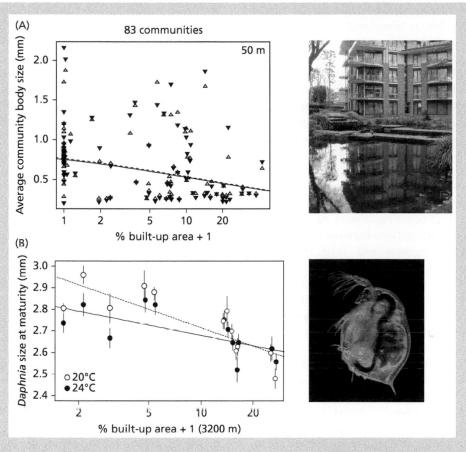

Figure 11.1 (A) (Left) Average body size of 83 zooplankton communities along a multicity urbanization gradient (urbanization as percentage built-up area in 50 m around the pond) across Flanders (Belgium) decreases with increasing urbanization. (Right) A typical urban pond in the sampling region. (B) (Left) *Daphnia magna* body size assessed in a common garden study at both 20 and 24 °C for 12 populations across the same urbanization gradient shows that urban *Daphnia* evolved to mature at a smaller size with increasing urbanization (percentage built-up area in a radius of 3200 m around the pond; patterns at 50 m are the same; more details are given in Brans et al. 2017a). (Right) *Daphnia magna* (photo by Joachim Mergeay).

They showed that intraspecific variation in the large-bodied species *D. magna* accounted for more than 97 per cent of the changes in zooplankton community body size in the set of ponds along the urbanization gradient in which this species occurred, whereas species sorting accounted for less than 3 per cent of the community trait change. Most of the intraspecific trait variation, however, was not linked to genetic differences or plastic responses to temperature, but likely to differences in age structure among the *D. magna* populations (referred to as ontogeny-driven plasticity). This example (1) shows both evolution-mediated and community-wide trait shifts along the same urbanization gradients, (2) demonstrates that intraspecific trait variation is very important, but also (3) shows that in this case genetic variation was only a minor component to both intraspecific and community trait variation along the urbanization gradients. We note that the data are more complex than the simple patterns suggested at first glance by Figure 11.1. More specifically, a subset of the communities shown in panel A of this figure are dominated by large species, and this subset shows a slight increase in average body size with urbanization (calculated at a 3200-m radius around the pond). *Daphnia magna* is the largest *Daphnia* species, and the 12 populations studied by Brans et al. (2017a, 2017b) belong to the subset of communities that shows an increase in body size. Genetic variation in body size in *D. magna* is therefore opposite in direction to the intraspecific non-genetic trait variation in that species (larger individuals with increasing urbanization) and community-wide trait variation for the communities in which it occurs.

gradients in the focal water flea species *Daphnia* (Brans et al. 2017a; Brans and De Meester 2018) (Box 11.1).

In another study, Dahirel et al. (2017, 2018) documented within-species and community-wide trait variation for web-building behaviour in orb web-spider communities to change in the same direction along urbanization gradients. The parallel processes structuring metapopulations and metacommunities combined with shared sets of traits can foster eco-evolutionary interactions at the metacommunity level. The recognition that evolutionary and ecological timescales can converge inspired the development of the field of *eco-evolutionary dynamics*, which focuses on the interactions between ecological and evolutionary processes and their implications (Thompson 1998; Hendry 2017). Localized evolutionary trait change of the species within the metacommunity can influence metacommunity structure and thus community-wide trait variation along environmental gradients. On the other hand, evolution can directly impact community-wide trait change along the gradient as it changes patch-specific trait values of the individual species (Jung et al. 2014; Lajoie and Vellend 2015; Govaert et al. 2016; Brans et al. 2017a). In addition, evolution can impact community assembly trajectories (Jung et al. 2014; Pantel et al. 2015). In order to obtain improved insight into how environmental gradients such as urbanization shape metacommunities, a joint analysis of responses at both levels of ecological organization and the integration of evolution in metacommunity ecology is needed.

The evolving metacommunity concept (Urban and Skelley 2006) aims at integrating evolutionary insights into metacommunity ecology. In an evolving metacommunity framework, one strives to understand how both metacommunities and the metapopulations of its member species are structured in landscapes, and how they influence each other. Evolving metacommunity analyses explicitly take into account gradients in both environment and space (spatial distances, connectedness), and aim at integrating data on both the population (within-species distribution of genetic variation at the genome level and in trait values) and community level (species occurrences and abundances) and variation in

ecologically relevant traits at both the population and community level. A comparison of patterns of variation across the landscape for functional traits with those obtained for neutral genetic markers or taxonomic composition can provide indications for the role of selection in structuring metapopulations and metacommunities, respectively. In addition, analyses of functional traits can inform on potential *eco-evolutionary feedback* loops and ecosystem consequences. Accounting for evolution might dramatically change our predictions of responses to environmental change, including the different dimensions of human-induced global change (Urban et al. 2016). In the context of climate change, for instance, it has been predicted that evolution might increase the importance of local responses compared to dispersal in determining local population persistence and community composition (Van Doorslaer et al. 2009; Urban et al. 2012; De Meester et al. 2018). Van Doorslaer et al. (2009) allowed a resident *D. magna* population isolated from a pond located in the UK to genetically adapt to a warming treatment (ambient +4 °C) in a mesocosm experiment. They then showed that this evolution resulted in a reduced establishment success of immigrant French genotypes that were pre-adapted to warmer conditions. The fact that some species might respond to climate warming by local adaptation whereas others might migrate will also result in altered competitive interactions (Urban et al. 2012, 2016).

The research agenda set out by the evolving metacommunity framework is highly challenging, as it requires assessing the relative importance of both local and regional responses, as well as the relative importance of ecological (e.g., species sorting) and evolutionary (genetic trait change in individual species) responses, and integrating them in one vision (Alberti 2015; De Meester et al. 2019). There is increased attention on theoretical simulation models that include both ecological and evolutionary dynamics in spatially explicit settings (Govaert et al. 2019). Such models can provide important insights into the consequences of eco-evolutionary interactions for responses to global change. For example, a spatially explicit eco-evolutionary model by Norberg et al. (2012) showed that the ecological and evolutionary responses induced by climate change result in diversity loss even long after the climate stabilized. This is due to the continuing evolution of species and its subsequent effects on species interactions. Other models have found that the presence of competitors can reduce a species' evolutionary response to environmental change, which can result in diversity loss of species in the metacommunity (e.g., de Mazancourt et al. 2008). Making the connection between such theoretical models and empirical work will be crucial to better understand the implications of evolving metacommunities for evolutionary and ecological responses to global change, including urbanization. We know of no empirical studies that achieved the goal of characterizing evolving metacommunities in natural landscapes in their full extent, especially given that this needs an explicit multispecies approach (De Meester et al. 2019). The example of zooplankton illustrated in Box 11.1 does integrate community and population responses in the context of urbanization, but has important limitations. One such limitation is the fact that evolution along the urbanization gradient was studied in only one species.

We here suggest that the repeatability and strength of the urbanization signal might offer unique opportunities to disentangle the different drivers structuring evolving metacommunities across multiple organism groups. In section 11.3, we discuss how this field can be developed. We start by providing a hypothetical example that is inspired by the case study on zooplankton community dynamics and evolution in one focal species (*D. magna*—Box 11.1), but explicitly consider evolutionary responses in multiple species. We then outline approaches to obtain insight into the structure of evolving metacommunities in urbanized landscapes both based on survey data and through experiments.

11.3 Urban evolving metacommunities: a hypothetical example

The following hypothetical example is outlined to illustrate the importance of studying the evolving metacommunity structure in differentially urbanized landscapes. While it is hypothetical, it is inspired by our observations on the metacommunity structure of zooplankton (Gianuca et al. 2018) and evolutionary

responses of one of its key species, the water flea *Daphnia* (Brans et al. 2017a, 2017b, 2018; Brans and De Meester 2018; Box 11.1). Our hypothetical example does, however, take evolution in all species of the community into account. Imagine a metacommunity in a heterogeneous landscape with multiple urban patches (Figure 11.2A). Urban patches select for specific trait values with respect to a given trait (cf. trait under selection in Figure 11.2B). One example would be the selection for smaller body size reported for zooplankton (Gianuca et al. 2018) and a number of other organism groups (Merckx et al. 2018). Urbanization-mediated shifts in species composition linked to trait values have been reported repeatedly (e.g., Concepción et al. 2015; Barnum et al. 2017; Piano et al. 2017). In our hypothetical example, selection for specific trait values results in shifts in species composition (Figure 11.2B,D). However, the urban populations of the member species of the communities might also show a shift in trait values in response to the selection pressure imposed by urbanization (Figure 11.2B, Urban 1–4; Brans et al. 2017a, 2017b; Brans and De Meester 2018; Winchell et al. 2018). This trait shift might be achieved through phenotypic plasticity or evolution, but we here assume that it has an important genetic component. In the case that individual species show trait evolution in the same direction as the community-wide trait shift (Figure 11.2B, Urban 1–4), species composition in urban areas might be less impacted by local species sorting in the presence compared to the absence of trait evolution (Figure 11.2B,D, compare Urban 1–4 with Urban no evolution). The expected difference in species composition between rural and urban areas might be limited by the fact that the different species making up the community themselves already show an evolutionary shift in trait values. Through local evolution, species that would otherwise have individuals with very low fitness in urban areas might be able to persist (Figure 11.2D).

In our example, we assumed that selection for the trait that we focus on in Figure 11.2B is strong. The different communities in urban centres have a very similar value for this trait, different from the value of the community inhabiting the rural area. In the scenarios with evolution (Urban 1–4), however,

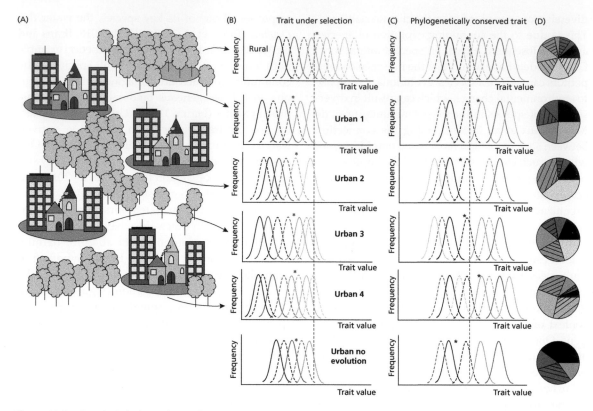

Figure 11.2 A hypothetical scheme showing how evolution can impact evolving metacommunity structure in an urbanized landscape. (A) We assume a landscape with natural, rural habitat as a matrix and several urban patches. (B) Trait distributions of species in one rural (top row) and four urban patches (Urban 1–4) for a trait under selection along the urbanization gradient. Different species are represented by different grey scales and line types, and are characterized by different distributions of trait values. We assume that in urban sites, only species can persist that have a mean trait value that is lower than a certain threshold depicted by the hatched vertical line. Community-wide average trait values are indicated by an asterisk. The plots illustrate that in all four urban sites, identity of the species that survive can be influenced by trait evolution of member species of the community. In all sites, surviving species and their trait distributions are different due to differential evolutionary trajectories. In the lower row, we depict the situation for a community in which the species do not show evolution for the trait under selection. Here trait values for all species are identical to those in the rural site, but only a few species can survive. (C) Trait distributions in rural and urban sites for a phylogenetically conserved trait, i.e., a trait that shows values that are fixed within species. Here there is no evolution of trait value in member species, and changes in community-wide trait values as one moves from rural to urban sites depend on which species are selected against based on their values for the trait under selection (see B). (D) Species composition in the rural and four urban sites. In the lower row, we also plot the species composition that would have resulted for the urban sites in the absence of evolution. The fact that member species of the community can evolve substantially alters species composition in the urban sites, and also results in a different species composition in the different urban sites.

the community-wide distribution for the trait under selection is partially mediated by population responses rather than only by shifts in species composition (Figure 11.2B). This might have far-reaching consequences. For instance, some species that would otherwise become very rare or disappear from the community might persist thanks to evolution (cf. *evolutionary rescue* (Gomulkiewicz and Shaw 2013)). Their higher densities might

potentially have an important buffering impact on the ecosystem consequences of urbanization-mediated selection.

In the zooplankton example outlined in Box 11.1, urbanization selects for smaller species (Brans et al. 2017a; Gianuca et al. 2018) associated with the urban heat island effect. Cladocerans of the genus *Daphnia* tend to be relatively large, and are thus partially selected against (Gianuca et al. 2018). They are, however, also the most efficient grazers on phytoplankton (Gianuca et al. 2016). To the extent that *Daphnia* species can persist in urbanized areas thanks to evolution, this might have important consequences for the top-down control of algae in urban ponds. Given that grazing rates in zooplankton are a function of body size, top-down control of algae might be reduced in urban compared to rural ponds. However, the fact that *Daphnia* can persist would enhance top-down control in urban ponds compared to the situation where large *Daphnia* would have disappeared in the absence of evolution. This example illustrates that ignoring evolution would result in erroneous predictions not only on species composition of communities in urban centres, but also on the ecosystem consequences of the shifts in community composition associated with urbanization. This is illustrated in Figure 11.2C, where we show shifts in trait values for a phylogenetically conserved trait, i.e., a trait that tends to be fixed within lineages and shows an increased divergence as taxa become phylogenetically more distant. It also has methodological consequences. For instance, studies on metacommunity structure that calculate urbanization-induced community-wide shifts in trait values based on species average trait values reported in the literature (e.g., Merckx et al. 2018) might underestimate the actual trait change associated with urbanization.

Obviously, the hypothetical example outlined in Figure 11.2 is idealized and highly simplified. For instance, we assumed that all communities in the natural matrix surrounding the urban centres are identical. Also, we assumed that selection in all urban centres was identical and equally effective. Many real-life situations will be complex mixtures of ecological (species sorting) and evolutionary trait shifts in multiple traits simultaneously. Yet, to the extent that this simplified example grasps key aspects of reality, there are a number of predictions we can make (Figure 11.2):

(1) Evolution might buffer shifts in community composition along urbanization gradients (Figure 11.2D). In its extreme, the absence of a shift in community composition along an urbanization gradient might be due to evolution of member species. From a metacommunity perspective, this will result in a reduced match between species occurrences and abundances with the environment (Figure 11.3G).

(2) Changes in community-wide trait values for those traits that are under selection in urban environments should match predictions stemming from knowledge on how environmental gradients translate into selection on particular traits (Figure 11.3B). From a trait-based metacommunity perspective, the match between trait values and the environmental gradients is expected to be high (Figure 11.3H). Because of evolutionary trait change within species, the association between trait values and species identity is, however, relaxed. Note that we here assume genetic trait shifts and adaptive responses. If responses are maladaptive, for instance due to genetic drift, the pattern can be entirely different. In addition, in real life, part of the observed trait shift might be due to phenotypic plasticity.

(3) Changes in community-wide trait values for traits that are not under selection in urban environments should be buffered compared to a setting in which no evolution occurred. This is particularly true for phylogenetically conserved traits (Figure 11.2C and Figure 11.3I). Evolutionary rescue allows species to persist in urban environments, along with their phylogenetically conserved traits. Phylogenetic signals in a metacommunity analysis (Knapp et al. 2008) might thus show a reduced response to urbanization in the presence of evolution, congruent with the buffered impact on species composition.

(4) With respect to ecosystem functions and associated ecosystem services, evolutionary rescue of key species in urban centres might contribute to buffer ecosystem-level changes, especially if the ecosystem functions and services are linked to phylogenetically conserved traits that are not directly under selection in the urban environment (Figure 11.2C and Figure 11.3I).

From the above, it follows that the degree to which traits are phylogenetically conserved or labile (i.e., can change independently of the phylogenetic background) might make an important

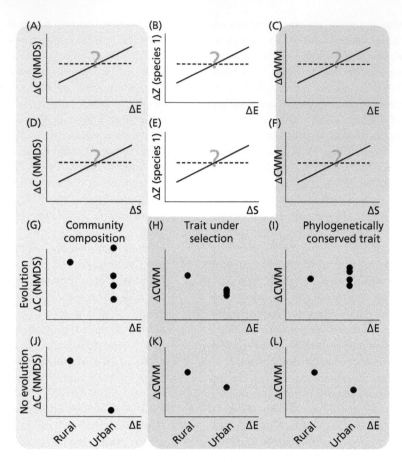

Figure 11.3 (A–F) For community composition (A,D), trait change within a species (B,E) and community-wide trait values (C,F), one might expect an association between differences among communities or populations occupying two patches and the difference in environmental conditions (ΔE; A–C) or the distance in space (ΔS; D–F) between patches. The strength of these relationships will depend among others on the importance of selection, dispersal, and drift. ΔC(NMDS) = change in coordinates in a non-metric multidimensional scaling ordination of community composition; Δz(species 1) = change in trait value of species 1; Δ(CWM) = change in community-weighted mean trait value. (G) and (J) illustrate how the association between pairwise differences in community composition and pairwise differences in environmental conditions differ in the presence (G) and absence (J) of evolution. (H) and (K) illustrate how the association between pairwise differences in community-weighted mean trait value for a trait under selection and pairwise differences in environmental conditions differ in the presence (H) and absence (K) of evolution, and (I) and (L) do the same but for a phylogenetically conserved trait. Data that are plotted in (G–L) are derived from the hypothetical example illustrated in Figure 11.2.

difference (Knapp et al. 2008; Cadotte et al. 2013). As a matter of fact, if the traits that are under selection in the urban environment are phylogenetically conserved, then phylogenetic constraints will limit the capacity for evolutionary rescue. The previously outlined buffering effect of evolution on community composition, community-wide trait values for traits that are not under selection, and ecosystem functions that are not dependent on traits under selection only applies if the urban environment mainly selects traits that are phylogenetically labile.

11.4 Approaches to study evolving metacommunities across urbanization gradients

In order to obtain insight into the evolving metacommunity responses to urbanization, we need to integrate information at both the community and population level, and quantify the impact of evolution on community features. Urban evolution can, through trait change at the level of local populations, directly influence trait value distributions at the community level (Figure 11.2B,C). Another way evolution might impact community features is through its impact on the relative abundances of species and thus on community composition (Figure 11.2D). In sections 11.4.1 and 11.4.2, we provide a brief overview of some approaches to quantify the contribution of evolution to community trait change and community composition along urbanization gradients. We discuss approaches that start from field observations as well as manipulative experiments. Overall, field surveys can allow one to assess the direct influence of intraspecific trait variation on community-wide trait distribution patterns. Snapshot field data in general do not allow one, however, to quantify how evolution impacts species composition. To assess this contribution, manipulative experiments are needed.

11.4.1 Community trait change: eco-evolutionary partitioning metrics

As a first approximation to assess metacommunity structure, it is helpful to quantify the degree to which variation in community structure is associated with environmental gradients (Figure 11.3A) or spatial distances (Figure 11.3D). This can be achieved through an analysis of the species abundance × site matrix as a function of local environmental conditions or spatial distances. An association with environmental gradients independent of space in such an analysis suggests that selection by the environment is a key determinant of community composition (species sorting (Leibold and Chase 2017)). As mentioned in section 11.2.1, landscape genetic analyses similarly relate patterns of genetic variation to environmental and spatial distances (Chapter 4; Orsini et al. 2013). To study the features of the evolving metacommunity, patterns at both the community and population level need to be analysed, and integrated through a trait-based approach. Traits represent the common currency driving both population and community responses to selective gradients. Having information on both community composition and species trait values allows the

quantification of community-weighted mean (CWM) trait values (e.g., Palma et al. 2017; Piano et al. 2017; Merckx et al. 2018). CWM trait values are obtained by multiplying relative abundances of species by their average trait values. They allow us to characterize the average trait value and the trait value distribution of a community.

Figure 11.3 illustrates that community composition (cf. metacommunity ecology; Figure 11.3A,D), intraspecific trait values of member species (cf. evolutionary ecology; Figure 11.3B,E), and CWM trait values (Figure 11.3C,F) can be associated with either environmental gradients (Figure 11.3A,B,C) or spatial distances (Figure 11.3D,E,F). These associations will depend on the relative importance of selection, dispersal (gene flow), and drift, but can also be influenced by eco-evolutionary interactions. Panels G–L of Figure 11.3 show how community composition (G,J), community-wide trait values of a trait under selection (H,K), and community-wide trait values of a phylogenetically conserved trait (H,L) differ in their association with urbanization depending on whether or not evolution occurred. The patterns that are depicted in Figure 11.3 are derived from the hypothetical example illustrated in Figure 11.2. For community composition, evolutionary trait change is expected to often reduce the match between species abundances and the environmental gradient. A low match of species composition to urbanization cannot, however, be used to infer an impact of evolution. This is because there are several other mechanisms, such as ecological drift, functional redundancy, and priority effects, that can also result in a low match between species composition and environmental gradients. To gain insight into the role of evolution on community composition, we need to carry out common garden or transplant experiments (see section 11.4.2). To quantify the impact of evolution on trait change along the environmental gradient, however, one can use eco-evolutionary partitioning metrics such as the ones developed by Lepš et al. (2012) and Govaert et al. (2016).

The method developed by Lepš et al. (2012) quantifies the contribution of intraspecific trait variation to observed community trait change along an environmental gradient using field survey data (Figure 11.4A,B,C). To that purpose, Lepš et al. (2012) calculate for each patch the CWM trait values

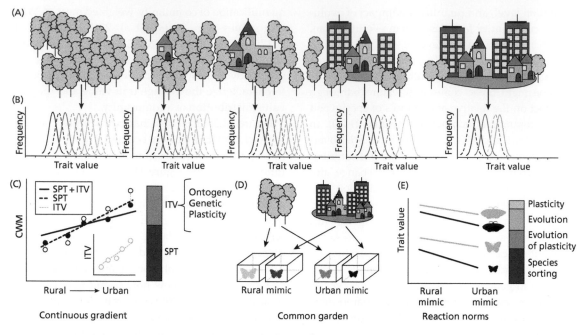

Figure 11.4 Quantifying the contribution of evolutionary trait change to CWM trait change along the urbanization gradient. (A) Urbanization gradient. (B) Trait distributions of member species of communities that occur along the gradient. (C) Regression analysis to quantify the contribution of interspecific (SPT, linked to community composition) and intraspecific (ITV) trait variation to community-wide trait change (SPT+ITV) following Lepš et al. (2012). Solid symbols in (C) give CWM trait values when metacommunity average trait values are used for each species, whereas empty symbols do the same for the setting in which local trait values are used for the different species. If population-specific trait values are measured on individuals directly isolated from nature and for traits for which it is not known that the phenotypes are independent of environmental conditions, this approach does not allow us to quantify the contribution of evolutionary trait change, but instead integrates intraspecific trait change resulting from differences age distribution, phenotypic plasticity, and evolutionary trait change. (D) To quantify the contribution of evolutionary trait change, one needs to carry out a common garden experiment in which one isolates a representative number of genotypes from populations along the urbanization gradient and quantifies their phenotype after purging from maternal effects in common garden conditions. (D) depicts how rural (lighter grey) and urban (darker grey) populations are cultured under environmental conditions mimicking urban and rural settings (e.g., mimicking the temperature gradient associated with the urban heat island effect). (E) With data gathered in the common garden experiment outlined in (D), one can apply a reaction norm approach to quantify relative importance of phenotypic plasticity and evolution to total trait change within each species. The reaction norm plot shows how trait values of the different populations and species (symbols) vary over the urbanization-associated environmental gradient (x-axis). If one carries out this analysis on multiple species isolated along the urbanization gradient, the reaction norm approach allows quantifying the relative importance of phenotypic plasticity, genetic trait change, and changes in species composition to community-wide trait change along the urbanization gradient.

(1) using local species abundances and local population-specific trait values, and (2) using local species abundances and mean species trait values taken from the literature or calculated as the mean values across the metacommunity; in addition (3) they calculate the difference between the two previously described CWMs. These three sets of values obtained are then separately used in regression analyses along the environmental gradient of interest (Figure 11.4C). In the regression analysis

using the first CWM trait value, trait variation along the environmental gradient integrates the effect of both changes in species composition (species sorting, SPT) and patch-specific variation in the trait values of all member species of the community (intraspecific trait variation, ITV). Using the second CWM trait value or the third set of values in a regression analysis only captures the effect of changes in species composition (SPT) or local trait change within species (ITV), respectively. From these regression analyses, one obtains regression sums of squares to quantify the relative importance of species sorting (i.e., changes in community composition) and intraspecific trait variation (genetic and non-genetic variation) to total variation in CWM trait values (Lajoie and Vellend 2015; Dahirel et al. 2017). If the genetic component to phenotypic variation in local species trait values can be assessed in common garden or transplant experiments, such an approach can also be used to assess the relative importance of genetic (i.e., genotypic trait variation, GTV) and non-genetic variation in trait values across localities (Brans et al. 2017a; Govaert et al. 2018).

The previously described metrics partition the total variation in community-wide trait values across a gradient into inter- and intraspecific contributions (Figure 11.4C). There are also partitioning metrics that quantify contributions of ecology and evolution to mean trait change rather than to trait variation. These eco-evolutionary partitioning metrics can be applied to any pair of communities and do not necessarily require an environmental gradient. Examples of such metrics involve the Price equation (Price 1970; Collins and Gardner 2009) and metrics using reaction norms, such as the metric developed by Ellner et al. (2011) and Govaert et al. (2016). All these metrics were originally developed to quantify the genetic and non-genetic contributions to intraspecific trait change (Price 1970; Ellner et al. 2011; Stoks et al. 2016), but they have been extended so that they can also be applied at the community level (Collins and Gardner 2009; Govaert et al. 2016).

Provided that the right data are collected, they allow quantification of the relative contribution of individual (phenotypic plasticity), population (evolution), and community (changes in relative abundances of species) level contributions to observed community-wide trait change (Figure 11.4E; Govaert et al. 2016; Govaert 2018). These approaches can suffer strong limitations, however. First, the amount of data that need to be collected is substantial. For reaction norm approaches (Figure 11.4E) one needs to collect common garden data from multiple populations of multiple species and test responses to relevant gradients (Figure 11.4D). For the Price equation, one needs to track genetic lineages within

species, and as a result this approach in practice is largely limited to asexually reproducing taxa. In sexually reproducing species, one would need detailed genetic relationship matrices. Second, these methods are developed to quantify the contribution of evolution to trait change between two consecutive time points. They assume that there is an ancestral state that is used as a starting point for the calculation of the relative contribution of plasticity, evolution, and changes in species abundances to trait change. They are therefore less suited for studies that collect data along spatial gradients than studies that collect data through time. For spatial data, one can either assume that one kind of environment is the ancestral state or apply the metrics to calculate contributions to trait change relative to an average community obtained by averaging species abundances and trait values across the whole metacommunity (Govaert et al. 2019). In the context of urbanization, one can, for instance, assume that the populations and communities in the rural areas represent the ancestral state, and then one can calculate the eco-evolutionary contributions to trait change as one moves from rural areas to urbanized sites (Figure 11.4E).

11.4.2 The dynamics of community change: common gardening experiments

The eco-evolutionary partitioning metrics described in section 11.4.1, when applied to field survey data that represent a snapshot in time, can quantify the relative contribution of phenotypic plasticity, evolution, and changes in species abundances to the observed trait change along the urbanization gradient. The actual dynamics of trait change and changes in species composition through time cannot, however, be reconstructed from single snapshot surveys. A space-for-time approach, where the local communities and populations at rural sites are viewed as the ancestral state that changed into the observed communities and species attributes in urban sites, is likely to fail to grasp the true dynamics. One reason is that not all rural sites are identical (Figure 11.2 is an oversimplification of reality in this respect). One way to obtain insight into the dynamics of change would be to apply the partitioning metrics to data collected over time as urbanization

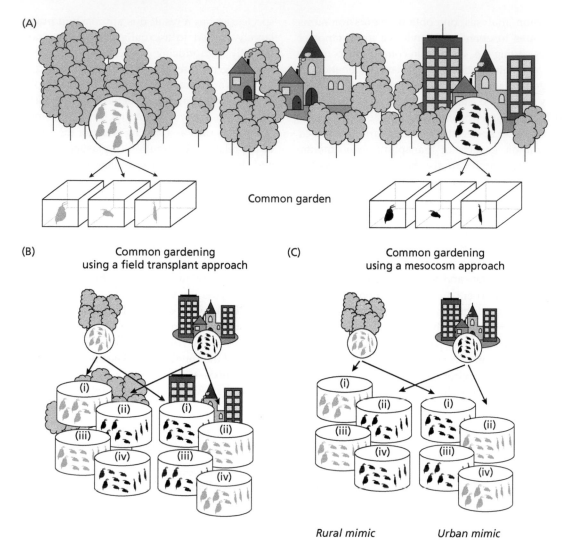

Figure 11.5 Quantifying the contribution of evolution to community composition and ecosystem features. (A) Isolation of genotypes of different species along the urbanization gradient and culturing a common garden environment to purge maternal effects. Note that urban and rural sites differ in both genotypes and relative abundances of species in the community. We here depict one common garden environment, but one might create two common garden environments, mimicking either the rural or urban conditions. In this case, however, both populations need to be grown in both common garden environments. (B) and (C) Common gardening experiments using populations reared in common garden environments where genotypes are transplanted across sites (B) and it is quantified how genotypic differences affect features of the local ecosystems or (C) where genotypes are inoculated in mesocosms that are used as common gardens whose features are monitored. Here we depict a setting in which we create two common gardens to which we inoculate either urban or rural genotypes of multiple species, and quantify to what extent genetic differentiation then impacts community composition or ecosystem features. In both the field and mesocosm approaches, we schematically depict four treatments all involving communities of multiple species according to the following: (i) reconstructing local community composition using local genotypes; (ii) reconstructing local community composition but using genotypes of the population inhabiting the contrasting habitat; (iii) reconstructing community composition of the contrasting habitat but using local genotypes; (iv) reconstructing community composition of the contrasting habitat using genotypes of the contrasting habitat.

increases locally or at the level of entire landscapes (see also Chapter 2).

Yet, even analyses that involve repeated surveys through time only report on changes in community composition and in species' and community-wide trait values as urbanization progresses. They do not allow assessing how evolutionary trait change impacts community composition. To this end, one must carry out manipulative experiments (Figure 11.5A–C). Following common garden experiments to isolate the genetic component to trait change (Figure 11.5A), one might engage in common gardening experiments to quantify the impact of evolutionary trait change on community composition (Figure 11.5B,C). In a common gardening experiment, one inoculates phenotypically divergent genotypes (e.g., urban vs rural *Daphnia* genotypes) in a common garden (e.g., a mesocosm mimicking an ecosystem, or in field enclosures) to then quantify how that impacts the features of the garden (Matthews et al. 2011). In the context of urbanization, one could use urban versus rural populations and inoculate them in common gardens mimicking an urban or rural setting (e.g., by mimicking a key gradient, such as temperature in mesocosms), and reconstruct community composition of the target habitat. One can then use the eco-evolutionary partitioning metric developed by Ellner et al. (2011) to quantify the relative contribution of urban evolution and of the urban-induced environmental gradient itself to the observed change in community composition (e.g., change in NMDS coordinates; Pantel et al. 2015) and ecosystem functioning (e.g., for freshwater-related parameters: chlorophyll a and phycocyanin concentrations, both linked to phytoplankton blooms). As an alternative to common gardening experiments using mesocosms, one can engage in reciprocal transplants of populations across habitats and use *in situ* enclosures in a common gardening approach (Figure 11.5B). So far we know of no experiments that in this way quantified the impact of urban evolution on community composition in urban systems. Ideally, one would carry out such experiments manipulating multiple dominant species separately and simultaneously, to capture the impact of urban evolution in single species and their combined effect on community composition in urban ecosystems (De Meester et al. 2019).

11.5 Eco-evolutionary feedbacks of urban evolution on ecosystem features

As it integrates ecological and evolutionary dynamics in a spatially explicit context, the evolving metacommunity framework provides important added value to the analysis of *urban eco-evolutionary dynamics*. Urban evolution can profoundly impact the ecology of urban ecosystems: evolution of functional response traits that mediate the response to environmental change can buffer the effect of environmental change on population densities and community composition; evolution of functional interaction traits that mediate interactions with competitors, predators, preys, hosts, parasites, and/or mutualists can change species interactions, community structure, and ecosystem functioning by changing food webs and the fluxes of energy and matter (see also Chapter 7); and evolution of functional effect traits that directly impact ecosystem functions such as production, respiration, and decomposition might directly impact ecosystem features. Urban evolution therefore has the potential to profoundly influence key such as pollination, soil fertility, and top-down control of pest species or, inspired by our example, harmful toxic algae.

In one example, Theodorou et al. (2018) found several loci (including genes associated with metabolism, heat stress, and oxidative stress) under directional selection in urban compared to rural populations of the red-tailed bumblebee (*Bombus lapidarius*), an important pollinator. Such adaptive changes can be crucial for its persistence in urban areas, and consequently its pollinating services (see also Chapter 7 on the evolutionary ecology of mutualisms in the urban context). With regard to pest species, the evolution of pesticide resistance in rats, lice, mice, bed bugs, and cockroaches in urban centres can have important consequences. For example, Rost et al. (2009) showed evolved resistance to warfarin in commensal rodents, whereas Combs et al. (2018) reported on the spatial pattern of relatedness and dispersal distances of rats across multiple cities, identifying barriers and filters to among-population gene flow. The combination of such studies is an important avenue for future control of urban pests, invasive plants, disease-spreading vectors such as mosquitos, and others (Rivkin et al. 2018). In a third example, evolution-mediated persistence of large zooplankton species such as the water flea *Daphnia* might facilitate top-down control of algae blooms in urban ponds (Brans et al. 2017a, 2017b).

Locally adapted urban populations might also impact ecological responses in the surrounding rural matrix. Once urban populations have evolved, and under the assumption of regular or occasional gene flow, they can act as a source of genetic variants to support rapid adaptation of populations in peri-urban areas as these are faced with increasing human pressures. This might influence how populations and communities assemble in expanding urban centres or respond to other human pressures. Intensive agriculture or mass recreation (e.g., tourist activities) might induce selection pressures similarly to urbanization, via drivers such as increased disturbances, pollution (e.g., chemical, plastic, noise), and higher nutrient levels. It would be an interesting avenue of research to quantify to what extent local populations of species inhabiting nature reserves neighbouring urban centres are impacted by urban-derived populations.

On a larger scale, adaptation to urban heat island effects might pre-adapt species to cope with future climate change, as suggested by Brans and De Meester (2018). While it is often suggested that urban populations are sink populations fuelled by rural dispersers (e.g., Mueller et al. 2018), it is conceivable, but so far unstudied, that urban populations might act as sources for rural populations (e.g., Björklund et al. 2010) as they adapt to climate warming and rural areas undergo habitat and climatic alterations.

From the above, it is clear that the interplay between evolution and community and ecosystem ecology along urbanization gradients can yield highly interesting dynamics that are so far largely unexplored. In many cases, such as the dynamics of community assembly in peri-urban areas and responses to climate change, the spatial setting and thus evolving metacommunity structure likely is an important determinant of the dynamics. The potentially important consequences of eco-evolutionary dynamics for our understanding and predictions of future responses to urbanization as well as global change warrant a focus on using cities as model systems to examine the dynamics of evolving metacommunities in heterogeneous and rapidly changing landscapes.

If one can link trait change to population, community, or ecosystem consequences, then one can extend the use of eco-evolutionary partitioning metrics to assess potential ecological consequences of trait evolution. Ellner et al. (2011) developed a regression method to quantify the effect size of evolutionary trait change in a species on ecological endpoints. This can be population growth rate (Ellner et al. 2011), community composition (e.g., coordinates in an NMDS analysis as depicted in Figure 11.3A (Pantel et al. 2015)), or ecosystem features. Similarly, one can quantify ecosystem functions or ecosystem services as endpoints in mesocosm common gardening or reciprocal transplant experiments to quantify the impact of urban evolution on ecosystem features (Figure 11.5).

11.6 Future directions

11.6.1 Multispecies approach

For a proper assessment of the structure and dynamics of urban evolving metacommunities, one needs a multispecies approach, also in the study of evolution. So far, no study has achieved integrating urban evolution in multiple species of target communities or food webs and quantifying their joint effect on metacommunity structure along urbanization gradients. Eco-evolutionary feedbacks of different species might interfere with each other, such that their integrated effect is different from that expected if their effects would be additive. This might lead to cryptic eco-evolutionary dynamics (Kinnison et al. 2015). There are three aspects of a multispecies approach that need to be taken into account: (1) the presence of other species can impact evolutionary trajectories (Barraclough 2015), (2) the eco-evolutionary feedbacks of urban evolution might strongly depend on the identity of the evolving species, and (3) it might depend on whether or not the other species in the community also evolve in response to the urbanization gradient (De Meester et al. 2019).

11.6.2 Urban niches

One interesting avenue of research would be to explore how the niches of species shift along urbanization gradients, as recently addressed by De León et al. (2019). As species adapt to urbanization, this might result in narrowing their niche for some environmental variables, but a widening of their niche for other environmental variables compared to niche use in rural settings. Especially changes in ecological resource availability and distributions, likely to occur across urbanization gradients, could alter the shape of adaptive landscapes underpinning niche segregation. Niche segregation might reflect plastic or genetic changes in niche use. Here too, partitioning methods could be used to assess which part of the community-wide shift in niche use along the urbanization gradient is due to ecology or evolution. The change in niche use might be the result of competitive release (e.g., if other species were not capable of surviving in urban settings) or because the urban environment provides a different selective environment where other tradeoffs operate (Brans et al. 2018). In one example, De León et al. (2019) reported urbanization and bird feeding behaviour by humans leading to a wider niche breadth within Darwin finch species. This resulted in larger overlaps between species in urban sites, eroding the ecological niche differences that used to stabilize adaptive radiations in Darwin finches on the Galápagos Islands. As there are no systematic studies yet on niche use shifts in species along the urbanization gradient, it is difficult to generalize. One prediction is that urban populations are expected to show broadened niches with respect to some typical urban stressors, such as increased noise tolerance, pollution tolerance, or heat tolerance. As stress resistance often involves energetic costs, this might be traded off with other responses. One expectation is that urban populations might develop narrower niches with respect to their capacity to co-occur with predators or parasites, as there might be some degree of enemy release at least in some urban settings (see Fisher et al. 2012: the *predation paradox* on vertebrate predators). These are, however, expectations that need to be tested.

11.6.3 Reconstructing urban evolution and its consequences: resurrection ecology and historical data

A key limitation of metacommunity survey data is that they provide snapshots of patterns, from which it is difficult to derive processes. Changes in the relative abundances of species and evolution are processes that develop as time progresses, and snapshot survey data fail to provide insight into how they reciprocally influence each other. Community and evolutionary trajectories might strongly influence each other. For instance, a species might have failed to show urban evolution in one urban centre while it did so in another urban centre, because it was rapidly reduced to very low numbers by other species present in the community in the first and not in the second urban centre. Conversely, urban evolution of an already dominant species might lead to exclusion of other species in a given urban centre even though the local populations did show significant urban evolution.

Datasets on community ecology and population-specific traits of member species across multiple snapshots through time are highly valuable to document how evolving metacommunities develop. These reconstructions through time can make use of old records of species occurrences of in sites subject to urbanization or of museum collections (Magurran et al. 2010; Silvertown et al. 2011). While many of these records are often mere species lists, they yield highly valuable information on species losses and gains as urbanization developed (e.g., Silvertown et al. 2011; Kern and Langerhans 2018). For organisms that produce dormant stages, one can use resurrection ecology to reconstruct evolution through time (Stoks et al. 2016; Franks et al. 2018) by hatching old populations and comparing them to more recent ones following urbanization in the surroundings of the habitat. If one can hatch whole assemblages of species, such as the guild of plant species occurring at one or a number of sites that experienced well-documented increases in urbanization, then this offers the possibility to reconstruct the evolving community as it developed during urbanization. We are, however, not aware of any study that achieved this. Evolving metacommunity data can also be obtained without hatching, through paleolimnological approaches and paleogenomics (Jeppesen et al. 2001; Orsini et al. 2013). A key challenge with reconstructing the past through the use of sediment cores or dormant egg banks might be the disturbance of biological archives in urban areas. For example, repeated cleaning of detritus and sediment during yearly management is likely to disturb the formation of layered sediment banks over time in urban ponds. For seedbanks in terrestrial systems, a similar difficulty might arise as soils are disturbed upon urbanization, e.g., through construction works.

11.6.4 Forward-looking empirical work on urban evolving metacommunities

Given the predictability of the spread of urban centres, urbanization offers a unique opportunity to develop forward-looking approaches. One can collect baseline data and material from sites that likely are to become urbanized in the near future. Through follow-up research, one can then develop a thorough understanding of how population and community responses to urbanization interact to shape metacommunities in partially urbanized landscapes (cf. Project Baseline (Etterson et al. 2016)). A first step would be the regular monitoring of community and population genetic changes as urbanization proceeds. A research programme that captures pre- and post-urbanization population and community features would, however, be crucially powerful. Again, taxa that produce dormant stages, and notably plants (Franks et al. 2018), provide unique opportunities, as one might engage in collecting seeds of multiple plant species in areas along urban peripheries that are expected to become urbanized soon. Specifically, one could include sites in long-term monitoring programmes that are likely to become urbanized in the near future, next to sites that are likely to remain pristine or rural, and urban sites. Regular monitoring of these sites would then allow reconstructing community and population genetic changes through time, whereas hatching pre- and post-urbanization populations of key species would allow quantifying evolutionary trait change in response to urbanization via common garden experiments. Moreover, hatching pre-urban populations would allow transplant experiments in time. Transplanting pre-urban populations and communities in post-urban settings will allow to quantify to what extent local evolution affected population, community, and ecosystem features in urbanized settings. If this can be done for the key species dominating specific communities, it would yield unprecedented insight into how urbanization structures both communities and populations, and how these two responses interact with each other.

Such an approach might also generate insights into local dynamics that are difficult to study using metacommunity and landscape genetics, such as how 'local' responses to urbanization develop. As mentioned in sections 11.2.1 and 11.2.2, some landscape genetic analyses indicate that evolutionary responses to urbanization are local, i.e., urban populations are genetically more related to neighbouring rural populations than to distant urban populations (e.g., Combs et al. 2018; Mueller et al. 2018). Yet, once there is a locally adapted urban population, this population might take over from rural populations as peri-urban populations are exposed to increasing urbanization. So far, we do not have much insight into the extent to which urban populations might be sources of genetic material for peri-urban areas (Björklund et al. 2010) or vice versa (Evans et al. 2009; Tüzün et al. 2017a). This might also have implications for community composition, as it is conceivable that species that underwent local adaptation in a given urban centre might take over in associated peri-urban settings, even though other species in principle had the evolutionary potential to also adapt locally. This would yield evolution-mediated priority effects (De Meester et al. 2016), where genetic adaptation of an early arriving species gives it a fitness advantage over later arriving species. Such eco-evolutionary dynamics might result in different species of a guild (e.g., a plant community) dominating in different urban centres, and thus determine metacommunity structure.

Acknowledgements

We thank Jason Munshi-South, Anne Charmantier, and Marta Szulkin for the opportunity to contribute to this book. We furthermore thank the reviewers and editors for their thoughtful and constructive comments on the first versions of this chapter. This work was supported by KU Leuven Research Fund project C16/2017/002 and FWO project G0B9818. K.I.B. acknowledges postdoctoral funding via KU Leuven (PDM/18/112). L.G. was supported by the University of Zurich Research Priority Programme on 'Global Change and Biodiversity'.

References

Alberti, M. (2015). Eco-evolutionary dynamics in an urbanizing planet. *Trends in Ecology & Evolution*, 30(2), 114–26.

Alberti, M., Correa, C., Marzluff, J.M., et al. (2017). Global urban signatures of phenotypic change in animal and plant populations. *Proceedings of the National Academy of Sciences of the United States of America*, 114(34), 8951–6.

Aronson, M.F.J., La Sorte, F.A., Nilon, C.H., et al. (2014). A global analysis of the impacts of urbanisation on bird and plant diversity reveals key anthropogenic drivers. *Proceedings of the Royal Society B: Biological Sciences*, 281(1780), 20133330.

Atkinson D. (1994) Temperature and organism size—a biological law for ectotherms? In: Begon, M. and Fitter, A.H. (eds) *Advances in Ecological Research*. Academic Press, New York.

Balbi, M., Ernoult, A., Poli, P., et al. (2018). Functional connectivity in replicated urban landscapes in the land snail (*Cornu aspersum*). *Molecular Ecology*, 27(6), 1357–70.

Baldissera, R., Rodrigues, E.N.L., and Hartz, S.M. (2012). Metacommunity composition of web-spiders in a fragmented neotropical forest: relative importance of environmental and spatial effects. *PLOS ONE*, 7(10), e48099.

Barnum, T.R., Weller, D.E., and Williams, M. (2017). Urbanisation reduces and homogenizes trait diversity in stream macroinvertebrate communities. *Ecological Applications*, 27(8), 2428–42.

Barraclough, T.G. (2015). How do species interactions affect evolutionary dynamics across whole communities? *Annual Review of Ecology, Evolution, and Systematics*, 46(1), 25–48.

Beninde, J., Feldmeier, S., Werner, M., et al. (2016). Cityscape genetics: structural vs. functional connectivity of an urban lizard population. *Molecular Ecology*, 25(20), 4984–5000.

Beninde, J., Feldmeier, S., Veith, M., and Hochkirch, A. (2018). Admixture of hybrid swarms of native and introduced lizards in cities is determined by the cityscape structure and invasion history. *Proceedings of the Royal Society B: Biological Sciences*, 285(1883).

Björklund, M., Ruiz, I., and Senar, J.C. (2010). Genetic differentiation in the urban habitat: the great tits (*Parus major*) of the parks of Barcelona city. *Biological Journal of the Linnean Society*, 99, 9–19.

Bourassa, A.L., Fraser, L., and Beisner, B.E. (2017). Benthic macroinvertebrate and fish metacommunity structure in temperate urban streams. *Journal of Urban Ecology*, 3(1).

Brans, K.I. and De Meester, L. (2018). City life on fast lanes: urbanisation induces an evolutionary shift towards a faster lifestyle in the water flea *Daphnia*. *Functional Ecology*, 32(9), 2225–40.

Brans, K.I., Govaert, L., Engelen, J.M.T., et al. (2017a). Eco-evolutionary dynamics in urbanised landscapes: evolution, species sorting and the change in zooplankton body size along urbanisation gradients. *Philosophical Transactions of the Royal Society B: Biological Sciences*, 372(1712).

Brans, K.I., Jansen, M., Vanoverbeke, J., et al. (2017b). The heat is on: genetic adaptation to urbanisation mediated by thermal tolerance and body size. *Global Change Biology*, 23(12), 5218–27.

Brans, K.I., Stoks, R., and De Meester, L. (2018). Urbanisation drives genetic differentiation in physiology and structures the evolution of pace-of-life syndromes in the water flea *Daphnia magna*. *Proceedings of the Royal Society B: Biological Sciences*, 285(1883), 20180169.

Cadotte, M., Albert, C.H., and Walker, S.C. (2013). The ecology of differences: assessing community assembly with trait and evolutionary distances. *Ecology Letters*, 16(10), 1234–44.

Collins, S. and Gardner, A. (2009). Integrating physiological, ecological and evolutionary change: a Price equation approach. *Ecology Letters*, 12(8), 744–57.

Combs, M., Byers, K.A., Ghersi, B.M., et al. (2018). Urban rat races: spatial population genomics of brown rats (*Rattus norvegicus*) compared across multiple cities. *Proceedings of the Royal Society B: Biological Sciences*, 285(1880), 20180245.

Concepción, E.D., Moretti, M., Altermatt, F., Nobis, M.P., and Obrist, M.K. (2015). Impacts of urbanisation on biodiversity: the role of species mobility, degree of specialisation and spatial scale. *Oikos*, 124(12), 1571–82.

Crispo, E., Moore, J.-S., Lee-Yaw, J.A., Gray, S.M., and Haller, B.C. (2011). Broken barriers: human-induced changes to gene flow and introgression in animals. *BioEssays*, 33(7), 508–18.

Dahirel, M., Dierick, J., De Cock, M., and Bonte, D. (2017). Intraspecific variation shapes community-level behavioral responses to urbanisation in spiders. *Ecology*, 98(9), 2379–90.

Dahirel, M., De Cock, M., Vantieghem, P., and Bonte, D. (2018). Urbanisation-driven changes in web building and body size in an orb web spider. *Journal of Animal Ecology*, 88(1), 79–91.

De León, L.F., Sharpe, D.M.T., Gotanda, K.M., et al. (2019). Urbanisation erodes niche segregation in Darwin's finches. *Evolutionary Applications*, 12(7), 1329–43 de Mazancourt, C., Johnson, E., and Barraclough, T.G. (2008). Biodiversity inhibits species' evolutionary responses to changing environments. *Ecology Letters*, 11(4), 380–88.

De Meester, L., Vanoverbeke, J., Kilsdonk, L.J., and Urban, M.C. (2016). Evolving perspectives on monopolization and priority effects. *Trends in Ecology & Evolution*, 31(2), 136–46.

De Meester, L., Stoks, R., and Brans, K.I. (2018). Genetic adaptation as a biological buffer against climate change: potential and limitations. *Integrative Zoology*, 13(4), 372–91.

De Meester, L., Brans, K.I., Govaert, L., et al. (2019) Analysing eco-evolutionary dynamics—the challenging complexity of the real world. *Functional Ecology*, 33(1), 43–59.

Donihue, C.M. and Lambert, M.R. (2015). Adaptive evolution in urban ecosystems. *Ambio*, 44(3), 194–203.

Ellner, S.P., Geber, M.A., and Hairston, N.G. (2011). Does rapid evolution matter? Measuring the rate of contemporary evolution and its impacts on ecological dynamics. *Ecology Letters*, 14(6), 603–14.

Etterson, J.R., Franks, S.J., Mazer, S.J., et al. (2016). Project Baseline: an unprecedented resource to study plant evolution across space and time. *American Journal of Botany*, 103(1), 164–73.

Evans, K.L., Gaston, K.J., Frantz, A.C., et al. (2009). Independent colonization of multiple urban centres by a formerly forest specialist bird species. *Proceedings of the Royal Society B: Biological Sciences*, 276(1666), 2403–10.

Fisher, J., Cleeton, S., Lyons, T., and Miller, J. (2012). Urbanisation and the predation paradox: the role of trophic dynamics in structuring vertebrate communities. *BioScience*, 62(9), 809–18.

Franks, S.J., Hamann, E., and Weis, A.E. (2018). Using the resurrection approach to understand contemporary evolution in changing environments. *Evolutionary Applications*, 11(1), 17–28.

Gianuca, A.T., Pantel, J.H., and De Meester, L. (2016). Disentangling the effect of body size and phylogenetic distances on zooplankton top-down control of algae. *Proceedings of the Royal Society B: Biological Sciences*, 283(1828), 20160487.

Gianuca, A.T., Engelen, J., Brans, K.I., et al. (2018). Taxonomic, functional and phylogenetic metacommunity ecology of cladoceran zooplankton along urbanisation gradients. *Ecography*, 41(1), 183–94.

Gomulkiewicz, R. and Shaw, R.G. (2013). Evolutionary rescue beyond the models. *Philosophical Transactions of the Royal Society B*, 368(1610), 20120093.

Gorton, A.J., Moeller, D.A., and Tiffin, P. (2018). Little plant, big city: a test of adaptation to urban environments in common ragweed (*Ambrosia artemisiifolia*). *Proceedings of the Royal Society B: Biological Sciences*, 285(1881), 20180968.

Govaert, L. (2018). Eco-evolutionary partitioning metrics: a practical guide for biologists. *Belgian Journal of Zoology*, 148(2), 167–202.

Govaert, L., Pantel, J.H., and De Meester, L. (2016). Eco-evolutionary partitioning metrics: assessing the importance of ecological and evolutionary contributions to population and community change. *Ecology Letters*, 19, 839–53.

Govaert, L., Pantel, J.H., and De Meester, L. (2019). Quantifying eco-evolutionary contributions to trait divergence in spatiall structured systems. *bioRxiv*. Doi: https://doi.org/10.1101/677526.

Harris, S.E. and Munshi-South, J. (2017). Signatures of positive selection and local adaptation to urbanisation in white-footed mice (*Peromyscus leucopus*). *Molecular Ecology*, 26(22), 6336–50.

Hendry, A.P. (2017). *Eco-Evolutionary Dynamics*. Princeton University Press, New Jersey.

Jeppesen, E., Jensen, J.P., Leavitt, P., and De Meester, L. (2001). Functional ecology and palaeolimnology: using cladoceran remains to reconstruct anthropogenic impact. *Trends in Ecology & Evolution*, 16(4), 191–8.

Johnson, M.T.J. and Munshi-South, J. (2017). Evolution of life in urban environments. *Science*, 358(6363), eaam8327.

Johnson, M.T.J., Prashad, C.M., Lavoignat, M., and Saini, H.S. (2018). Contrasting the effects of natural selection, genetic drift and gene flow on urban evolution in white clover (*Trifolium repens*). *Proceedings of the Royal Society B: Biological Sciences*, 285(1883), 20181019.

Johnson, P.T.J., Hoverman, J.T., McKenzie, V.J., Blaustein, A.R., and Richgels, K.L.D. (2013). Urbanisation and wetland communities: applying metacommunity theory to understand the local and landscape effects. *Journal of Applied Ecology*, 50(1), 34–42.

Jung, V., Albert, C.H., Violle, C., and Kunstler, G. (2014). Intraspecific trait variability mediates the response of subalpine grassland communities to extreme drought events. *Journal of Ecology*, 102, 45–53.

Kern, E.M.A. and Langerhans, R.B. (2018). Urbanisation drives contemporary evolution in stream fish. *Global Change Biology*, 24(8), 3791–803.

Kinnison, M.T., Hairston, N.G., and Hendry, A.P. (2015). Cryptic eco-evolutionary dynamics. *Annals of the New York Academy of Sciences*, 1360(1), 120–44.

Knapp, S., Kühn, I., Schweiger, O., and Klotz, S. (2008). Challenging urban species diversity: contrasting phylogenetic patterns across plant functional groups in Germany. *Ecology Letters*, 11(10), 1054–64.

Lajoie, G. and Vellend, M. (2015). Understanding context dependence in the contribution of intraspecific variation to community trait–environment matching. *Ecology*, 96(11), 2912–22.

Leibold, M.A. and Chase, J.M.C. (2017). *Metacommunity Ecology*. Princeton University Press, New Jersey.

Lepš, J., de Bello, F., Šmilauer, P., and Dole˘zal, J. (2012). Community trait response to environment: disentangling species turnover vs intraspecific trait variability effects. *Ecography*, 34(5), 856–63.

Liu, Z., He, C., Zhou, Y., and Wu, J. (2014). How much of the world's land has been urbanised, really? A hierarchical framework for avoiding confusion. *Landscape Ecology*, 29(5), 763–71.

Magurran, A.E., Baillie, S.R., Buckland, S.T., et al. (2010). Long-term datasets in biodiversity research and monitoring: assessing change in ecological communities through time. *Trends in Ecology & Evolution*, 25(10), 574–82.

Manel, S., Joost, S., Epperson, B.K., et al. (2010). Perspectives on the use of landscape genetics to detect genetic adaptive variation in the field. *Molecular Ecology*, 19(17), 3760–72.

Matthews, B., Narwani, A., Hausch, S., et al. (2011). Toward an integration of evolutionary biology and ecosystem science. *Ecology Letters*, 14(7), 690–701.

McKinney, M.L. (2008). Effects of urbanisation on species richness: a review of plants and animals. *Urban Ecosystems*, 11, 161–76.

Merckx, T., Souffreau, C., Kaiser, A., et al. (2018). Body-size shifts in aquatic and terrestrial urban communities. *Nature*, 558(7708), 113–16.

Miles, L.S., Johnson, J.C., Dyer, R.J., and Verrelli, B.C. (2018). Urbanisation as a facilitator of gene flow in a human health pest. *Molecular Ecology*, Jul 4. Doi: 10.1111/mec.14783. (Epub ahead of print.).

Mueller, J.C., Partecke, J., Hatchwell, B.J., Gaston, K.J., and Evans, K.L. (2013). Candidate gene polymorphisms for behavioural adaptations during urbanisation in blackbirds. *Molecular Ecology*, 22(13), 3629–37.

Mueller, J.C., Kuhl, H., Boerno, S., et al. (2018). Evolution of genomic variation in the burrowing owl in response to recent colonization of urban areas. *Proceedings of the Royal Society B: Biological Sciences*, 285(1878), 20180206.

Munshi-South, J. and Kharchenko, K. (2010). Rapid, pervasive genetic differentiation of urban white-footed mouse (*Peromyscus leucopus*) populations in New York City. *Molecular Ecology*, 19(19), 4242–54.

Niemelä, J. and Kotze, D.J. (2009). Carabid beetle assemblages along urban to rural gradients: a review. *Landscape and Urban Planning*, 92(2), 65–71.

Norberg, J., Urban, M.C., Vellend, M., Klausmeier, C.A., and Loeuille, N. (2012). Eco-evolutionary responses of biodiversity to climate change. *Nature Climate Change*, 2, 747–51.

Oke, T.R. (1973). City size and the urban heat island. *Atmospheric Environment*, 7, 769–79.

Orsini, L., Schwenk, K., De Meester, L., et al. (2013). The evolutionary time machine: forecasting how populations can adapt to changing environments using dormant propagules. *Trends in Ecology & Evolution*, 28, 274–82.

Palma, E., Catford, J.A., Corlett, R.T., et al. (2017). Functional trait changes in the floras of 11 cities across the globe in response to urbanisation. *Ecography*, 40(7), 875–86.

Pantel, J.H., Duvivier, C., and De Meester, L. (2015). Rapid local adaptation mediates zooplankton community assembly in experimental mesocosms. *Ecology Letters*, 18, 992–1000.

Parris, K.M. (2016). *Ecology of Urban Environments*. Wiley-Blackwell, Chichester.

Piano, E., De Wolf, K., Bona, F., et al. (2017). Urbanisation drives community shifts towards thermophilic and dispersive species at local and landscape scales. *Global Change Biology*, 23(7), 2554–64.

Price, G.R. (1970). Selection and covariance. *Nature*, 227(5257), 520–21.

Reid, N.M., Proestou, D.A., Clark, B.W., et al. (2016). The genomic landscape of rapid repeated evolutionary adaptation to toxic pollution in wild fish. *Science*, 354(6317), 1305–8.

Rivkin, L.R., Santangelo, J.S., Alberti, M., et al. (2018). A roadmap for urban evolutionary ecology. *Evolutionary Applications*, 2(3), 384–98.

Rost, S., Pelz, H.-J., Menzel, S., et al. (2009). Novel mutations in the VKORC1 gene of wild rats and mice—a response to 50 years of selection pressure by warfarin? *BMC Genetics*, 10(1), 4.

Seto, K.C., Güneralp, B., and Hutyra, L.R. (2012). Global forecasts of urban expansion to 2030 and direct impacts on biodiversity and carbon pools. *Proceedings of the National Academy of Sciences of the United States of America*, 109(40), 16083–8.

Silvertown, J., Cook, L., Cameron, R., et al. (2011). Citizen science reveals unexpected continental-scale evolutionary change in a model organism. *PLOS ONE*, 6(4), e18927.

Stoks, R., Govaert, L., Pauwels, K., Jansen, B., and De Meester, L. (2016). Resurrecting complexity: the interplay of plasticity and rapid evolution in the multiple trait response to strong changes in predation pressure in the water flea *Daphnia magna*. *Ecology Letters*, 19(2), 180–90.

Theodorou, P., Radzevičiūtė, R., Kahnt, B., et al. (2018). Genome-wide single nucleotide polymorphism scan suggests adaptation to urbanisation in an important pollinator, the red-tailed bumblebee (*Bombus lapidarius* L.). *Proceedings of the Royal Society B: Biological Sciences*, 285(1877), 20172806.

Thompson, J. (1998). Rapid evolution as an ecological process. *Trends in Ecology & Evolution*, 13(8), 329–32.

Thompson, K.A., Renaudin, M., and Johnson, M.T.J. (2016). Urbanisation drives the evolution of parallel clines in plant populations. *Proceedings of the Royal Society of London B: Biological Sciences*, 283(1845), 20162180.

Turrini, T. and Knop, E. (2015). A landscape ecology approach identifies important drivers of urban biodiversity. *Global Change Biology*, 21(4), 1652–67.

Tüzün, N., Op de Beeck, L., and Stoks, R. (2017a). Sexual selection reinforces a higher flight endurance in urban damselflies. *Evolutionary Applications*, 10(7), 694–703.

Tüzün, N., Op de Beeck, L., Brans, K.I., Janssens, L., and Stoks, R. (2017b). Microgeographic differentiation in thermal performance curves between rural and urban populations of an aquatic insect. *Evolutionary Applications*, 10(10), 1067–75.

Urban, M.C. and Skelly, D.K. (2006). Evolving metacommunities: toward an evolutionary perspective on metacommunities. *Ecology*, 87(7), 1616–26.

Urban, M.C., De Meester, L., Vellend, M., Stoks, R., and Vanoverbeke, J. (2012). A crucial step toward realism: responses to climate change from an evolving metacommunity perspective. *Evolutionary Applications*, 5(2), 154–67.

Urban, M.C., Bocedi, G., Hendry, A.P., et al. (2016). Improving the forecast for biodiversity under climate change. *Science*, 353(6304), aad8466.

Van Doorslaer, W., Vanoverbeke, J., Duvivier, C., et al. (2009). Local adaptation to higher temperatures reduces immigration success of genotypes from a warmer region in the water flea *Daphnia*. *Global Change Biology*, 15, 3046–55.

Vellend, M. (2015). *The Theory of Ecological Communities*. Princeton University Press, New Jersey.

Winchell, K.M., Maayan, I., Fredette, J.R., and Revell, L.J. (2018). Linking locomotor performance to morphological shifts in urban lizards. *Proceedings of the Royal Society B: Biological Sciences*, 285(1880), 20180229.

Zhao, L., Oppenheimer, M., Zhu, Q., et al. (2018). Interactions between urban heat islands and heat waves. *Environmental Research Letters*, 13(3), 034003.

CHAPTER 12

Terrestrial Locomotor Evolution in Urban Environments

Kristin M. Winchell, Andrew C. Battles, and Talia Y. Moore

Winchell, K.M., Battles, A.C. and Moore, T.Y., *Terrestrial Locomotor Evolution in Urban Environments* In: *Urban Evolutionary Biology*.
Edited by Marta Szulkin, Jason Munshi-South and Anne Charmantier, Oxford University Press (2020). © Oxford University Press.
DOI: 10.1093/oso/9780198836841.003.0012

12.1 Introduction

How animals move through and interact with their habitats is directly influenced by their ecology, behaviour, physiology, and morphology. The structure and properties of the substrates through and on which animals move, the functional demands of an animal's life, and morphological structures involved in movement all contribute to different locomotor strategies specific to the organism in a given environment. Thus, while some generalizations can be made regarding the types of locomotor adaptations we may see in terrestrial urban species, the consequences of urban life differ among species to varying extents, depending on how each interacts with the environment. Since most traits are not universally beneficial and are instead context-dependent, we can connect habitat use and phenotypic change in an environment by examining the target of natural selection: performance in context (Arnold 1983). The functional gain of a phenotype leads to fitness advantages, for example if a trait enables an organism to more effectively escape predators, capture prey, or exploit novel resources. This provides a mechanistic basis by which adaptive morphological traits are shaped by natural selection in novel habitats. In other words, major changes to a habitat could favour adaptation through altered performance, which may be achieved via adaptive plastic and genetic morphological changes. In the case of locomotion, alterations of the habitat's spatial structure or the properties of substrates encountered by an animal could shift locomotion strategies, and thereby alter demands on locomotor morphology.

Urbanization transforms natural landscapes in multiple dimensions, most notably the habitat structure, leading to novel habitat spaces that are fundamentally different from less disturbed, natural landscapes. These differences in abiotic and biotic elements create challenges for animals occupying urban landscapes, resulting in behavioural, morphological, and physiological shifts (reviewed in Johnson and Munshi-South 2017). For many species, including species that do particularly well in urban forest patches, the open urban landscape presents insurmountable challenges to movement, leading to population fragmentation and reduced gene flow (Chapter 4; Munshi-South 2012; Beninde et al. 2016). For others that are able to move throughout the urban landscape, the shift in spatial distribution of habitat elements presents novel challenges to be overcome for terrestrial movement. Scansorial species (those that use terrestrial habitat and are also specialized for climbing) face the additional challenge of contending with altered substrate properties of anthropogenic structures. These environmental differences create novel demands on functionally relevant traits. Some animals may be pre-adapted to the locomotory challenges of urban life, possessing traits that allow them to exploit urban habitats, while others may be filtered from the urban environment because they lack key traits necessary for urban habitat use. For other species, occupying urban habitats may promote morphological change through plasticity, evolutionary adaptation, or both.

In this chapter, we discuss the key elements of urban habitats that affect locomotion for terrestrial and scansorial animals. Examples of evolutionary adaptation in urban environments abound (Johnson and Munshi-South 2017), but locomotion-specific instances of evolution in urban habitats are not as well studied. First, we examine how urbanization alters the spatial arrangement of habitat elements and how this can alter behaviour and morphology by changing the adaptive landscape of traits related to terrestrial locomotion. Then we describe how urbanization results in a dominance of substrates with very different physical properties and how this change impacts behaviour and morphology associated with climbing. Throughout this chapter, we draw on existing literature on locomotion in more natural environments to predict what types of behavioural, biomechanical, and morphological changes we might expect in species in urban environments. Since this research area is currently being developed, we present hypotheses and suggestions for future studies that we expect will significantly broaden the field. We realize that the consequences of interaction with urban environments can be very different across species, but we aim to deal with the most common conditions faced by species throughout this chapter. We focus on terrestrial systems here, but acknowledge that the selective pressures altered by urbanization are not limited to affecting terrestrial organisms. For example, altered hydrology has implications for urban aquatic organisms (Chapter 10) and the altered wind currents caused by the arrangement of buildings, convection from car parks, and active airflow systems affect flight in birds and insects.

12.2 Spatial organization of habitats

One of the most easily recognized characteristics of urban landscapes is the structural simplification and fragmentation of natural habitat. Trees form a non-continuous canopy in urban environments, with percentages of 0–55 typical at the landscape scale (Nowak et al. 1996). Vegetation is reduced to small fragmented patches and heavily manicured urban green spaces, commonly composed of non-native plants and intermixed with patches of built environment (Forman 2014). Concomitant with this loss and isolation of vegetative habitat is the increase in impervious surfaces (e.g., concrete and pavement), which contributes to elevated and heterogeneous thermal profiles in urban environments and creates an inhospitable matrix with little to no refuges (Forman 2014). Together, these features result in urban habitats that are typically fragmented by roads and buildings, have large open areas with small clusters of vegetation, and have little to no vegetative cover (see Figure 12.1 and Chapter 2). Even in urban parks, the distance between trees and the lack of canopy cover is strikingly different from exurban locations.

12.2.1 Behaviourally mediated habitat use

Heterogeneous (or 'patchy') distribution of vegetation and resources can act as either a barrier or a conduit for organism movement, in both naturally occurring and urban environments. Even if urban forest patches provide sufficient resources to sustain a small population, the artificial spacing of these patches may make it difficult for a scansorial community to expand its habitat use to more urbanized habitats or migrate to a more suitable patch across the urban matrix. Behavioural affinity for specific microhabitats, based on type of locomotion, preferred substrate, or preferred vegetation density, can filter which animals are able to pass through or occupy urbanized areas (Prevedello and Vieira 2010). Moreover, behaviour, particularly whether a species restricts habitat use to vegetative elements within the urban environment (e.g., urban parks) or fully exploits novel urban habitat space (e.g., use of anthropogenic resources), has implications for how natural selection may shape adaptive morphological changes (Wong and Candolin 2015). Two metrics of behaviour, open-field anxiety and flight initiation distance, reflect how an animal's terrestrial movement can depend on the distribution of plants and animals within a habitat. Such metrics are particularly insightful for understanding how behaviour influences urban habitat use, which in turn influences how natural selection shapes traits related to locomotion.

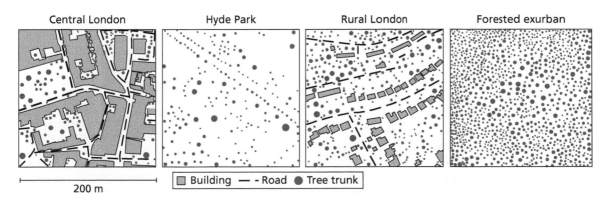

Central London   Hyde Park   Rural London   Forested exurban

├────────────────┤
   200 m

□ Building   — - Road   ● Tree trunk

Figure 12.1 The urban environment is fragmented and dominated by anthropogenic structures with large spans of open space and non-continuous canopy cover. In each 200-m square we show building footprints, tree trunks (proportional to canopy size), and roads. From left to right (increasing in distance from the city centre): an intensely urbanized section of central London; Hyde Park, a large urban park in central London; Caterham, a rural suburb of London; and a forested area outside Greater London.

Open-field anxiety, also called 'thigmotaxis', is the behavioural affinity for shelter (Simon et al. 1994). This behaviour has been selectively bred in laboratory animals and is preserved even in laboratory-reared animals that have never experienced predation, suggesting that these traits are heritable and genetically based (Leppänen et al. 2005; Moore et al. 2017a). For animals that have historically experienced strong predation pressure, open-field anxiety is inversely correlated with predator-evasion ability, due to the fitness tradeoff between remaining safe in a shelter and exploring new areas to find food. While open-field anxiety is a baseline behaviour independent of the presence of a predator, flight-initiation distance (FID) is a behavioural response to a perceived predation threat. The risk humans or other potential predators pose as perceived by the animal is approximated by measuring the proximity at which human presence induces a typical fear or flight response (reviewed in Cooper and Blumstein 2015). These responses are mainly associated with perceived predator characteristics and distance to refuge, but are also influenced by interacting factors such as time of day, amount of ground cover, degree of crypsis, and ability to flee (Stankowich and Blumstein 2005).

In urban environments, where human encounters are frequent, species that are able to tolerate human presence are more likely to thrive. Thus, open-field anxiety and flight response influence whether and how an animal will explore an urban environment. Open-field anxiety may act as a filter on urban colonization: animals with high open-field anxiety are likely to have a high FID regardless of habitat occupied, and therefore are unlikely to persist in urban areas (Figure 12.2, species A). If, in fact, species with high open-field anxiety and high FID do persist in an urban environment, they are likely to remain in areas with vegetation and ample refuges. The tendency to find secluded and protected spaces, a characteristic of species with high open-field anxiety, determines the suitability of potential refuges, whether natural or anthropogenic (e.g., underground tunnels, attics). In contrast, animals more capable of evading predators, or with fewer natural predators, are naturally less anxious in exposed areas such as urban environments (Figure 12.2, species B and C; Lima and Dill 1990). Among these

species, those that flee less readily (low FID) are more likely to leave areas with ample refuges and explore areas occupied by humans (e.g., pavements, buildings, fences), with habituation playing a role in increased behavioural divergence between urban and natural populations. As animals with low open-field anxiety and low FID explore more of the urban habitat, they will be affected by novel selection pressures favouring morphological adaptations associated with shifts in locomotor demands.

In general, urban animals, such as the brown anole lizard (*Anolis sagrei*) (Lapiedra et al. 2017), eastern grey squirrel (*Sciurius carolinensis*) (Engelhardt and Weladji 2011), and several species of birds (Møller 2008), tend to have lower FID than their forest counterparts (i.e., they allow a closer approach before fleeing). This difference may arise within urban environments through habituation (i.e., fear responses and FID decrease with increased human exposure)

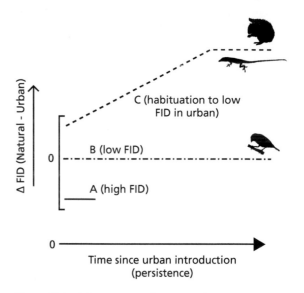

Figure 12.2 FID is an important determinant of persistence in an urban environment. Species with high FID regardless of whether they are in urban or non-urban sites flee early when predators approach, and so may not be able to establish and grow populations (A). Some species have relatively low FID in general and so individuals in urban environments may easily persist (B). Some species may have initial high FID in natural habitats, but through either plasticity or evolutionary adaptation, urban populations may develop lower FID relative to natural populations (C). Silhouettes (from PhyloPic) represent example species cited in this chapter. In these hypothetical examples, species start with no difference in FID between natural and urban populations (0 on y-axis).

or natural selection on genetically based behaviours such as open-field anxiety. As with open-field anxiety, some species may be predisposed to have increased tolerance of humans. Birds that have shorter FIDs in their ancestral habitat tend to be more tolerant of urbanization and have even shorter FIDs in urban environments (Møller 2014). Moreover, through habituation these innate fear responses may be modified, leading to increased boldness and decreased neophobia in urban environments. For example, urban *Anolis* lizards have been shown to be bolder and spend more time exploring new environments than their forest counterparts (Lapiedra et al. 2017) (similar to Figure 12.2, species C).

Future studies should take an integrative approach to examine heritable behaviours as well as the influence of habitat use and predation pressures in shaping behavioural responses in urban environments. Such studies will provide a more precise understanding of the genetic and plastic behavioural preferences that determine which species persist and thrive in changing environments.

12.2.2 Mechanisms of locomotion in urban habitats

The environmental differences described at the beginning of section 12.2, combined with increased human–wildlife interaction, create a suite of novel demands on movement within the urban environment. Long-distance movement may become less relevant for some species as range size and dispersal distances are limited, leading to more sedentary individuals (e.g., Etter et al. 2002). Yet large spans of open space typical of urban environments necessitate increased terrestrial movement and reduced arboreal locomotion. Little overhead cover or refuges, clustered resources, and shifts in predator abundance and communities require that this over-ground movement be quick. Additional threats associated with human interaction, in terms of both direct and indirect persecution (Koenig et al. 2002), may also require behavioural adjustments to minimize stress and mortality. Escape behaviour in response to perceived predation threat, locomotion across open areas, and shifts from one form of locomotion to another all depend on the spatial organization of suitable refuges and substrates in urban environments.

Individuals in urban areas may escape in different ways than those in natural habitats. Due to the heritability of thigmotaxis (Leppänen et al. 2005; Moore et al. 2017a), increased open space in urban areas would limit habitat use between refuges, potentially leading prey animals to evolve reduced FID. Indeed, urban Indian rock agamas (*Psammophilus dorsalis*) exhibited shorter FID but used perches closer to refuges than individuals in rural areas (Batabyal et al. 2017). In other words, although urban lizards allowed researchers to approach more closely, this was likely because suitable refuges existed in close proximity. However, this pattern is not universal. For instance, garden skinks (*Lampropholis guichenoti*) in Sydney (Australia) were found further from refuges (in modified environments which had reduced ground cover) and had greater FIDs than in natural areas (Prosser et al. 2006). In addition, alternative escape strategies may be employed. Sparkman et al. (2018) found that western fence lizards (*Sceloporus occidentalis*) on university campuses in California (USA) were more likely to remain tonically immobile rather than flee when confronted with a potential predator compared to rural lizards. Certainly, factors other than the habitat structure, such as predation pressure, influence escape behaviour, and more research is needed to understand how urban habitats alter escape strategies.

When traversing these open areas, animals may alter their trajectories in ways that deter predation. These responses depend on the predation strategies encountered: ambush, pursuit, or ballistic interception (when a predator follows a trajectory that should lead to the future location of the prey) (Moore and Biewener 2015). Prey that mainly encounter ambush predators are likely to increase their ability to rapidly respond to a predation attempt. Pursued prey are likely to increase their locomotor economy and escape speed, with a few agile manoeuvres to throw off the predator. Prey that experience ballistic interception predation are likely to change their behaviour to make it difficult for predators to predict their future positions and facilitate evasive escape manoeuvres. Such unpredictable escape behaviour, also known as 'protean behaviour', includes the tendency to escape both away from and towards a predator, alternating between different escape strategies, or using rapid escape manoeuvres (reviewed by Domenici et al. 2011). The unpredictability of escape trajectories in three-dimensional space can be directly measured using the entropy equation to predict predator evasion ability and affinity for open spaces (Moore et al. 2017a). Since protean behaviours often correspond to an affinity for open environments, we would expect that these animals would be predisposed to persist as environments become more urbanized, and suggest future research examine this relationship.

Despite being specialized for movement on a particular substrate, animals may alter their locomotor behaviour to overcome barriers to movement or adjust to other features of the urban environment that differ from their natural habitat. Studies show that animals alter their locomotor strategies when the spatial distribution of suitable habitats changes, even in the absence of urbanization. Although apparently not tolerant of urbanization, gibbons, which have evolved specialized anatomical features for brachiating (swinging from branch to branch at high speeds) have been documented to shift from this typical form of locomotion when faced with habitat fragmentation. Instead of reducing their home range size in response to fewer suitable trees in modified habitats, brachiating gibbons voluntarily cross gaps in the canopy up to 9 m wide and will bipedally walk across larger gaps when necessary (Cannon and Leighton 1994). This example demonstrates that even in natural environments, habitat structure can have a behavioural effect on locomotion that, if sustained, may favour certain morphologies that can accumulate over time. Other animals could have a less drastic, but still evolutionarily significant, context-dependent change in their locomotion. For instance, degus (*Octodon degus*), small diurnal rodents, run faster and pause for longer when moving between refuges in open habitats compared to more vegetated, 'sheltered' habitats (Vasquez et al. 2006), and urban *Anolis* lizards sprint faster than individuals from forest habitats (Winchell et al. 2018b). Such behavioural modifications may drive morphological changes to maintain movement in habitats undergoing anthropogenic modifications.

Quite a few studies have investigated FID in urban animals, but few have connected this aspect

of behaviour with ecological and evolutionary consequences. We still know relatively little about how innate behaviours and habituation impact habitat use in the urban environment, although there is growing evidence that behavioural modifications reducing stress have genetically based physiological consequences for urban species (Chapter 15; van Dongen et al. 2015). We know even less about how anthropogenic changes to habitat structure impact locomotion strategies. Linking these behavioural and physiological shifts with ecological shifts in urban species provides an important context for predicting and interpreting evolutionary adaptations.

12.2.3 Shifts in locomotion drive morphological change

As described in section 12.2.2, species with a propensity to use open urban environments are expected to experience novel selective pressures to effectively use such habitat while minimizing risks. Depending on the ecology and behaviour of a species, different morphologies may become favourable. Although few studies have explicitly examined morphological differences in urban animals and even fewer have examined the effects of morphological shifts on locomotor performance, we can make predictions based on functional–morphological relationships known from natural environments across a range of taxa.

A shift from arboreal or scansorial lifestyles toward more terrestrial habitat use in urban environments is expected to have morphological consequences, particularly related to speed and acceleration. Since there are direct tradeoffs between morphological specializations for speed and acceleration, the propensity to evolve specialization for either depends on the natural history of the species in addition to environmental alterations and ecological interactions. Generally, predators associated with open grassland environments tend to use a pounce-pursuit strategy (Van Valkenburgh 1985). In contrast, many forest environments are dominated by specialist ambush or interception predators. In urban areas, predator communities tend to consist of generalist, often domestic, mesopredators that pursue their prey, such as cats, raccoons, dogs, coyotes, and foxes (Faeth et al. 2005). This shift in predator communities may lead to increased pounce-pursuit predation, or alternatively, to lower rates of predation if animals are only predated upon by ambush or interception predators that are ineffective in urban environments. If mortality from predation is reduced (through either reduced predation pressure or decreased predator efficiency), the possibility also exists that selection on anti-predator traits may instead be relaxed in urban environments.

Since predation strategies shape prey escape strategies, we expect that many prey species will be subject to selective pressures favouring morphologies that enable fast sprinting and agile movement as animals move quickly within patchy habitat. For example, kangaroo rats (*Dipodomys merriami*) are specialized for rapid acceleration (Biewener and Blickhan 1988) and are successful in moderately urbanized habitats (California State University San Bernardino campus), but disappear in completely urbanized habitats (downtown San Bernardino; T.Y. Moore, personal observations). Their rapid responses and high manoeuvrability make them particularly suited to exploit the newly formed open space, but as human density increases, open space decreases, and they have fewer refuges for escape. However, these predicted changes assume that predator and prey communities are abundant enough to justify such strategies. Trophic dynamics in urban areas can be drastically different from more natural habitats; vertebrate prey community size and diversity are reduced, and animals opportunistically scavenge anthropogenic food sources with high frequency (Faeth et al. 2005).

If high-speed movement is favoured in urban environments, we would expect urban species to evolve more slender and elongate limbs, which increase stride length and decrease the moment of inertia of each limb (Hildebrand 1982). In particular, longer distal limb elements (i.e., tibia/fibula, radius/ulna) and larger metatarsal to femur ratios (i.e., relatively longer feet) are correlated with high-speed terrestrial locomotion (reviewed in Foster et al. 2015). Such limbs have a lower moment of inertia, which reduces the energy needed to redirect the limb in every stride during sustained locomotion. Speed can be further enhanced by adjusting the posture of the limb. Animals with more upright postures and erect limbs have higher effective

mechanical advantage and transmit a greater proportion of muscle force into movement, rather than into tendon and bone (Biewener 2005). On the other hand, limb elements that are generally held in flexed positions, such as species with crouched postures, can be rapidly extended to elongate the duration of force production, resulting in a single accelerative event or leap. Crouched postures are likely to be more useful for cryptic prey animals evading ambush or interception predation, as well as arboreal species that cling to vertical substrates, such as frogs.

Osteological differences in response to the selective pressures in urban environments likely extend beyond simple changes in limb length or proportions. In mammals that sustain running at higher speeds, species have vertebral morphologies that enhance flexion and extension of the spine. Elongate pelvic ischia and symphyses are associated with faster running speeds, quicker acceleration, and changes in direction at high speeds, while the location of the tibial tuberosity affects knee extension for fast acceleration and propulsion (Álvarez et al. 2013). In mammals, pounce-pursuit predation is associated with elbow joints that reduce forearm rotation, fore limbs with the hand fixed in a prone position (e.g., relatively long metacarpals), relatively long hind limbs, and limb movement restricted to the parasagittal plane (Van Valkenburgh 1985). While some changes in bone length and density can occur in response to forces encountered during ontogeny (Losos et al. 2000; La Croix 2011), consistent changes in bone shape within a population are accumulated through adaptive evolution.

A few studies have documented some of these hypothesized shifts in open natural environments and in urban environments. Lizard species (e.g., *Anolis* sp. and *Niueoscincus* sp.) occupying open natural habitats have longer limbs and attain faster sprint speeds, while those occupying more cluttered or arboreal habitats tend to have shorter limbs and move more slowly (Losos and Irschick 1996; Melville and Swain 2000). In urban environments, we see similar shifts, where urban lizards (*A. cristatellus*) have relatively long limbs compared to nearby forest populations of the same species (Winchell et al. 2016). These differences in limb length in urban *A. cristatellus* were maintained in common garden rearing, supporting a genetic basis of this trait shift (Winchell et al. 2016). Of course, these predictions assume the directionality of habitat-use shifts to be from arboreality to terrestriality, but in some species the manufactured human environment may provide novel opportunities leading to increased scansoriality. For example, populations of the Aegean wall lizard (*Podarcis erhardii*) inhabiting human-built rock walls had relatively longer limbs, particularly the distal elements, compared to populations inhabiting flat, loosely consolidated terrestrial habitat (Donihue 2016).

Survival in urban environments is likely dependent on both genetic and plastic aspects of locomotor morphology, and probably in complex and interactive ways. In addition to osteological changes, quick and agile locomotion to enhance prey evasion in more open urban environments could be achieved with plastic changes in neuromuscular and muscle-tendon architecture. Although no studies to date have examined muscular morphology of urban animals in an adaptive context, natural selection may act on this aspect of locomotor morphology. Increasing muscle fibre length or cross-sectional area increases the speed

of muscle length change and the force produced by the muscle, respectively (Zajac 1989). Muscle fibre type can also change plastically, altering the force–length characteristics of the muscle (Aryan and Alnaqeeb 2002). Less-compliant tendons transmit muscle contraction more directly into limb movement, since minimal muscle contraction is used to stretch the tendon, providing greater control of motion and faster response time at the cost of elastic energy savings (Figure 12.3; Biewener 2005). A faster response time could also be achieved neuromechanically by increasing the sensitivity of sensory organs, and reflexive responses to specific stimuli (Brown and Loeb 2000). For example, kangaroo rats will reflexively jump in the air when they hear a specific vibration frequency produced by owl wings (Webster 1962). Such reflexes evolve via shortening the neural pathways between stimuli and motor control centres. These neuromuscular changes have the potential to enhance prey and predator manoeuvrability in the open spaces created by the onset of urbanization, but in highly urbanized areas with fewer open spaces, animals require additional adaptations to access manufactured substrates as refuges.

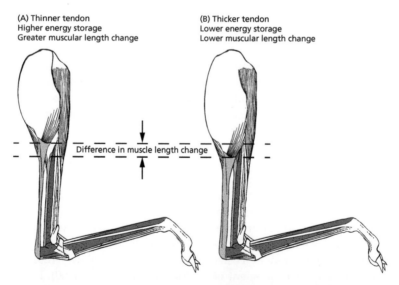

(A) Thinner tendon
Higher energy storage
Greater muscular length change

(B) Thicker tendon
Lower energy storage
Lower muscular length change

Difference in muscle length change

Figure 12.3 Tendon stiffness affects cost of transport and transmission of muscular contraction into joint movement. (A) Compliant and thin tendons elastically store energy that can be returned to decrease the total energy required for a motion. However, muscle length must change greatly to compensate for tendon stretch and achieve a given angular deflection in the limb. (B) Stiffer and thicker tendons absorb less energy because they stretch less. Thus, less muscle length change is required to match the same angular deflection as in (A). This higher stiffness facilitates more direct transmission of muscular contraction into locomotory changes, such as acceleration and deceleration. (Figure adapted from Moore et al. (2017).)

Building upon extensive biomechanical research on the functional significance of neural, muscular, and skeletal morphologies, future functional morphology studies should examine these features in the context of urbanization, as they are key to evasive success.

12.3 Substrate properties

Urbanization drastically changes the types of perches available to arboreal urban dwellers by replacing existing vegetation with anthropogenic structures, which differ dramatically in their structural, mechanical, and surface properties (Meyers et al. 2008). Anthropogenic perches tend to be less structurally complex compared to trees: unbranching, flat, and broad in diameter (e.g., Winchell et al. 2016, 2018a). In addition, urban animals must contend with entirely novel substrates, such as metal, painted surfaces, and glass. These surfaces are more rigid (less compliant), smoother, and harder (increased surface strength) than many natural substrates (Figure 12.4). For arboreal species, the urban environment thus presents a second set of challenges relevant to functional morphology. As described in section 12.2.1, species must either restrict their habitat use to natural vegetative habitat elements (effectively reducing their available habitat space) or adapt/be pre-adapted to use the anthropogenic structures that dominate the environment. Many species across a range of taxa use anthropogenic structures to varying extents. For example, raccoons and squirrels climb buildings to nest in attics and chimneys, and frogs and geckos are commonly found on buildings, particularly near lights.

12.3.1 Climbing behaviour on urban substrates

Just as the layout and distribution of habitat elements influence locomotor behaviour (section 12.2.2), the properties of the substrate, and the challenges they pose, often dictate the locomotor strategy required. When responding to a predator, climbing animals must either flee vertically, or move quickly to the other side of the structure where it is out of sight, or jump from the structure. Predators are able to approach an animal perched on a broad substrate more closely than on a narrow trunk, because evasive movement is more effective on broad trunks compared to those that might not fully hide the animal (Regalado 1998). Prey behavioural responses reflect this: *Anolis* lizards on thin trunks are more likely to jump when responding to a simulated approaching predator rather than move to the opposite side of the structure as they would on broader trunks (Losos and Irschick 1996). Walls, though they

Figure 12.4 Arboreal species such as these *Anolis cristatellus* in urban areas must be able to move across and between structures that are smoother, harder, and broader than those found in natural environments. The urban lizard on the left clings to a smooth, flat, painted concrete wall, which differs drastically from the typical arboreal habitat used by the lizard on the right. (Photos by K.M. Winchell.)

are broad in diameter, often do not have another side to which an animal can move to, therefore changing how animals react to perch breadth as a substrate property. For example, Avilés-Rodríguez and Kolbe (2019) found that urban *A. cristatellus* lizards perched on walls had significantly longer FIDs compared to conspecifics on other substrates, and that lizards sprinted more often as an escape strategy on walls than they did on trees and other urban substrates.

If sprinting across an anthropogenic substrate is not a viable escape strategy, for example if a species is unable to rapidly move or effectively cling on smooth surfaces, then jumping may be necessary. Many arboreal animals tend to jump from rigid substrates (to produce high ground-reaction forces) onto more compliant substrates (to absorb excess forces and cushion their landing) (Gilman and Irschick 2013). Other animals, such as tree frogs, use the energy stored in a compliant substrate to minimize energy loss during jumping (Astley et al. 2015). It may not always be possible to find appropriate substrates for jumping in urban environments, where natural substrates are clustered and anthropogenic substrates dominate the landscape (Figure 12.1). In nature, the same plant often has several portions that differ in compliance (boughs and branches versus twigs and leaves), while in urban areas, man-made structures tend to be uniform in material and compliance. A lack of suitable launching or landing structures in urban habitats could result in injury (Winchell et al. 2019) and thus would necessitate a shift in escape strategy. One possible strategy to deal with the paucity of compliant substrates is to adjust in-flight movement. For example, gliding mammals tend to land on rigid and large-diameter branches and trunks of trees (Byrnes and Spence 2011).

Alternatively, animals that are unable to effectively climb or move across anthropogenic substrates may be 'filtered out' of urban environments or may persist by avoiding these structures entirely and restrict their habitat use to urban vegetation. By avoiding habitats where performance is submaximal, animals may avoid associated fitness costs (this is known as the habitat constraint hypothesis (Irschick and Losos 1999)) and impede evolutionary change associated with using manufactured aspects of the urban environment. We have some evidence that this may occur. Urban lizards (specifically *A. cristatellus*) may choose to avoid smooth anthropogenic substrates on which they struggle to sprint effectively (Battles et al. 2018; Winchell et al. 2018b), but this pattern does not hold for all species or contexts (Kolbe et al. 2015; Battles et al. 2018). Habitat avoidance (constrained by suboptimal performance) may help explain why all animals that possess the capabilities of climbing anthropogenic structures do not always do so. For anole lizards, and likely many other taxa, species that use anthropogenic structures may be exposed to different selective pressures than those that restrict their habitat to vegetation (Winchell et al. 2018a). The extent to which this is generally true within and between species has important implications for the repeatability and convergence of urban morphological shifts.

Although the biomechanics of jumping and landing have been well studied in non-urban contexts, the consequences of locomotion on anthropogenic substrates on bone development, injury risk, and behavioural strategies for locomotion in the urban context are poorly understood. For example, the inertial data collected by Byrnes et al. (2008) could be applied to animals that occupy urban and natural environments to provide insight into whether animals are encountering higher forces or altering their behaviour to minimize the force encountered during impact. To understand the evolutionary impacts of compliance on jumping in urban habitats, further studies should apply the experimental frameworks cited here, and others, in urban contexts.

12.3.2 Mechanisms of climbing in the urban habitat

The differences in physical properties of anthropogenic substrates described in section 12.3 combined with the resultant behavioural shifts likely to occur (section 12.3.1) create an altered adaptive landscape for climbing locomotion in urban environments. When animals traverse the urban environment without descending to the ground, they must be able to quickly and adeptly move across and between broad, hard, and smooth anthropogenic surfaces that offer few places to hide. Vertical movement relies primarily on the ability to cling to a substrate, which is essential for climbing animals to evade predators, locate food, and reproduce. Since animal movement strategies are greatly determined by the roughness of a surface, smooth surfaces typical in urban habitats may limit certain forms of animal movement. Many species may not possess suitable attachment mechanisms to climb anthropogenic structures at all and may be filtered out of the urban habitat.

While long limbs are advantageous for quick terrestrial movement (section 12.2.3), climbing requires a different suite of traits. For example, animals occupying more cluttered or arboreal habitats encounter many perches of smaller diameter, upon which longer limbs result in reduced stability, speed, and acceleration capacity (Losos and Irschick 1996, Sathe and Husak 2015). The stability of climbing animals is inversely proportional to their body size and distance from their centre of mass to the surface (Cartmill 1974). Thus, smaller animals and animals that are able to maintain their centre of mass closer to a surface would be at an advantage when climbing in general. Animals with shorter limbs, a flattened body, and reduced head height are able to maintain their centre of mass closer to the surface, providing better stability when climbing and reducing backward pitching effects that could result in falls (Cartmill 1985; Vanhooydonck and VanDamme 1999; Zaaf et al. 1999). Furthermore, actively controlled tails can be used to constantly press the body centre of mass close to the surface and can act as a kickstand to prevent backwards pitching moments (Jusufi et al. 2008).

Of course, these arboreal movements require that a species be able to cling to the smooth surfaces in urban environments in the first place. Different approaches and mechanisms for adhesion or attachment may be employed by a single species, with varying effectiveness depending on the organism, habitat use, and substrate. Grasping (Figure 12.5B) is related to hand, foot, and digit morphology, involving highly integrated structures that vary widely among species (reviewed in Sustaita et al. 2013). Increasing the length of the phalanges or increasing the length of a prehensile tail can enhance grasping performance. Many animals (especially arthropods and lizards, but also some bats, hyraxes, marsupials, and primates) supplement their grasping with adhesive tarsal and toe padding (hereafter we use

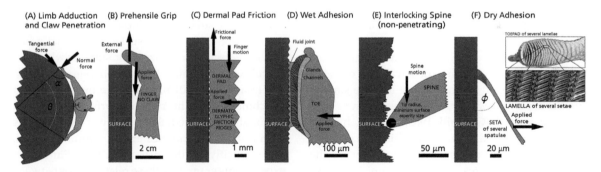

Figure 12.5 Attachment mechanisms used by animals as surface asperities decrease in size, from left to right. (A) Tarsal pads and claws are enhanced by adducting opposing limbs across an arched surface, when their limbs subtend a large angle θ with respect to centre of surface arc. Penetrating claws are stable if angle of penetration (α) is greater than 90°. This is commonly seen in semi-arboreal mammals, such as squirrels and lemurs. (Figure adapted from Cartmill (1974).) (B) Prehensile appendages, such as fingers and tails, can wrap around large asperities or narrow objects to grip, as demonstrated by primate phalanges and tails, chameleon tails, and opossum tails. (C) Friction of tarsal pads can be slightly enhanced with dermatoglyphic ridges (fingerprints), which conform to small asperities in substrate. Primates, koalas, and lemurs make extensive use of fingerprints for enhanced grip. (D) Wet adhesion requires forming a fluid joint between the appendage and substrate, which resists shearing forces. In many frogs and insects, fluid is produced in glands and delivered via channels in surface of toepad. (Figure adapted from Green (1981).) (E) Insect and crab spines interlock with small asperities in surface of substrate that are larger than tip radius of spine. (Figure adapted from Asbeck et al. (2006).) (F) Geckoes and other animals rely on cumulative van der Waals forces between atoms of substrate and atoms of numerous spatulae at terminal ends of setae to adhere to smooth substrates. Tokay gecko setae automatically detach when angle φ is greater than critical detachment angle for that species (between 15 and 40°) (Hagey et al. 2017)). These attachment mechanisms are not exclusive. That is, an animal could utilize a combination of mechanisms. (Figure adapted from Autumn and Peattie (2002).)

tarsal padding to refer to both tarsal and toe pad-ding, unless otherwise specified), and most amphib-ians rely primarily on toepads for adhesion (Federle et al. 2006; Labonte and Federle 2015). Compliant toepads and dermatoglyphic friction ridges con-form to the surface of the substrate to provide fric-tional forces that scale with grip strength (Sustaita et al. 2013). The steep vertical inclines typical of anthropogenic substrates (e.g., walls) pose chal-lenges to vertical locomotion, as climbing animals must overcome increased tangential forces with respect to their coefficient of friction (Cartmill 1985). Friction can be enhanced across a curved substrate (i.e., tree branch, telephone pole) by adducting opposing limbs (Figure 12.5A), especially when the opposing limbs form a large angle (θ) with respect to the centre of the surface arc (Cartmill 1974). Flatter surfaces, such as walls, or a narrower grip decrease θ, making adduction less effective. Thus, evolution-ary changes increasing limb length and range of adduction, as well as potentially plastic changes in strength of adduction, can enhance gripping ability on manufactured substrates.

On substrates that are too broad and flat for adduction (Figure 12.5A) to enhance frictional attachment forces, adhesion must rely on alterna-tive mechanisms. Where grasping is not sufficient, claws (Figure 12.5E) play a crucial role in adhesion. On smoother surfaces, claws that have evolved to be sharper, more curved, shorter, and higher (i.e., taller at the base) provide the advantage of being able to penetrate the surface or fit into the smaller surface asperities with substantial strength to sup-port the animal (Cartmill 1985). Insects, lizards, mammals, and birds all use a combination of tarsal padding and rigid claws to attach to surfaces—the claw either penetrates (Figure 12.5A) or interlocks (Figure 12.5E) with minute asperities in the surface of the substrate and is pulled towards the body (Cartmill 1974). Thus, the diameter, surface strength, and size of surface asperities determine whether an animal can effectively cling using a combination of tarsal pads and claws (Asbeck et al. 2006). In add-ition to friction, myriad mechanisms, including wet adhesion (the secretion of adhesive fluid from the smooth tarsal pad to form a fluid joint with the sub-strate), capillary action, and van der Waals forces, have evolved to enable tarsal pads to adhere to smoother surfaces.

The effectiveness of attachment mechanisms depends on multiple factors related to how exactly the digit interacts with the substrate to enable cling-ing (Figure 12.5). In contrast to section 12.2.3, the majority of our understanding about biological attachment mechanisms is based on interaction with manufactured surfaces (in laboratory settings) that are common to urban environments, providing a foundation for understanding the effectiveness of different climbing morphologies of urban species. In certain lizards (anoles, geckos, and some skinks) and some arthropods, adhesive toepads are hier-archical structures comprised of specialized scales (known as lamellae) resembling fringes, which are comprised of rows of hair-like setae, which divide at the distal end into hundreds of microscopic spat-ular tips. The spatulae are so numerous that the total adhesion produced on vertical glass by a sin-gle toe is more than sufficient to support the body weight of a tokay gecko. An added benefit of this dry form of adhesion is the immediate detachment of setae from a surface at a specific angle, which confers the ability to rapidly locomote while effect-ively adhering, and a self-cleaning property that maintains adhesion from step to step (Federle 2006). Furthermore, this dry toepad adhesion can be effective in both wet and dry environments, although the strength of adhesion also depends on the hydrophobicity of the surface, such that adhe-sion is ineffective on hydrophilic surfaces when wet as well as on extremely hydrophobic surfaces when dry (Stark et al. 2013). While the mechanisms of attachment on anthropogenic surfaces have been thoroughly examined, the majority of this work has been done in laboratories and in natural contexts, not explicitly in the urban context. Functional experiments examining these relationships between morphology and performance *in situ* in urban environments will provide insight into how com-plex changes to the physical structures within a habitat alter selective pressures that shape climbing morphologies in urban taxa.

12.3.3 Morphological changes associated with climbing urban substrates

For species that exploit anthropogenic structures, the precise nature of adaptive urban morphological differences depends on the extent to which a species also uses terrestrial habitat and on the attachment mechanisms employed. As with terrestrial locomotion (section 12.2.3), few studies have investigated climbing performance in urban animals outside of *Anolis* lizards, but we draw on functional–morphological relationships examined in natural environments to inform our broad predictions.

Since the number of perches is reduced and distributed in a way that likely requires increased terrestrial movement in urban environments (section 12.2), quick over-ground movement may be more critical than quick climbing for scansorial animals that extensively use terrestrial habitat. If long limbs are a favourable adaptive solution to challenges posed by terrestrial urban habitat use (section 12.2.3), this morphological shift may come at a cost to climbing performance. Indeed, in urban *A. cristatellus*, longer-limbed lizards sprint faster on horizontal surfaces but slower than short-limbed lizards at steep inclinations (Winchell et al. 2018b). Urban animals that climb must therefore balance the competing selective pressures of both terrestrial and arboreal habitats. In non-urban habitats, analogous habitat use favours relatively long limbs but also flattened bodies and other traits that enable the organism to maintain its centre of mass close to the surface (e.g., low-rock-dwelling lizards (*Urosaurus ornatus*) and cave-specialist lizards (*Anolis bartschi*) (Herrell et al. 2001; Revell et al. 2007). Similarly, Winchell et al. (2018b) found that urban *A. cristatellus* have trait combinations that enable effective climbing: lizards with long limbs but also large toepads with many lamellae were able to sprint as quickly on steeply inclined anthropogenic surfaces as lizards with shorter limbs. Enhanced clinging of the toepads, particularly on fore feet, would enable animals to pull their centre of mass closer to the surface (Autumn et al. 2006). This coevolution of traits that optimize both terrestrial and scansorial movement is probably not uncommon in urban animals.

For purely arboreal species that use vertically inclined perches and infrequently descend to the ground, urban species may tend to converge on the morphologies of their cliff-dwelling relatives, since walls superficially resemble cliffs. In addition, while this has not been studied in terrestrial species (so far as we are aware), many lizard species regularly traverse smooth substrates in their natural habitats (e.g., waxy leaves and smooth bark, such as

on agave plants and palm trees), which may be similar to urban habitat elements. Species that use these habitats tend to have dorsally compressed bodies with relatively short limbs and highly specialized adhesive toepads, which confer greater acceleration on smooth substrates compared to rough substrates (Autumn and Peattie 2002; Autumn et al. 2006). This may help explain why so many geckos, such as the globally distributed *Hemidactlyus* sp., as well as species in their native ranges such as *Aristelliger praesignis*, are so regularly found on buildings.

In addition to maintaining stability and moving quickly on anthropogenic structures, urban animals must first and foremost be able to effectively adhere to these surfaces. Evidence is mounting that smooth substrates limit maximal sprint speed for anoles, even for urban individuals that should encounter these substrates more frequently (Kolbe et al. 2015; Winchell et al. 2018b; Battles et al. 2019). The mechanics of this phenomenon are not yet well understood, but Battles et al. (2019) show that foot traction and the ability to execute a complete stride may greatly determine locomotor performance. Failure to effectively adhere to anthropogenic surfaces can have severe fitness consequences for urban animals. The selective pressures shaping adhesion morphologies may well be strong, even if anthropogenic structures are used infrequently. For example, a single fall can result in injury or predation, and returning to a safe perch is energetically costly. Moreover, if an individual is not able to effectively locomote on these surfaces, a substantial proportion of habitat space in the urban environment is rendered unusable and results in lost ecological opportunity, with fitness consequences. Animals that rely on high surface roughness (such as mammals that use primarily frictional forces and claws) are likely to have decreased performance on smoother surfaces (imagine a dog trying to move quickly across a hardwood floor). In contrast, animals that use multiple attachment mechanisms that function at multiple scales of surface roughness are more likely to persist in urban areas and fully exploit a variety of anthropogenic spaces. This use of multiple scales of adhesion mechanisms explains why so many arthropods and lizards are capable of rapidly invading a house and persist even in highly urbanized habitats.

Animals in many taxa that use smooth vertical substrates in natural environments generally have claws that are shorter, higher, and more curved compared to terrestrial species (Tulli et al. 2009; Zani 2000; Yuan et al. 2019). With the increased use of smooth anthropogenic substrates in cities (Figure 12.4), we would similarly expect an increase in the sharpness, curvature, and height and a decrease in the length of claws of species exploiting this niche space. As surfaces decrease in roughness, the maximum asperity size of a surface decreases, which reduces the ability for a claw to interlock (Figure 12.5E). Evidence from *Anolis* lizards suggests that claw curvature and toepad size coevolve (larger toepads and more curved claws) to enable more effective adhesion to smooth perches higher in the arboreal habitat (Yuan et al. 2019). Preliminary evidence from urban *Anolis* lizards similarly demonstrates variation in claw shape between urban and forest populations in the directions predicted (K.M. Winchell, personal observations). However, on extremely smooth substrates primarily found in anthropogenic habitats, claws are no longer an effective attachment mechanism and animals must rely entirely on wet or dry adhesion (Figure 12.5D,F). Because wet adhesion does not allow rapid detachment, it is less likely to be a successful attachment mechanism for urban animals compared to dry adhesion, which can allow rapid locomotion and is self-cleaning. Moreover, producing the fluid necessary for wet adhesion may be especially costly to animals in many urban environments, which are often hotter and drier than natural environments.

Animals with dry adhesion may be pre-adapted to cling to smooth anthropogenic surfaces. If adhering to anthropogenic surfaces provides a selective advantage, then we expect natural selection to favour increased surface area (toepad size and width), a larger number of lamellae, or changes to the arrangement or microscopic structure of setae (Irschick et al. 1996; Zani 2000). Urban *A. cristatellus* lizards tend to have larger toepads and toepads with more lamellae than nearby forest populations (Winchell et al. 2016; Winchell et al. 2018b). As squamation is fixed early in ontogeny and these differences were maintained in common garden rearing, these adaptive differences are likely genetically based (Winchell et al. 2016). Variation in the setae

array such as density, arrangement, and height, as well as individual setal properties such as branching pattern, length, diameter, and resting angle all impact cling force and may be additional targets of natural selection in urban animals that employ this mechanism of adhesion (Peattie and Full 2007). Furthermore, relationships between toepad morphology and other morphological traits are important for performance on smooth substrates. For instance, among urban anoles, *Anolis stratulus*, a species that has a greater toepad area relative to its body size, demonstrated greater stability on a smooth, vertical racetrack compared to a large species with relatively smaller toepads (Kolbe et al. 2015).

Due to the harder and less-compliant substrates present in urban environments, we would expect that the limb bones of urban animals would be exposed to repeated high-stress loadings that can lead to fatigue damage accumulation and bone failure (Biewener 2005). This could lead to higher rates of traumatic injury in urban animals (as observed in urban *A. cristatellus* (Winchell et al. 2019), with obvious fitness costs. The high-stress loadings are likely to select for stiffer tendons and denser bone trabeculae to enhance force absorption, much as we see in animals that are specialized for high accelerations and jumping (Biewener 2005). Smaller body sizes may also reduce injury risk from falls, since loading stresses increase with body size (Biewener 2005). Thus, we would expect that urban-adapted animals would evolve stiffer tendons, denser bones, and smaller body sizes than their close relatives living in natural habitats. Furthermore, small jumping animals may be pre-adapted to inhabit habitats dominated by manufactured substrates, while larger animals that are susceptible to fall injuries may be excluded from urban environments.

Alternatively, the force of impacts can be reduced by enhancing specializations for gliding in terrestrial animals that increase aerial manoeuvrability. Although our focus in this chapter is not on aerial locomotion, the ability to partially control aerial descent is a viable option for some climbing animals. Gliding is enhanced by manipulating how fluid (e.g., water or air) moves around a structure, which is influenced by the spacing and size of non-winged gliding structures in small animals. Larger animals, which require a solid structure for gliding, have evolved morphological specializations to increase the gliding surfaces of their body. Arboreal snakes, for example, flatten their bodies dorsoventrally to enhance gliding manoeuvrability (Socha 2011). In addition, many gliding mammals have evolved a membranous fold of skin between the fore and hind legs, which enables them to reduce the velocity and impact of their descent (Byrnes et al. 2008).

Current efforts have barely scratched the surface towards understanding morphological adaptations for climbing anthropogenic structures. Research on *Anolis* lizards has provided a solid starting point to guide future research questions by integrating evolutionary and ecological approaches. Studies of adhesion should extend beyond lamellae number and toepad size and investigate the microscopic aspects of dry adhesion that may be under selection, as well as the costs and effectiveness of wet adhesion in urban environments. More research is needed for other types of locomotion in which foot traction is important, such as jumping (and landing). In addition, comparative studies examining anatomical changes in cliff-dwelling, terrestrial, and arboreal species in response to urbanization will provide valuable insight into how perch structure affects attachment morphology. Finally, researchers should conduct studies that directly integrate performance, preference, and fitness on smooth, anthropogenic substrates across a range of urban-dwelling taxa.

12.4 Conclusions

The precise morphological shifts that will occur in species in urban environments depend on a species' evolutionary history, its habitat use, and how it interacts with the environment in natural habitats. For purely terrestrial animals, selective pressures on locomotion will be related primarily to over-ground movement. For scansorial animals that use both terrestrial and arboreal habitat, we might expect traits related to over-ground movement to face stronger selection pressures compared to traits related to climbing because of the reduction and sparse distribution of trees and structures. Secondarily then, traits related to climbing would be selected

for in a compensatory manner, as many traits that enhance terrestrial movement (e.g., longer limbs and a laterally compressed body) may decrease stability at vertical inclinations. Finally, for animals that have purely arboreal habitat use, traits related to climbing should be selected for most strongly, even if they would reduce performance of terrestrial locomotion. While our focus here is on terrestrial locomotion, aerial and aquatic locomotion are also affected by changes in the urban environment that may favour shifts in traits related to manoeuvrability and acceleration in fluid environments.

Although many factors are involved in shaping the precise morphology necessary to achieve an optimal locomotor performance, we can make some generalized conclusions about the most common types of morphological shifts we are likely to observe across terrestrial, scansorial, and arboreal species in urban environments (Figure 12.6). In purely terrestrial species, we hypothesize urban animals will exhibit trait shifts related to acceleration: longer slender limbs (especially the distal limb elements), more crouched posture, an elongated pelvis, limited elbow supination, increased muscle cross-sectional area, and increased neurosensitivity. In scansorial species that use both arboreal and terrestrial habitat, these same traits would enhance over-ground movement but may be costly on vertical substrates. Traits that act to bring the centre of mass closer to the climbing surface and improve clinging ability will be strongly selected for, perhaps coevolving with traits that enhance terrestrial movement. In particular, we hypothesize that scansorial animals will have longer limbs but also wider bodies, dorsoventrally compressed bodies, and enhanced clinging abilities. Clinging abilities depend on the mechanism of attachment by the organism, but in general we would expect sharper, more curved, higher, and shorter claws, larger toepads with more adhesive lamellae, and alterations in the density, arrangement, and properties of the setae. Lastly, for purely arboreal species, traits that enhance stability, clinging, and inter-perch aerial movement will be favoured. Specifically, we hypothesize that urban animals will have cliff-like morphologies (e.g., a dorsoventrally compressed body, shortened limbs), enhanced clinging morphologies, smaller bodies, and adaptations that either increase bone safety

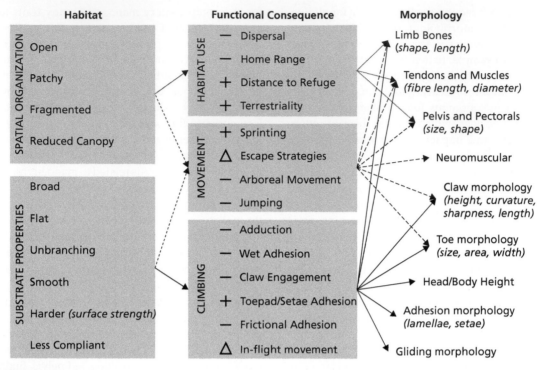

Figure 12.6 Differences in structural habitat in urban areas in both spatial organization and properties of climbing substrates influence habitat use, terrestrial movement, and climbing movement. This in turn creates selective pressures on morphology specific to the locomotor demands of the urban environment. In the functional consequence column, + indicates increased demand for the function, − reduced demand, and Δ a non-directional change.

factor (e.g., greater bone density, smaller body size, stiff high-stress tendons) or enhance control during gliding.

12.4.1 Future directions

We acknowledge that these conclusions are simplistic generalizations and that terrestrial locomotion involves extremely complex integration of morphological structures at multiple levels. Many other bones, muscles, and structures are relevant to locomotion with effects that are species- and context-dependent. We have attempted here to present what we consider to be likely locomotor adaptations in urban environments, based on biomechanical principles. We hope that this chapter provides a jumping-off point for readers to find avenues for future investigation into the many ways animals are morphologically adapting to urban areas. With this in mind, we provide some suggestions for future study to advance this field.

We emphasize the importance of studying phenotypic adaptation to urbanization from a functional morphological perspective. Trait shifts in urban populations only make sense in light of their functional consequences relating to habitat differences from natural environments. While the functions of locomotor and attachment morphologies are generally well characterized, we have the opportunity to test our understanding of their adaptive benefits by observing how these traits change in response to the selective pressures encountered in urban environments. Future research on locomotion in urban areas should start with hypotheses based on spatial or surface differences in the urban environment, should consider how species interact with habitat structures in non-urban environments, and ideally should connect morphological differences to functional locomotor consequences in urban environments. By following this simple habitat use, morphology, performance framework, and whenever possible fitness output, observed phenotypic shifts will be more easily interpretable and hypotheses will be more robust. This will prevent wasted research efforts. For example, if examining limb length in a wall-dwelling urban species that occupies cliff-like habitats in its natural environment, you may not find any differences—but, as described above, such changes may not be expected. More importantly, this framework will prevent erroneous conclusions based on post-hoc interpretation of phenotypic shifts. This is especially important because traits relevant to locomotion may be shaped by selection on unrelated traits (as discussed in Box 12.1). In addition, where possible, researchers should investigate the genetic and plastic contributions to morphological differences detected through common garden experiments to fully understand the evolutionary implications of adaptive phenotypic shifts (see Chapter 5). This integrated approach is key to understanding broadly how urbanization affects animal evolution and ecology.

As we consider the structural and environmental changes with the urbanization of a habitat, we gain insight into the new biomechanical and behavioural

challenges that animals encounter. Urbanization can act as either a filter barrier or a conduit for animal movement, depending on their ability to behaviourally or evolutionarily grapple with these changes. These phenomena will, no doubt, have profound changes on the gene flow within these populations and the speed of local adaptation to the urbanized environment. A mechanistic understanding of how locomotion and patchiness of micro-habitats affect gene flow would likely enhance the study of filter barriers and niche models to predict the presence and absence of animals in ecosystems encountering urbanization. This is just one example of how a functional perspective can facilitate the application of existing tools and approaches for the study of adaptation in urban ecosystems.

Acknowledgements

We thank our colleagues who read early versions of this manuscript: Natalie Holt, Jonathan Losos, Duncan Irschick, Travis Hagey, and the Mississippi University for Women Fall 2018 Biomechanics class. We also thank Marta Szulkin, Anne Charmantier, Jason Munshi-South, Brian Langerhans, and an anonymous reviewer for providing helpful comments to guide our revisions.

Box 12.1 Indirect effects of urban habitats on terrestrial locomotion

Species that persist in urban habitats may experience evolutionary changes that are not induced by directional selection on biomechanical function but have effects on locomotor traits, nonetheless. Such trait shifts are outside the scope of the current chapter but provide fascinating systems for future investigation. Here we provide a few key examples of how selection on unrelated traits may impact locomotor morphology and performance:

- Ants in urban environments evolve longer limbs to avoid the hot substrates, which increases stride length (Chapter 6; Diamond et al. 2017).
- Male white-footed mice have decreased bone density over the past 85 years, concomitant with a greater concentration of xenoestrogens in urban environments, which impacts bone safety factor and likely affects agility (Nakkula et al. 2016).
- One feature of domestication syndrome is a reduction in vertebral number, specifically caudal vertebrae, leading to the characteristic curled tails of many domesticates (Darwin 1868), which may affect agility.

References

Álvarez, A., Ercoli, M.D., and Prevosti, F.J. (2013). Locomotion in some small to medium-sized mammals: a geometric morphometric analysis of the penultimate lumbar vertebra, pelvis and hindlimbs. *Zoology*, 116(6), 356–71.

Arnold, S.J. (1983). Morphology, performance and fitness. *American Zoologist*, 23(2), 347–61.

Aryan, F.A. and Alnaqeeb, M.A. (2002). Effect of immobilization and underload on skeletal muscle in the hindlimb of the Jerboa. *Kuwait Journal of Science & Engineering*, 29(1), 83–97.

Asbeck, A.T., Kim, S., Cutkosky, M.R., Provancher, W.R., and Lanzetta, M. (2006). Scaling hard vertical surfaces with compliant microspine arrays. *International Journal of Robotics Research*, 25(12), 1165–79.

Astley, H.C., Haruta, A., and Roberts, T.J. (2015). Robust jumping performance and elastic energy recovery from compliant perches in tree frogs. *Journal of Experimental Biology*, 218(21), 3360–63.

Autumn, K. and Peattie, A.M. (2002). Mechanisms of adhesion in geckos. *Integrative and Comparative Biology*, 42(6), 1081–90.

Autumn, K., Hsieh, S.T., Dudek, D.M., et al. (2006). Dynamics of geckos running vertically. *Journal of Experimental Biology*, 209(2), 260–72.

Avilés-Rodríguez, K.J. and Kolbe, J.J. (2019). Escape in the city: urbanization alters the escape behavior of *Anolis* lizards. *Urban Ecosystems*, 22(4), 733–42.

Batabyal, A., Balakrishna, S., and Thaker, M. (2017). A multivariate approach to understanding shifts in escape strategies of urban lizards. *Behavioral Ecology and Sociobiology*, 71(5), 83.

Battles, A.C., Moniz, M., and Kolbe, J.J. (2018). Living in the big city: preference for broad substrates results in niche expansion for urban *Anolis* lizards. *Urban Ecosystems*, 21(6), 1087–95.

Battles, A.C., Irschick, D.J., and Kolbe J.J. (2019). Do structural habitat modifications associated with urbanization influence locomotor performance and limb kinematics in *Anolis* lizards? *Biological Journal of the Linnean Society*, 127(1), 100–12.

Beninde, J., Feldmeier, S., Werner, M., et al. (2016). Cityscape genetics: structural vs. functional connectivity of an urban lizard population. *Molecular Ecology*, 25(20), 4984–5000.

Biewener, A.A. (2005). Biomechanical consequences of scaling. *Journal of Experimental Biology*, 208(9), 1665–76.

Biewener, A.A. and Blickhan, R. (1988). Kangaroo rat locomotion: design for elastic energy storage or acceleration? *Journal of Experimental Biology*, 140(1), 243–55.

Brown, I.E. and Loeb, G.E. (2000). A reductionist approach to creating and using neuromusculoskeletal models. In: Winters, J.M. and Crago, P.E. (eds) *Biomechanics and Neural Control of Posture and Movement*, pp. 148–63. Springer, New York.

Byrnes, G. and Spence, A.J. (2011). Ecological and biomechanical insights into the evolution of gliding in mammals. *Integrative and Comparative Biology*, 51(6), 991–1002.

Byrnes, G., Lim, N.T.L., and Spence, A.J. (2008). Take-off and landing kinetics of a free-ranging gliding mammal, the Malayan colugo (*Galeopterus variegatus*). *Proceedings of the Royal Society B: Biological Sciences*, 275(1638), 1007–13.

Cannon, C.H. and Leighton, M. (1994). Comparative locomotor ecology of gibbons and macaques: selection of canopy elements for crossing gaps. *American Journal of Physical Anthropology*, 93(4), 505–24.

Cartmill, M. (1974). Pads and claws in arboreal locomotion. In: Jenkins, F.A. Jr (ed.) *Primate Locomotion*, pp. 45–83. Academic Press, Cambridge, MA.

Cartmill, M. (1985). Climbing. In: Hildebrand, M., Bramble, D., Liem, K., and Wake, D.B. (eds) *Functional Vertebrate Morphology*, pp. 73–88. The Belknap Press, Cambridge, MA.

Cooper, W.E. and Blumstein, D.T. (eds) (2015). *Escaping from Predators: An Integrative View of Escape Decisions*. Cambridge University Press, Cambridge.

Darwin, C., (1868). *The Variation of Animals and Plants under Domestication*. John Murray, London.

Diamond, S.E., Chick, L., Perez, A.B.E., Strickler, S.A., and Martin, R.A. (2017). Rapid evolution of ant thermal tolerance across an urban-rural temperature cline. *Biological Journal of the Linnean Society*, 121(2), 248–57.

Domenici, P., Blagburn, J.M., and Bacon, J.P. (2011). Animal escapology II: escape trajectory case studies. *Journal of Experimental Biology*, 214(15), 2474–94.

Donihue, C. (2016). Microgeographic variation in locomotor traits among lizards in a human-built environment. *PeerJ*, 4, e1776.

Engelhardt, S.C. and Weladji, R.B. (2011). Effects of levels of human exposure on flight initiation distance and distance to refuge in foraging eastern gray squirrels (*Sciurus carolinensis*). *Canadian Journal of Zoology*, 89(9), 823–30.

Etter, D.R., Hollis, K.M., Van Deelen, et al. (2002). Survival and movements of white-tailed deer in suburban Chicago, Illinois. *Journal of Wildlife Management*, 66(2), 500–510.

Faeth, S.H., Warren, P.S., Shochat, E., and Marussich, W.A. (2005). Trophic dynamics in urban communities. *BioScience*, 55(5), 399–407.

Federle, W. (2006). Why are so many adhesive pads hairy? *Journal of Experimental Biology*, 209(14), 2611–21.

Federle, W., Barnes, W.J.P., Baumgartner, W., Drechsler, P., and Smith, J.M. (2006). Wet but not slippery: boundary friction in tree frog adhesive toe pads. *Journal of the Royal Society Interface*, 3(10), 689–97.

Forman, R.T.T. (2014). *Urban Ecology: Science of Cities*. Cambridge University Press, Cambridge.

Foster, K.L., Collins, C.E., Higham, T.E., and Garland, T. Jr (2015). Determinants of lizard escape performance: decision, motivation, ability, and opportunity. In: Cooper, W.E. and Blumstein, D.T. (eds) *Escaping from Predators: An Integrative View of Escape Decisions*, pp. 287–321. Cambridge University Press, Cambridge.

Gilman, C.A. and Irschick, D.J. (2013). Foils of flexion: the effects of perch compliance on lizard locomotion and perch choice in the wild. *Functional Ecology*, 27(2), 374–81.

Green, D.M. (1981). Adhesion and the toe-pads of treefrogs. *Copeia*, 4, 790–96.

Hagey, T.J., Harte, S., Vickers, M., Harmon, L.J., and Schwarzkopf, L. (2017). There's more than one way to climb a tree: limb length and microhabitat use in lizards with toe pads. *PLOS ONE*, 12(9), e0184641.

Herrel, A., Meyers, J.J., and Vanhooydonck, B. (2001). Correlations between habitat use and body shape in a phrynosomatid lizard (*Urosaurus ornatus*): a population-level analysis. *Biological Journal of the Linnean Society*, 74(3), 305–14.

Hildebrand, M. (1982). *An Analysis of Vertebrate Structure*. Wiley, New York.

Irschick, D.J. and Losos, J.B. (1999). Do lizards avoid habitats in which performance is submaximal? The relationship between sprinting capabilities and structural habitat use in Caribbean anoles. *The American Naturalist*, 154(3), 293–305.

Irschick, D.J., Austin, C.C., Petren, K., et al. (1996). A comparative analysis of clinging ability among pad-bearing lizards. *Biological Journal of the Linnean Society*, 59(1), 21–35.

Johnson, M.T. and Munshi-South, J. (2017). Evolution of life in urban environments. *Science*, 358(6363), eaam8327.

Jusufi, A., Goldman, D.I., Revzen, S., and Full, R.J. (2008). Active tails enhance arboreal acrobatics in geckos. *Proceedings of the National Academy of Sciences of the United States of America*, 105(11), 4215–19.

Koenig, J., Shine, R., and Shea, G. (2002). The dangers of life in the city: patterns of activity, injury and mortality in suburban lizards (*Tiliqua scincoides*). *Journal of Herpetology*, 36(1), 62–9.

Kolbe, J.J., Battles, A.C., and Aviles-Rodriguez, K.J. (2016). City slickers: poor performance does not deter *Anolis* lizards from using artificial substrates in human-modified habitats. *Functional Ecology*, 30(8), 1418–29.

Labonte, D. and Federle, W. (2015). Scaling and biomechanics of surface attachment in climbing animals. *Philosophical Transactions of the Royal Society B*, 370(1661), 20140027.

La Croix, S. (2011). *Morphological and Behavioral Development in a Top North American Carnivore, the Coyote*. PhD thesis. Michigan State University, East Lansing.

Lapiedra, O., Chejanovski, Z., and Kolbe, J.J. (2017). Urbanization and biological invasion shape animal personalities. *Global Change Biology*, 23(2), 592–603.

Leppänen, P.K., Ewalds-Kvist, S.B.M., and Selander, R.K. (2005). Mice selectively bred for open-field thigmotaxis: life span and stability of the selection trait. *Journal of General Psychology*, 132(2), 187–204.

Lima, S.L. and Dill, L.M. (1990). Behavioral decisions made under the risk of predation: a review and prospectus. *Canadian Journal of Zoology*, 68(4), 619–40.

Losos, J.B. and Irschick, D.J. (1996). The effect of perch diameter on escape behaviour of *Anolis* lizards: laboratory predictions and field tests. *Animal Behaviour*, 51(3), 593–602.

Losos, J.B., Creer, D.A., Glossip, D., et al. (2000). Evolutionary implications of phenotypic plasticity in the hindlimb of the lizard *Anolis sagrei*. *Evolution*, 54(1), 301–5.

Melville, J. and Swain, R.O.Y. (2000). Evolutionary relationships between morphology, performance and habitat openness in the lizard genus *Niveoscincus* (Scincidae: Lygosominae). *Biological Journal of the Linnean Society*, 70(4), 667–83.

Meyers, M.A., Chen, P.Y., Lin, A.Y.M., and Seki, Y. (2008). Biological materials: structure and mechanical properties. *Progress in Materials Science*, 53(1), 1–206.

Møller, A.P. (2008). Flight distance of urban birds, predation, and selection for urban life. *Behavioral Ecology and Sociobiology*, 63, 63–75.

Møller, A.P. (2014). Behavioral and ecological predictors of urbanization. In: Gil, D. and Brumm, H. (eds) *Avian Urban Ecology*, pp. 54–67. Oxford University Press, Oxford.

Moore, T.Y. and Biewener, A.A. (2015). Outrun or outmaneuver: predator–prey interactions as a model system for integrating biomechanical studies in a broader ecological and evolutionary context. *Integrative and Comparative Biology*, 55(6), 1188–97.

Moore, T.Y., Cooper, K.L., Biewener, A.A., and Vasudevan, R. (2017a). Unpredictability of escape trajectory explains predator evasion ability and microhabitat preference of desert rodents. *Nature Communications*, 8(1), 440.

Moore, T.Y., Rivera, A.M., and Biewener, A.A. (2017b). Vertical leaping mechanics of the Lesser Egyptian Jerboa reveal specialization for maneuverability rather than elastic energy storage. *Frontiers in Zoology*, 14(1), 32.

Munshi-South, J. (2012). Urban landscape genetics: canopy cover predicts gene flow between white-footed mouse (*Peromyscus leucopus*) populations in New York City. *Molecular Ecology*, 21(6), 1360–78.

Nakkula, R.W. Jr, Smith, L.M., Bigelow, E.M., et al. (2016). Secular changes in bone morphology and sexual dimorphism of white-footed mice over 85 years. Poster 1607. In: *Orthopaedic Research Society Program Book*, 5–8 March, Orlando, FL.

Nowak, D.J., Rowntree, R.A., McPherson, E.G., et al. (1996). Measuring and analyzing urban tree cover. *Landscape and Urban Planning*, 36(1), 49–57.

Peattie, A.M. and Full, R.J. (2007). Phylogenetic analysis of the scaling of wet and dry biological fibrillar adhesives. *Proceedings of the National Academy of Sciences of the United States of America*, 104(47), 18595–600.

Prevedello, J.A. and Vieira, M.V. (2010). Does the type of matrix matter? A quantitative review of the evidence. *Biodiversity and Conservation*, 19(5), 1205–23.

Prosser, C., Hudson, S., and Thompson, M.B. (2006). Effects of urbanization on behavior, performance, and morphology of the garden skink, *Lampropholis guichenoti*. *Journal of Herpetology*, 40(2), 151–60.

Regalado, R. (1998). Approach distance and escape behavior of three species of Cuban *Anolis* (Squamata: Polychrotidae). *Caribbean Journal of Science*, 34, 211–17.

Revell, L.J., Johnson, M.A., Schulte, J.A., Kolbe, J.J., and Losos, J.B. (2007). A phylogenetic test for adaptive convergence in rock-dwelling lizards. *Evolution*, 61(12), 2898–912.

Sathe, E.A. and Husak, J.F. (2015). Sprint sensitivity and locomotor trade-offs in green anole (*Anolis carolinensis*) lizards. *Journal of Experimental Biology*, 218(14), 2174–9.

Simon, P., Dupuis, R., and Costentin, J. (1994). Thigmotaxis as an index of anxiety in mice. Influence of dopaminergic transmissions. *Behavioural Brain Research*, 61(1), 59–64.

Socha, J.J. (2011). Gliding flight in chrysopelea: turning a snake into a wing. *Integrative and Comparative Biology*, 51(6), 969–82.

Sparkman, A., Howe, S., Hynes, S., Hobbs, B., and Handal, K. (2018). Parallel behavioral and morphological divergence in fence lizards on two college campuses. *PLOS ONE*, 13(2), e0191800.

Stankowich, T. and Blumstein, D.T. (2005). Fear in animals: a meta-analysis and review of risk assessment. *Proceedings of the Royal Society B: Biological Sciences*, 272(1581), 2627–34.

Stark, A.Y., Badge, I., Wucinich, N.A., et al. (2013). Surface wettability plays a significant role in gecko adhesion underwater. *Proceedings of the National Academy of Sciences of the United States of America*, 110(16), 6340–45.

Sustaita, D., Pouydebat, E., Manzano, A., et al. (2013). Getting a grip on tetrapod grasping: form, function, and evolution. *Biological Reviews*, 88(2), 380–405.

Tulli, M.J., Cruz, F.B., Herrel, A., Vanhooydonck, B., and Abdala, V. (2009). The interplay between claw morphology and microhabitat use in neotropical iguanian lizards. *Zoology*, 112(5), 379–92.

van Dongen, W.F., Robinson, R.W., Weston, M.A., Mulder, R.A., and Guay, P.J. (2015). Variation at the DRD4 locus is associated with wariness and local site selection in urban black swans. *BMC Evolutionary Biology*, 15(1), 253.

Vanhooydonck, B. and Van Damme, R. (1999). Evolutionary relationships between body shape and habitat use in lacertid lizards. *Evolutionary Ecology Research*, 1(7), 785–805.

Van Valkenburgh, B. (1985). Locomotor diversity within past and present guilds of large predatory mammals. *Paleobiology*, 11(4), 406–28.

Vásquez, A.R., Grossi, B., and Natalia Márquez, I. (2006). On the value of information: studying changes in patch assessment abilities through learning. *Oikos*, 112(2), 298–310.

Webster, D.B. (1962). A function of the enlarged middle-ear cavities of the kangaroo rat, *Dipodomys*. *Physiological Zoology*, 35(3), 248–55.

Winchell, K.M., Reynolds, R.G., Prado-Irwin, S.R., et al. (2016). Phenotypic shifts in urban areas in the tropical lizard *Anolis cristatellus*. *Evolution*, 70(5), 1009–22.

Winchell, K.M., Carlen, E.J., Puente-Rolón, A.R., and Revell, L.J. (2018a). Divergent habitat use of two urban lizard species. *Ecology and Evolution*, 8(1), 25–35.

Winchell, K.M., Maayan, I., Fredette, J.R., and Revell, L.J. (2018b). Linking locomotor performance to morphological shifts in urban lizards. *Proceedings of the Royal Society B: Biological Sciences*, 285(1880), 20180229.

Winchell, K.M., Briggs, D., and Revell, L.J. (2019). The perils of city life: patterns of injury and fluctuating asymmetry in urban lizards. *Biological Journal of the Linnean Society*, 126(2), 276–88.

Wong, B. and Candolin, U. (2015). Behavioral responses to changing environments. *Behavioral Ecology*, 26(3), 665–73.

Yuan, M.L., Wake, M.H., and Wang, I.J. (2019). Phenotypic integration between claw and toepad traits promotes microhabitat specialization in the *Anolis* adaptive radiation. *Evolution*, 73(2), 231–44.

Zaaf, A., Herrel, A., Aerts, P., and De Vree, F. (1999). Morphology and morphometrics of the appendicular musculature in geckoes with different locomotor habits (Lepidosauria). *Zoomorphology*, 119(1), 9–22.

Zajac, F.E. (1989). Muscle and tendon: properties, models, scaling, and application to biomechanics and motor control. *Critical Reviews in Biomedical Engineering*, 17(4), 359–411.

Zani, P.A. (2000). The comparative evolution of lizard claw and toe morphology and clinging performance. *Journal of Evolutionary Biology*, 13(2), 316–25.

CHAPTER 13

Urban Evolutionary Physiology

Caroline Isaksson and Frances Bonier

Isaksson, C. and Bonier, F., *Urban Evolutionary Physiology* In: *Urban Evolutionary Biology*. Edited by Marta Szulkin, Jason Munshi-South and Anne Charmantier, Oxford University Press (2020). © Oxford University Press.
DOI: 10.1093/oso/9780198836841.003.0013

13.1 Why physiology?

Physiology is the underlying cornerstone for adaptive responses to environmental change, and is often considered to be the 'black box' that connects the environment to fitness. By measuring individual- and population-level variation in physiological markers, we can gain insight into species resilience and sensitivity to current urbanization, and also future threats of expanding urban sprawl. However, physiological responses can vary among species, populations, and individuals for many reasons; thus, ecophysiology is best studied within an evolutionary framework to fully understand the sources of variation and potential impacts of urbanization. For example, are the physiological responses of successful urban species a consequence of natural selection acting over time or a precolonization sorting of certain individuals, or have physiological responses acclimated to urban challenges over an individual's lifetime? Beyond these adaptationist explanations, we also expect some physiological responses to reflect constraints or maladaptation in urban individuals. This failure to exhibit adaptive physiological responses to urban challenges could explain why some species are absent from urban habitats and why some urban populations are sinks, whereas other populations persist and even thrive in urban habitats (e.g., Fischer et al. 2015).

Urban ecophysiology is an active and expanding area of research, and we aim to cover examples from most animal taxa, although most research has so far been conducted on birds. To date, one of the main aims of urban ecophysiology research has been to link physiological markers to a cost of urban life—for example, measuring oxidative damage and telomere attrition in urban and non-urban populations (e.g., Gillis et al. 2014; Herrera-Dueñas et al. 2014; Isaksson 2015, 2018 and references therein; Salmón et al. 2018a). Increased levels of these markers are expected to be detrimental for cellular functioning, and thus also lifespan. This approach is highly valuable for assessing costs and identifying species, populations, or life-stages of particular concern, and also for identifying current targets of selection. However, physiological systems are highly interconnected; thus, to use

ecophysiological approaches to go beyond describing costs of urban living, to understanding the role of physiology in mediating responses to urban habitat, it is also valuable to investigate physiological systems/networks and how they differ or change in response to urbanization, rather than to focus solely on a single marker linked to a cost.

13.2 Challenges of studying evolution of plastic physiological traits

Evolutionary physiology has its early beginnings in the 1970s (reviewed in Garland and Carter 1994), whereas urban evolutionary physiology only started to appear in the literature in the beginning of 2000 (e.g., Partecke et al. 2004; Bonier et al. 2006). Early evolutionary physiologists aimed to identify general physiological principles linked, for example, to flight, climate (e.g., arid or cold conditions), or broad allometric patterns between body size and metabolic rate. Following a similar approach, and based on the fact that all living cells share some highly conserved processes and traits through shared ancestry, one could predict general evolutionary responses or principles to apply to the adaptive urban landscape (at least on the macroscale). However, although urban evolutionary physiology is in its early days, there seems to be an intriguing complexity in the physiological responses to urbanization. In other words, there appear to be multiple solutions or pathways to solve the same problem, at both the physiological and genetic level (Cohen et al. 2017; Perrier et al. 2018).

If we seek to move beyond categorizing physiological responses to urbanization and understanding mechanisms that explain them (e.g., artificial night light causes X change), toward understanding if these observed changes are adaptive and if they are driven by plasticity or evolutionary change, we need to approach these questions within an evolutionary framework. This can be challenging for highly plastic traits—particularly traits involved in presumed adaptive plastic responses to challenging environments (Isaksson 2015; Bonier and Martin 2016), which encompasses almost all physiological traits. Consider, for example, the case of glucocorticoid hormone concentrations or endogenously synthe-sized antioxidants. In a cross-sectional sample, we inevitably find individual variation in glucocortic-oids/antioxidants. Within a population, we can ascribe this variation to differences in the challenges facing individuals, differences in the ways individ-uals respond to challenges, and/or stable differ-ences in mean glucocorticoid/antioxidant levels among individuals (perhaps reflecting heritable variation and/or lasting effects of developmental plasticity). But how can we determine if some of this variation reflects phenotypes that are more or less favoured by natural selection in an urban set-ting? Also, how can we determine if there is herit-able variation in the degree to which individuals' physiological phenotypes or their repertoire of physiological plasticity are adapted to the urban environment? Common garden studies and genomic analyses are two relevant approaches that future studies should employ to better characterize the physiological response to urbanization and to understand whether such a response results from physiological acclimation or selection (Box 13.1; see also Chapter 5).

Box 13.1 Measuring selection on physiological traits

- Trait–fitness correlations are often used to characterize selection, but might also simply reveal variation in the degree of environmental challenge confronting individ-uals (Bonier and Martin 2016). However, variation in these trait–fitness relationships between urban and non-urban populations, when assessed in careful field studies with rigorous controls for other sources of variation (e.g., life-history stage), might provide a good indication of how selection on physiological traits varies among urban and non-urban populations.
- Common garden and translocation studies can also pro-vide insight into the degree to which population-level variation reflects heritable differences. For example, upon finding differences in physiological phenotypes among an urban and non-urban population of the same species, individuals could be cross-fostered in the wild or bred and reared in captivity (to reduce parental and early environmental effects that can per-sist throughout an individual's lifetime) to determine if differences observed in the field persist in the new environment (e.g., Donihue and Lambert 2015; Salmón et al. 2018b).
- Experiments that seek to block a physiological response could reveal whether the response mitigates or exacer-bates fitness costs of urban challenges, providing insight into the adaptive value of the intact response. For example, one could inhibit the synthesis of the main intracellular antioxidant by providing the animal with buthionine sul-foximine (BSO) to investigate the role of cellular oxidative protection for fitness in urban and non-urban habitats (Romero-Haro and Alonso-Alvarez 2015).
- Genomic or candidate gene analyses can be useful to identify key physiological systems under selection. For example, serotonin transporter gene (*SERT/SLC6A*) and other genes linked to neurological function and development are promising targets of selection (e.g., Mueller et al. 2013). Other likely targets are genes linked to cellular repair (e.g., nucleases, *EXO1* and *FEN1*) and protection such as genes in the cytochrome (*CYP*)-family, ceruloplasmin (*CP*), and metallothioneins (*B5G2T6_TAEGU* and *MT4*) (Watson et al. 2017). However, physiological systems are quantita-tive traits and the end-result is likely to be highly poly-genic, i.e., selection acting on different genes in different populations could generate similar physiological (pheno-typic) responses (e.g., Perrier et al. 2018). Thus, in most cases, single gene mutations are unlikely to be associated with a change in the urban physiological phenotype, and identification of genetic networks might be more fruitful.
- There are also several strong approaches which cannot eas-ily be conducted in the field or in most vertebrates, but are feasible in captivity with organisms with a short generation time. These approaches include experimental evolution and artificial selection, which could involve exposing organ-isms to conditions associated with the urban environment (e.g., artificial light at night) or selecting on physiological traits thought to be important to urban adaptation over sufficient generations to detect an evolved response.
- Longitudinal studies or studies using historic data (collec-tions, archived samples, feather or fur for longitudinal hormone comparisons) can assess evidence of evolution in populations that have become established and success-ful in cities (e.g., Snell-Rood and Wick 2013).

13.3 Urban stressors or stimulators

Stressors are factors that challenge homeostasis and trigger a physiological response, causing either short- or long-term negative effects at the molecular, cellular, and/or organismal level. Any factor, even water and music, can be a stressor at high concentrations/levels. Thus, no factor is a stressor in itself; rather, it is the concentration/level, exposure time, and species/individual sensitivity that differentiates a stressor from a beneficial stimulator (see below). This is old knowledge in ecotoxicology and was already stated in the sixteenth century by Paracelsus. Although less explored, a similar toxicological reasoning can apply to our understanding of the response not only to substances but also to other urban factors, such as light at night, noise, and food quality and quantity.

Some factors show a linear dose-response relationship in severity (toxicity) (Figure 13.1A, *i* and *ii*). However, others (most) show a non-linear dose-response relationship (Figure 13.1 A, *iii*). In these non-linear responses, a negative effect is often not evident until a certain threshold concentration/level is reached (which could vary across species and life-stages). Other factors can be as detrimental at low concentrations/level as they are at high concentrations/level (Figure 13.1B, *iv*). This can, for example, be the case for essential metals and dietary antioxidants (such as carotenoids) that are crucial for cellular functions. Thus, deficiency at low concentrations and toxicity at high concentrations can lead to similar negative effects on fitness. As a result, the selective pressures associated with physiological responses to different factors can be expressed very differently.

Other factors, however, could show a stimulatory and even beneficial effect at low concentrations/levels, before becoming a stressor (Figure 13.1B, *v*). Thus, a 'stimulator' is here defined as a factor that triggers a physiological response at low concentrations/levels that gives rise to a positive effect at the molecular, cellular, and/or organismal level. This response is referred to as a hormetic response. Perhaps the most well-known hormetic response is the intake of alcohol, where small amounts have positive effects on cardiovascular systems, but a high intake is detrimental. An initially low exposure to a factor can also lead to a preconditioned

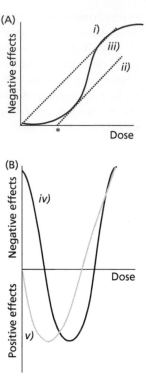

Figure 13.1 Theoretical dose-response curves. These physiological dose-response curves can be applicable to many different urban factors, i.e., not only the traditional toxic chemical substances, but also factors such as noise and artificial night light, where wavelength spectrum, amplitude, and frequency could show a dose-response relationship. (A) Two linear dose-response curves—one without (i) and one with (ii) a threshold for when the stressor starts to cause negative effects on the individual. However, the most common dose-response curve is non-linear and with a threshold value (iii). (B) Two dose-response curves that can have both negative and positive effects on the individual. In iv (black line) the factor has negative effects at low and high levels (a stressor), whereas there is a positive effect at intermediate doses. In v (grey line) the factor acts as a stimulator with positive effects at low doses; however, if the dose is too high there will be negative effects. This type of dose-response curve is often referred to as a hormetic response.

(protecting) hormetic effect when subsequently exposed to a more intense level of the same factor (Calabrese et al. 2007). In terms of toxic substances, this effect is mediated through the stimulation of the detoxifying systems by low non-toxic levels, which then is protective when a more massive insult occurs. Well-known hormetic pathways include antioxidant systems, heat-shock proteins, and anti-apoptotic proteins (Arumugam et al. 2006)—all likely to be important in urban habitats. Positive fitness effects of preconditioning hormetic effects have been observed in essentially all organisms studied—from plants to vertebrates (Calabrese et al. 2007). Regarding most pollutants and other urban stressors, we largely ignore which dose-response curve applies to urban-dwelling species, especially because there is a cocktail of pollutants and stressors that could act in additive or synergistic ways. This means that the threshold level for induction of fitness costs, and thereby selection, could be reached at a much lower concentration compared to a single substance exposure (Chapter 2; Watson et al. 2015).

13.4 Urban habitats and detoxification of xenobiotics

Human-introduced chemical substances in the environment can affect wildlife as well as humans. Strong selection on specific detoxifying and receptor genes has been documented in response to heavy metals and pesticides with a targeted mode of action (e.g., Whitehead et al. 2016). Less is known about the evolutionary outcomes, in terms of adaptive physiological responses to the 'cocktail' of urban pollutants. In this section, we review some of the common pollutants, their route of transmission to absorption, and some of the key physiological responses involved in detoxification, which are potential targets for selection.

13.4.1 Urban pollution

In cities, pollution levels are relatively high in air, soil, and water. The most common urban pollutants are different forms of nitrogen oxides (NO_x), sulfur dioxide (SO_2), polynuclear aromatic hydrocarbons (PAHs), carbon monoxide (CO), tropospheric ozone (O_3), particulate matter ($PM_{2.5}$–PM_{10}), and dioxin (which is a collective name for chlorinated dibenzo-p-dioxin and dibenzofuran congeners), but also heavy metals such as lead and mercury. In urban environments, diesel- and gasoline-powered motor vehicles are the main source of emission for many of the pollutants; thus, the urban pollution level generally increases with human population size, i.e., traffic-load. However, the level of industrialization of the city will also affect the overall pollution level, in particular for heavy metals and dioxins.

Although negative health effects are well documented for most of these pollutants and in most animals (e.g., Austin 1998; Acevedo-Whitehouse and Duffus 2009; Koivula and Eeva 2010), the specific biochemical or physiological pathways involved for a potential evolutionary response are still a rather unknown area. The main reason for this is because urban pollutants are not acutely toxic and exposure in nature is via a 'cocktail' of chemicals; thus, the pathways through which urban pollution affects physiology are complex and multifactorial. For example, cadmium and lead affect different components of the immune system (e.g., Krocova et al. 2000). This is in contrast to pesticides, which are targeted and acute in their design, and hence intense selection can lead to rapid evolution of tolerance, often through a single gene mutation (Whitehead et al. 2016). This has been commonly seen in cytochrome P450 genes and glutathione S-transferases in response to common classes of pesticides, such as dichlorodiphenyltrichloroethane (DDT), neonicotinoids, and organophosphates. In terms of urban evolutionary ecotoxicology and to understand the mechanistic underpinnings of selection, it could be more fruitful to use a genomic approach and network analyses rather than focusing on specific genes under selection. Likewise, most evolutionary ecophysiologists focus on physiology as characterized by markers in the blood stream, which can indeed be a good proxy for whole-body physiology and can be collected somewhat less invasively; however, other tissues might be more informative (such as the lungs, liver, and kidney) as to the actual mechanistic effects caused by urban stressors.

13.4.2 Air pollution and its consequences: a case study of birds

The main emission route of pollutants in urban areas is via air. The air pollutants can then deposit and accumulate in soil and water. Thus, the intake routes of pollutants are via the respiratory tract (inhalation), mouth (intake via food and water), and skin. Among terrestrial vertebrates, birds are probably the most sensitive group when it comes to inhalation of air pollutants. This sensitivity derives from their faster metabolic rate and the more efficient gas-exchange in their unidirectional air-sac lung, relative to other terrestrial vertebrates. When pollutants are inhaled, the first lines of defence in the respiratory tract are the mucus and cilia in the trachea, which prevent larger aerosols (> 10 μm) from entering deeper into the respiratory tract. This mucociliary clearance can be negatively affected by controlled exposures to air pollutants such as O_3 and SO_2, by reducing the length and number of cilia (reviewed in Sanderfoot and Holloway 2017). These changes have consequences for respiration by causing hypertrophy (swelling) of cells in the bronchi, inflammation in capillary networks, pulmonary oedema, and necrosis of epithelial cells: in short, respiratory illness of the bird. Comparisons of lungs from urban and non-urban birds (e.g., laughing doves (*Streptopelia senegalensis*), house sparrows (*Passer domesticus*), and Cape glossy starlings (*Lamprotornis nitens*)) confirm these effects shown in controlled exposure experiments (Sanderfoot and Holloway 2017).

Apart from this first line of defence, uptake and distribution of pollutants are similar whether they are inhaled, digested, or absorbed through the skin; it is the pollutants' chemical properties that are more important in determining how they affect an organism. Gases and vapours are readily absorbed by diffusion to the blood stream, whereas particulate matter (PM) and droplets can be absorbed either through simple diffusion or by pinocytosis, if not accumulated in the lung or excreted with the faeces. While in the circulatory system, the pollutants can be transferred to all organs in the body. Because the liver and kidneys have a higher volume of blood flow, they are also exposed to higher levels of pollutants, and thus are a target tissue for toxicological effects and selection. In addition, the metabolic and excretory functions of these organs make them further targets for accumulation of pollutants as well as exposure to breakdown products, which can be more toxic than the original substance. Furthermore, lipid-soluble toxicants and PM can pass through and/or impair the blood–brain barrier and cause damage to the brain. For example, diesel-fuel-generated PM has been shown to penetrate this barrier when living in polluted cities, causing neuro-inflammation, neuro-degeneration, and oxidative stress (Hartz et al. 2008).

Toxic gases, such as ozone (O_3), are highly potent pro-oxidants and can cause cellular damage. This damage can be avoided if the detoxifying antioxidant system is upregulated. When damage does occur, it can have severe implications, requiring the cell to either repair, engulf, or excrete these damaged molecules. The most severe damage is if the DNA itself is impaired; then transcription will be affected and enzymes and proteins might not function properly. Inhaled PM and PAH can also have mutagenic and genotoxic effects (e.g., Samet et al. 2004). The mutagenic effects not only affect somatic cells but sometimes also the germline, with implications for following generations. DNA changes have been shown in many different urban-dwelling organisms, but the links to fitness are often missing (see review in Isaksson 2018).

Regarding oxidative stress, there is no simple equation that suggests that the more antioxidants the merrier, in terms of fitness. Redox reactions are a normal part of cell life, and reactive oxygen/nitrogen species are required as signalling molecules. Thus, there is a fine line, and an evolutionary constraint, in terms of maximizing defence systems to combat the oxidative pollutants (Cohen et al. 2017). Despite this, oxidative damages have proven to be some of the best biomarkers for assessing costs associated with urban pollution. Other good candidates for evolutionary potential in urban and polluted environments are metallothionein, cytochrome P450, and the aryl hydrocarbon receptors (AHRs) (e.g., Box 13.1; Posthuma and van Straalen 1993; Good et al. 2014). For example, AHRs have a 250-fold decrease in affinity to dioxin and PAH in aquatic birds compared to chickens, which reduces the impact of these toxicants in aquatic birds (Karcher et al. 2006); see also studies in fish (e.g., Doering et al. 2013; Reitzel et al. 2014). This variation across species is a promising target for finding physiological adaptations to specific urban pollutants, and potential targets for ongoing selection in urban habitat.

13.5 Urban habitats and the endocrine regulation of reproduction

Reproduction is central to an individual's ability to succeed in any habitat. As such, the ways that the urban environment influences reproduction are likely to present critical selective pressures. These challenges might act as selective filters—allowing only pre-adapted individuals and species to persist in cities—or they might induce adaptive, or maladaptive, plastic responses or lead to adaptive evolution in urban populations. These varied outcomes will determine the success or failure of individuals and populations confronted by urbanization.

Environmental factors can influence two broad reproductive phenotypes that are regulated by endocrine signals: breeding phenology and investment in reproduction. Across vertebrates, several interacting endocrine axes play a role in mediating this, with two systems playing a primary role. (1) Photoperiod-associated activation of the pineal gland induces release of melatonin, which acts as a central regulator of both daily (circadian) and seasonal patterns of behaviour and physiology, including reproduction, in many vertebrates (Arendt 1998; Kumar et al. 2010). (2) Activation of the hypothalamic–pituitary–gonadal (HPG) axis, which is also associated with melatonin secretion patterns, induces release of gonadotropins (luteinizing hormone (LH) and follicle stimulating hormone (FSH)), androgens, and oestrogens, regulating both the timing and level of reproductive investment. Other endocrine signals, such as progestogens, prolactin, and glucocorticoids (see section 13.6.1), are also involved in regulation of reproduction (Norris and Carr 2013).

Differences exist in reproductive physiology among some urban and non-urban bird populations, but very few studies have sought to determine to what extent these differences derive from plastic responses to the environment or evolved differences among populations. The few that have sought to address this question point to an important role for plasticity. For example, in Germany, urban European blackbirds (*Turdus merula*) breed earlier than forest-dwelling birds. Partecke and colleagues (2004) hand-raised urban- and forest-derived birds from the nestling stage in a common garden captive setting. When the birds reached reproductive maturity 1 year after they were brought into captivity, urban-derived males exhibited a surge in LH earlier and urban-derived females' LH levels declined earlier than forest-derived birds. Despite these endocrine differences, timing of testis

and follicle development was similar in both urban and forest birds. Birds were followed through a second year, and by then no differences persisted in LH. Similarly, an urban population of dark-eyed juncos (*Junco hyemalis*) in California has a longer breeding season than nearby non-urban mountain populations, and urban males exhibit elevated circulating testosterone over a correspondingly longer period (Atwell et al. 2014). However, when juvenile birds were raised in a common garden setting, differences in circulating testosterone did not persist. These differences among free-ranging birds that are diminished or disappear entirely in a common garden setting suggest a role for phenotypic plasticity in either creating the initial differences among birds (e.g., in response to natal site environmental or maternal effects) or causing differences to disappear (e.g., in response to the common garden environment). For other urban-dwelling animals, there are several documented effects of urbanization on both timing and investment in reproduction (e.g., bats: Russo and Ancillotto 2015; insects: Chick et al. 2019). However, there is a large gap of knowledge about the physiological mechanisms underlying timing and investment.

In sections 13.5.1 and 13.5.2, we review how two specific urban challenges (night light pollution and altered food availability) impact reproductive physiology, focusing on experimentally documented effects of environmental factors on endocrine regulators of reproduction and fitness. While other aspects of the urban environment might influence the reproductive physiology of urban organisms, we focus on these two challenges because they are well documented and associated with important determinants of reproductive phenotypes.

13.5.1 Night light pollution and reproductive endocrinology

Changing day length is a primary cue for timing of reproduction for seasonal breeders, including some tropical species. Artificial light at night (ALAN) is a hallmark of cities across the globe and generates light pollution that is most intense in urban centres. The introduction of ALAN and the resulting disruption of natural photoperiod has clearly documented effects on reproductive physiology in an array of organisms. Relative to control environments with normal photoperiodic conditions and dark nights, even conservative exposure to low levels of ALAN can suppress melatonin and alter timing and level of activation of the HPG axis in seasonally breeding vertebrates (e.g., Dominoni et al. 2013; Robert et al. 2015; reviewed in Ouyang et al. 2018). For example, Dominoni and colleagues brought adult male European blackbirds into captivity and exposed half of them to low levels of ALAN and half to dark nights, with both groups experiencing increasing day length that simulated local conditions. The blackbirds exposed to ALAN exhibited increased circulating testosterone and advanced development of testes by an average of 26 days relative to control birds (Dominoni et al. 2013). Interestingly, blackbirds derived from an urban population responded differently to the ALAN treatment than birds from a non-urban area. Urban birds increased their testosterone concentrations and developed testes in response to ALAN before non-urban birds, but they also regressed their testes earlier than non-urban birds. This population-level difference could be driven by an evolutionary or a plastic response to the urban environment and might be adaptive in birds that have experience with high levels of light pollution. While most studies find the expected suppressive effect of ALAN on melatonin, and a stimulatory effect on reproductive axes, Schoech et al. (2013) found that ALAN unexpectedly increased melatonin and suppressed reproductive hormones in captive western scrub-jays (*Aphelacoma californica*). This species thrives in urban areas, and indeed, study subjects were sourced from urban/suburban areas, suggesting some potentially adaptive mechanism for maintaining reproductive function in free-ranging birds despite exposure to ALAN. Alternatively, findings from some captive experiments could be confounded by the stress of captivity itself, with elevated glucocorticoid hormones potentially stimulating melatonin secretion.

Captive experiments provide evidence that ALAN can influence the endocrine axes that regulate reproduction. However, many species will not breed in captivity, and so while physiology can be monitored, effects of ALAN on reproduction itself often cannot be assessed, and there is potential for confounding effects of the captive environment to confuse results. Field experiments that expose free-ranging organisms to ALAN have the advantage of preserving the potentially important signal of natural, full-spectrum lighting during the day and other cues (like changing food availability), and have provided advances in our understanding of how ALAN affects reproductive phenotypes that likely reflect underlying physiological responses. For example, when breeding great tits (*Parus major*) were experimentally provided with a light outside their nest box after their clutches had hatched, females increased their provisioning rate during later stages of the nestling period (potentially due to stimulation of prolactin or glucocorticoids, which can regulate parental effort), relative to controls (Titulaer et al. 2012). However, correlative work comparing blue tits (*Cyanistes caeruleus*) nesting close to and far from street lamps found that birds in territories with more ALAN did not fledge more offspring, but did begin laying eggs slightly earlier than females with less ALAN exposure. Males in dark territories were more successful at securing extrapair matings, a phenotype known to be influenced by testosterone (Kempanaers et al. 2010). As described in section 13.5, timing of reproduction is largely regulated by the interplay between melatonin and the HPG axis.

A limitation of many of these field studies, including some of the experimental ones, is that birds are not randomly assigned to treatment groups—that is, even when lights are placed following a randomized experimental design, birds choose where to settle and breed. Thus, when interpreting results, we must consider potential non-random differences between treatment groups (e.g., in bird age, condition, quality, or dominance status) as potential confounds. Overall, we currently lack clear evidence of fitness effects of light pollution, which would be required for this challenge to be a strong driver of evolutionary change in physiological responses to ALAN in urban populations.

13.5.2 Food availability/quality and reproductive endocrinology

Many organisms use food availability as an important cue for timing of reproduction, and synchronizing breeding with suitable food availability for feeding offspring can be critical for successful breeding. Urbanization can dramatically alter overall food availability or its timing. In addition, food quality can be altered in urban habitats. These changes can influence reproductive physiology.

A large body of literature provides evidence that experimental food supplementation can stimulate the primary endocrine axis that regulates reproduction (i.e., the HPG axis), advance onset of reproduction, increase investment in reproduction, and increase reproductive success across habitats, although effects vary, depending on the reproductive ecology and seasonality of the focal population (Boutin 1990; Ruffino et al. 2014). The most consistently observed effect of food on reproductive phenotypes found across experimental food supplementation studies is advanced timing of breeding, as well as delayed breeding in response to restricted food (reviewed in Davies and Deviche 2014). Notably, many urban birds breed earlier than non-urban conspecifics, pointing to a potential role for urban food subsidies, although other factors (e.g., ALAN, social environment, and temperature) might also contribute to this pattern. Additionally, despite advanced laying dates, many urban birds lay smaller clutches of eggs and fledge fewer young per nesting attempt (Chamberlain et al. 2009), suggesting that urban food availability does not stimulate reproductive physiology and investment at every level.

Additional food might stimulate the HPG axis, but the quality of food provided can also affect the timing of and investment in reproduction (e.g., Plummer et al. 2013; Toledo et al. 2016). For example, Schoech and Bowman (2003) have found that Florida scrub-jays (*Aphelocoma coerulescens*) in suburban habitats breed earlier than birds in natural habitats, which they posited might reflect anthropogenic food subsidies, usually rich in fats and proteins (e.g., pet food and sunflower seeds). Indeed, male scrub-jays provided with supplemental food high in both fat and protein had elevated circulating testosterone concentrations relative both to unsupplemented controls and birds provided with supplemental food that was high in fat but low in protein content. Interestingly, female oestradiol concentrations were unaffected by treatment, but

birds from both food-supplement groups advanced timing of clutch initiation (Schoech et al. 2004).

Given the species-specific nature of foraging ecology, and likely city-specific effects of urbanization on food availability and quality, we should not expect general rules regarding effects of urban food on reproductive endocrinology. Whether selection on the HPG axis might be an important factor driving urban adaptation remains to be revealed. Timing of reproduction is often directly linked to reproductive success (but see Caizergues et al. 2018), and therefore fitness, although these relationships might reflect the influence of adaptive plasticity and variation in female condition, rather than natural selection on heritable variation in phenology (Price et al. 1988). A likely target for selection in urban environments would be the fit of endocrine responsiveness to cues of food availability and/or quality such that phenology matches availability of appropriate resources for raising offspring. For example, if anthropogenic food subsidies cause animals to upregulate activation of the HPG axis and breed early in cities, but abundance of insects—an important food item for provisioning offspring for many birds—is not similarly increased or temporally shifted, mismatches could occur between reproduction and resource availability (but see Seress et al. 2018). Thus, natural selection in cities might favour dampened endocrine sensitivity to unreliable cues, such as anthropogenic food, if they do not predict availability of appropriate resources for breeding.

13.6 Urban habitats and endocrine responses to challenges

The HPA axis is central in regulating behavioural and physiological responses to challenges and, unsurprisingly, has been the focus of several studies aimed at understanding the role of endocrine traits in responses to urbanization. Studies investigating the effects of urbanization on the HPA axis have largely been descriptive, comparing urban and non-urban conspecific populations. However, experimental studies have isolated some of the environmental factors that change with urbanization, and documented their effects on the HPA axis. The findings of studies employing these two approaches, however, reveal few clear patterns.

Comparative studies in the field find no consistent patterns in responses of the HPA axis to urban habitat. For example, populations in cities have been found to have higher, lower, or similar circulating concentrations of glucocorticoid hormones (the end product of activation of the HPA axis). This variation is evident depending on species, life-history stage, and sex of the individuals being sampled, with variation often evident even within the same study (reviewed in Bonier 2012). Clearly, we cannot currently draw broad conclusions from this growing literature. Here, we focus on the effects of two challenges associated with urbanization—altered food availability and human activity. These two urban challenges are most directly relevant to HPA axis regulation because of the two distinct functions of activation of this endocrine axis: maintenance of energetic balance and response to acute challenges.

13.6.1 Altered food availability and the HPA axis

At baseline concentrations, glucocorticoid hormones are central regulators of metabolic processes (Landys et al. 2006). Glucocorticoids increase in response to elevated energetic demands, and facilitate mobilization of resources (e.g., by stimulating gluconeogenesis) and resource acquisition (e.g., by stimulating foraging). These hormones also interact with the HPG axis, acting as important modulators of reproduction, and contributing to the matching of reproductive investment and timing with resource availability (Acevedo-Rodriguez et al. 2018). As such, fitness effects of altered food availability in the urban environment could be mediated, in part, by HPA axis responses.

Ample evidence points to a relationship between food availability and circulating glucocorticoid concentrations. For example, adult barn swallows (*Hirundo rustica*) have higher circulating corticosterone concentrations when weather conditions result in shortages in the aerial insect prey that they rely on to feed themselves and their offspring (Jenni-Eiermann et al. 2008). Lendvai and colleagues (2014) found that changes in corticosterone in captive house sparrows experimentally exposed to moderate food restriction (60 per cent daily food intake) depended on changes in body mass. Birds that lost more body

mass also increased corticosterone more, indicating a link between the degree of energetic challenge the bird experienced during food restriction and its HPA axis response. These increases in glucocorticoids in response to food shortage may be adaptive, as they can spur increases in foraging activity and mobilization of stored resources (e.g., Crossin et al. 2012; reviewed in Landys et al. 2006; Vera et al. 2017).

In cities, where organisms might confront novel foods of differing types and abundance relative to what is found in more pristine environments, appropriate HPA responses to these challenges, which contribute to adaptive metabolic, behavioural, and reproductive responses, should be favoured by selection. To date, we have little direct empirical evidence to assess this prediction. Florida scrub-jays breeding in suburban habitats with access to ample anthropogenic food have lower circulating corticosterone than jays in non-urban habitats and, in some years, food supplementation can reduce corticosterone in non-urban jays (Schoech et al. 2004). In this same population, food supplementation increases reproductive success, through increased clutch size and offspring survival (Schoech et al. 2008). Given that suburban scrub-jays breed earlier than non-urban conspecifics, Schoech and colleagues hypothesized that increased food in suburban habitats might cause lower corticosterone levels, which in turn drives earlier breeding. Food supplementation experiments provide some support for a causal link between food and corticosterone in this species, and a link between food and timing of breeding, but experimentally increased corticosterone (through oral doses of exogenous corticosterone in a food item) did not influence timing of reproduction (Schoech et al. 2007).

Overall, we see that links exist between the urban environment (food availability) and the HPA axis, and between the environment and components of fitness (timing and success of reproduction). The degree to which the fitness effects associated with variation in food availability are modulated or mediated by the HPA axis, in any environment, is as yet unclear. Further work similar to that of Schoech and colleagues that attempts to link the environment to an endocrine response and to traits important for fitness (and variation in those links) is needed to determine the degree to which the HPA response to varied urban food availability is shaped by natural selection.

13.6.2 Human activity and the HPA axis

For an organism to succeed in urban environments, it must tolerate increased levels of human activity. Encounters with people and our affiliates (e.g., our vehicles and pets) might constitute acute stressors for some animals, and thereby activate an acute HPA axis response, resulting in rapid elevation of circulating glucocorticoids and redirection of activities and resources towards coping with the challenge at hand. Because of this temporary reallocation of resources, repeated induction of an acute HPA response could interfere with successful reproduction and survival. As such, what distinguishes a successful urban species or individual from an unsuccessful one could hinge strongly on the HPA response to human disturbance. Despite the potential importance of appropriate HPA responsiveness to human activity in determining the success of organisms in urban environments, very few studies have attempted to address this question.

Human disturbance around nesting sites has been linked to both breeding failure and corticosterone. For example, nests of American kestrels (*Falco sparverius*) in disturbed sites (characterized by greater human development and/or near busy, high-speed roads) were almost ten times more likely to fail than nests in less-disturbed sites (Strasser and Heath 2013). Furthermore, female kestrels nesting in disturbed sites had higher corticosterone concentrations, and nests of females with higher corticosterone were more likely to fail. Overall, this study provides correlative evidence linking human disturbance to both corticosterone and reproductive success, but we need experimental evidence to establish causal links, and to relate the importance of those links to success or failure in cities. In addition, studies of the HPA response to human disturbance in species with similar or even higher reproductive success in cities (e.g., Ancillotto et al. 2016; Corsini et al. 2017) could help shed light on adaptive HPA reactivity in the urban environment.

Some correlative evidence shows that corticosterone can be elevated in populations in more heavily urbanized or human-disturbed habitats.

For example, Zhang and colleagues (2014) found a positive association between mean corticosterone concentrations in tree sparrow (*Passer montanus*) populations and an urbanization score applied to several sampling sites across an urbanization gradient in China. This suggests that birds in more highly urbanized areas might be experiencing more challenges, but whether the pattern is driven by human disturbance or other factors such as food availability (see section 13.6.1), and whether it is associated with fitness effects, is unknown. Furthermore, in a recent experimental study on zebra finches (*Taeniopygia guttata*), an acoustic stressor (noise, which is elevated in urban areas due to human activities) induced a similar increase in both metabolic rate and corticosterone (Jimeno et al. 2018), which is suggestive that a key function of elevated corticosterone in the urban environment could be metabolic regulation. However, many studies find no difference or even lower glucocorticoid concentrations in urban populations (e.g., Lyons et al. 2017; reviewed in Bonier 2012).

Overall, the effects of high levels of human activities on fitness of urban organisms are likely to be mediated by their HPA responses, but this is currently a wide-open area for future work, given the paucity of experimental studies directly investigating these questions to date. One defining characteristic of the HPA axis response is that it is generalized—that is, the axis is activated by challenges, broadly defined, rather than a more specified response to particular challenges. Important differences across individuals, and perhaps targets for selection in the urban environment, could lie in the perception of challenges (e.g., what constitutes a challenge worthy of an HPA response and what does not), recovery from challenges, and acclimation to challenges. However, relatively few studies have described differences among urban and non-urban individuals in sensitivity and regulation of the HPA response to challenges. Curve-billed thrashers (*Toxostoma curvirostre*) in urban populations show greater elevation of circulating corticosterone in response to injection with pituitary peptides (arginine vasotocin and adrenocorticotropin) than desert birds (Fokidis and Deviche 2011). In contrast, urban and desert thrashers increased corticosterone similarly in response to injection with a hypothalamic peptide (corticotropin-releasing factor), and reduced corticosterone similarly in response to injection with a synthetic glucocorticoid that induces negative feedback mechanisms (dexamethasone). Thus, at least in these populations, urban birds exhibit increased sensitivity to some components of the HPA endocrine cascade, but it is unclear if they also exhibit differential responses to the environmental challenges that can activate this axis, such as human activity. We also do not know the relative role of plasticity versus evolution in explaining these population-level differences. One common garden study found that urban European blackbirds had similar baseline corticosterone but dampened responses to an acute stressor compared to non-urban birds (Partecke et al. 2006). These differences might reflect local adaptation, although early environmental and maternal effects are also plausible, given birds were brought into captivity at the late nestling stage.

An element that has been lacking in most work to date involving HPA responses to urbanization is consideration of the degree to which species are challenged by cities. Most studies focus on one species, comparing HPA traits among urban and non-urban populations or along a gradient of urbanization. Perhaps some of the variation in findings among such comparisons is driven by broader differences among species that determine the degree to which cities represent challenges or opportunities. For species that persist but are not thriving in urban habitats, we would not have the same expectations in terms of how selection might act on the HPA axis in urban populations as we would for species that are thriving. Uncovering these species-level differences might hold some of the keys to understanding urban adaptation. For example, future work characterizing trait–fitness correlations (Box 13.1) across species that are challenged to varying degrees by the urban environment might provide some evidence to determine if the responses of these different species to urban challenges constrain or facilitate their ability to cope with urbanization.

13.7 Urban habitats and metabolic responses

The conversion of food to energy to build up biological molecules and to eliminate waste products is the cornerstone of metabolism. Intra- and interspecific variation in the different components of metabolism (e.g., metabolic rate, catabolism (breaking down molecules), and anabolism (building up molecules)) is driven by a variety of environmental factors such as climate, diet, and pollution, but also activity patterns (e.g., awake versus sleep), many of which are highly relevant to the urban context (Isaksson 2018 and references therein). Thus, genes underlying different metabolic responses are likely targets of selection.

In evolutionary physiology, one key component to understanding metabolism is the measure of basal metabolic rate (BMR). BMR is the minimal rate of energy expenditure required to sustain life, which then dictates the species/population/individuals' pace-of-life. Broadly, a high BMR is associated with a fast-pace of life, and vice versa for a low BMR. The results for urban species/populations are mixed; some seem to follow the slow pace-of-life path, with higher survival and lower yearly reproductive output, whereas others show the opposite (e.g., Charmantier et al. 2017; Brans et al. 2018). Many of these life-history studies have not measured BMR but instead other physiological traits such as breath rate, immune responses, antioxidants, and oxidative damage, which are linked to pace-of-life syndromes and more easily measured in wild populations. To date, only one study has investigated the BMR of urban and non-urban animals (great tits) using a common garden approach; however, there was no overall effect of population of origin on BMR (section 13.7.1; Andersson et al. 2018).

A much less explored, but emerging field in relation to metabolism is the measure of various metabolites, i.e., metabolomics. These metabolites can be both substrates and products of anabolic and catabolic processes such as sugars, vitamins, organic acids, and lipids, thus indicating either an intrinsic or extrinsic change that can affect survival and reproduction. As mentioned in section 13.5, metabolism is under hormonal regulation; however, it is becoming increasingly apparent that the metabolic substrates and products themselves (such as fatty acids and glucose) can be part of a feedback process in either a feed-forward or feedback manner (Widmaier 1992). Thus, studying selection on metabolism requires a holistic approach to physiological systems and networks; otherwise it will be more like finding a needle in a haystack.

In terms of studies of metabolic responses in urban habitats, very little to almost nothing has yet been explored. Given this dearth of data, in section 13.7.1 we illustrate the *potential* rather than the *evidence* for metabolism to be a key target of selection based on documented differences in urban and non-urban habitats and selection on some of these traits.

13.7.1 Food quality and metabolic responses

Food type and quality can clearly affect many different physiological systems, with implications for urban-dwelling populations and individuals (see sections 13.5.2 and 13.6.1). The composition of many macro- and micronutrients will change for urban wildlife when feeding on anthropogenic food sources. This includes, for example, increased intake of sugar and salt, altered fatty acid composition, and reduced intake of dietary antioxidants and vitamins (Isaksson and Andersson 2007; Andersson et al. 2018; Shulte-Hostedde et al. 2018). Some micronutrients are essential, and so deficiencies will lead to dramatic effects (Figure 13.1). One example is thiamine deficiencies, which have detrimental effects on growth and embryonic development in birds, mussels, and salmonines (Balk et al. 2016). Although the underlying reason for the declining availability of thiamine in nature is still unknown (e.g., Balk et al. 2016), it seems to be related to increased exploitation and human population growth.

An evolutionary solution to these extreme effects of nutrient deficiency is unlikely; thus, from an evolutionary perspective, dietary shifts that are not acutely detrimental are more interesting. From hominin evolution, we know that there have been several dietary shifts, with corresponding evolutionary adaptations. For example, human populations have evolved in response to increased intake of starch and changed fatty acid composition of diet. A comparison between isolated human populations on naturally high versus low starch diets revealed significantly more copies of the amylase gene (*AMY1*), the salivary enzyme that breaks down starch to sugars, in the population on high starch diet (Perry et al. 2007). In non-human vertebrates, selection on genes encoding for starch digestion and fat metabolism has also been identified in dogs, as part of domestication (Axelsson et al. 2013).

In wild animals, there has been no corresponding evolutionary study in relation to the dietary shifts caused by urbanization. However, wild urban yellow baboons (*Papio cynocephalus*) and raccoons (*Procyon lotor*) that have shifted their diet to include or even entirely consist of anthropogenic waste food are heavier and have a changed glucose metabolism, i.e., insulin resistance, and increased glycated serum proteins (Banks et al. 2003; Schulte-Hostedde et al. 2018). Furthermore, in northern Europe and North America, birds commonly use bird feeders during the winter months. In a common garden experiment of urban and non-urban great tits, urban birds had a consistently lower BMR compared to non-urban birds when on an unsaturated fat diet (UFA) (sunflower seed oil) and not when on a saturated fat diet (SFA) (coconut butter) (Andersson et al. 2018). Thus, different diets seem to affect BMR of birds from different habitats differently, which could be a result of a preconditioned hormetic response (see Figure 13.1) or selection to cope with a sunflower seed diet in the urban habitat (Watson et al. 2017). In line with this, urban and non-urban great tits show a difference in the expression of genes encoding for lipid metabolism, more specifically fatty acid synthesis and elongation of polyunsaturated fatty acids (PUFA) (Watson et al. 2017). Although, the link between PUFAs and fitness has been well documented in poultry and in the fish industry, it was only recently confirmed in a wild vertebrate (Twining et al. 2016). Surprisingly, the study by Andersson and colleagues (2018) on BMR is the only study to date on BMR of urban and non-urban living organisms. Future studies should reveal whether these metabolic responses in urban animals are selected or plastic responses to dietary shifts or to other urban factors.

13.8 Unanswered questions and concluding remarks

The field of urban evolutionary physiology can add something unique to the study of urban evolution, namely, an increased understanding of the mechanistic underpinnings of adaptation, and its constraints.

An understanding of the role of phenotypic plasticity and preconditioned hormetic responses in urban adaptation will be crucial to understanding the often large variation found in physiological responses within and across populations. However, to take the field further, we might also need to take one step back and investigate the dose-response relationship with single and multiple urban factors to be able to establish and model future scenarios for adaptive potential versus constraints (Watson et al. 2015; Cohen et al. 2017): for example, micronutrient deficiencies caused by human activities that cannot be circumvented by evolution (such as thiamine deficiency), but instead require policy changes and conservation interventions.

Most urban factors are, however, not acutely toxic. Instead, it is the balance between the stimulating (beneficial) and detrimental factors that cause a challenge. Where are the threshold points that distinguish an urban population that thrives from one that declines? What is the underlying mechanism(s) causing this switch, and explaining variation among species and individuals? Which urban factors affect the same physiological systems, and do these effects act in opposition to each other or in an additive or synergistic manner? This chapter sought to illustrate that several urban challenges, for example food abundance and quality, affect a variety of physiological systems, and thereby different components of fitness. However, we lack an understanding of how natural selection is shaping the physiological mechanisms that mediate the links between urban challenges and fitness. This is a promising area for future work on urban evolution.

Overall, we see that air pollution, noise, light pollution, altered food availability, and human activity can impact physiological processes and that the responses are more interconnected than previously appreciated. Thus, we call for a more integrative physiological approach in future research, investigating physiological networks and the strengths of different paths. In addition, it is largely unknown if the ways that urban organisms alter their physiology in response to these challenges translate into important fitness effects, and the degree to which plasticity versus evolution plays a role in response to these urban challenges. Studies that map the selective landscape associated with these factors,

describe the ways that selection acts on the urban physiological phenotype (Box 13.1), and disentangle the relative roles of plasticity versus evolutionary change offer promising avenues for significant progress in this area. Although we did not present a comprehensive review of the field, encompassing all physiological responses and systems and how they respond to all urban factors, our broad conclusions can be generalized across all physiological systems, taxonomic groups, and urban challenges. The field of urban evolutionary physiology is in its early days, and given the current interest in mechanistic biology in an ecological context, we predict exciting times and burgeoning research activity in this field in the near future.

Acknowledgements

We would like to thank the three editors, Jason Munshi-South, Anne Charmantier, and Marta Szulkin, for putting together this book and inviting us to write this chapter. We also thank the reviewers for constructive feedback, which has improved the breadth of this chapter.

References

Acevedo-Whitehouse, K. and Duffus, A.L.J. (2009). Effects of environmental change on wildlife health. *Philosophical Transactions of the Royal Society B*, 364, 3429–38.

Acevedo-Rodriguez, A., Kauffman, A.S., Cherrington, B.D., et al. (2018). Emerging insights into hypothalamic–pituitary–gonadal axis regulation and interaction with stress signalling. *Journal of Neuroendocrinology*, 30, e12590.

Ancillotto, L., Tomassini, A., and Russo, D. (2016). The fancy city life: Kuhl's pipistrelle, *Pipistrellus kuhlii*, benefits from urbanisation. *Wildlife Research*, 42, 598–606.

Andersson, M.N., Nilsson, J., Nilsson, J-Å., and Isaksson, C. (2018). Diet and ambient temperature interact to shape plasma fatty acid composition, basal metabolic rate, and oxidative stress in great tits. *Journal of Experimental Biology*, 221, 186759.

Arendt, J. (1998). Melatonin and the pineal gland: influence on mammalian seasonal and circadian physiology. *Reviews of Reproduction*, 3, 13–22.

Arumugam, T.V., Gleichmann, M., Tang, S.C., and Mattson, M.P. (2006). Hormesis/preconditioning mechanisms, the nervous system and ageing. *Ageing Research Reviews*, 5, 165–78.

Atwell, J.W., Cardoso, G.C., Whittaker, D.J., Price, T.D., and Ketterson, E.D. (2014). Hormonal, behavioral, and life-history traits exhibit correlated shifts in relation to population establishment in a novel environment. *The American Naturalist*, 184, 147–60.

Austin, B. (1998). The effects of pollution on fish health. *Journal of Applied Microbiology*, 85, 234S–42.

Axelsson, E., Ratnakumar, A., Arendt, M.-J., et al. (2013). The genomic signature of dog domestication reveals adaptation to a starch-rich diet. *Nature*, 495, 360–64.

Balk, L., Hägerroth, P.-Å., Gustavsson, H., et al. (2016). Widespread episodic thiamine deficiency in northern hemisphere wildlife. *Scientific Reports*, 6, 38821.

Banks, W.A., Altmann, J., Sapolsky, R.M., Phillips-Conroy, J.E., and Morley, J.E. (2003). Serum leptin levels as a marker for a syndrome X-like condition in wild baboons. *Journal of Clinical Endocrinology and Metabolism*, 88, 1234–40.

Bonier, F. (2012). Hormones in the city: endocrine ecology of urban birds. *Hormones and Behavior*, 61, 763–72.

Bonier, F. and Martin, P.R. (2016). How can we estimate natural selection on endocrine traits? Lessons from evolutionary biology. *Proceedings of the Royal Society B: Biological Sciences*, 283, 20161887.

Bonier, F., Martin, P.R., Sheldon, K.S., et al. (2006). Sex-specific consequences of life in the city. *Behavioral Ecology*, 18, 121–9.

Boutin, S. (1990). Food supplementation experiments with terrestrial vertebrates: patterns, problems, and the future. *Canadian Journal of Zoology*, 68, 203–20.

Brans, K.I., Stoks, R., and De Meester, L. (2018). Urbanization drives genetic differentiation in physiology and structures the evolution of pace-of-life syndromes in the water flea *Daphnia magna*. *Proceedings of the Royal Society B: Biological Sciences*, 285, 20180169.

Caizergues, A.E., Grégoire, A., and Charmantier, A. (2018). Urban versus forest ecotypes are not explained by divergent reproductive selection. *Proceedings of the Royal Society B: Biological Sciences*, 285, 20180261.

Calabrese, E.J., Bachmann, K.A., Bailer, A.J., et al. (2007). Biological stress response terminology: integrating the concepts of adaptive response and preconditioning stress within a hormetic dose-response framework. *Toxicology and Applied Pharmacology*, 222, 122–8.

Chamberlain, D.E., Cannon, A.R., Toms, M., et al. (2009). Avian productivity in urban landscapes: a review and meta-analysis. *Ibis*, 151, 1–18.

Charmantier, A., Demeyrier, V., Lambrechts, M., Perret, S., and Grégoire, A. (2017). Urbanization is associated with divergence in pace-of-life in great tits. *Frontiers in Ecology and Evolution*, 5, 53.

Chick, L.D., Strickler, S.A., Perez, A., Martin, R.A., and Diamond, S.E. (2019). Urban heat islands advance the timing of reproduction in a social insect. *Journal of Thermal Biology*, 80, 119–25.

Cohen, A., Isaksson, C., and Salguero Gomez, R. (2017). Co-existence of multiple trade-off currencies shapes evolutionary outcomes. *PLOS ONE*, 12, e0189124.

Corsini, M., Dubiec, A., Marrot, P., and Szulkin, M. (2017). Humans and tits in the city: quantifying the effects of human presence on great tit and blue tit reproductive trait variation. *Frontiers in Ecology and Evolution*, 5, 82.

Crossin, G.T., Trathan, P.N., Phillips, R.A., et al. (2012). Corticosterone predicts foraging behavior and parental care in macaroni penguins. *The American Naturalist*, 180, E31–41.

Davies, S. and Deviche, P. (2014). At the crossroads of physiology and ecology: food supply and the timing of avian reproduction. *Hormones and Behavior*, 66, 41–55.

Doering, J.A., Giesy, J.P., Wiseman, S., and Hecker, M. (2013). Predicting the sensitivity of fishes to dioxin-like compounds: possible role of the aryl hydrocarbon receptor (AhR). ligand binding domain. *Environmental Science and Pollution Research*, 20, 1219–24.

Dominoni, D., Quetting, M., and Partecke, J. (2013). Artificial light at night advances avian reproductive physiology. *Proceedings of the Royal Society B: Biological Sciences*, 280, 20123017.

Donihue, C.M. and Lambert, M.R. (2015). Adaptive evolution in urban ecosystems. *Ambio*, 44, 194–203.

Fischer, J.D., Schneider, S.C., Ahlers, A.A., and Miller, J.R. (2015). Categorizing wildlife responses to urbanization and conservation implications of terminology. *Conservation Biology*, 29, 1246–8.

Fokidis, H.B. and Deviche, P. (2011). Plasma corticosterone of city and desert curve-billed thrashers, *Toxostoma curvirostre*, in response to stress-related peptide administration. *Comparative Biochemistry and Physiology Part A: Molecular & Integrative Physiology*, 159, 32–8.

Garland, T. Jr. and Carter, P.A. (1994). Evolutionary physiology. *Annual Review of Physiology*, 56, 579–621.

Gillis, P.L., Higgins, S.K., and Jorge, M.B. (2014). Evidence of oxidative stress in wild freshwater mussels (*Lasmigona costata*) exposed to urban-derived contaminants. *Ecotoxicology and Environmental Safety*, 102, 62–9.

Good, R.T., Gramzow, L., Battlay, P., et al. (2014). The molecular evolution of cytochrome P450 genes within and between *Drosophila* species. *Genome Biology and Evolution*, 6, 1118–34.

Hartz, A.M.S., Bauer, B., Block, M.L., Hong, J.-S., and Miller, D.-S. (2008). Diesel exhaust particles induce oxidative stress, proinflammatory signaling, and P-glycoprotein up-regulation at the blood–brain barrier. *The FASEB Journal*, 22, 2723–33.

Herrera-Dueñas, A., Pineda, J., Antonio, M.T., and Aguirrea, J.I. (2014). Oxidative stress of house sparrow as bioindicator of urban pollution. *Ecological Indicators*, 42, 6–9.

Isaksson, C. (2015). Urbanisation, oxidative stress and inflammation: a question of evolving, acclimatizing or coping with urban environmental stress. *Functional Ecology*, 29, 913–23.

Isaksson, C. (2018). Impact of urbanization on birds. In: Tietze, D.T. (ed.) *Birds Species: How they Arise, Modify and Vanish*, pp. 235–57. Springer Nature, Cham, Switzerland.

Isaksson, C. and Andersson, S. (2007). Carotenoid diet and nestling provisioning in urban and rural great tits *Parus major*. *Journal of Avian Biology*, 38, 564–72.

Jenni-Eiermann, S., Glaus, E., Grüebler, M., Schwabl, H., and Jenni, L. (2008). Glucocorticoid response to food availability in breeding barn swallows (*Hirundo rustica*). *General and Comparative Endocrinology*, 155, 558–65.

Jimeno, B., Hau, M., and Verhulst, S. (2018). Corticosterone levels reflect variation in metabolic rate, independent of 'stress'. *Scientific Reports*, 8, 13020.

Karcher, S.I., Franks, D.G., Kennedy, S.W., and Hahn, M.E. (2006). The molecular basis for differential dioxin sensitivity in birds: role of the aryl hydrocarbon receptor. *Proceedings of the National Academy of Sciences of the United States of America*, 103, 6252–7.

Kempenaers, B., Borgström, P., Loës, P., Schlicht, E., and Valcu, M. (2010). Artificial night lighting affects dawn song, extra-pair siring success, and lay date in songbirds. *Current Biology*, 20, 1735–9.

Koivula, M.J. and Eeva, T. (2010). Metal-related oxidative stress in birds. *Environmental Pollution*, 158, 2359–70.

Krocova, Z., Macela, A., Kroca, M., and Hernychova, L. (2000). The immunomodulatory effect(s) of lead and cadmium on the cells of immune system in vitro. *Toxicology in Vitro*, 14, 33–40.

Kumar, V., Wingfield, J.C., Dawson, A., et al. (2010). Biological clocks and regulation of seasonal reproduction and migration in birds. *Physiological and Biochemical Zoology*, 83, 827–35.

Landys, M.M., Ramenofsky, M., and Wingfield, J.C. (2006). Actions of glucocorticoids at a seasonal baseline as compared to stress-related levels in the regulation of periodic life processes. *General and Comparative Endocrinology*, 148, 132–49.

Lendvai, A.Z., Ouyang, J.Q., Schoenle, L.A., et al. (2014). Experimental food restriction reveals individual differences in corticosterone reaction norms with no oxidative costs. *PLOS ONE*, 9, e110564.

Lyons, J., Mastromonaco, G., Edwards, D.B., and Schulte-Hostedde, A.I. (2017). Fat and happy in the city: eastern chipmunks in urban environments. *Behavioral Ecology*, 28, 1464–71.

Mueller, J.C., Partecke, J., Hatchwell, B.J., Gaston, K.J., and Evans, K.L. (2013). Candidate gene polymorphisms for behavioural adaptations during urbanization in blackbirds. *Molecular Ecology*, 22, 3629–37.

Norris, D.O. and Carr, J.A. (2013). *Vertebrate Endocrinology*, 5th edn. Academic Press, London.

Ouyang, J.Q., Davies, S., and Dominoni, D. (2018). Hormonally mediated effects of artificial light at night on behavior and fitness: linking endocrine mechanisms with function. *Journal of Experimental Biology*, 221, 156893.

Partecke, J., Van't Hof, T., and Gwinner, E. (2004). Differences in the timing of reproduction between urban and forest European blackbirds (*Turdus merula*): result of phenotypic flexibility or genetic differences? *Proceedings of the Royal Society B: Biological Sciences*, 271, 1995–2001.

Partecke, J., Schwabl, I., and Gwinner, E. (2006). Stress and the city: urbanization and its effects on the stress physiology in European blackbirds. *Ecology*, 87, 1945–52.

Perrier, C., Lozano del Campo, A., Szulkin, M., et al. (2018). Great tits and the city: distribution of genomic diversity and gene–environment associations along an urbanization gradient. *Evolutionary Applications*, 11, 593–613.

Perry, G.H., Dominy, N.J., Claw, K.G., et al. (2007). Diet and the evolution of human amylase gene copy number variation. *Nature Genetics*, 39, 1256–60.

Plummer, K.E., Bearhop, S., Leech, D.I., Chamberlain, D.E., and Blount, J.D. (2013). Winter food provisioning reduces future breeding performance in a wild bird. *Scientific Reports*, 3, 2002.

Posthuma, L. and van Straalen, N.M. (1993). Heavy-metal adaptation in terrestrial invertebrates: a review of occurrence, genetics, physiology and ecological consequences. *Comparative Biochemistry and Physiology—Part C: Comparative Pharmacology and Toxicology*, 106, 11–38.

Price, T., Kirkpatrick, M., and Arnold, S.J. (1988). Directional selection and the evolution of breeding date in birds. *Science*, 240, 798–9.

Reitzel, A.M., Karchner, S.I., Franks, D.G., et al. (2014). Genetic variation at aryl hydrocarbon receptor (AHR) loci in populations of Atlantic killifish (*Fundulus heteroclitus*) inhabiting polluted and reference habitats. *BMC Evolutionary Biology*, 14, 6.

Robert, K.A., Lesku, J.A., Partecke, J., and Chambers, B. (2015). Artificial light at night desynchronizes strictly seasonal reproduction in a wild mammal. *Proceedings of the Royal Society B: Biological Sciences*, 282, 20151745.

Romero-Haro, A.A. and Alonso-Alvarez, C. (2015). The level of an intracellular antioxidant during development determines the adult phenotype in a bird species: a potential organizer role for glutathione. *The American Naturalist*, 185, 390–405.

Ruffino, L., Salo, P., Koivisto, E., Banks, P.B., and Korpimäki, E. (2014). Reproductive responses of birds to experimental food supplementation: a meta-analysis. *Frontiers in Zoology*, 11, 80.

Russo, D. and Ancillotto, L. (2015). Sensitivity of bats to urbanization: a review. *Mammalian Biology*, 80, 205–12.

Salmón, P., Stroh, E., Herrera-Dueñas, A., von Post, M., and Isaksson, C. (2018a). Oxidative stress in birds along a NO$_x$ and urbanization gradient: an interspecific approach. *Science of the Total Environment*, May 1, 622–3, 635–43.

Salmón, P., Watson, H., Nord, A., and Isaksson, C. (2018b). Effects of the urban environment on oxidative stress in early life: insights from a cross-fostering experiment. *Integrative and Comparative Biology*, 58, 986–94.

Samet, J.M., DeMarini, D.M., and Malling, H.V. (2004). Do airborne particles induce heritable mutations? *Science*, 304, 971–2.

Sanderfoot, O.V. and Holloway, T. (2017). Air pollution impacts on avian species via inhalation exposure and associated outcomes. *Environmental Research Letter*, 12, 083002.

Schoech, S.J. and Bowman, R. (2003). Does differential access to protein influence differences in timing of breeding of Florida scrub-jays (*Aphelocoma coerulescens*) in suburban and wildland habitats? *The Auk*, 120, 1114–27.

Schoech, S.J., Bowman, R., and Reynolds, S.J. (2004). Food supplementation and possible mechanisms underlying early breeding in the Florida scrub-jay (*Aphelocoma coerulescens*). *Hormones and Behavior*, 46, 565–73.

Schoech, S.J., Bowman, R., Bridge, E.S., et al. (2007). Corticosterone administration does not affect timing of breeding in Florida scrub-jays (*Aphelocoma coerulescens*). *Hormones and Behavior*, 52, 191–6.

Schoech, S.J., Bridge, E.S., Boughton, R.K., et al. (2008). Food supplementation: a tool to increase reproductive output? A case study in the threatened Florida scrub-jay. *Biological Conservation*, 141, 162–73.

Schoech, S.J., Bowman, R., Hahn, T.P., et al. (2013). The effects of low levels of light at night upon the endocrine physiology of western scrub-jays (*Aphelocoma californica*). *Journal of Experimental Zoology Part A: Ecological Genetics and Physiology*, 319, 527–38.

Seress, G., Hammer, T., Bókony, V., et al. (2018). Impact of urbanization on abundance and phenology of caterpillars and consequences for breeding in an insectivorous bird. *Ecological Applications*, 28, 1143–56.

Shulte-Hostedde, A.I., Mazal, Z., Jardine, C.M., and Gagnon, J. (2018). Enhanced access to anthropogenic food waste is related to hyperglycemia in raccoons (*Procyon lotor*). *Conservation Physiology*, 6(1), coy026.

Snell-Rood, E.C. and Wick, N. (2013). Anthropogenic environments exert variable selection on cranial capacity in mammals. *Proceedings of the Royal Society B: Biological Sciences*, 280, 20131384.

Strasser, E.H. and Heath, J.A. (2013). Reproductive failure of a human-tolerant species, the American kestrel, is associated with stress and human disturbance. *Journal of Applied Ecology*, 50, 912–19.

Titulaer, M., Spoelstra, K., Lange, C.Y., and Visser, M.E. (2012). Activity patterns during food provisioning are affected by artificial light in free living great tits (*Parus major*). *PLOS ONE*, 7, e37377.

Toledo, A., Andersson, M.N., Wang, H-L., et al. (2016). Fatty acid profiles of great tit (*Parus major*). eggs differ between urban and rural habitats, but not between coniferous and deciduous forests. *The Science of Nature*, 103, 55.

Twining, C.W., Brenna, J.T., Lawrence, P., et al. (2016). Omega-3 long-chain polyunsaturated fatty acids support aerial insectivore performance more than food quantity. *Proceedings of the National Academy of Sciences of the United States of America*, 113, 10920–25.

Vera, F., Zenuto, R., and Antenucci, C.D. (2017). Expanding the actions of cortisol and corticosterone in wild vertebrates: a necessary step to overcome the emerging challenges. *General and Comparative Endocrinology*, 246, 337–53.

Watson, H., Cohen, A.A., and Isaksson, C. (2015). A theoretical model of the evolution of actuarial senescence under environmental stress. *Experimental Gerontology*, 71, 80–88.

Watson, H., Videvall, E., Andersson, M.N., and Isaksson, C. (2017). Transcriptome analysis of a wild bird reveals physiological responses to the urban environment. *Scientific Reports*, 7, 44180.

Whitehead, A., Clark, B.W., Reid, N.M., Hahn, M.E., and Nacci, D. (2016). When evolution is the solution to pollution: key principles, and lessons from rapid repeated adaptation of killifish (*Fundulus heteroclitus*) populations. *Evolutionary Applications*, 10, 762–83.

Widmaier, E.P. (1992). Metabolic feedback in mammalian endocrine systems. *Hormones and Metabolic Research*, 24, 147–53.

Zhang, S., Chen, X., Zhang, J., and Li, H. (2014). Differences in the reproductive hormone rhythm of tree sparrows (*Passer montanus*) from urban and rural sites in Beijing: the effect of anthropogenic light sources. *General and Comparative Endocrinology*, 206, 24–9.

CHAPTER 14

Urban Sexual Selection

Tuul Sepp, Kevin J. McGraw, and Mathieu Giraudeau

Sepp, T., McGraw, K.J. and Giraudeau, M., *Urban Sexual Selection* In: *Urban Evolutionary Biology*. Edited by Marta Szulkin, Jason Munshi-South and Anne Charmantier, Oxford University Press (2020). © Oxford University Press.
DOI: 10.1093/oso/9780198836841.003.0014

14.1 Introduction

Urbanization is an ongoing and worldwide phenomenon that strongly affects the environmental conditions encountered by wild animals. The rapid expansion of urbanized landscapes disrupts natural environments and increases the exposure of organisms to multiple stressors (e.g., pollutants, artificial light at night, noise, human presence) (Seto et al. 2012), but also can increase resource availability and provide safe havens for breeding and escaping predators (Fischer et al. 2012; Birnie-Gauvin et al. 2017). Whether positive or negative, these urban effects act as selection pressures that uniquely shape phenotypic and genetic diversity of many animal species (e.g., Bonier et al. 2007; Giraudeau et al. 2014; Meillere et al. 2015).

Knowledge on *natural selection* pressures experienced by urban animals and resulting evolutionary changes is scarce (Johnson and Munshi-South 2017), and even less is known about the directions and degree to which cities impact sexual selection in animals (but see Yeh 2007; Tüzün et al. 2017; Halfwerk et al. 2018). Given the importance of sexual selection as a major force of evolution, the scarcity of studies on the effects of urban disturbances on sexual selection is surprising (Candolin et al. 2008). Studying

sexual selection in urban settings could lead to a better understanding of the role of this process in adaptation to a novel environment and in the early stages of speciation (Thompson et al. 2018).

Urbanization could influence sexual selection through both biotic and abiotic mechanisms (Figure 14.1). Abiotic changes include, among other things, noise, light, and chemical pollution, habitat structure and fragmentation, and weather. Biotic changes include modifications in food availability, population densities, predator/parasite/pathogen communities, and exposure to humans. All these factors can influence life-history strategies, mating behaviours, and resource competition. Resulting shifts in selection pressures can affect the development of sexual signals, competitive interactions, mate preferences, and ultimately pairing/mating success in urban habitats at the individual and population levels (Delhey and Peters 2017), with many complex interactions among the various different processes.

The study of urban sexual selection has focused so far on few model systems (mainly birds) and particular types of signals—largely acoustic and visual. Moreover, the attention has been mainly on signal senders and not receivers. While this is a rapidly growing field of study, we are still in relatively early

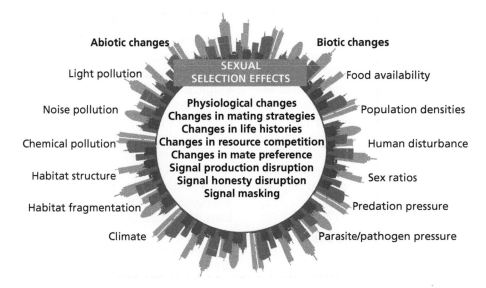

Figure 14.1 Sexual selection pressures can shift in urban environments based on changes in abiotic and biotic characteristics. Possible proximate and ultimate effects on sexual selection are depicted in the middle.

stages of understanding how the urban environment modifies sexual selection. Presently, we are in need of a systematic overview of what is known about the changes in sexual signals in response to urbanization in different animal groups (with differing life histories and trophic levels), and among populations exposed to different phases of the urbanization process. In addition, we need to understand (1) how sexually selected signals are received and processed, (2) possible genetic bases of these trait shifts, and (3) the impact of shifts in sexual signal expression on the pairing and mating success of individuals. Based on the available literature, it appears that a large variety of sexually selected traits, such as colouration (e.g., Yeh 2007; Giraudeau et al. 2018), song (e.g., Slabbekoorn and Peet 2003; Halfwerk et al. 2018), and timing of reproduction (e.g., Partecke et al. 2004), are strongly impacted by urban environmental characteristics, but the consequences of these modifications on sexual selection and fitness are seldom considered. In other words, our understanding of how mate choice and competition for mates are influenced by urban conditions is very limited.

In this chapter, we start by defining sexual selection, and provide perspectives on how this process is expected to change in urban environments compared to natural habitats. We continue with an explanation of how urbanization may alter sexual selection pressures. We examine the impact of urbanization on sexual signal senders and receivers, and the consequences of these responses for mating and reproductive behaviour. We then briefly review evidence for evolutionary change, and finish with describing the potential role of sexual selection for speciation in urban environments. Although evidence of urbanization shaping species through sexual selection is so far scarce, our current understanding of sexual selection theory and urban ecology suggests a number of potential impacts, opening up new research avenues.

14.2 Sexual selection and fitness in urban environments

In sexually reproducing organisms, competition among individuals over access to copulations and fertilizations is responsible for many of the more

extravagant traits found in nature and determining the relative fitness of an organism (Andersson 1994). We define relative fitness here as the number of descendants produced by an individual compared to the average produced by other individuals in the population (Hunt et al. 2004). As changes in the environment potentially affect who reproduces, and thereby influence sexual selection, changes in sexual selection pressures are expected to occur in response to urbanization. For example, it has been suggested that melanin colouration in birds might be under fluctuating sexual selection depending on parasites and other environmental conditions, which could result in spatiotemporal and frequency-dependent variability in mate choice (Côte et al. 2018). While this hypothesis is not supported by strong direct evidence so far, a study in lark buntings (*Calamospiza melanocorys*) showed that populations living in spatiotemporally variable environments can indeed substantially differ in their preferences for melanin-based colouration (Chaine and Lyon 2008). As an additional example of environmental effects on sexual selection pressures, the reduction of white in the tail and behavioural responses to territorial song observed in urban dark-eyed juncos (*Junco hyemalis*) have been related to reduced territorial competition resulting from milder climate, increased resource availability, and longer breeding seasons in urban environments (Newman et al. 2006; Yeh 2007).

While we now have decades worth of studies on urban-dependent changes in sexual signals, few of these allow us to assess whether these changes are adaptive, since a comprehensive understanding of the associated fitness consequences is lacking. The numerous studies published so far on urban effects on sexual signals can be used as an early, descriptive step in studying urban evolution through sexual selection. Most of these studies have focused on visual and auditory signals, and specific urban-associated factors that are suggested to affect these traits. The visual signals of an animal can be affected by a multitude of urban factors, including reduced availability of dietary pigments or other nutrients (Isaksson and Andersson 2007), increased pollutant levels (Chatelain et al. 2017), changes in parasite pressure (Jacquin et al. 2013), health status (Sumasgutner et al. 2018), stress level, and hormonal state

Table **14.1** Examples of recent studies of traits related to sexual selection in an urbanization context.

Higher taxon	Species	Urbanization context	Aspect of sexual selection	Main finding	Evidence for evolution?	Reference
Vertebrates: Aves	Song sparrow *Melospiza melodia*	Urban vs rural populations	Colouration, behaviour (territoriality), hormones	Urbanization alters the relationship between colouration and territorial behaviour	Not tested	Beck et al. 2018
Vertebrates: Aves	Oriental magpie-robin *Copsychus saularis*	Urban–rural variation	Acoustic signals (bird song)	Urban birds sing longer and slower songs to maximize transmission of important song information	Not tested	Hill et al. 2018
Vertebrates: Aves	Northern cardinal *Cardinalis cardinalis*	Urban–rural variation	Song, male body size, territoriality	The association between male quality and song characteristics that is present in rural areas is disrupted in urban areas	Not tested	Narango and Rodewald 2017
Vertebrates: Aves	Northern cardinal *Cardinalis cardinalis*	Urbanization gradient	Song	Higher song frequencies are related to higher levels of background noise, but temporal attributes of song (e.g., syllable rate, length) are best explained by conspecific densities, which are substantially greater in urban than rural landscapes	Not tested	Narango and Rodewald 2016
Vertebrates: Aves	White-crowned sparrow *Zonotrichia leucophrys nuttalli*	Noise pollution	Song	Males on louder territories produce songs at higher minimum frequencies but also with reduced bandwidth and lower vocal performance	Not tested	Luther et al. 2015
Vertebrates: Aves	House finch *Haemorhous mexicanus*	Urban vs rural populations	Colouration	Urban finches are less colourful (carotenoid-based colouration) than desert birds at capture, but the difference disappears in captivity	Probably not: differences disappear in common garden	Giraudeau et al. 2014
Vertebrates: Aves	House finch *Haemorhous mexicanus*	Urban vs rural populations	Colouration, behaviour	The association between colouration and aggressiveness varies between urban and rural populations. More colourful urban birds are less aggressive	Not tested	Hasegawa et al. 2014
Vertebrates: Aves	Dark-eyed junco *Junco hyemalis*	Urban vs rural populations	Hormones, behaviour, colouration	Urban birds have lower extra-pair paternity and reduced colouration (melanic), but with testosterone levels elevated for longer time	Maybe: colour differences persisted in common garden, while differences in testosterone levels did not	Atwell et al. 2014
Vertebrates: Aves	Great tit *Parus major*	Urban vs rural populations	Colouration	Forest great tits with wider black ties survive better, but the reverse was found for urban birds	Maybe: fitness effects tested	Senar et al. 2014
Vertebrates: Aves	Burrowing owl *Athene cunicularia*	Urban vs rural populations	Extra-pair matings, brood parasitism	Urban birds breed at higher densities, but no urban–rural differences in the frequency at which extra-pair matings or brood parasitism occurs	Not tested	Rodriguez-Martinez et al. 2014
Vertebrates: Aves	Domestic canary *Serinus canaria*	Noise pollution	Song	Sexual preference for low- over high-frequency songs fades because of urban noise	No: same-source individuals	Des Aunay et al. 2014

(continued)

Table 14.1 Continued

Higher taxon	Species	Urbanization context	Aspect of sexual selection	Main finding	Evidence for evolution?	Reference
Vertebrates: Aves	European robin *Erithacus rubecula*	Noise pollution	Song	Males on noisy territories sing less complex songs, at a higher minimum frequency, with a narrower song frequency bandwidth, and shorter songs at lower rates than males on quiet territories	Mostly not: noise-exposure experiment revealed plasticity in most song attributes	Montague et al. 2013
Vertebrates: Amphibia	Túngara frog *Engystomops pustulosus*	Urban vs rural populations	Calls	Urban frogs have increased call conspicuousness	Maybe: translocation experiment	Halfwerk et al. 2018
Vertebrates: Reptilia	Common wall lizard *Podarcis muralis*	Urban vs rural populations	Fluctuating asymmetry	Higher fluctuating asymmetry in a sexually selected trait (femoral pores) in urban lizards	Not tested	Lazić et al. 2013
Invertebrates: Odonata	Damselfly *Coenagrion puella*	Urban vs rural populations	Mating success (matings during lifespan)	Urban males have lower lifetime mating success	Maybe: common garden	Tüzün and Stoks 2017
Invertebrates: Hymenoptera	Acorn ant *Temnothorax curvispinosus*	Adaptation to urban heat islands	Offspring production	Urban ants have more offspring at high temperatures	Maybe: common garden, F1 offspring tested	Diamond et al. 2018
Invertebrates: Odonata	Damselfly *Coenagrion puella*	Urban vs rural populations	Mate preference for flight endurance	Sexual selection for higher flight endurance is detected only in urban populations	Maybe: fitness effects measured	Tüzün et al. 2017
Invertebrates: Orthoptera	Cricket *Teleogryllus commodus*	Light pollution	Mating behaviour, song	Probability of mating is higher under light pollution, but light disrupts precopulatory mating behaviour. No effect on song	No: same-source individuals	Botha et al. 2017
Invertebrates: Orthoptera	Grasshopper *Chorthippus biguttulus*	Noise pollution	Song	Males from roadside habitats produce songs with higher local frequency maximum under standardized, quiet recording conditions	Maybe: differences remained in standardized conditions	Lampe et al. 2012
Invertebrates: Lepidoptera	Moth *Mamestra brassicae*	Light pollution	Pheromones	Artificial night lighting reduces sex pheromone production and alters the chemical composition of the pheromone blend	No: same-source individuals	Van Geffen et al. 2015

(Beck et al. 2018). Similar to visual signals, several urban factors also affect the auditory signals of animals. Anthropogenic noise is probably the most relevant factor affecting auditory signals (Slabbekoorn and Peet 2003; Parris et al. 2009; Halfwerk et al. 2011; Costello and Symes 2014; Lampe et al. 2014), but additional factors like conspecific densities (Narango and Rodewald 2016) or morphological changes in urban animals (Badyayev et al. 2008; Giraudeau et al. 2014) have also been considered. While many studies have found differences in signal quality between animals from urban and rural populations (reviewed in Hutton and McGraw 2016; Sepp et al. 2018), most of the described links between sexual signals and urban environmental factors are correlational, and the possibilities for evolutionary change are very rarely tested (Table 14.1).

Adaptation to environmental change can occur via phenotypic plasticity or genetic change (Hoffmann and Sgro 2011), and most of the studies of urbanization and changes in sexual signals do not distinguish between plastic vs genetic responses. Phenotypic plasticity allows an animal to adjust its morphology or behaviour to the conditions of its immediate environment, thus potentially increasing its fitness (Thibert-Plante and Hendry 2011), and a lack of phenotypic plasticity has been postulated as contributing to the exclusion of some species from altered environments (Badyaev 2005; Lowry et al. 2012). Accordingly, we can expect urban species to exhibit considerable phenotypic plasticity also in sexually selected traits. At the same time, sexual signals not only are phenotypically plastic, but also can indicate the benefits that the signallers can offer in the form of 'good genes' (Andersson 1994). However, the relative fitness of different genotypes is strongly dependent on the environment; in other words, what genes can be considered 'good' depends upon local/variable conditions (Hunt et al. 2004). The extent to which the genetic benefits of mating with the most attractive males can be generalized over different environments (e.g., non-urban and urban environments) is not well understood (Hunt et al. 2004). Accordingly, true fitness might be decoupled from the quality and expression of signalling traits in urban habitats (Hutton and McGraw 2016), suggesting the possibility of an evolutionary trap arising from urbanization (Sih et al. 2011). If

signals do not reliably mirror the bearer's fitness in urban areas, then an increase in the frequency of mating events between choosers and low-quality mates is expected (Halfwerk et al. 2011), unless choosers are able to adjust and choose a partner based on more reliable information (Candolin et al. 2008). Since vulnerability to evolutionary traps is suggested to be one of the factors determining the ability of species to colonize cities (Sih et al. 2011), we can expect that, in urban-adapted species, the signal-response relationship for sexually selected traits is more flexible than in urban avoiders. However, this hypothesis needs further empirical support.

14.3 Changes in sexual selection pressures

The novel conditions typical of urban environments can constitute changes in selection pressures and thereby affect organism fitness (Johnson and Munshi-South 2017). The strength of sexual selection is affected not only by changes in reproductive potential and strategies, but also by changes in survival probability. For example, sexual selection for more elaborate displays is counterbalanced by natural selection pressures imposed by resource availability, predators, and parasites. Whether the urban environment increases or decreases survival seems to be highly species specific. For example, many bird species survive better in cities (reviewed by Sepp et al. 2018), as do urban-adapted small mammals (Łopucki et al. 2013), while the opposite is shown for many reptile species (reviewed by French et al. 2018). Urbanization might by analogy increase sexual selection pressures in some species, leading to more elaborate sexual signal traits and displays, but weaken sexual selection in other species, leading to less pronounced secondary sexual characteristics. For example, while the role of changed nutrient availability probably plays an important role in colour production, the generally drabber plumage of urban birds compared to rural birds (reviewed by Hutton and McGraw 2016; Sepp et al. 2018) may also be explained by relaxed sexual selection, if cities perturb resource or competitive balances to render traits either easier to produce or less valuable as a signal to a mate or rival.

For example, in northern cardinals (*Cardinalis cardinalis*), a disassociation between brightness of male plumage and territory attributes was recorded in urbanized areas (Rodewald et al. 2009). Additionally, in urban dark-eyed juncos, the size of a sexually signalling trait—amount of white in the tail—has declined by approximately 22 per cent compared to mountain-dwelling juncos, and this potential relaxation of selection has been attributed to decreased need for territorial behaviour in urban habitats (Yeh 2007).

Although a decreased need to advertise a fitness-related trait, such as the ability to hold territories, may weaken sexual selection, a decreased cost, such as weaker predation or decreased cost of pigment intake, can in turn strengthen selection on a sexual signal trait. As it becomes less costly to display extravagant sexual signals under changed environmental conditions, we can expect that, in order to maintain the honesty of a sexual signal, it needs to become even more conspicuous (Zahavi 1975). A change in the preferences of the signal receiver could lead to a similar outcome. As has been repeatedly demonstrated (i.e., Breden et al. 1987; Forsgren 1992; Johnson and Basolo 2003), the presence of predators can make females less choosy or even reverse their preferences for normally attractive mates. Accordingly, lower predation risk (which has been often demonstrated in urban habitats (Sepp et al. 2018)) should increase the choosiness of females, and thereby also increase selection pressure on sexual signals. This possibility was recently described in an urban context in a study of túngara frogs (*Engystomops pustulosus*), which indicated that males have increased the conspicuousness of their calls in urban environments (Halfwerk et al. 2018). This result was associated with lower rates of predation and parasitism in urban sites compared with forest sites (Halfwerk et al. 2018). A previous study in the same species indicated that female choosiness was indeed affected by perceived predation risk, with weaker expression of preferences under high rates of predation (Bonachea and Ryan 2011).

The associations between survival and sexual selection can be complex. Milder conditions in urban environments (reduced predation and parasite pressure and increased resource availability) can lead to a decrease in variation in survival probability. The amount of variance in relative fitness in a population is termed the 'opportunity for selection', and small variation means little opportunity for selection to act (Crow 1958). For example, a recent study in white-crowned sparrows (*Zonotrichia leucophrys*) indicated that, while birds had higher survival in urban areas, their average body condition was lower (Phillips et al. 2018). Whether this survival of birds with lower body condition is also related to relaxed selection on condition-dependent sexual signals is not known. However, it is also possible that changes in survival probability increase the strength of sexual selection, when high mortality before reaching reproductive age 'hides' traits related to reproduction from selection. Reduced adult mortality has been suggested to facilitate sexual selection by favouring the evolution of sexual signals, especially if these signals are age-dependent (Adamson 2013). While we are still largely missing studies comparing the age structure of urban and rural animal populations, a study in European blackbirds (*Turdus merula*) indicated a lower proportion of first-year birds in urban populations (Evans et al. 2009). Future studies could relate these differences in age structures to variation in the expression of age-dependent sexual signals.

Additional links between survival probability and sexual selection could arise from tradeoffs between investment in traits related to survival or sexual signals. These tradeoffs can be described through the concept of pace-of-life syndrome (Reale 2010; Tarka et al. 2018). This concept describes variation in life-history traits along a slow-to-fast pace-of-life continuum, with long lifespans, low reproductive and metabolic rates, and elevated somatic defences at the slow end of the continuum and the opposite traits at the fast end (Gaillard 1989). Pace-of-life can vary in relation to local environmental conditions (e.g., latitude, altitude), and it has been recently proposed that this variation may also occur along an urbanization gradient (Sepp et al. 2018). Accordingly, by comparing intraspecific variation in pace-of-life related traits, the authors concluded that urban animals may exhibit a slower pace-of-life compared to those from rural areas and thus invest more in self-maintenance and less in annual reproduction (Sepp et al. 2018). For species that demonstrate higher survival in urban habitats, urban colonization can therefore lead to the evolution of

Box 14.1 Stages of urban colonization and strength of sexual selection

Existing studies on urban changes in sexual selection pressures show contradictory results. While some studies indicate that animals in cities have lower reproductive output, decreased investment in sexual signals, and relaxed mate choice (Sepp et al. 2018), others indicate higher investment in sexual signalling and increased reproductive effort among urban organisms (Minias 2016; Halfwerk et al. 2018). As stronger sexual selection could be a mechanism that facilitates adaptation to novel environmental conditions (Candolin and Heuschele 2008; Martinez-Ruiz and Knell 2017), it is possible that sexual selection pressures are dependent on urban colonization stage (Figure 14.2). Accordingly, we hypothesize that, in early stages of urban invasion, populations experiencing strong sexual selection are favoured, and high investment in sexual signal traits is accentuated by higher fitness of genotypes with a fast pace-of-life. Once adaptation has occurred, (sexually selected) alleles that facilitate urban adaptation are widely spread in a population, survival probability increases, and a slower pace-of life may be favoured. Under these conditions, sexual selection pressures could either decrease (when the opportunity for selection decreases) or increase (in the case of age-dependent sexual signals). Alternatively, low survival in the early stages of urban invasion could reduce sexual selection pressures due to high random (offspring) mortality (Pischedda et al. 2015), or favoured investment in survival-enhancing traits over sexual signals. Which trends are more likely to prevail in specific species might depend on the trophic level of the species, since predation risk is a major ecological factor shaping the evolution of reproductive tactics in prey species (Lima and Dill 1990). For prey species with fewer natural predators in urban environments, we might expect increased choosiness of females and stronger sexual selection pressures in urban-adapted populations (i.e., Halfwerk et al. 2018). For predator species, the opposite might hold true, and due to relaxed mate competitions, individuals may be less choosy. For example, a study in crested goshawks (*Accipiter trivirgatus*) indicated that mixed-aged pairs were more common in urban than rural areas, a result the authors attributed to the fact that urban areas provide more food and nesting sites compared to rural areas, creating more opportunities for immature birds (Lin et al. 2015).

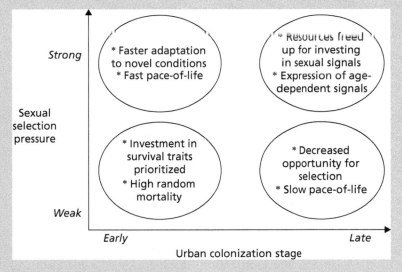

Figure 14.2 Hypothetical links between the recency of urban colonization and strength of sexual selection.

slow life histories, with the downstream effect of evolutionary pressures towards investing in fewer offspring per year, but maximizing lifetime fitness given the greater survival rates. A related effect would be lower investment in sexual signals (reviewed in Sepp et al. 2018).

Studies have shown that strong sexual selection can lead to faster adaptation to environmental change (e.g., Parrett and Knell 2018), and rapidly drive adaptive sexual signal change when changes in sexual and natural selection pressures align. A good example of this is the study on túngara frogs described above (Halfwerk et al. 2018), where changes in both natural selection (lower predator and parasite pressures) and sexual selection (increased female choosiness) facilitated the development of more extravagant male sexual signals. Whether changes in sexual selection pressures increase or decrease the likelihood of adapting to the urban environment remains to be studied. Indeed, it has been shown that sexual selection can lead to both increases and decreases in population-level fitness measures, such as extinction probability and adaptation rate (Martinez-Ruiz and Knell 2017). According to the theoretical model developed by Martinez-Ruiz and Knell (2017), the direction of the link between fitness and sexual selection depends on a variety of environmental and demographic factors, including the nature of environmental change, the carrying capacity of the environment, the average fecundity of the population in question, and the strength of condition dependence. The model states that, when the environmental change is directional and similar conditions are maintained for long periods of time, as is often the case with urbanization, populations in which sexual selection pressures are strong are less vulnerable to extinction than populations mating randomly.

Under these circumstances, sexual selection leads to less-adapted males acquiring relatively few mates, whereas those males who happen to be best adapted to the new environment will achieve much higher mating success than they would in a random mating system, allowing the populations in which sexual selection pressures are stronger to adapt faster. According to the good-genes model, mating success is positively correlated with genetic quality, and sexual selection therefore increases the proportion of alleles that are beneficial under the novel (urban) conditions (Candolin and Heuschele 2008). A good example of this process is the study on damselflies (*Coenagrion puella*), where natural selection for better locomotor performance and flight endurance necessary in isolated and fragmented urban habitats in males was reinforced by sexual selection (measured as mating success (Tüzün et al. 2017)). It is therefore intriguing to hypothesize that sexual selection fosters rapid adaptation in early stages of urban colonization (Box 14.1). However, whether sexual selection pressures weaken or strengthen in a population or species that is already

well adapted to urban habitats probably depends on the specific characteristics of the population/ species, i.e., survival probability, life-history strategy, or mating strategy.

14.4 Responses of signal senders to urban changes

Changes in sexual signals are probably the most studied aspect of sexual selection in the context of urbanization (reviewed in Hutton and McGraw 2016). Urbanization has been shown to affect all types of sexually selected traits, including auditory, visual, behavioural, and olfactory, in a wide variety of animal species, although the main focus has been on birds and insects (Table 14.1). Expression of sexual signals can reflect the genetic quality of the signal sender, and offspring are expected to inherit both the genes underlying choice and the genes for quality (Lande 1981; Hamilton and Zuk 1982). However, as fitness is inevitably the product of how genotypes interact with the environment (Hunt et al. 2004), the definition of 'genetic quality' may differ between natural and urban habitats. Accordingly, the information content of sexual signals can be affected by urbanization (Senar et al. 2014; Narango and Rodewald 2017; Beck et al. 2018). For example, it has been suggested that the adaptive value of traits that are advertised by colouration in birds (i.e., competitiveness, foraging efficiency, predator avoidance) might differ between wild and urbanized habitats (Isaksson et al. 2018). As an example of the changed information content of sexual signals in urban settings, a recent study on song sparrows (*Melospiza melodia*) showed that the relationship between chest spotting, a sexually selected trait, and territorial behaviour differed between urban and natural habitats (Beck et al. 2018). Rural males with more extensive spotting were less territorially aggressive than males with less extensive spotting, but this relationship between colouration and aggressiveness was absent in urban males. In contrast, the darkness of chest spots was associated with aggressiveness in urban, but not rural, males.

Information content change is not the only factor affecting sexual signals in cities. The physiological mediator between the signal trait and condition might also be disrupted by habitat characteristics.

To illustrate, Corbel et al. (2016) demonstrated a disruption of the link between stress responses and melanin colouration in urban feral rock pigeons (*Columba livia*) compared to rural pigeons. In addition, the masking effects of urban environmental characteristics can induce a change in the signal, independent from the physiological mechanism and information content. An emblematic masking study, and one of the first published on changes in urban sexual signals, revealed that songbirds sing at higher frequencies in urban environments polluted with traffic noise, reducing the overlap in frequencies between their song and low-frequency anthropogenic noise (Slabbekoorn and Peet 2003). This result has been later replicated in a number of studies (see Table 14.1 for some examples). The understanding that both the adaptive value of the behaviour advertised with sexual signals and signal production mechanisms can be affected by habitat type is important to keep in mind when interpreting differences between urban and rural animals. Future studies that include measurements of true fitness for individuals with different signal expressions will be key for understanding the evolutionary context and optimal trait values for individuals living in urban environments.

14.5 Responses of signal receivers to urban environmental changes

Although most urban sexual selection studies have focused on the signal sender, potential changes in the responses of signal receivers (competitive rivals or prospective mates) are equally important to consider. Alterations to mate preferences have been investigated more frequently than those of resource competitors. Signal perception might be affected by urban conditions that impact signal transmission: for example, changes in light conditions affect visual detection/discrimination of colour, changes in background noise affect hearing of auditory signals, and air pollution affects sensing of chemical signals. In a study of canaries (*Serinus canaria domestica*), female preference for low-frequency songs faded away under urban-noise conditions (des Aunay et al. 2014). In another example of male–male interactions, male house sparrows (*Passer domesticus*)

were more sensitive and responded more aggressively to acoustic signals in noisier urban conditions (Phillips and Derryberry 2018). On the other hand, urban shifts in mate preference from the viewpoint of the signal receiver might be caused by differing fitness optima of behavioural and physiological traits in the city that are advertised with sexual signals. For instance, a study in humans showed that preferences for sexual facial dimorphism (more masculine faces), and perceptions that masculinity signals aggressiveness, are stronger in urban societies and in groups that have low disease, fertility, and homicide rates (Scott et al. 2014). The authors explain this finding based on the fact that urban individuals are exposed to large numbers of unfamiliar faces, providing novel opportunities and more motivation to discern subtle relationships between facial appearance and other traits (Scott et al. 2014). This study also fits the aforementioned theory of reduced adult mortality as a factor that facilitates sexual selection by favouring the evolution of sexual signals (Adamson 2013). Another study in damselflies (*Coenagrion puella*) indicated that the better flight performance of urban males is reinforced by sexual selection (Tüzün et al. 2017).

Taking into account the rapid shifts in trait optima that are possible under sexual selection, urban changes in expression of sexual signals should not be considered maladaptive before the (mal)adaptiveness is tested in a mate-preference (or intrasexual-competition) context. We can again expect that changes in mate preference are more pronounced in species and populations that have had more time to adapt to urban environments, although changes in preference at the population level may be considerably slowed by ongoing dispersal from non-urban to urban habitats. As reciprocal effects between signal trait and receiver preference can be expected (Houde 1994), changes in preference can themselves lead to changes in signal characteristics. For example, in the presence of high levels of background sound, female tree frogs (*Hyla ebraccata*) abandoned discrimination for low-frequency calls and reverted to the task of detecting signals with modal properties for the population (Wollerman and Wiley 2002). This process can trigger evolutionary change in sexual signals without the need for any primary changes to the signals produced.

14.6 Consequences for mating and reproductive strategies

Reproductive strategies are different approaches through which members of a population can achieve reproductive success. These include the number of mates, behavioural tactics used in acquiring these mates, and patterns of parental care (Emlen and Oring 1977). In a review of anthropogenic impacts on animal mating systems, Lane et al. (2011) suggested four possible pathways through which human activities may impact reproductive strategies.

First, urban *habitat fragmentation* leads to changes in temporal and spatial distribution of resources, but also to changes in population densities and dispersal. Mate encounter rates may influence mate choice, multiple mating, female resistance to male mating attempts, mate searching, mate guarding, parental care, and the probability of divorce (Kokko and Rankin 2006). Cities can be fragmented into small separated patches like parks or ponds or into large areas with impervious surfaces (e.g., car parks, roads) (Liu et al. 2016). Among other things, fragmentation can change the availability and distribution of receptive females, leading to changes in reproductive strategies. For example, comparison of two neighbouring brushtail possum (*Trichosurus cunninghami*) populations, one living in a forest and the other in a fragmented roadside habitat, indicated that a polygynous mating system prevailed in the second, depending mainly on resource distribution (Martin and Martin 2007). A study on spotted towhees (*Pipilo maculatus*) showed that nests located closer to human food subsidies contained more extra-pair offspring (Smith et al. 2016). Also, the high density in urban populations is suggested to increase extra-pair paternity in raccoons (*Procyon lotor*) (Hauver et al. 2010, but see Rodriguez-Martínez et al. 2014 for a study of burrowing owls (*Athene cunicularia*), where density did not affect the probability of either extra-pair matings or brood parasitism). In insects, high attraction rates to artificial light sources can also lead to the formation of fragmented, small, dense communities of individuals (Degen et al. 2016), which may in turn affect their reproductive strategies.

Second, *chemical pollution* can act as an endocrine disruptor or as interference in sexual communication. A review article by Lürling and Scheffer (2007) concluded that even low, non-toxic concentrations of chemicals, ranging from heavy metals and pesticides to seemingly harmless substances such as surfactants, can disrupt the transfer of chemical information and thereby induce maladaptive responses in both signal senders and receivers. To illustrate, non-lethal doses of pesticides disrupted the courtship behaviour of lynx spider (*Oxyopes salticus*) (Hanna and Hanna, 2014).

Third, *selective harvesting* (hunting) can remove variation in sexually selected traits (Coltman, 2008, Edeline et al. 2009). This is probably less relevant in an urban context, because the harvesting of animals based on their expression of sexually selected traits, such as horns or antlers, but also body size, is rarer in urban habitats compared to wild habitats. However, rapid growth of cities in the forested wilderness of Amazonia and Congo, with increasing urban demand for wildlife as food (Parry et al. 2014), indicates that this may become an issue in some parts of the urbanizing world.

Fourth, *climate change* affects the timing of reproduction in many animals. Urban heat islands are one of the most widely cited ecological effects of urbanization, since cities typically experience higher temperatures due to the elevated prevalence of heat-absorbing and -emitting surfaces and additional heat production, compared to rural areas (Kaiser et al. 2016). Higher temperatures, changes in lighting regimes, and seasonal food availability can affect the wintering or migratory behaviour of birds (Robb et al. 2008; Plummer et al. 2015), which could be one explanation for the common finding that urban animals start breeding earlier in the season (Chamberlain et al. 2009; Sepp et al. 2018). Length of the breeding season and availability of mates could in turn affect the decision of whether to produce one or several broods during the season (Breedevald et al. 2017). This is another aspect of reproduction that could be affected by recency of urban colonization, since the number of broods has been suggested to affect the success of organisms in novel environments when there is adaptive mismatch (Sol and Maspons 2016).

As with other traits linked to sexual selection, studies to date of urban impacts on reproductive success have mainly focused on birds. These studies have generally indicated that urban birds have lower productivity per nesting attempt (reviewed by Chamberlain et al. 2009; Sepp et al. 2018). The extent to which this reproductive pattern is due to unsuitable habitat or changed reproductive strategies in cities is still open for discussion. A common weakness of many bird studies of reproductive success and urbanization is that lifetime reproductive success is not measured. Therefore, the possibility that lower investment in current reproduction is a habitat-specific strategy for increasing future reproductive success cannot be excluded. It is interesting to note that an increasing number of studies report nonlinearity of wildlife responses across the urbanization gradient (reviewed for birds by Batáry et al. 2018), with highest abundance, diversity, and reproductive success in transition zones between rural and urban environments, possibly due to a combination of beneficial characteristics from both rural and urban habitats (e.g., mosaic environment with abundant greenery, supplementary feeding stations, low predation). For example, in a study on little brown bats (*Myotis lucifugus*), it was found that reproductive success was highest in the transition zones, compared to rural and urban habitats (Coleman and Barclay 2011), which the authors attributed to lower population densities of bats compared to urban habitats.

At present, it is mostly unclear whether urban-associated changes in mating systems and reproductive strategies are plastic or heritable responses to habitat characteristics, as the mating behaviours of individuals will depend on both current-day ecological conditions and the evolutionary history of the species (Bennett and Owens 2002). Mating-system plasticity is likely species-specific, depending on how variable in space and time the distribution of mates and resources has been historically (Hernaman and Munday 2007). Plasticity is less likely to evolve if the ecological conditions favouring one particular mating system tend to be relatively stable in space and time. Future studies should therefore aim to include more information about the evolutionary history of the species, and about environmental plasticity in mating strategies in natural populations, to understand the role of plasticity in the urban mating strategies of different species.

14.7 Evidence for evolutionary changes

To detect urban-associated evolutionary changes that are driven by sexual selection, a three-step approach can be used (Donihue et al. 2014). First,

there must be documented *variation* in sexually selected traits and reproductive success between urban and rural populations. Many past studies of sexual selection and urbanization stop at this first step, unfortunately, and do not demonstrate an evolutionary response *per se*. Second, there must be a *heritable basis* to the phenotypic change; this is currently the field with the most knowledge gaps. Third, *experimental manipulation* is needed to identify specific urban drivers of trait variation. Although still scarce, studies using experimental manipulation of either environmental conditions or sexual selection pressures are slowly starting to accumulate. There are also other approaches that can be used to study adaptive microevolution in sexually selected traits in urban settings, for example genomic approaches that can target candidate genes for the given trait (if known), or correlative approaches that compare the direction and strength of sexual selection across the urban gradient (Chapter 5).

Urban and rural populations of the same species commonly differ genetically (Gortat et al. 2013; Lourenço et al. 2017; Perrier et al. 2017; Mueller et al. 2018) and epigenetically (McNew et al. 2017). However, such differences are rarely linked to changes in sexual selection or signals, and are mainly proposed to be the result of habitat fragmentation and genetic drift. However, some of these differences could still be associated with changes in sexual selection. For example, the epigenetic differences between urban and rural Darwin's finches found by McNew et al. (2017) were associated with genes related to melanin (feather pigment) production. In addition, when analysing variation in bill shape and song characteristics in house finches (*Haemorhous mexicanus*), Badyaev et al. (2008) also found some (albeit weak) support for genetic divergence between urban and rural populations. However, this study, like most urban evolution studies to date, used microsatellite markers, which are suitable for basic estimates of genetic diversity, differentiation, and gene flow, but not for identifying genes and pathways under selection (Johnson and Munshi-South 2017). Future genetic studies should apply methods like whole genome sequencing or genome-wide SNPs, or specifically target genes linked to sexual signal expression, in order to shed light on the potential for sexual selection to drive urban-associated evolutionary change.

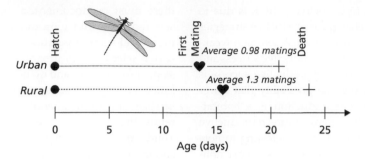

Figure 14.3 When urban and rural damselflies were reared in common garden conditions, urban males still had significantly lower mating success, possibly mediated by shorter lifespan, which indicates fitness costs of adjusting to an urban lifestyle. (Data source: Tüzün and Stoks 2017.)

Experimental studies of urban sexual selection are becoming more common. Experiments that demonstrate more permanent changes in sexual signalling strategies between urban and rural populations could be indicative of an early evolutionary response. The study on túngara frogs (section 14.3; Halfwerk et al. 2018) included a translocation experiment, which indicated that urban frogs can decrease the complexity of their calls to reduce risk of predation and parasitism when moved to the forest, but that forest frogs are not capable of this kind of plasticity, since they do not increase their sexual attractiveness when moved to the city. Another study, where urban and rural damselflies (*Coenagrion puella*) were reared in a 'common garden', found lower lifetime mating success for urban males (Figure 14.3; Tüzün and Stoks 2017). These findings suggest the possibility that environmental matching plays an important role in shaping the phenotypic traits or adaptations that ensure maximal mating success in urban habitats. This study also hints at an opposite possibility compared to the túngara frog study: once an organism is habituated to urban conditions, it might be difficult for these organisms to return to their 'natural' habitat and to be able to compete with local conspecifics for maximal fitness outcomes.

14.8 Potential role for speciation

While the study of urban speciation is in its infancy, city environments could provide ideal conditions for studying fundamental questions about speciation, since urban wild animal populations may be in the initial stages of divergence (Thompson et al. 2018). First examples of contemporary urban speciation have already been described. For example, a new species of leopard frog (*Anura: Ranidae*) was recently discovered from the New York City metropolitan area and surrounding coastal regions (Freinberg et al. 2014), and a new form of the mosquito *Culex pipiens*, the so-called *molestus* form, was found in the London Underground railway system (Byrne and Nichols 1999). Rates of recent phenotypic change, especially in sexually selected traits, seem to be greater in urban areas than in other habitat types (Alberti et al. 2017b), also suggesting that human-mediated speciation might be particularly potent in urban areas. As an example, advanced breeding onset is one of the most consistent findings in urban habitats (reviewed for birds in Chamberlain et al. 2008; Sepp et al. 2018), and this phenomenon might lead to assortative mating and facilitate divergence between urban and rural animals (Thompson et al. 2018).

Most urban-related changes in sexual signal expression (e.g., vocalizations, colouration) are probably plastic (Slabbekoorn 2013). However, plastic behavioural change can still contribute to speciation, since plastic responses in one trait can result in novel selection pressures on other traits (Price et al. 2003). For example, a plastic increase in song frequency in response to urban noise can lead to heritable changes in female preference (Montague et al. 2013). Since there are various ways in which urbanization can affect sexual selection, it is intriguing to hypothesize that these changes could lead to early stages of speciation. Future studies should aim to determine whether urban-induced divergence in sexual selection pressures and secondary sexual characteristics contributes to premating isolation between urban and non-urban populations.

14.9 Conclusions and future directions

We are in the midst of an exciting and blossoming era for urban ecological and evolutionary research. As cities and human populations expand globally, so too has the literature on many intricate aspects of phenotypic and genotypic change in organisms that avoid, tolerate, or thrive near human development. Alongside this growth have emerged many creative field (i.e., experimental) and technical (e.g., transcriptomic) approaches we can employ to investigate responses of urban populations (Alberti et al. 2017a; Watson et al. 2017; Isaksson et al. 2018). Though emphases were initially placed on naturally selected adaptations and plastic responses to human-induced environmental changes (Alberti et al. 2017b), in recent years increasing attention has been paid to the diverse sexually selected pressures and responses that may be unique in urbanized habitats (Parris 2016).

In this chapter, we have summarized the literature and theory to date on this relatively new body of work, which indicates that sexual selection is indeed altered in urban environments and that some of these patterns are highly consistent (i.e., changes in sexual signalling and reproductive strategies). However, few of the studies conducted so far allow us to assess whether these changes are adaptive. We have suggested here that the changes in the strength of sexual selection pressure in urban habitats may be species-specific, depending on the strength of survival selection and recency of urban colonization. Depending on the species, urbanization can affect sexual signals from both the sender and receiver perspective, shaping the information content of the signals, physiological mediators between signals and internal state, and female preference. Many of these changes described so far are probably plastic responses, but this does not mean that they are irrelevant in the context of urban evolution, since plastic responses in one trait can result in novel selection pressures on other traits.

We suggest a series of neglected areas and compelling new directions as we continue to explore the relative importance of sexual selection in shaping urban animal dynamics and biodiversity. From an evolutionary perspective, it is critical to identify the mechanism(s)—genetic adaptation or behavioural/ developmental plasticity—by which organisms respond to anthropogenic activities (Hendry et al. 2008). A recent large-scale taxonomic review (Alberti et al. 2017b) suggests that most evidence points to short-term plastic responses, but these studies too are dominated by traits subject to natural selection more than sexual selection. Given the relative recency of human urban development on an evolutionary timescale and often an overemphasis of research on longer-lived, larger creatures (i.e., with slower generational turnover), it makes sense that present evidence would more likely point to a preponderance of plastic responses. For only a few sexually selected traits has the idea of plasticity vs adaptation been considered, and it is high time that these evolutionary mechanisms be carefully studied, both from an adaptive genetic/locus perspective and causally using common garden experiments. Understanding these mechanisms will provide a clear predictive framework for the speed and direction in which organisms with different life histories may respond to anthropogenic changes. Moreover, given that sexual selection pressures can outpace those of natural selection for certain traits/taxa, one might predict that sexual selection pressures would be more likely/quickly to reveal adaptive signatures in urban populations (Candolin et al. 2008). In instances where plasticity is the expected or dominant mechanism for sexual-trait responsiveness to urbanization, it will be interesting to probe whether this is driven by flexibility in early life (e.g., ontogenetic conditions) and/or extended plasticity into adulthood, perhaps due to differing exposures to urban conditions at different life-stages.

As has been true through most of the history of sexual selection research (Andersson 1994), the majority of work on urban sexual selection has been centred on male traits and female responses, with less attention to intrasexual competition (Hasegawa et al. 2014; Beck et al. 2018) and even less on female traits, female–female competition, and male mate choice. The relative scarcity of data on male–male competition, for example, is somewhat surprising, given the vast literature on resource competition in general in urban animals (Anderies et al. 2007). Even in the detailed studies that have been conducted on urban sexual signals, age demographics are typically ignored, which can be quite problematic if the urban selective suite differentially culls (an) age-cohort(s) (i.e., comparatively more juveniles than adults in urban raccoons (Prange et al. 2003); a lower proportion of first-year birds in urban populations of European blackbirds (Evans et al. 2009)). Thus, apparent differences in ornament expression may not be due to urbanization *per se* but simply due to a comparison of ornaments from different age classes between urban and rural animals.

Urban environments vary along a number of different axes (e.g., space, time, size, types of habitat modification—see Chapter 2), and as such we should not expect equal effects of all city developments on even a single species. At present, the literature is dominated by single-city and time-point studies that often focus on just one anthropogenic variable, so we have little sense of the consistency of effects of any given urban variable (let alone any combination of urban variables), within or among cities, on a sexually selected trait (or suite of traits) in any species. Moreover, most urban sexual selection research has been dominated by studies in developed North American and European cities, for example, which themselves differ, on average, in age, but also illustrate how comparatively little we know about urban impacts in southern hemispheric cities (Sumasgutner et al. 2018), including in rich biodiverse tropical regions in which there is clear operational uniqueness in sexual selection systems (Macedo and Machado 2013). Finally, for most taxa investigated to date, we altogether lack experimental approaches to studying urban sexual selection. These include controlled mate-choice (Giraudeau et al. 2018) or competition (Hasegawa et al. 2014) tests, as well as manipulations of various urban-related stimuli (e.g., noise (Bermúdez-Cuamatzin et al. 2010)), which are critical for isolating the particular organismal or environmental traits that drive sexually selected responses in cities.

In summary, we use this opportunity as a call to diversify yet bolster our research coverage of different sexes, ages, taxa, cities, continents, and types of environmental modification in order to advance the relatively few published studies of urban sexual selection to date. Some model systems and taxa (namely, acoustic and colour signalling in birds) continue to serve as excellent paradigm- and prediction-forming research opportunities, but these by no means should exclusively inform how biodiverse sexual-selection schemes worldwide may be shifting or be unaffected by the increasing rates and types of urbanization expected in the coming decades. Urban sexual selection remains surprisingly unstudied in some clades, such as mammals. However, even human sexual selection systems are experiencing rapid changes (see Chapter 16) due to our often-hidden and elevated encounter rates with unique potential partners in the cities we create (McGraw 2002), and we too can learn about our own behaviours from those of the wildlife around us, in areas that we now massively mould and will continue to mould with and for all of our cohabitants.

Acknowledgements

We thank Daniel Sol, Wouter Halfwerk, and the editors of this book for excellent and constructive comments on the earlier versions of this chapter.

References

Adamson, J.J. (2013). Evolution of male life histories and age-dependent sexual signals under female choice. *PeerJ*, 1, e225.

Alberti, M., Marzluff, J., and Hunt, V.M. (2017a). Urban driven phenotypic changes: empirical observations and theoretical implications for eco-evolutionary feedback. *Proceedings of the Royal Society B: Biological Sciences*, 372, 20160029.

Alberti, M., Correa, C., Marzluff, J.M., et al. (2017b). Urban signatures of phenotypic change. *Proceedings of the National Academy of Sciences of the United States of America*, 114, 8951–6.

Anderies, J.M., Katti, M., andShochat, E. (2007). Living in the city: resource availability, predation, and bird population dynamics in urban areas. *Journal of Theoretical Biology*, 247, 36–49.

Andersson, M. (1994). *Sexual Selection*. Princeton University Press, New Jersey.

Atwell, J.W., Cardoso, G.C., Whittaker, D.J., Price, T.D., and Ketterson, E.D. (2014). Hormonal, behavioral, and life-history traits exhibit correlated shifts in relation to population establishment in a novel environment. *The American Naturalist*, 184, E147–60.

Badyaev, A.V. (2005). Stress-induced variation in evolution: from behavioural plasticity to genetic assimilation. *Proceedings of the Royal Society B: Biological Sciences*, 272, 877–86.

Badyaev, A.V., Young, R.L., Oh, K.P., and Addison, C. (2008). Evolution on a local scale: developmental, functional, and genetic bases of divergence in bill form and associated changes in song structure between adjacent habitats. *Evolution*, 62, 1951–64.

Batáry, P., Kurucz, K., Suarez-Rubio, M., and Chamberlain, D.E. (2018). Non-linearities in bird responses across urbanization gradients: a meta-analysis. *Global Change Biology*, 24, 1046–54.

Beck, M.L., Davies, S., and Sewall, K.B. (2018). Urbanization alters the relationship between coloration and territorial aggression, but not hormones, in song sparrows. *Animal Behaviour*, 142, 119–28.

Bennett, P.M. and Owens, I.P.F. (2002). *Evolutionary Ecology of Birds: Life History, Mating System and Extinction*. Oxford University Press, Oxford.

Bermúdez-Cuamatzin, E., Ríos-Chelén, A.A., Gil, D., and Garcia, C.M. (2010). Experimental evidence for real-time song frequency shift in response to urban noise in a passerine bird. *Biology Letters*, rsbl20100437.

Birnie-Gauvin, K., Peiman, K.S., Raubenheimer, D., and Cooke, S.J. (2017). Nutritional physiology and ecology of wildlife in a changing world. *Conservation Physiology*, 5, cox030.

Bonachea, L.A. and Ryan, M.J. (2011). Simulated predation risk influences female choice in túngara frogs, *Physalaemus pustulosus*. *Ethology*, 117, 400–407.

Bonier, F., Martin, P.R., Sheldon, K.S., et al. (2007). Sex-specific consequences of life in the city. *Behavioral Ecology*, 18, 121–9.

Botha, L.M., Jones, T.M., and Hopkins, G.R. (2017). Effects of lifetime exposure to artificial light at night on cricket (*Teleogryllus commodus*) courtship and mating behaviour. *Animal Behaviour*, 129, 181–8.

Breden, F. and Stoner, G. (1987). Male predation risk determines female preference in the Trinidad guppy. *Nature*, 329, 831–3.

Breedevald, M.C., San-Jose, L.M., Romero-Diaz, C., Roldan, E.R.S., and Fitze, P.S. (2017). Mate availability affects the trade-off between producing one or multiple annual clutches. *Animal Behaviour*, 123, 43–51.

Byrne, K. and Nichols, R.A. (1999). *Culex pipiens* in London Underground tunnels: differentiation between surface and subterranean populations. *Heredity*, 82, 7–15.

Candolin, U. and Heuschele, J. (2008) Is sexual selection beneficial during adaptation to environmental change? *Trends in Ecology and Evolution*, 23, 446–52.

Chaine, A.S. and, Lyon, B.E. (2008). Adaptive plasticity in female mate choice dampens sexual selection on male ornaments in the lark bunting. *Science*, 319, 459–62.

Chamberlain, D.E., Cannon, A.R., Toms, M.P., et al. (2009). Avian productivity in urban landscapes: a review and meta-analysis. *Ibis*, 151, 1–18.

Chatelain, M., Pessato, A., Frantz, A., Gasparini, J., and Leclaire, S. (2017). Do trace metals influence visual signals? Effects of trace metals on iridescent and melanic feather colouration in the feral pigeon. *Oikos*, 126, 1542–53.

Coleman, J.L. and Barclay, R.M.R. (2011). Influence of urbanization on demography of little brown bats (*Myotis lucifugus*) in the prairies of North America. *PLOS ONE*, 6(5), e20483.

Coltman, D.W. (2008). Evolutionary rebound from selective harvesting. *Trends in Ecology & Evolution*, 23, 117–18.

Corbel, H., Legros, A., Haussy, C., et al. (2016). Stress response varies with plumage colour and local habitat in feral pigeons. *Journal of Ornithology*, 157, 825–37.

Costello, R.A. and Symes, L.B. (2014). Effects of anthropogenic noise on male signalling behaviour and female phonotaxis in *Oecanthus* tree crickets. *Animal Behaviour*, 95, 15–22.

Côte, J., Boniface, A., Blanchet, S., et al. (2018). Melanin-based coloration and host–parasite interactions under global change. *Proceedings of the Royal Society B: Biological Sciences*, 285(1879).

Crow, J.F. (1958). Some possibilities for measuring selection intensities in man. *Human Biology*, 30, 1–13.

Degen, T., Mitesser, O., Perkin, E.K., et al. (2016). Street lighting: sex-independent impacts on moth movement. *Journal of Animal Ecology*, 85, 1352–60.

Delhey, K. and Peters, A. (2017). Conservation implications of anthropogenic impacts on visual communication and camouflage. *Conservation Biology*, 31, 30–39.

des Aunay, G.H., Slabbekoorn, H., Nagle, N., Passas, F., and Draganoiu, T.I. (2014). Urban noise undermines female sexual preferences for low-frequency songs in domestic canaries. *Animal Behaviour*, 87, 67–75.

Diamond, S.E., Chick, L.D., Perez, A., Strickler, S.A., and Martin, R.A. (2018). The evolution of city life: evolution of thermal tolerance and its fitness consequences: parallel and non-parallel responses to urban heat islands across three cities. *Proceedings of the Royal Society B: Biological Sciences*, 285, 20180036.

Donihue, C.M. and Lambert, M.R. (2014). Adaptive evolution in urban ecosystems. *Ambio*, 44, 194–203.

Edeline, E., Le Rouzic, A., Winfield, I.J., et al. (2009). Harvest-induced disruptive selection increases variance in fitness-related traits. *Proceedings of the Royal Society B: Biological Sciences*, 276, 4163–71.

Emlen, S.T. and Oring, L.W. (1977). Ecology, sexual selection and the evolution of mating systems. *Science*, 197, 215e223.

Evans, K.L., Gaston, K.J., Sharp, S.P., et al. (2009). Effects of urbanisation on disease prevalence and age structure in blackbird *Turdus merula* populations. *Oikos*, 118, 774–82.

Feinberg, J.A., Newman, C.E., Watkins-Colwell, G.J., et al. (2014). Cryptic diversity in metropolis: confirmation of a new leopard frog species (Anura: Ranidae) from New York City and surrounding Atlantic coast regions. *PLOS ONE*, 9, e108213.

Fischer, J.D., Cleeton, S.H., Lyons, T.P., and Miller, J.R. (2012). Urbanization and the predation paradox: the role of trophic dynamics in structuring vertebrate communities. *BioScience*, 62, 809–18.

Forsgren, E. (1992). Predation risk affects mate choice in a gobiid fish. *The American Naturalist*, 140, 1041–9.

French, S.S., Webb, A.C., Hudson, S.B., and Virgin, E.E. (2018). Town and country reptiles: a review of reptilian responses to urbanization. *Integrative and Comparative Biology*, 58, 948–66.

Gaillard, J.M., Pontier, D., Allaine, D., et al. (1989). An analysis of demographic tactics in birds and mammals. *Oikos*, 56, 59–76.

Giraudeau, M., Nolan, P.M., Black, C.E., et al. (2014). Song characteristics track bill morphology along a gradient of urbanization in house finches (*Haemorhous mexicanus*). *Frontiers in Zoology*, 11, 83.

Giraudeau, M., Toomey, M.B., Hutton, P., and McGraw, K.J. (2018). Expression of and choice for condition-dependent carotenoid-based color in an urbanizing context. *Behavioral Ecology*, ary093.

Gortat, T., Rutkowski, R., Gryczynska-Siemiatkowska, A., Kozakiewicz, A., and Kozakiewicz, M. (2013). Genetic structure in urban and rural populations of *Apodemus agrarius* in Poland. *Mammalian Biology*, 78, 171–7.

Halfwerk, W., Bot, S., Buikx, J., et al. (2011). Low-frequency songs lose their potency in noisy urban conditions. *Proceedings of the National Academy of Sciences of the United States of America*, 108, 14549–54.

Halfwerk, W., Blaas, M., Kramer, L., et al. (2018). Adaptive changes in sexual signalling in response to urbanization. *Nature Ecology & Evolution*, 3(3), 374–380.

Hamilton, W.D. and Zuk, M. 1982). Heritable true fitness and bright birds: a role for parasites? *Science*, 218, 384–7.

Hanna, C. and Hanna, C. (2014). Sublethal pesticide exposure disrupts courtship in the striped lynx spider, *Oxyopes salticus* (Araneae: Oxyopidae). *Journal of Applied Entomology*, 138, 1–2.

Hasegawa, M., Ligon, R.A., Giraudeau, M., Watanabe, M., and McGraw, K.J. (2014). Urban and colorful male house finches are less aggressive. *Behavioral Ecology*, 25, 641–9.

Hauver, S.A., Gehrt, S.D., Prange, S., and Dubach, J. (2010). Behavioral and genetic aspects of the raccoon mating system. *Journal of Mammalogy*, 91, 749–57.

Hendry, A.P., Farrugia, T.J., and Kinnison, M.T. (2008). Human influences on rates of phenotypic change in wild animal populations. *Molecular Ecology*, 17, 20–29.

Hernaman, V. and Munday, P.L. (2007). Evolution of mating systems in coral reef gobies and constraints on mating system plasticity. *Coral Reefs*, 26, 585.

Hill, S.D., Aryal, A., Pawley, M.D.M., and Ji, W.H. (2018). So much for the city: urban–rural song variation in a widespread Asiatic songbird. *Integrative Zoology*, 13, 194–205.

Hoffmann, A.A. and Sgro, C.M. (2011) Climate change and evolutionary adaptation. *Nature*, 470, 479–85.

Houde, A.E. (1994). Effect of artificial selection on male colour patterns on mating preference of female guppies. *Proceedings of the Royal Society B: Biological Sciences*, 256(1346), 125–30.

Hunt, J., Bussiere, L.F., Jennions, M.D., and Brooks, R. (2004). What is genetic quality? *Trends in Ecology & Evolution*, 19, 329–33.

Hutton, P. and McGraw, K.J. (2016). Urban impacts on oxidative balance and animal signals. *Frontiers in Ecology and Evolution*, 19 May.

Isaksson, C. and Andersson, S. (2007). Carotenoid diet and nestling provisioning in urban and rural great tits *Parus major*. *Journal of Avian Biology*, 38, 564–72.

Isaksson, C., Rodewald, A.D., and Gil, D. (eds) (2018). *Behavioural and Ecological Consequences of Urban Life in Birds*. Frontiers Media, Lausanne, Switzerland.

Jacquin, L., Récapet, C., Prévot-Julliard, A.C., et al. (2013). A potential role for parasites in the maintenance of color polymorphism in urban birds. *Oecologia*, 173, 1089–99.

Johnson, J.B. and Basolo, A.L. (2003). Predator exposure alters female mate choice in the green swordtail. *Behavioral Ecology*, 14, 619–25.

Johnson, M.T. and Munshi-South, J. (2017). Evolution of life in urban environments. *Science*, 358, eaam8327.

Kaiser, A., Merckx, T., and Van Dyck, H. (2016). The urban heat island and its spatial scale dependent impact on survival and development in butterflies of different thermal sensitivity. *Ecology and Evolution*, 6, 4129–40.

Kokko, H. and Rankin, D. J. (2006). Lonely hearts or sex in the city? Density-dependent effects in mating systems. *Philosophical Transactions of the Royal Society B*, 361, 319–34.

Lampe, U., Reinhold, K., and Schmoll, T. (2014). How grasshoppers respond to road noise: developmental plasticity and population differentiation in acoustic signalling. *Functional Ecology*, 28, 660–68.

Lande, R. (1981). Models of speciation by sexual selection on polygenic traits. *Proceedings of the National Academy of Sciences of the United States of America*, 78, 3721–5.

Lane, J.E., Forrest, M.N.K., and Willis, C.K.R. (2011). Anthropogenic influences on natural animal mating systems. *Animal Behaviour*, 81, 909–17.

Lazic, M.M., Kaliontzopoulou, A., Carretero, M.A., and Crnobrnja-Isailović, J. (2013). Lizards from urban areas are more asymmetric: using fluctuating asymmetry to evaluate environmental disturbance. *PLOS ONE*, 8(12), e84190.

Lima, S.L. and Dill, L.M., (1990). Behavioral decisions made under the risk of predation: a review and prospectus. *Canadian Journal of Zoology*, 68, 619–40.

Lin, W.-L., Lin, S.-M., Lin, J.-W., Wang, Y., and Tseng, H.-T. (2015). Breeding performance of crested goshawk *Accipiter trivirgatus* in urban and rural environments of Taiwan. *Bird Study*, 62, 177–84.

Liu, Z., He, C., and Wu, J. (2016). The relationship between habitat loss and fragmentation during urbanization: an empirical evaluation from 16 world cities. *PLOS ONE*, 11(4), e0154613.

Łopucki, R., Mróz, I., Berliński, L., and Burzych, M. (2013). Effects of urbanization on small-mammal communities and the population structure of synurbic species: an example of a medium-sized city. *Canadian Journal of Zoology*, 91, 554–61.

Lourenço, A., Álvarez, D., Wang, I.J., and Velo-Antón, G. (2017). Trapped within the city: integrating demography, time since isolation and population-specific traits to assess the genetic effects of urbanization. *Molecular Ecology*, 26, 1498–1514.

Lowry, H., Lill A., and Wong, B.B.M. (2012). Behavioural responses of wildlife to urban environments. *Biological Reviews*, 88, 537–49.

Lürling, M. and Scheffer, M. (2007). Info-disruption: pollution and the transfer of chemical information between organisms. *Trends in Ecology & Evolution*, 22, 374–9.

Luther, D.A., Phillips, J., and Derryberry, E.P. (2015). Not so sexy in the city: urban birds adjust songs to noise but compromise vocal performance. *Behavioral Ecology*, 27, 332–40.

Macedo, M. and Machado, G. (2013). *Sexual Selection. Perspectives and Models from the Neotropics*. Academic Press, Cambridge, MA.

Martin, J.K. and Martin, A.A. (2007). Resource distribution influences mating system in the bobuck (*Trichosurus cunninghami*: Marsupialia). *Oecologia*, 154, 227–36.

Martínez-Ruiz, C. and Knell, R.J. (2017). Sexual selection can both increase and decrease extinction probability: reconciling demographic and evolutionary factors. *Journal of Animal Ecology*, 86, 117–27.

McGraw, K.J. (2002). Environmental predictors of geographic variation of human mating preferences. *Ethology*, 108, 303–18.

McNew, S.M., Beck, D., Sadler-Riggleman, I., et al. (2017). Epigenetic variation between urban and rural populations of Darwin's finches. *BMC Evolutionary Biology*, 17, 183.

Meillere, A., Brischoux, F., Parenteau, C., and Angelier, F. (2015). Influence of urbanization on body size, condition, and physiology in an urban exploiter: a multi-component approach. *PLOS ONE*, 10, e0135685.

Minias, P. (2016). Reproduction and survival in the city: which fitness components drive urban colonization in a reed-nesting waterbird? *Current Zoology*, 62, 79–87.

Montague, M.J., Danek-Gontard, M., and Kunc, H.P. (2013). Phenotypic plasticity affects the response of a sexually selected trait to anthropogenic noise. *Behavioral Ecology*, 24, 342–8.

Mueller, J.C., Kuhl, H., Boerno, S., et al. (2018). Evolution of genomic variation in the burrowing owl in response to recent colonization of urban areas. *Proceedings of the Royal Society B: Biological Sciences*, 285(1878).

Narango, D.L. and Rodewald, A.D. (2016). Urban-associated drivers of song variation along a rural–urban gradient. *Behavioral Ecology*, 27, 608–16.

Narango, D. and Rodewald, A.D. (2017). Signal information of bird song changes in human-dominated landscapes. *Urban Ecosystems*, 21, 41–50.

Newman, M.M., Yeh, P.J., and Price, T.D. (2006). Reduced territorial responses in dark-eyed juncos following population establishment in a climatically mild environment. *Animal Behaviour*, 71, 893–9.

Parrett, J.M. and Knell, R.J. (2018). The effect of sexual selection on adaptation and extinction under increasing temperatures. *Proceedings of the Royal Society B: Biological Sciences*, 25, 20100000.

Parris, K.M. (2016). *Ecology of Urban Environments*. Wiley-Blackwell, Chichester.

Parris, K.M., Velik-Lord, M., and North, J.M.A. (2009). Frogs call at a higher pitch in traffic noise. *Ecology and Society*, 14, 25.

Parry, L., Barlow, L., and Pereira, H. (2014). Wildlife harvest and consumption in Amazonia's urbanized wilderness. *Conservation Letters*, 7, 565–74.

Partecke, J., Van't Hof, T., and Gwinner, E. (2004). Differences in the timing of reproduction between urban and forest European blackbirds (*Turdus merula*): result of phenotypic flexibility or genetic differences? *Proceedings of the Royal Society B: Biological Sciences*, 271, 1995–2001.

Perrier, C., del Campo, A.L., Szulkin, M., et al. (2017). Great tits and the city: distribution of genomic diversity and gene–environment associations along an urbanization gradient. *Evolutionary Applications*, 11, 593–613.

Phillips, J.N. and Derryberry, E.P. (2018). Urban sparrows respond to a sexually selected trait with increased aggression in noise. *Scientific Reports*, 8, 7505.

Phillips, J.N., Gentry, K.E., Luther, D.A., and Derryberry, E.P. (2018). Surviving in the city: higher apparent survival for urban birds but worse condition on noisy territories. *Ecosphere*, 9, e02440.

Pischedda, A., Friberg, U., Stewart, A.D., Miller, P.M., and Rice, W.R. (2015). Sexual selection has minimal impact on effective population sizes in species with high rates of random offspring mortality: an empirical demonstration using fitness distributions. *Evolution*, 69, 2638–47.

Plummer, K.E., Siriwardena, G.M., Conway, G.J., Risely, K., and Toms, M.P. (2015). Is supplementary feeding in gardens a driver of evolutionary change in a migratory bird species? *Global Change Biology*, 21, 4353–63.

Prange, S., Gehrt. S.D., and Wiggers, E.P. (2003). Demographic factors contributing to high raccoon densities in urban landscapes. *Journal of Wildlife Management*, 67, 324–33.

Price, T.D., Qvarnström, A., and Irwin, D.E. (2003). The role of phenotypic plasticity in driving genetic evolution. *Proceedings of the Royal Society B: Biological Sciences*, 270, 1433–40.

Reale, D., Garant, D., Humphries, M.M., et al. (2010). Personality and the emergence of the pace-of-life syndrome concept at the population level. *Philosophical Transactions of the Royal Society B*, 365, 4051–63.

Robb, G.N., McDonald, R.A., Chamberlain, D.E., and Bearhop, S. (2008). Food for thought: supplementary feeding as a driver of ecological change in avian populations. *Frontiers in Ecology and Environment*, 6, 476–84.

Rodriguez-Martinez, S., Carrete, M., Roques, S., Rebolo-Ifrán, N., and Tella, J.L. (2014). High urban breeding densities do not disrupt genetic monogamy in a bird species. *PLOS ONE*, 9, e91314.

Scott, I.M., Clark, A.P., Josephson, S.C., et al. (2014). Human preferences for sexually dimorphic faces may be evolutionarily novel. *Proceedings of the National Academy of Sciences of the United States of America*, 111, 14388–93.

Senar, J.C., Conroy, M.J., Quesada, J., and Mateos-Gonzalez, F. (2014). Selection based on the size of the black tie of the great tit may be reversed in urban habitats. *Ecology and Evolution*, 4, 2625–32.

Sepp, T., McGraw, K.J., Kaasik, A., and Giraudeau, M. (2018). A review of urban impacts on avian life-history evolution: does city living lead to slower pace of life? *Global Change Biology*, 24, 1452–69.

Seto, K.C., Guneralp, B., and Hutyra, L.R. (2012). Global forecasts of urban expansion to 2030 and direct impacts on biodiversity and carbon pools. *Proceedings of the National Academy of Sciences of the United States of America*, 109, 16083–8.

Sih, A., Ferrari, M.C., and Harris, D.J. (2011). Evolution and behavioural responses to human-induced rapid environmental change. *Evolutionary Applications*, 4, 367–87.

Slabbekoorn, H. (2013). Songs of the city: noise-dependent spectral plasticity in the acoustic phenotype of urban birds. *Animal Behaviour*, 85, 1089–99.

Slabbekoorn, H. and Peet, M. (2003). Ecology: birds sing at a higher pitch in urban noise. *Nature*, 424, 267.

Smith, S.B., McKay, J.E., Murphy, M.T., and Duffield, D.A. (2016). Spatial patterns of extra-pair paternity for spotted towhees *Pipilo maculatus* in urban parks. *Journal of Avian Biology*, 47, 815–23.

Sol, D. and Maspons, J. (2016). Life history, behaviour and invasion success. In: Weis, J.S. and Sol, D. (eds) *Biological Invasions and Animal Behaviour*, pp. 64–82. Cambridge University Press, Cambridge.

Sumasgutner, P., Rose, S., Koeslag, A., and Amar, A. (2018). Exploring the influence of urbanization on morph distribution and morph-specific breeding performance in a polymorphic African raptor. *Journal of Raptor Research*, 52, 19–30.

Tarka, M., Guenther, A., Niemelä, P.T., Nakagawa, S., and Noble, D.W.A. (2018). Sex differences in life history, behavior, and physiology along a slow-fast continuum: a meta-analysis. *Behavioural Ecology and Sociobiology*, 72, 132.

Thibert-Plante, X. and Hendry, A.P. (2011). The consequences of phenotypic plasticity for ecological speciation. *Journal of Evolutionary Biology*, 24, 326–42.

Thompson, K.A., Rieseberg, L.H., and Schluter, D. (2018). Speciation and the city. *Trends in Ecology & Evolution*, 33, 815–26.

Tüzün, N. and Stoks, R. (2017). Pathways to fitness: carry-over effects of late hatching and urbanisation on lifetime mating success. *Oikos*, 127, 949–59.

Tüzün, N., de Beeck, L.O., and Stoks, R. (2017). Sexual selection reinforces a higher flight endurance in urban damselflies. *Evolutionary Applications*, 10, 694–703.

Van Geffen, K.G., Groot, A.T., Grunsven, R.H.A., et al. (2015). Artificial night lighting disrupts sex pheromone in a noctuid moth: moth sex pheromone in illuminated nights. *Ecological Entomology*, 40, 401–408.

Watson, H., Videvall, E., Andersson, M.N., and Isaksson, C. (2017). Transcriptome analysis of a wild bird reveals physiological responses to the urban environment. *Scientific Reports*, 7, 44180.

Wollerman, L. and Wiley, H. (2002). Background noise from a natural chorus alters female discrimination of male calls in a Neotropical frog. *Animal Behaviour*, 63, 15–22.

Yeh, P.J. (2007). Rapid evolution of a sexually selected trait following population establishment in a novel habitat. *Evolution*, 58, 166–74.

Zahavi, A. (1975). Mate selection—a selection for a handicap. *Journal of Theoretical Biology*, 53, 205–14.

CHAPTER 15

Cognition and Adaptation to Urban Environments

Daniel Sol, Oriol Lapiedra, and Simon Ducatez

Sol, D., Lapiedra, O. and Ducatez, S., *Cognition and Adaptation to Urban Environments* In: *Urban Evolutionary Biology*. Edited by Marta Szulkin, Jason Munshi-South and Anne Charmantier, Oxford University Press (2020). © Oxford University Press.
DOI: 10.1093/oso/9780198836841.003.0015

15.1 Introduction

Urbanization is one of the most rapid and extreme forms of environmental alteration, causing a reduction and fragmentation of natural habitats, along with profound changes in resource availability and disturbance regimes (McKinney 2002; Shochat et al. 2006). These abrupt alterations can create mismatches between organisms' phenotypes and the environment, causing maladaptation and local extinction. Not surprisingly, intensively urbanized areas generally maintain lower species richness than the surrounding natural vegetation (Aronson et al. 2014; Sol et al. 2014). This loss of biodiversity has general consequences—given that biodiversity is rapidly diminishing at a global scale—and local consequences—because local biodiversity sustains fundamental ecosystem functions and provides key services to human societies. Since urbanization is predicted to continue expanding rapidly during the twenty-first century (Seto et al. 2012), there is an increasing need to understand whether and how organisms can adapt to urban environments (Sih 2013).

The realization that evolution can occur on contemporary timescales suggests that natural selection should be able to generate a new phenotype–environment match after the environment suddenly changes (Gonzalez et al. 2012; Bell 2017; Hendry et al. 2018). There is evidence that the rates of phenotypic change are greater in urbanized systems than in natural and non-urban anthropogenic systems (Alberti et al. 2017). However, demonstrations that these changes reflect rapid adaptive evolution are rare (Johnson and Munshi-South 2017). While the lack of evidence can in part reflect the paucity of studies, it may also indicate that urban contexts are particularly difficult for adaptation. Cities often confront organisms with novel challenges that are drastically different from those found in nature, including the need to exploit artificial foods or avoid risks associated with traffic and buildings (Sol et al. 2013). These challenges often act simultaneously and vary over time and space, making local adaptation difficult. Urban populations are also likely to show reduced genetic variation for selection to act upon (Evans et al. 2009; Johnson and Munshi-South 2017) as a result of bottlenecks—if individuals are relics

that pre-dated urbanization—or founder effects—if they established from a small number of colonizers after the city was formed. In the latter situation, adaptation may be further impeded by source-sink dynamics whenever individuals from nearby non-urban populations—presumably not well adapted to the urban environment—significantly contribute to growth and recruitment of urban populations.

Evolutionary adaptation to urban contexts may be particularly difficult for organisms with longer generation times. A long generation time reduces the accumulation of beneficial mutations and slows down changes in allele frequencies, delaying evolutionary responses (Rosenheim and Tabashnik 1991; Vander Wal et al. 2012). However, some long-lived animals, notably vertebrates, can attain high success in urban environments. Examples include monkeys, foxes, skunks, crows, parakeets, and gulls (Sol et al. 2014; Santini et al. 2019). The fact that some long-lived animals are able to thrive in urban areas is at first sight puzzling because their life history is based on ensuring a high survival rate every year, which is challenging when the animal is not fully adapted to the environment (but see Sol and Maspons 2016; Maspons et al. 2019). This is less puzzling, however, if we consider that many urban dwellers are ecological generalists (Evans et al. 2011; Sol et al. 2014; Ducatez et al. 2018). By definition, ecological generalists exhibit a more relaxed phenotype–environment match, which allows them to proliferate under a variety of environmental contexts. The urban environment can be particularly benign for generalist animals, offering a wide array of ecological opportunities such as artificial feeders, exotic fruits, and waste generated by human activities (Rodewald and Arcese 2017). However, the urban environment is so different from natural environments that it seems unlikely that even an ecological generalist is already well equipped for an urban life. Thus, the long-term persistence of a generalist population in an urban environment may still depend on improving the phenotype–environment match through natural selection.

In this chapter, we argue that to understand how animals evolve in cities we need to recognize the central role of cognition in shaping the way individuals interact with their environment (Figure 15.1). Cognition—the neural processes that regulate

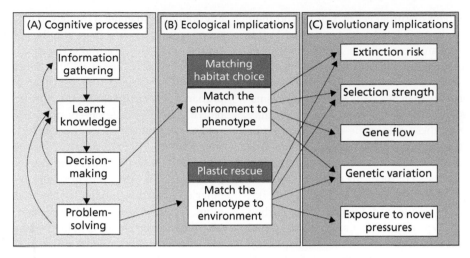

Figure 15.1 Schematic representation of how cognition can affect evolutionary processes by altering the relationship between individuals and their environment. Cognition allows animals to gather environmental information and retain it in the memory (A). The acquired knowledge is then used to make decisions (e.g., where and when to reproduce) and solve problems (e.g., how to exploit a new food). This generates new learned knowledge that further improves skills in acquiring new information and using it to make decisions and solve new problems. These processes reduce phenotype–environment mismatch either by facilitating matching habitat choice or by allowing the construction of new behavioural responses. Once the new phenotype–environment match is established, learning can contribute to maintain it through environmental and sexual imprinting. An evolutionary implication of these processes (C) is a reduction of the strength of selection. However, they can also facilitate evolutionary change by reducing the risk of extinction and by maintaining standing genetic variation for selection to act upon. Construction of new behavioural responses may in addition expose the population to new environmental pressures, driving evolution toward novel directions. Finally, matching habitat choice can reduce gene flow through environmental or sexual imprinting.

how animals gather, preserve, and use information (Shettleworth 2010)—allows individuals to improve decision-making and problem-solving when confronted with novel challenges. In doing so, it can affect eco-evolutionary processes through a variety of ways (Wyles et al. 1983; Huey et al. 2003; Price et al. 2003; Sol et al. 2005; Verzijden et al. 2012; Dukas 2013; Kawecki 2013), particularly in long-lived generalists (Sol et al. 2016; Maspons et al. 2019). The most obvious way is by weakening the strength of selection on heritable phenotypic variation, thereby inhibiting adaptive evolution. However, cognitive solutions may also facilitate evolutionary adaptation to the urban environment in the long term by reducing the risk of population extinction by maladaptation, by ensuring that individuals are more gradually exposed to the new conditions, and by allowing the population to move close to the realm of attraction of a new adaptive peak. Cognitive processes can also help preserve genetic variation on which natural selection can act and facilitate adaptive divergence by limiting the entrance of

genes from non-urban populations through environmental imprinting.

Although our concern in this chapter is primarily with the processes through which cognition influences evolutionary adaptation of animals to the urban environments, cognition itself is expected to evolve as a response to the selective pressures exerted by the urban environment. Thus, we also discuss our current knowledge on the evolution of cognition in urban environments. Admittedly, empirical evidence is still insufficient to fully understand the eco-evolutionary implications of cognition in urban environments, reflecting the fact that urban evolution is a young field. However, theoretical advances allow us to discuss the research needs that might improve our understanding of the process in the future.

15.2 Cognition and phenotype–environment mismatch

The challenges that face animals in urban environments include the need to exploit artificial foods, confront frequent disturbances by people and their pets, avoid risks associated with traffic and buildings, and communicate under noisy conditions (reviewed in Sol et al. 2013). These challenges are drastically different from those found in nature, often creating mismatches between the organisms' phenotypes and the environment. Cognition may compensate for these phenotype–environment mismatches via two main mechanisms (Figure 15.1; Maspons et al. 2019). The first is by helping individuals to choose those parts of the environment that better match their phenotype (matching habitat choice, *sensu* Nicolaus and Edelaar 2018). Urban environments typically exhibit substantial spatial and temporal heterogeneity (see also Chapter 2). This heterogeneity implies that the use of adequate stimuli (i.e., environmental and/or social cues) to decide where and when to live and reproduce can have important fitness benefits (Stamps and Swaisgood 2007; Sih 2013).

Cognition may also compensate for adaptive mismatches caused by urbanization through a 'plastic rescue' mediated by learning (Chevin et al. 2010; Snell-Rood et al. 2018). Learning is the neuronal representation of new information (Dukas 2012) and may enhance fitness in novel environments by allowing animals to create new associations between events. These associations can be between stimuli, between stimuli and behavioural responses, or between behavioural responses (Dickinson 2012). Exposure to certain stimuli can, for instance, increase subsequent preferences for habitats that contain those same stimuli, a process known as natal habitat preference induction (Stamps et al. 2010). In this way, learning can contribute to match the phenotype to the new environment by improving matching habitat choice. By creating new associations between stimuli and behaviours, learning can also increase fitness in urban environments by allowing individuals to use the environment in new ways. For example, a major hurdle for animals in intensively urbanized environments is the need to change their diet (reviewed in Sol et al. 2013). This requires the animal to recognize a new food item as edible, evaluate its profitability, and develop behavioural strategies to efficiently exploit it. Evidence shows that a higher tendency to invent feeding behaviours increases the likelihood of persisting in novel

environments in general (reviewed in Sol 2007) and urban environments in particular (Møller 2009; Carrete and Tella 2011; Maklakov et al. 2011 ; but see Kark et al. 2007; Evans et al. 2011; Sol et al. 2014).

Empirical evidence suggests that some species primarily rely on matching their phenotypes to the native habitats, while others primarily exploit the urban environment in novel ways. In *Anolis* lizards from Puerto Rico, for example, Winchell et al. (2018) found that *A. stratulus* tends to use more 'natural' portions of the urban environment (i.e., trees and other cultivated vegetation) whereas *A. cristatellus* more frequently uses anthropogenic structures. Matching habitat choice is in theory an efficient way to reduce phenotype–environment mismatches, but it is limited by the types of environments available and by the existence of reliable cues for habitat choice to avoid ecological traps (Edelaar et al. 2017; Maspons et al. 2019). Learning seems to be a better alternative in environments that are too novel to provide reliable cues—such as urban environments (Maspons et al. 2019). Thus, high learning abilities may be a feature of ecological generalists, who frequently expose themselves to novel challenges (Ducatez et al. 2015; Sol et al. 2016). Learning is not exempt of costs, however, such as the need to invest in expensive neural structures and to allocate time to gather information and devise behavioural solutions (Mery and Kawecki 2003; Dukas 2004). Yet, these costs are likely to be more assumable in animals with long generation times (Kokko and Sutherland 2001; Sol and Maspons 2016; Sol et al. 2016), because they are less constrained by time and are more likely to benefit from learned behaviours during their long lives. Thus, whether an animal relies more on learned or innate responses is expected to depend on its ecology and life history.

15.3 Is cognition facilitating or inhibiting adaptive evolution in urban environments?

The notion that cognition can protect individuals against maladaptation through learning, matching habitat choice, or a combination of both is often referred to as the cognitive buffer (Allman 1999; Sol 2009). The existence of a cognitive buffer may have important evolutionary consequences (Figure 15.1). The most obvious consequence is to relax or weaken selection on phenotypic traits (including cognition itself), thereby slowing evolutionary responses. If individuals tend to select the habitats and resources that better match their phenotypes, selection should become weaker because intraspecific variation in fitness is reduced. Likewise, if learning is enough to move the population close to a new adaptive peak, the strength of selection should also decrease because even individuals with inappropriate phenotypes should have chances to survive and reproduce (Huey et al. 2003; Price et al. 2003). This latter scenario is similar to the one considering that phenotypic plasticity can delay adaptive responses by shifting the distribution of phenotypes, shielding the population from natural selection (Lahti et al. 2009; Fox et al. 2019). While cognition may weaken the strength of selection in novel environments, empirical evidence suggests that it rarely inhibits completely the effect of selection. For instance, translocation experiments in *Anolis* lizards from the Caribbean show that individuals alter their habitat use in the presence of an introduced predator, but that these behavioural shifts do not prevent natural selection from changing morphology (Losos et al. 2004) and risk-taking behaviour (Lapiedra et al. 2018).

Despite potentially weakening the strength of selection, cognitive solutions may still facilitate evolutionary adaptation to the urban environment through a variety of mechanisms. The first of these mechanisms is by reducing the risk of population extinction due to maladaptation, a form of Baldwin effect (Baldwin 1896; Maynard-Smith 1987) facilitated by the cognitive buffer (Allman 1999; Sol 2009). Theoretical and empirical evidence provides ample support for the role of cognition in facilitating population persistence in novel environments (Sol et al. 2013; Maspons et al. 2019). Dark-eyed juncos (*Junco hyemalis*), for instance, established in an urban area in San Diego (California) over 40 years ago. Unlike their ancestors from the mountains, which mainly nest on the ground, the new population predominantly showed off-ground nesting. This behavioural shift substantially increased reproductive success, contributing to the persistence of the population in the new environment (Yeh et al. 2007). Persistence in a new environment is crucial for adaptive evolution

because it provides time for selection to act on heritable variation of any trait that improves fitness.

Second, although environmental changes associated with urbanization occur abruptly, habitat matching choice can ensure that individuals are more gradually exposed to the new conditions. A more gradual exposure to new selective pressures is important to facilitate adaptive evolution because abrupt changes can severely compromise population persistence (Gonzalez et al. 2012). One example is suggested in studies of urban pigeons (*Columba livia*), the free-living descendants of domesticated rock pigeons (Figure 15.2). Feral pigeons are an interesting case because artificial selection has altered hind-limb morphology for efficient locomotion (Johnston 1990; Sol 2008). The lack of an efficient locomotor system is not problematic in captivity, but it may reduce searching efficiency and survival in the wild (Sol 2008). Interestingly, pigeons with morphologies more suitable for locomotion tend to allocate more time to search for their own food, while those with morphologies closer to their domestic ancestor primarily rely on food provided by humans (Sol 2008). Thus, habitat matching choice

Figure 15.2 Feral pigeons—the free-living descendants from domesticated, artificially selected rock pigeons (*Columba livia*)—are one of the most successful urban dwellers and a model study in cognitive sciences (Bouchard et al. 2007). Their evolutionary dynamics have also been investigated. Contemporary feral pigeons are significantly closer in morphology and plumage colour to wild rock pigeons than to their more direct domestic ancestors (Johnston 1990), suggesting that natural selection has been reconstituting their wild phenotype (Sol 2008; Rutz 2012; Santos et al. 2015). (Photography credits: Oriol Lapiedra/Cesar González-Lagos.)

may have facilitated adaptive evolution by exposing the population to less abrupt changes that otherwise could have compromised the population's persistence (Gonzalez et al. 2012).

Third, new learned behaviours can expose the population to new adaptive peaks (Price et al. 2003), driving evolution toward new directions. This process is an expression of the Baldwin effect (Baldwin 1896; Maynard-Smith 1987), which is usually referred to as behavioural drive (Wyles et al. 1983; Huey et al. 2003). In urban house finches (*Carpodacus mexicanus*), for instance, the decision to adopt human-derived food resources has placed new selective pressures on the shape of their beaks (Badyaev et al. 2008). Selection has favoured longer and wider beaks, which increase bite force to break the larger and harder sunflower seeds upon which they depend in the city. More generally, the propensity to solve problems by inventing new behaviours has been suggested as one of the main factors promoting evolutionary divergence in vertebrates. Comparative studies in birds have, for instance, revealed that lineages that have diversified more taxonomically and morphologically are those that have larger brains (a surrogate of learning ability) and a higher propensity to solve problems by inventing new behaviours (e.g., Sol et al. 2005; Sol and Price 2008). Thus, cognition has a great potential for facilitating adaptive divergence of urban populations.

Fourth, cognition may influence adaptive evolution by preventing genetic variation from being eroded by natural selection. This is a direct consequence of the cognitive buffer provided by learning and matching habitat choice, which reduces the strength of selection on heritable traits. In situations where gene flow between urban and non-urban populations is reduced, the preservation of genetic variation is essential to provide the raw material for selection to act upon in the future. Although empirical evidence is lacking, the process can be of particular relevance for urban evolution, considering the high probability that populations have experienced important losses of genetic variation through bottlenecks or founder effects (Johnson and Munshi-South 2017).

Finally, cognition can limit gene flow through environmental and sexual imprinting (ten Cate and

Vox 1999; Irwin and Price 1999). Animals commonly learn at a young age which environmental features are important in surviving or mating, a process known as imprinting (Greggor et al. 2014). It follows that if individuals born in the city have a learned preference for settling in urban areas, this should favour assortative mating between urban individuals (Evans et al. 2010). Assortative mating can promote evolutionary divergence by reducing gene flow with nearby non-urban populations (Irwin and Price 1999; Beltman et al. 2004). Other mechanisms that might reduce gene flow with non-urban populations by favouring assortative mating include cultural selection for acoustic signals less masked by urban noise (Moseley et al. 2018) and changes in the breeding schedule (Mueller et al. 2013).

15.4 Evolution of cognition in urban environments

If learning and matching habitat choice help individuals to cope with the challenges of urban environments, the cognitive traits underlying these processes are expected to be selected. These cognitive traits can also potentially evolve, if heritable, under the influence of the mechanisms discussed in section 15.3. Unfortunately, the lack of urban-focused studies assessing the fitness consequences of heritable variation in cognition limits our actual understanding of the evolution of cognition in urban contexts.

Most previous studies on divergence in cognition driven by urbanization have involved a comparison between urban and either suburban or rural populations of the same species on tasks requiring habituation, discrimination, reversal learning, and problem-solving. The usual prediction is that because urban animals encounter a broader range and a higher frequency of novel challenges, they should have superior cognitive abilities (i.e., be better at habituation, reversal learning, and solving novel problems; Box 15.1).

Table 15.1 summarizes the main results of this literature (updated from Sol et al. 2013; Griffin et al. 2017). Strikingly, the majority of studies suggest higher cognitive performance in urban animals in cognitive processes related to the foraging domain. For example, urban individuals tend to be better at solving an extractive foraging problem (e.g.,

Box 15.1 Experimental approaches to study cognition

Cognitive ecologists have devised a number of experimental approaches to assess cognitive performance of individuals. Below we provide a few examples relevant to urban contexts. Other cognitive processes such as spatial learning, inhibition, and causal reasoning may affect animals' ability to respond to the challenges encountered in urban environments, though more studies are required to better comprehend their potential importance (see Table 15.1 for more examples and references).

Habituation: This involves the reduction of a preprogrammed response as a result of repeated exposure to a normally eliciting stimulus. Cities are characterized by the increased presence of stimuli that would be interpreted as a threat in a natural environment and elicit an escape behaviour (e.g., motor vehicles, people, and human-associated predators such as cats and dogs). Some of these threats occur at very high rates with little consequences, and animals might reduce their response and habituate to these stimuli, allowing them to focus on rewarding activities instead of consistently flying away from non-risky situations. For example, in Hungary, urban sparrows (*Passer domesticus*) habituated to human disturbance (by decreasing their reactivity to human intrusion after repeated exposure) faster than their rural counterparts (Vincze et al. 2016), suggesting higher habituation performances in urban populations.

Operant conditioning: This is a learning process in which the likelihood of a behaviour is strengthened (increases in frequency) or weakened (decreases in frequency) in response to consequences (reinforcement or punishment, respectively), occurring after the behaviour is exhibited (Papageorgi 2018). The urban environment offers both new opportunities and new risks, and operant conditioning is one way for animals to adapt their response to these new conditions. The adoption of new foraging techniques allowing animals to exploit resources that would otherwise remain inaccessible is, for example, likely to involve operant conditioning processes, whereby individuals learn by trial and error to access new resources (e.g., to open

take-away containers or rubbish bins, to pierce plastic bags). It has been found that animals from urban areas are often faster than their rural counterparts at solving problems requiring the removal of an obstacle or pulling of a stick to access a food reward (see Table 15.1), suggesting higher operant conditioning performances in urban populations (though other cognitive and non-cognitive processes are likely to affect problem-solving performance).

Associative learning: This is a learning process by which reliable and consistent associative relationships are formed between environmental conditions or between environmental and individual conditions (see Figure 15.3). Associating specific cues (a colour, smell, or noise, a human behaviour, etc.) with a danger or a reward could optimize responses to the urban environment by eliciting adaptive behaviours based on relatively simple associations. For example, being able to associate artificial street lights (environmental cue) with the presence of flying insects (food reward) will likely help some insectivorous species to thrive in urban areas. Likewise, recognizing human faces may allow animals to discriminate between humans that provide food from those that pose risks. Being able to unlearn associations that are not meaningful anymore or to switch to new ones can also be rewarding. For example, being able to ignore street lights during seasons where flying insects are absent around them will be crucial to optimize foraging. Performance at reversal learning, where a previously rewarded stimulus becomes unrewarded and a previously unrewarded one is now rewarded, is often measured in the literature to estimate this type of flexibility. Recently, Batabyal and Thaker (2019) trained delicate skinks (*Psammophilus dorsalis*, a tropical agamid lizard) to associate refuges in a Y-maze with their 'safe' or 'unsafe' characteristics. They showed that individuals from suburban areas were quicker to learn and unlearn the location of the safe refuge than rural individuals.

Figure 15.3 Associative learning is often measured using tests of colour discrimination in captivity, whereby an individual is trained to associate a rewarded colour (e.g., the unicolour circle on the picture) with food, and to avoid a non-rewarded colour (e.g., the bicolour circle). The picture shows associative learning experiments conducted by Leal and Powell (2012) on a tropical *Anolis*. (Photography credit: Manuel Leal.)

Table 15.1 Studies comparing cognition in urban (U) and non-urban (R) populations. NS, Non-significant difference between urban and non-urban populations; U > R, urban individuals significantly outperform non-urban ones.

Behaviour	Species	Trend	Reference
Innovation and problem-solving			
Consumer innovation	Common myna (*Acridotheres tristis*)	NS	Sol et al. (2011)
Problem solving	House finch (*Haemorhous mexicanus*)	NS	Cook et al. (2017)
Problem solving	House finch (*Haemorhous mexicanus*)	NS	Arnold (2013)
Problem solving	Mountain chickadee (*Poecile gambeli*)	U > R	Kozlovsky et al. (2017)
Problem solving (four tests)	House sparrow (*Passer domesticus*)	U > R in 1/4 tests	Papp et al. (2015)
Problem solving	House sparrow (*Passer domesticus*)	U > R	Liker and Bókony (2009)
Problem solving	Common myna (*Acridotheres tristis*)	U > R	Sol et al. (2011)
Problem solving (two tests)	Great tit (*Parus major*)	U > R	Preiszner et al. (2017)
Problem solving (two tests)	Barbados bullfinch (*Loxigilla barbadensis*)	U > R	Audet et al. (2016)
Habituation and learning			
Habituation to human disturbance	House sparrow (*Passer domesticus*)	U > R	Vincze et al. (2016)
Learning	Delicate skink (*Lampropholis delicata*)	NS	Kang et al. (2018)
Learning	Barbados bullfinch (*Loxigilla barbadensis*)	NS	Audet et al. (2016)
Learning and reversal learning speed	Common myna (*Acridotheres tristis*)	U > R	Federspiel et al. (2017)
Learning to use a problem-solving task	Zenaida dove (*Zenaida aurita*)	U > R	Boogert et al. (2010)
Learning/shaping at a problem-solving task	Zenaida dove (*Zenaida aurita*)	U > R	Seferta et al. (2001)
Long-term memory retention	Mountain chickadee (*Poecile gambeli*)	U > R	Kozlovsky et al. (2017)
Spatial memory acquisition	Mountain chickadee (*Poecile gambeli*)	NS	Kozlovsky et al. (2017)
Reversal learning	Barbados bullfinch (*Loxigilla barbadensis*)	NS	Audet et al. (2016)
Caching rate			
Food caching rate	Mountain chickadee (*Poecile gambeli*)	NS	Kozlovsky et al. (2017)
Sociality			
Social cohesion	Black-capped chickadees (*Poecile atricapillus*)	NS	Jones et al. (2019)
Anatomy			
Hippocampus volume	Mountain chickadee (*Poecile gambeli*)	NS	Kozlovsky et al. (2017)
Telencephalon volume	Mountain chickadee (*Poecile gambeli*)	U > R	Kozlovsky et al. (2017)
Total number of hippocampus neurons	Mountain chickadee (*Poecile gambeli*)	NS	Kozlovsky et al. (2017)

removing an obstacle hiding food) than rural or suburban individuals. However, most current evidence is biased toward birds. Moreover, most experiments from Table 15.1 do not isolate any specific cognitive process or examine potential neural mechanisms. One of the few exceptions is the study by Kozlovsky et al. (2017) in black-capped chickadees (*Poecile atricapillus*), which reported that urban individuals were better problem-solvers and had a larger telencephalon, but did not show a larger

hippocampus (a brain area specialized in spatial learning) nor a better spatial memory than forest birds. While much effort has been devoted to study cognitive performance in extractive foraging problems, other important cognitive processes remain little studied. This is the case of cognitive processes involved in social interactions. Despite being crucial in facilitating the spread of new behaviours throughout the population (Aplin et al. 2015), evidence for differences between urban and non-urban individuals in cognitive traits that structure social networks and facilitate the transmission of learned behaviours is notoriously absent (but see Jones et al. 2019).

The degree to which the observed differences in cognitive performance between urban and non-urban individuals result from rapid evolution driven by natural selection or other processes remains mostly unknown. Classic alternatives to selection include drift and a sorting process by which only individuals with certain cognitive skills are able to colonize urban environments. Other alternatives also potentially explaining differences in cognition between urban and non-urban individuals include motivation, personality, and previous experience. In Australia, for example, common mynas (*Sturnus tristis*) from highly urbanized areas tend to solve a foraging task (removing a lid from a well hiden food) faster than mynas from less urbanized areas (Sol et al. 2011). However, these differences do not reflect divergence in cognitive skills. Rather, the principal factors underlying differences in problem-solving ability of mynas are differences across habitats in neophobic–neophilic responses.

Demonstrating natural selection in cognition requires showing that fitness differences are associated with cognitive performance in the wild (Morand-Ferron et al. 2016). However, there is currently little evidence to establish a link between cognitive traits and their function in the wild (Cole et al. 2012; Cauchard et al. 2013; Sonnenberg et al. 2019). Moreover, while selection acts on phenotypic variation, adaptive evolutionary responses also require that this variation is heritable. Yet, the majority of studies investigating the heritability of cognitive traits or of the underlying neural structures have focused on humans and captive animals. Studies on wild animals are much scarcer and the results are highly variable (e.g., heritability of 0.04 for problem-solving performance in wild great tits (*Parus major*) (Quinn et al. 2016); heritability of 0.60–0.75 for the total cranial volume of wild Rhesus macaques (*Macaca mulatta*) (Cheverud et al. 1990)). As far as we know, no study has examined the existence of additive genetic variance and/or heritable variation in cognition in the urban context. More investigation is thus needed to better assess the

evolutionary potential of cognitive traits (Croston et al. 2015; Morand-Ferron et al. 2016).

15.5 Future studies to investigate the role of cognition in urban evolution

Unravelling the evolution of cognition and its influence on eco-evolutionary dynamics in urban habitats requires demonstrating that variation in cognition affects fitness in ways that alter the frequency of genetically based traits (including cognitive traits themselves). The basic tools to investigate adaptive evolution in cities have been nicely reviewed by Donihue and Lambert (2015). Here we will focus on three of the main challenges specific to the study of the eco-evolutionary consequences of cognition in urban contexts.

The first challenge is the availability of reliable, individual-level data on variation in cognition. Assessing intraspecific variation in cognition is not straightforward for several reasons (Morand-Ferron et al. 2016; Boogert et al. 2018). One reason is that cognition cannot be directly measured. Rather, it is commonly estimated through its effect on behaviour. Yet, behaviour can be affected by other factors, such as motivation and personality. Therefore, experiments are commonly necessary to characterize individual cognitive performance (see definitions in Box 15.1). To be informative, these experiments need to be ecologically relevant for wild ranging animals, and they need to be conducted in standardized conditions to minimize environmental effects. The experiments should also be replicated to confirm that interindividual variation in cognition is consistent (Lihoreau et al. 2018).

The second relevant challenge is to understand the consequences of cognition for fitness (Croston et al. 2015; Morand-Ferron et al. 2016). Ideally, the way to establish a link between cognition and fitness is by designing experiments in which environmental factors expected to select for cognition are manipulated. Conducting experimental translocations is potentially a powerful tool to assess the survival of different phenotypes in areas with different selective forces (e.g., Lapiedra et al. 2018). Brady (2012), for instance, used a reciprocal transplant experiment to show that populations of spotted salamanders (*Ambystoma maculatum*) adjacent to roads have become locally adapted. In many cases, however, experiments might not be feasible in urban habitats. In these cases, one alternative is to explore natural variation in cognition in replicated, well-established gradients of urbanization (Johnson and Munshi-South 2017). Another alternative is to use longitudinal analyses of populations directly confronted to urban development (or, alternatively, urban shrinkage) as an evolutionary quasi-experiment (Haase 2008; Ramalho and Hobbs 2012). However, even in these cases, studying the evolutionary consequences of cognition in urban environments can be challenging. For instance, aspects of field-work planning such as the establishment of permanent study sites or the need to leave field equipment in areas frequented by people might not be feasible in urban areas. Direct interaction with people can also be problematic (Dyson et al. 2019), hindering the ability to track individuals through time, a crucial aspect for studies of selection (Lapiedra 2018). To overcome these and other challenges, researchers will need to address logistic and ethical limitations associated with carrying out research in urban habitats (reviewed in Dyson et al. 2019). Despite these difficulties, studying natural selection in free-living urban animal populations is feasible (Yeh and Price 2004; Sol 2008; Charmantier et al. 2017) and it holds the largest potential to obtain biologically relevant information of the fitness consequences of variation in cognitive traits in urban areas.

The last challenge to understand the evolutionary implications of cognition is to evaluate how differences in fitness alter the frequency of genetically based cognitive and non-cognitive traits. Ideally, this requires long-term measures of natural selection on heritable traits that are ecologically relevant. In long-lived animals this type of study can take many decades. The alternative is to explore the nature of existing ecologically relevant phenotypic variation. For example, common garden and artificial selection experiments can be particularly useful to establish the genetic basis of differences in cognitive and non-cognitive traits between urban and non-urban populations. These studies can help in determining whether different geographic patterns are driven by evolutionary processes or, instead, whether they are the product of phenotypic plasticity (Chapter 5; Donihue and Lambert 2015). Combined

with evidence of fitness consequences, the studies can also suggest the action of natural selection (assuming that the patterns of selection we currently observe are the same that caused the observed phenotypic differences). Molecular analyses can provide additional insight to demonstrate the adaptive nature of phenotypic differences (see Chapter 5). Some studies have indeed provided evidence for changes in allelic frequencies of urban populations in genes presumed to affect cognition. In one of these studies, Mueller et al. (2013) found that urban blackbirds (*Turdus merula*) differ in allelic frequencies of the SERT gene, a gene related to harm-avoidance behaviour. In twelve of the fourteen cities studied, blackbirds had lower frequency of the most common SERT variant found in rural populations. In another study, urban black swans (*Cygnus atratus*) were reported to differ in the dopamine receptor DRD4. The authors suggested that this difference resulted in a decreased fear of humans (van Dongen et al. 2015).

While molecular approaches represent important avenues for future research, and are essential to investigate certain mechanisms (e.g., whether habitat imprinting reduces gene flow), a major limitation at present is our insufficient understanding of the genetic determinants of cognition (Audet et al. 2018). Snell-Rood and Wick (2013) used an alternative retrospective approach to investigate the evolution of cognition in urban environments. They measured museum specimens of ten mammal species to examine changes over time in relative cranial capacity—a surrogate for domain-general cognition—in urban and rural populations. For two species, the white-footed mouse (*Peromyscus leucopus*) and meadow vole (*Microtus pennsylvanicus*), urban populations had a larger cranial capacity than rural populations. However, no increase in cranial capacity was observed over time in urban populations.

15.6 Conclusions

According to Maynard-Smith (1989), avoiding extinction as a result of environmental change is a race between demography and adaptive evolution. In this chapter, we have developed the argument that animal cognition can represent a central nexus between demography and adaptive evolution in cities.

Cognition may favour adaptive divergence by facilitating population persistence, exposing individuals to novel selection pressures, maintaining genetic diversity in the population, and reducing gene flow through assortative mating. However, cognition can also inhibit adaptive evolution by reducing the strength of selection. Whether cognition facilitates or inhibits adaptation to urban environments is currently backed by insufficient evidence, and it will have to be elucidated by designing experimental and/or longitudinal studies (Donihue and Lambert 2015; Lapiedra 2018). The evolutionary consequences of cognition are likely to depend on the interplay between matching habitat choice and learning (see Maspons et al. 2019), and vary according to the life history and ecology of the species (Ducatez et al. 2015; Sol et al. 2016).

Cognition itself can be selected for and evolve if it shows enough heritable variation. Learning should be particularly selected in long-lived species, where the benefits of constantly updating information to make decisions outweigh the costs (Sol et al. 2016; Maspons et al. 2019). Moreover, as highlighted by Kokko and Sutherland (2001), learned preferences should be more useful than genetically determined preferences in long-lived species that respond more slowly to selection. Although current evidence suggests that cognition may be beneficial in the early stages of becoming urban dwellers, it is less clear whether it may still be useful in later stages (Snell-Rood and Wick 2013). As natural selection improves the phenotype–environment match, cognitive solutions can be disfavoured if they become too costly for the benefits they provide (Lahti et al. 2009). However, a main feature of the urban environment is its high spatial and environmental heterogeneity in the distribution of resources, enemies, and hazards. This heterogeneity may force urban animals to continuously update environmental information and adjust their behaviour conveniently. Thus, cognition may still be selected in long-lived species with opportunist–generalist lifestyles (Ducatez et al. 2015; Sol et al. 2016).

Given that by 2030 urban land cover is expected to nearly triple the global urban land area *circa* 2000 (Seto et al. 2012), understanding evolution in urban environments is becoming increasingly important (Johnson and Munshi-South 2017). Although we currently have more questions than answers, the interplay between cognitive responses and selection in reducing phenotype–environment mismatch and facilitating subsequent adaptive evolution is likely to be crucial to understand urban evolution in animals. The new tools of evolutionary ecology provide unprecedented opportunities to investigate how cognition affects eco-evolutionary dynamics in urban contexts. Because the urban environment exposes animals to a variety of pressures that require cognitive solutions, the study of urban evolution also offers excellent opportunities to investigate the functional significance and evolutionary origin of advanced cognitive processes such as general intelligence.

Acknowledgements

We thank the editors Anne Charmantier, Marta Szulkin, and Jason Munshi-South for the kind invitation to contribute to this volume, Louis Lefebvre, Alexis Chaine, Tuul Sepp, Mathieu Giraudeau, and the three editors for reviewing earlier versions of the chapter, and Manuel Leal for providing a picture of his experiments on *Anolis*. This work was supported by funds from the Spanish government (CGL2017-90033 P) to Daniel Sol. Oriol Lapiedra was supported by Agaur in the form of a Beatriu de Pinós postdoctoral fellowship 2016-BP00205. Simon Ducatez was supported by a McGill University–CREAF collaboration agreement partially funded by a Discovery grant from NSERC Canada to Louis Lefebvre.

References

Alberti, M., Marzluff, J., Hunt, and V.M. (2017). Urban driven phenotypic changes: empirical observations and theoretical implications for eco-evolutionary feedback. *Philosophical Transactions of the Royal Society B: Biological Sciences*, 372(1712), 20160029.

Allman, J.M. (1999). *Evolving Brains*. Scientific American Library, New York.

Aplin, L.M., Farine, D.R., Morand-Ferron, J., Cockburn, A., Thornton, A., and Sheldon, B.C. (2015). Experimentally induced innovations lead to persistent culture via conformity in wild birds. *Nature*, 518(754), 538–41.

Arnold, S. (2013). Beauty and brains: redder rural finches have it all. Undergraduate Honor's Thesis, Arizona State University, Tempe.

Aronson, M.F.J., La Sorte, F.A., Nilon, C.H., et al. (2014). A global analysis of the impacts of urbanization on bird and plant diversity reveals key anthropogenic drivers. *Proceedings of the Royal Society B: Biological Sciences*, 281(1780), 20133330.

Audet, J.N., Ducatez, S., and Lefebvre, L. (2016). The town bird and the country bird: problem solving and immunocompetence vary with urbanization. *Behavioral Ecology*, 27, 637–44.

Audet, J.-N., Kayello, L., Ducatez, S., et al. (2018). Divergence in problem-solving skills is associated with differential expression of glutamate receptors in wild finches. *Scientific Advances*, 4, eaao6369.

Badyaev, A.V., Young, R.L., Oh, K.P., and Addison, C. (2008). Evolution on a local scale: developmental, functional, and genetic bases of divergence in bill form and associated changes in song structure between adjacent habitats. *Evolution*, 62(8), 1951–64.

Baldwin, J.M. (1896). A new factor in evolution. *American Naturalist*, 30(354), 441–51.

Batabyal, A. and Thaker, M. (2019). Lizards from suburban areas learn faster to stay safe. *Biology Letters*, 15(2), 20190009.

Bell, G. (2017). Evolutionary rescue. *Annual Review of Ecology and Evolutionary Systems*, 48, 605–27.

Beltman, J.B., Haccou, P., and ten Cate, C. (2004). Learning and colonization of new niches: a first step toward speciation. *Evolution*, 58(1), 35–46.

Boogert, N.J., Monceau, K., and Lefebvre, L. (2010). A field test of behavioural flexibility in Zenaida doves (*Zenaida aurita*). *Behavioural Processes*, 85, 135–41.

Boogert, N.J., Madden, J.R., Morand-Ferron, J., and Thornton, A. (2018). Measuring and understanding individual differences in cognition. *Philosophical Transactions of the Royal Society B: Biological Sciences*, 373(1756), 20170280.

Bouchard, J., Goodyer, W., and Lefebvre, L. (2007). Social learning and innovation are positively correlated in pigeons (*Columba livia*). *Animal Cognition*, 10(2), 259–66.

Brady, S.P. (2012). Road to evolution? Local adaptation to road adjacency in an amphibian (*Ambystoma maculatum*). *Scientific Reports*, 2(1), 235.

Carrete, M. and Tella, J.L. (2011). Inter-individual variability in fear of humans and relative brain size of the species are related to contemporary urban invasion in birds. *PLOS ONE*, 6(4), e18859.

Cauchard, L., Boogert, N.J., Lefebvre, L., Dubois, F., and Doligez, B. (2013). Problem-solving performance is correlated with reproductive success in a wild bird population. *Animal Behaviour*, 85(1), 19–26.

Charmantier, A., Demeyrier, V., Lambrechts, M.M., Perret, S., and Grégoire, A. (2017). Urbanization is associated with divergence in pace-of-life in great tits. *Frontiers in Ecology and Evolution*, 5(May), 1–13.

Cheverud, J.M., Falk, D., Vannier, M., Konigsberg, L., Helmkamp, R.C., and Hildebolt, C. (1990). Heritability of brain size and surface features in rhesus macaques (*Macaca mulatta*). *Journal of Heredity*, 81(1), 51–7.

Chevin, L.-M., Lande, R., and Mace, G.M. (2010). Adaptation, plasticity, and extinction in a changing environment: towards a predictive theory. *PLOS Biology*, 8(4), e1000357.

Cole, E.F., Morand-Ferron, J., Hinks, A.E., and Quinn, J.L. (2012). Cognitive ability influences reproductive life history variation in the wild. *Current Biology*, 22(19), 1808–12.

Cook, M.O., Weaver, M.J., Hutton, P., McGraw, K.J. (2017). The effects of urbanization and human disturbance on problem solving in juvenile house finches (*Haemorhous mexicanus*). *Behavioral Ecology and Sociobiology*, 71, 85.

Croston, R., Branch, C.L., Kozlovsky, D.Y., Dukas, R., and Pravosudov, V.V. (2015). Heritability and the evolution of cognitive traits. *Behavioral Ecology*, 26(6), 1447–59.

Dickinson, A. (2012). Associative learning and animal cognition. *Philosophical Transactions of the Royal Society B: Biological Sciences*, 367(1603), 2733–42.

Donihue, C.M. and Lambert, M.R. (2015). Adaptive evolution in urban ecosystems. *Ambio*, 44(3), 194–203.

Ducatez, S., Clavel, J., and Lefebvre, L. (2015). Ecological generalism and behavioural innovation in birds: technical intelligence or the simple incorporation of new foods? *Journal of Animal Ecology*, 84(1), 79–89.

Ducatez, S., Sayol, F., Sol, D., and Lefebvre, L. (2018). Are urban vertebrates city specialists, artificial habitat exploiters, or environmental generalists? *Integrative and Comparative Biology*, 58(5), 929–38.

Dukas, R. (2004). Evolutionary biology of animal cognition. *Annual Review of Ecology and Evolutionary Systems*, 35(1), 347–74.

Dukas, R. (2012). Ecology of learning. In: Seel, N.M. (ed.) *Encyclopedia of the Sciences of Learning*, pp 1071–3. Springer, Boston.

Dukas, R. (2013). Effects of learning on evolution: robustness, innovation and speciation. *Animal Behaviour*, 85(5), 1023–30.

Dyson, K., Ziter, C., Fuentes, T.L., and Patterson, M.S. (2019). Conducting urban ecology research on private property: advice for new urban ecologists. *Journal of Urban Ecology*, 5(1), 1–10.

Edelaar, P., Jovani, R., Gomez-Mestre, I. (2017). Should I change or should I go? Phenotypic plasticity and matching habitat choice in the adaptation to environmental heterogeneity. *American Naturalist*, 190, 506–20.

Evans, K.L., Gaston, K.J., Frantz, A.C., et al. (2009). Independent colonization of multiple urban centres by a formerly forest specialist bird species. *Proceedings of the Royal Society B: Biological Sciences*, 276(1666), 2403–10.

Evans, K.L., Hatchwell, B.J., Parnell, M., and Gaston, K.J. (2010). A conceptual framework for the colonisation of urban areas: the blackbird *Turdus merula* as a case study. *Biological Reviews of the Cambridge Philosophical Society*, 85(3), 643–67.

Evans, K.L., Chamberlain, D.E., Hatchwell, B.J., Gregory, R.D., and Gaston, K.J. (2011). What makes an urban bird? *Global Change Biology*, 17(1), 32–44.

Federspiel, I.G., Garland, A., Guez, D., et al. (2017). Adjusting foraging strategies: a comparison of rural and urban common mynas (*Acridotheres tristis*). *Animal Cognition*, 20, 65–74.

Fox, R.J., Donelson, J.M., Schunter, C., Ravasi, T., and Gaitán-Espitia, J.D. (2019). Beyond buying time: the role of plasticity in phenotypic adaptation to rapid environmental change. *Philosophical Transactions of the Royal Society B: Biological Sciences*, 374(1768), 20180174.

Gonzalez, A., Ronce, O., Ferriere, R., and Hochberg, M.E. (2012). Evolutionary rescue: an emerging focus at the intersection between ecology and evolution. *Philosophical Transactions of the Royal Society B: Biological Sciences*, 368(1610), 20120404.

Greggor, A.L., Clayton, N.S., Phalan, B., and Thornton, A. (2014). Comparative cognition for conservationists. *Trends in Ecology and Evolution*, 29(9), 489–95.

Griffin, A.S., Netto, K., and Peneaux, C. (2017). Neophilia, innovation and learning in an urbanized world: a critical evaluation of mixed findings. *Current Opinions in Behavioural Science*, 16, 15–22.

Haase, D. (2008). Urban ecology of shrinking cities: an unrecognized opportunity? *Nature and Culture*, 3(1), 1–8.

Hendry, A.P., Schoen, D.J., Wolak, M.E., and Reid, J.M. (2018). The contemporary evolution of fitness. *Annual Review of Ecology and Evolutionary Systems*, 49, 457–76.

Huey, R.B., Hertz, P.E., and Sinervo, B. (2003). Behavioral drive versus behavioral inertia in evolution: a null model approach. *American Naturalist*, 161(3), 357–66.

Irwin, D.E. and Price, T. (1999). Sexual imprinting, learning and speciation. *Heredity*, 82, 347–54.

Johnson, M.T.J. and Munshi-South, J. (2017). Evolution of life in urban environments. *Science*, 358(6363), 607.

Johnston, R.F. (1990). Variation in size and shape in pigeons, *Columba livia*. *Wilson Bulletin*, 102, 213–25.

Jones, T.B., Evans, J.C., and Morand-Ferron, J. (2019). Urbanization and the temporal patterns of social networks and group foraging behaviors. *Ecology and Evolution*, 9(8), 4589–602.

Kang, C., Goulet, Y.E., and Chapple, Y. (2018). The impact of urbanization on learning ability in an invasive lizard. *Biology Journal of the Linnean Society*, 123, 55–62.

Kark, S., Iwaniuk, A., Schalimtzek, A., and Banker, E. (2007). Living in the city: can anyone become an 'urban exploiter'? *Journal of Biogeography*, 34(4), 638–51.

Kawecki, T.J. (2013). The impact of learning on selection-driven speciation. *Trends in Ecology and Evolution*, 28(2), 68–9.

Kokko, H. and Sutherland, W.J. (2001). Ecological traps in changing environments: ecological and evolutionary consequences of a behaviourally mediated Allee effect. *Evolutionary Ecology Research*, 3(5), 537–51.

Kozlovsky, D.Y., Weissgerber, E.A., Pravosudov, V.V. (2017). What makes specialized food-caching mountain chickadees successful city slickers? *Proceedings of the Royal Society B: Biological Sciences*, 284, 20162613.

Lahti, D.C., Johnson, N.A., Ajie, B.C., et al. (2009). Relaxed selection in the wild. *Trends in Ecology and Evolution*, 24(9), 487–96.

Lapiedra, O. (2018). Urban behavioral ecology: lessons from *Anolis* lizards. *Integregrative and Comparative Biology*, 58, 939–47.

Lapiedra, O., Schoener, T.W., Leal, M, Losos, J.B., and Kolbe, J.J. (2018). Predator-driven natural selection on risk-taking behavior in anole lizards. *Science*, 360(6392), 1017–20.

Leal, M. and Powell, B.J. (2012). Behavioural flexibility and problem-solving in a tropical lizard. *Biology Letters*, 8(1), 28–30.

Lihoreau, M., Doligez, B., Klein, S., et al. (2018). The repeatability of cognitive performance: a meta-analysis. *Philosophical Transactions of the Royal Society B: Biological Sciences*, 373(1756), 20170281.

Liker, A. and Bókony, V. (2009). Larger groups are more successful in innovative problem solving in house sparrows. *Proceedings of the National Academy of Sciences of the United States of America*, 106, 7893–8.

Losos, J.B., Schoener, T.W., and Spiller, D.A. (2004). Predator-induced behaviour shifts and natural selection in field-experimental lizard populations. *Nature*, 432(7016), 505–8.

Maklakov, A.A., Immler, S., Gonzalez-Voyer, A., Ronn, J., and Kolm, N. (2011). Brains and the city: big-brained passerine birds succeed in urban environments. *Biology Letters*, 7(5), 730–2.

Maspons, J., Molowny-Horas, R., and Sol, D. (2019). Behaviour, life history and persistence in novel environments. *Philosophical Transactions of the Royal Society B: Biological Sciences*, 374(1781), 20180056.

Maynard-Smith, J. (1987). When learning guides evolution. *Nature*, 329(6142), 761–2.

Maynard-Smith, J. (1989). The causes of extinction. *Philosophical Transactions of the Royal Society B: Biological Sciences*, 325(1228), 241–52.

McKinney, M.L. (2002). Urbanization, biodiversity and conservation. *Bioscience*, 52, 883–90.

Mery, F. and Kawecki, T.J. 2003. A fitness cost of learning ability in *Drosophila melanogaster*. *Proceedings of the Royal Society B: Biological Sciences*, 270(1532), 2465–9.

Møller, A.P. (2009). Successful city dwellers: a comparative study of the ecological characteristics of urban birds in the Western Palearctic. *Oecologia*, 159(4), 849–58.

Morand-Ferron, J., Cole, E.F., and Quinn, J.L. (2016). Studying the evolutionary ecology of cognition in the wild: a review of practical and conceptual challenges. *Biological Reviews of the Cambridge Philosophical Society*, 91(2), 367–89.

Moseley, D.L., Derryberry, G.E., Phillips, J.N., et al. (2018). Acoustic adaptation to city noise through vocal learning by a songbird. *Proceedings of the Royal Society B: Biological Sciences*, 285(1888), 20181356.

Mueller, J.C., Partecke, J., Hatchwell, B.J., Gaston, K.J., and Evans, K.L. (2013). Candidate gene polymorphisms for behavioural adaptations during urbanization in blackbirds. *Molecular Ecology*, 22(13), 3629–37.

Nicolaus, M. and Edelaar, P. (2018). Comparing the consequences of natural selection, adaptive phenotypic plasticity, and matching habitat choice for phenotype–environment matching, population genetic structure, and reproductive isolation in meta-populations. *Ecology and Evolution*, 8(8), 3815–27.

Papp, S., Vincze, E., Preiszner, B., Liker, A., and Bókony, V. (2015). A comparison of problem-solving success between urban and rural house sparrows. *Behavioural Ecology and Sociobiology*, 69, 471–80.

Preiszner, B., Papp, S., Pipoly, I., et al. (2017). Problem-solving performance and reproductive success of great tits in urban and forest habitats. *Animal Cognition*, 20, 53–63.

Price, T.D., Qvarnström, A., Irwin, D.E., Qvarnstrom, A., and Irwin, D.E. (2003). The role of phenotypic plasticity in driving genetic evolution. *Proceedings of the Royal Society B: Biological Sciences*, 270(1523), 1433–40.

Quinn, J.L., Cole, E.F., Reed, T.E., and Morand-Ferron, J., (2016). Environmental and genetic determinants of innovativeness in a natural population of birds. *Philosophical Transactions of the Royal Society B: Biological Sciences*, 371(1690), 20150184.

Ramalho, C.E. and Hobbs, R.J. (2012). Time for a change: dynamic urban ecology. *Trends in Ecology and Evolution*, 27(3), 179–88.

Rodewald, A.D. and Arcese, P. (2017). Reproductive contributions of cardinals are consistent with a hypothesis of relaxed selection in urban landscapes. *Frontiers in Ecology and Evolution*, 5, 1–7.

Rosenheim, J.A. and Tabashnik, B.E. (1991). Influence of generation time in the rate of response to selection. *American Naturalist*, 137(4), 527–41.

Santini, L., González-Suárez, M., Russo, D., Gonzalez-Voyer, A., von Hardenberg, A., and Ancillotto, L. (2019). One strategy does not fit all: determinants of urban adaptation in mammals. *Ecology Letters*, 22(2), 365–76.

Santos, C.D., Cramer, J.F., Pârâu L.G., Miranda, A.C., Wikelski, M., and Dechmann, D.K.N. (2015). Personality and morphological traits affect pigeon survival from raptor attacks. *Scientific Reports*, 5(1), 15490.

Seferta, A., Guay, P.J., Marzinotto, E., and Lefebvre, L. (2001). Learning differences between feral pigeons and zenaida doves: the role of neophobia and human proximity. *Ethology*, 107, 281–93.

Seto, K.C., Guneralp, B., Hutyra, L.R. (2012). Global forecasts of urban expansion to 2030 and direct impacts on biodiversity and carbon pools. *Proceedings of the National Academy of Sciences of the United States of America*, 109(40), 16083–8.

Shettleworth, S. (2010). *Cognition, Evolution, and Behavior*. Oxford University Press, Oxford.

Shochat, E., Warren, P., Faeth, S., McIntyre, N., and Hope, D. (2006). From patterns to emerging processes in mechanistic urban ecology. *Trends in Ecology and Evolution*, 21(4), 186–91.

Sih, A. (2013). Understanding variation in behavioural responses to human-induced rapid environmental change: a conceptual overview. *Animal Behaviour*, 85(5), 1077–88.

Snell-Rood, E.C. and Wick, N. (2013). Anthropogenic environments exert variable selection on cranial capacity in mammals. *Proceedings of the Royal Society B: Biological Sciences*, 280(1769), 20131384.

Snell-Rood, E.C., Kobiela, M.E., Sikkink, K.L., and Shephard, A.M. (2018). Mechanisms of plastic rescue in novel environments. *Annual Review of Ecology and Evolutionary Systems*, 49, 331–54.

Sol, D. (2007). Do successful invaders exist? Pre-adaptations to novel environments in terrestrial vertebrates. In: Nentwig, W. (ed.) *Biological Invasions*, pp. 127–41. Springer, Berlin.

Sol, D. (2008). Artificial selection, naturalization, and fitness: Darwin's pigeons revisited. *Biology Journal of the Linnean Society*, 93(4), 657–65.

Sol, D. (2009). The cognitive-buffer hypothesis for the evolution of large brains. In: Dukas, R. and Ratcliffe, J.M. (eds) *Cognitive Ecology II*, pp. 111–34. University of Chicago Press, Chicago.

Sol, D. and Maspons, J. (2016). Life history, behaviour and invasion success. In: Weis, J.S. and Sol, D. (eds) *Biological Invasions and Animal Behaviour*, pp. 63–81. Cambridge University Press, Cambridge.

Sol, D. and Price, T.D.D. (2008). Brain size and the diversification of body size in birds. *American Naturalist*, 172(2), 170–7.

Sol, D., Stirling, D.G., and Lefebvre, L. (2005). Behavioral drive or behavioral inhibition in evolution: subspecific diversification in Holarctic Passerines. *Evolution*, 59(12), 2669.

Sol D., Griffin, A.S, Bartomeus, I., and Boyce, H. (2011). Exploring or avoiding novel food resources? The novelty conflict in an invasive bird. *PLOS ONE*, 6(5).

Sol, D., Lapiedra, O., and González-Lagos, C. (2013). Behavioural adjustments for a life in the city. *Animal Behaviour*, 85(5), 1101–12.

Sol, D., González-Lagos, C., Moreira, D., Maspons, J., and Lapiedra, O. (2014). Urbanisation tolerance and the loss of avian diversity. *Ecology Letters*, 17(8), 942–50.

Sol, D., Sayol, F., Ducatez, S., and Lefebvre, L. (2016). The life-history basis of behavioural innovations. *Philosophical Transactions of the Royal Society B: Biological Sciences*, 371(1690), 20150187.

Sonnenberg, B.R., Branch, C.L., Pitera, A.M., Bridge, E., and Pravosudov, V.V. (2019). Natural selection and spatial cognition in wild food-caching mountain chickadees. *Current Biology*, 29(4), 670–676.e3.

Stamps, J.A. and Swaisgood, R.R. (2007). Someplace like home: experience, habitat selection and conservation biology. *Applied Animal Behaviour Science*, 102(3–4), 392–409.

Stamps, J.A., Krishnan, V.V., and Reid, M.L. (2010). Search costs and habitat selection by dispersers. *Ecology*, 86(2), 510–18.

ten Cate, C. and Vox, D. (1999). Sexual imprinting and evolutionary processes in birds: a reassessment. *Advances in the Study of Behavior*, 28, 1–32.

Vander Wal, E., Garant, D., Festa-Bianchet, M., and Pelletier, F. (2012). Evolutionary rescue in vertebrates.

evidence, applications and uncertainty. *Philosophical Transactions of the Royal Society B: Biological Sciences*, 368(1610), 20120090.

van Dongen, W.F.D., Robinson, R.W., Weston, M.A., Mulder, R.A., and Guay, P.-J. (2015). Variation at the DRD4 locus is associated with wariness and local site selection in urban black swans. *BMC Evolutionary Biology*, 15(1), 253.

Verzijden, M.N., ten Cate, C., Servedio, M.R., Kozak, G.M., Boughman, J.W., and Svensson, E.I. (2012). The impact of learning on sexual selection and speciation. *Trends in Ecology and Evolution*, 27(9), 511–19.

Vincze, E., Papp, S., Preiszner, B., Seress, G., Bókony, V., and Liker, A. (2016). Habituation to human disturbance is faster in urban than rural house sparrows. *Behavioral Ecology*, 27(5), 1304–13.

Winchell, K.M., Carlen, E.J., Puente-Rolón, A.R., and Revell, L.J. (2018). Divergent habitat use of two urban lizard species. *Ecology and Evolution*, 8(1), 25–35.

Wyles, J.S., Kunkel, J.G., and Wilson, A.C. (1983). Birds, behavior, and anatomical evolution. *Proceedings of the National Academy of Sciences of the United States of America*, 80(14), 4394–7.

Yeh, P.J. and Price, T.D. (2004). Adaptive phenotypic plasticity and the successful colonization of a novel environment. *American Naturalist*, 164(4), 531–42.

Yeh, P.J., Hauber, M.E., and Price, T.D. (2007). Alternative nesting behaviours following colonisation of a novel environment by a passerine bird. *Oikos*, 116(9), 1473–80.

CHAPTER 16

Selection on Humans in Cities

Emmanuel Milot and Stephen C. Stearns

The goal of much of modern medicine and culture is effectively to stop evolution. Is that happening?

(Michael Balter 2005)

Milot, E. and Stearns, S.C., *Selection on Humans in Cities* In: *Urban Evolutionary Biology*. Edited by Marta Szulkin, Jason Munshi-South and Anne Charmantier, Oxford University Press (2020). © Oxford University Press.
DOI: 10.1093/oso/9780198836841.003.0016

16.1 Introduction

Urbanization destroys natural habitats, forcing wild organisms to either move away or rapidly adapt by plastic or genetic changes (Alberti et al. 2016; Johnson and Munshi-South 2017; Santangelo et al. 2018). For *Homo sapiens*, however, the evolutionary logic may seem reversed to someone who looks at modern populations with a superficial eye: the organism (human) selects (builds) an environment that suits its needs rather than having the environment select individuals with certain traits. Modernization, the broader process in which urbanization is embedded, has increased our control over our ecosystem and biology to such an extent that some posit the decline of evolutionary potential in humans and even the outright end of natural selection (Box 16.1). Others, however, suggest that human evolution has been accelerating in the last thousand years, with no obvious signs of abating in contemporary times

(Cochran and Harpending 2009). Therefore, one question that inevitably arises about urbanization is how it changes the pace of human evolution.

If humans continue to evolve nowadays (we think they do), then further questions are what selective agents are specific to cities, and what are the traits that they target. Given the obvious differences between the environment in which citizens dwell today and that where their recent ancestors lived, selection is unlikely to have remained constant. Living in cities might indeed reshape selection in several ways, as we will show here. This leads to another question: what consequences could urban evolutionary processes have on human ecology and living conditions? Some consequences, such as selection on alleles modulating lifespan, might be more important than most people think (Mostafavi et al. 2017).

Humans both construct the urban habitat and experience its selective pressures. This adds layers

---

Box 16.1 The end of natural selection in humans?

Several arguments claim the end of selection and the unavoidable decline of evolutionary potential in humans. They are mostly heard in the media and rarely percolate into the scientific literature. Nonetheless, some bold statements made by prominent scientists and science journalists have prompted responses in the scientific community (e.g., see Wills 1998; Balter 2005; Stock 2008; Templeton 2010; Rickard 2013; Monosson 2014). Moreover, a Google search quickly identifies titles such as 'Are humans still evolving?', as if that were still an open question. We classify 'end of selection/evolution' arguments into four general categories defined by the mechanisms involved:

- *Promethean* arguments assert that humans are actively emancipating themselves from the influence of nature, and hence of selection, through cultural and techno-scientific progress. Because child survival is very high in societies that have gone through a transition to modernity (typically > 0.99 per cent), nearly everyone reaches sexual maturity and can reproduce (in developed societies). In addition, the correlated reduction in mean family size has presumably so reduced natural fertility that this component of fitness also no longer contributes to selection. Following this logic to its end, full escape from natural selection in the future would be conceivable with the

progress of individual medicine and the rise of transhumanism (see section 16.6.2).

- *Limitative* arguments view the decline in evolutionary forces as a by-product of progress that transforms our ecology. For example, it has been suggested that genetic inertia is too large in a population of billions of humans for either selection or genetic drift to have any significant impact on our species in the near-term future (e.g., as reported in Monosson 2014). Moreover, as the mobility of individuals—and hence of their genes—increasingly produces global mixing, the potential for local adaptation should shrink, particularly in cities that receive continuous migratory influx. Low fertility could further locally restrict the appearance and spread of new beneficial mutations.

- *Ontological* arguments claim that our evolved nature contains mechanisms that eliminate natural selection. Essentially, those who hold this view consider that cultural evolution has replaced biological evolution, thanks to the emergence of our cognitive capacities (e.g., Gould (2000) famously declared that no biological change occurred in humans in the past 40 000–50 000 years). Behind ontological arguments may also linger the old idea that humans are the end point of evolution (Balter 2005). More radically, Tort (1983) proposed that the evolution of social

continued

Box 16.1 *Continued*

instincts such as altruism and sympathy led humans to reject the 'destructive hierarchy of fitness' and to develop societal norms against natural selection.

- *Degeneration* arguments predict a genetic decline in physical and cognitive capacities when human actions impede natural selection. Note that the original degeneration theory (Morel 1857) was published before Darwin's *On the Origin of Species* and could not evoke natural selection. Here we refer to a different form of this theory that emerged around the mid-twentieth century, and which rests on the expectation that deleterious mutations will accumulate in genomes under relaxed selection (Muller 1950). Recent examples include Sykes (2004), who predicts the complete degeneration of the Y chromosome within 5000 years, and Lynch (2016), who worries that medical interventions will cause human fitness and cognitive abilities to deteriorate at a rate of ~ 1 per cent per generation.

Is evolutionary potential vanishing in modern populations?

While variation in survival offers less leverage to natural selection than it used to, the reverse is true for variation in fertility (see section 16.3.3), which clearly contradicts Promethean arguments. The factors evoked by limitative arguments can also serve to predict an increase in the pace of human evolution rather than the opposite (Templeton 2010). For instance, in large populations, selection is more efficient and more mutations occur, while large-scale mobility can help to spread beneficial ones. Ontological and degenerative arguments neglect the complex interplay between culture and biological evolution. Culture evolves continuously and rapidly and may keep phenotypes away from their fitness optimum, and hence maintain selective pressures over the long term. As individuals become freer in their reproductive choices, genetic variation in life history and behaviour has more room to be expressed (e.g., see section 16.5). Relaxed selection on a focal trait can have consequences on the expression of correlated traits and their fitness distribution, causing new selective pressures on these correlated traits. Cultural-based adaptations, including effective agricultural production, may not resist forthcoming climatic instability and prompt for genetic response to resulting environmental stress, such as resource depletion and competition (Stock 2008). Finally, 'decline or end of selection' arguments do not withstand a confrontation with data: there is mounting evidence that phenotype and fitness covary in modern human populations (e.g., Byars et al. 2010, 2017; Stulp et al. 2015; Beauchamp 2016; Kong et al. 2017).

---

of complexity for the analysis of urban evolution. Therefore, our aim is to present here the elements needed to grasp the scope of the problem. We focus on health traits to tighten our presentation with support from the literature on natural selection in humans that has emphasized health-related phenotypes or genotypes.

Our task is, however, further complicated by three limitations. First, the effects of urbanization per se (which started in the Fertile Crescent about 7500 BCE) on evolutionary processes are largely conflated with those of modernization in general, since the latter, which started in Europe about 1750 CE, involves urbanization both as a component and as a result. Thus, while we have some idea of the types of traits most likely to be under the scrutiny of natural selection in modern populations (Stearns et al. 2010; Field et al. 2016; Solomon 2016), we do not yet know what part of that selection is directly attributable to urbanization. The second limitation is that few studies have explicitly attempted to quantify selection on heritable human traits due to urban factors with the methods of evolutionary quantitative genetics. This is not unexpected given that selection and the genetic responses to it are just starting to be documented in contemporary populations, utilizing advances in genomics, analytical methods, and the increasing availability of multigenerational longitudinal data (e.g., Tropf et al. 2015; Beauchamp 2016; Byars et al. 2017). Quantifying the contribution of urban factors, including physical attributes of the habitat, stressors, and cultural norms, has yet to be done. The third limitation is that cities are connected to the countryside both via commuting, which blends the impacts of urban and rural selection, and by gene flow, which blends the urban and rural genetic responses to selection. For all three reasons, our attempt to identify selective agents specific to modern cities must remain partly speculative given current knowledge. While there are few data with which to address many of the most interesting questions, it is worth posing those questions to challenge research.

As does the rest of this book, this chapter focuses on contemporary populations. Nonetheless, we begin our survey with a brief tour back in time: cities have existed for millennia, and ancient populations can inform us about selective agents that have acted independently of modernization itself. They can also help identify traits or alleles deserving special attention in the study of contemporary populations. These topics are discussed in section 16.2. In section 16.3 we examine the impact of modernization on natural selection, a necessary step in understanding how urban populations continue to evolve. Then, bearing in mind the limitations mentioned above, in section 16.4 we dig into the question of contemporary evolutionary processes, examining the opportunity for selection, the selective agents, and the genetics of diseases in urban societies. We end the chapter (sections 16.5 and 16.6) with thoughts on two subjects that are likely to be keys to understanding the future evolution of urbanites but have received limited attention so far: eco-evolutionary dynamics in humans and potential interactions between selection and transhumanism.

16.2 Signals from the past

Diverse sources of information, including modern genomes, pedigrees, and comparative anthropological studies, provide insights into which traits or genes were selected in past environments and during major transitions in human history, and how rapidly traits can respond genetically to environmental changes, like urbanization, that occur quickly. We next briefly outline some elements of the past selective environment.

16.2.1 Old times, old friends, and selection at the dawn of urbanization

The transition to agriculture, which started about 12 000 years ago, abruptly modified the selective landscape for humans. Increases in food supply, changes in diet and physical activity, and new forms of socioeconomic organization arising from settlement and densification created novel environments, to which the genes and traits that had previously evolved in foraging societies were less adapted (Cochran and Harpending 2009). Behaviour and life history were likely among the traits most impacted by the shifts in selective pressures. For instance, settlement led to the growth and extension of capital (crops, livestock, tradeable products, etc.) that had previously been limited to the herds of animals owned by pastoralists, and modified the energy available for development and reproduction, as well as household economy and the fitness returns from labour (Kaplan 1996). The resulting adjustments to life-history strategies often involved the offspring quantity–quality tradeoff, as suggested by the associations among wealth, fertility, and parental investment in a broad range of human societies, from traditional pastoralists to urban populations (Mace 2000, 2008).

The transition to agriculture also had major impacts on health. Long before population settlement, humans were in contact with animal-associated microbes and parasites with oral–faecal transmission. These organisms played an important role in the development of immunoregulation (the 'old friends' hypothesis (Rook 2012)). Agricultural life, and later urban life, caused the advent of new pathogens such as influenza and cholera, which would later become epidemic and lead to a first epidemiological transition (Rook 2012). Some generalist and scavenger species, including mice, rats, and cockroaches, evolved as human-dependant pests, occupying the new urban niches and becoming vectors of infectious diseases (Johnson and Munshi-South 2017). Higher urban than rural mortality rates have been documented in historical populations, up to pre-industrial times, reflecting the effects of higher densities, competition, infection risks, and poorer sanitation (e.g., Gagnon and Mazan 2009; Walter and DeWitte 2017).

Evidence for genetic responses to selective pressures in ancient cities remains scarce and indirect. For example, the genomic study of Daub et al. (2013) suggests that population settlement catalysed the evolution of immune response. They found that most pathways in the human genome that are globally enriched for signals of selection are either directly or indirectly involved in immune responses. These responses are plausibly associated with the shift in exposure to infectious diseases that accompanied the agricultural revolution and the change from nomadic hunting and gathering to farming in settlements and then to living in cities. Deeper insights into urban-driven evolution can come from the comparison of genomic data across populations exhibiting different urban histories, as illustrated by a study by Barnes et al. (2010). Comparing seventeen populations across Eurasia, these authors found that the frequency of a tuberculosis-resistant allele at the *SLC11A1* gene increases as a function of the historical extent of urbanization (measured as the oldest recorded date of urban settlement in a population). The frequency of the resistant allele goes from ~ 0.75 in populations urbanized recently (< 200 years ago) to ~ 1 (fixation) where urban settlements have existed for > 2500 years, albeit with some deviations around this general pattern. These results support the hypothesis that selection for immune response has been more sustained in areas with a longer urban history, but should ideally be confirmed by sampling more populations.

16.2.2 How fast did traits respond genetically to past environmental changes?

The answers to this question have been sought at different timescales (Box 16.2). At the scale of thousands of years, signals are mostly associated with one or a few genes of known function in which rare alleles increased rapidly in frequency in response to selection, such as those conferring malaria resistance. They also allow the identification of common variants in genes shaping complex traits that played a historical role in human adaptation and could be the target of different selective pressures in modern societies. For instance, past positive selection has been detected on alleles modulating metabolism and that are nowadays associated with higher risks for common diseases like type II diabetes (Pickrell et al. 2009). In addition, selection on variants regulating gene expression may have been instrumental in inducing fast evolutionary responses to rapid changes in living conditions over the last thousand years (Johnson and Voight 2018).

At the scale of tens or hundreds of years, evidence is accumulating that small shifts in the frequencies of many genes, rather than large shifts in the frequencies of a few genes, can cause a rapid response to selection in polygenic traits (Boyle et al. 2017), including health, and morphological, physiological, and life-history traits (Box 16.2; Stearns et al. 2010). For example, Byars et al. (2017) found signatures of selection in the forty-plus genes associated with coronary artery disease by genome-wide association study (GWAS) and showed that all these genes had pleiotropic effects on reproductive traits. This result suggested that the selection that resulted in genomic signatures was caused by relatively recent changes in the contributions of the reproductive traits to lifetime reproductive success rather than by mortality effects on fitness later in life mediated by coronary artery disease. Finally, signals of selection in contemporary times are starting to be detected in genomic data, as shown in an innovative study that tracked changes in allele frequencies with age to document viability selection in the UK population (Mostafavi et al. 2017); this study found sets of genetic variants that delayed maturation and extended life.

The current evidence suggests that the evolutionary responses of humans to changed environments could be fairly rapid, with measurable changes within just a few (approximately two to ten) generations. The modern urban environment differs radically from the rural environments in which most humans lived until very recently. Urbanization may represent a rapid shift in selection that could elicit a genetic response. As signals of past selection teach us, whether such a response is likely, could occur rapidly, and could be measured depends on the answers to at least these questions: Is one looking at DNA sequences in the genome or at phenotypic traits? Is one examining a single gene or a polygenic trait, and how large are the effects of the genes on the traits? How strong is the impact of environmental change on reproductive success, on trait expression, and on the relationship between the traits and reproductive success? Human adaptation to the urban environment thus appears to be possible in principle, but whether it has actually happened during the transition to modernity and with urbanization depends on many details that vary

---

Box 16.2 Evidence for genetic response to selection on human traits

Signatures of selection in the genome at the scale of thousands of years

At the scale of thousands of years, many signatures of selection have been detected in the genome using methods reviewed by Sabeti et al. (2006) and Field et al. (2016). The signatures that have been most accurately dated are correlated with cultural changes, such as the domestication of dairy animals. This famously includes the evolution of lactase persistence (the ability to digest milk as adults), which started independently in Eurasia ca. 9000 years ago and in Africa ca. 4500 years ago (Tishkoff et al. 2007). Functions associated with other selection signals include malaria resistance, adaptation of skin pigmentation to reduced ultraviolet light, and adaptation to reduced oxygen at high altitude, all of which occurred at the scale of thousands to tens of thousands of years.

Polygenic traits could evolve more rapidly, but do they?

Early work on signals of selection focused on single genes with known function, but most traits of functional significance are determined by many genes. In humans, such traits include, for example, height, body mass index, age at menarche, blood pressure, blood lipid levels, glucose tolerance, and risk of coronary artery disease. They have fairly high heritabilities measured from correlations among relatives and are thus probably determined by many genes of small effect for each of which there is standing variation in the population. The response to selection in such cases can be both rapid and determined by small shifts in the frequencies of many genes rather than by large shifts in the frequencies of a few genes (Pritchard and Di Rienzo 2010; Jain and Stephan 2017). In such studies it is not yet possible to date accurately the era in which the response to selection happened, but methodological advances and massive sequencing could change the situation rapidly.

Tracking per-generation selection using genealogies

Studying the dynamics of traits in genealogies over several to tens of generations opens up another approach to estimating the rate at which responses to selection can occur in humans and to understanding the tradeoffs that constrain those responses. First, consider some of the tradeoffs that have been detected. Pettay et al. (2005), studying four generations of pre-industrial Finns, found that females who started to breed early or who bred frequently had reduced lifespans. Wang et al. (2013), analysing the Framingham Heart Study population, found that the genetic correlation between children-ever-born (CEB) and lifespan was large, significant, and negative, and the negative phenotypic relationship between CEB and lifespan was mostly contributed by women born after 1905, with the relationship becoming more strongly negative in cohorts born later in the twentieth century. These studies support the existence of a cost of reproduction in human females: those who reproduce earlier, more often, or at shorter intervals have shorter lives. Despite the constraints on response implied by such tradeoffs, evolutionary responses have sometimes been rapid. Milot et al. (2011), studying a population of farmers on Île aux Coudres, an island in the St Lawrence River in Quebec, found that mean age at first reproduction declined from 26 to 22 years old over a period of 140 years with a substantial change in the mean genetic value of the trait, supporting the hypothesis that the change is partly the result of a genetic response to selection. Kong et al. (2017), analysing a sample of nearly 130 000 Icelanders born between 1910 and 1990, found that the polygenic score for educational attainment was declining significantly at a rate of at least 0.01 standard units per decade, as a result of selection, providing evidence for a rapid genetic response in contemporary times.

---

from case to case. Nevertheless, human populations can have very large effective sizes that should increase the efficiency of natural selection compared to other long-lived, large-bodied species.

16.3 The transition to modernity

The transition to modernity (Corbett et al. 2018) began with the industrial revolution about 1800 CE in Europe and was soon accompanied by ecological,

demographic, and epidemiological transitions that have had profound impact on human populations. The industrial revolution led to changes in technology, lifestyle, transportation, and the built environment. The ecological transition changed the environments of many humans from rural to urban life; nutrition improved and, for many, food security was no longer an issue. The demographic transition involved first a drop in mortality rates, accompanied by a burst of population growth, then a drop in birth rates that stabilized populations that shifted from being dominated by the young to having a higher proportion of the old. The epidemiological transition shifted the causes of morbidity and mortality from infectious diseases like measles, smallpox, and malaria to non-communicable diseases like cancer, heart disease, neurodegenerative diseases, and autoimmune disorders. Thus the transition to modernity has been a complex process affecting vital rates, environmental interactions, age distributions, and changes in the causes of death and in the microbiota. It has had and is having many consequences for selection, health, and disease in humans.

As with the earlier urbanization, which started ca. 7500 BCE, the transition to modernity, which started ca. 1750 CE, caused a decline in fertility (see section 16.3.3) that is often seen as further impacting the offspring quantity–quality tradeoff (but see Lawson and Borgerhoff Mulder 2016; Stulp and Barrett 2016).

16.3.1 Antagonistic pleiotropy across the transition to modernity

The demographic shift in age distributions from populations dominated by the young to populations dominated by the old has revealed genetic effects in older people that were not easily detected in pre-industrial populations where life expectancy was only 30–40 years. Those costs late in life did not previously have to be paid because most individuals died for other reasons, such as infectious disease, but they now contribute to the global burden of diseases of the aging (Stearns and Medzhitov 2016). There is increasing evidence that prior evolution in pre-transition populations selected alleles that improved reproductive performance early in life but were linked via antagonistic pleiotropy to detrimental effects late in life, such as risk of cancer, coronary artery disease, and Alzheimer's disease (Corbett et al. 2018). *BRCA* genes, which are important for embryonic development and tumour suppression, are likely a good example. Germline mutations in those genes are associated with both a higher fertility and a high risk of contracting breast and ovarian cancers, including in women of reproductive age (e.g., DeSantis et al. 2017). Breast-

feeding and multiple pregnancies can, however, protect women against cancer. Plausibly, the fitness advantage provided by these variants overwhelmed their health costs in pre-industrial populations. A promising approach to test this hypothesis is to follow genes and fitness components down human pedigrees going back to pre-industrial generations. By linking molecular data collected on living individuals to genealogies, it is possible to assess how recent selection explains the frequency of genetic variants affecting health in contemporary populations (Milot et al. 2017; Peischl et al. 2018).

16.3.2 When modernity chases our old friends away

The move of the majority of the world's population into urban environments has also caused major shifts in our microbiota, leading to a second epidemiological transition (Rook 2012). Urbanized spaces increase the representation of human-associated microbes and decrease the representation of animal-associated microbes in our microbiota (Ruiz-Calderon et al. 2016). This loss and replacement of 'old friends' has contributed to the rise of outcomes associated with a deficit in immunoregulation, the most visible being allergies. Improved hygiene, in particular the reduction in helminth infections, is associated with increased risk of autoimmune diseases that include type I diabetes (Stearns and Medzhitov 2016). These problems are accentuated by modern lifestyles and medical interventions. For instance, Klein et al. (2018) provide evidence for the role of breast-feeding in the development of a microbiota and an acquired immunity that are tailored to the environment where the infant grows up. Reductions in breast-feeding (as well as increases in caesarean sections) change the gut and lung microbiota in ways that increase the risk of asthma and obesity.

16.3.3 Opportunity for natural selection across the demographic transition

The opportunity for selection—defined as the variance in relative fitness and denoted I (Crow 1958)—sets the upper limit to the strength of selection that can be exerted on phenotypes. In some populations, this parameter has decreased with the transition to modernity (Courtiol et al. 2013; Corbett et al. 2018). Specifically, the component of I created by variation among individuals in lifespan is now reduced to very low levels in developed countries, mainly through reductions in juvenile mortality, which reaches its lowest values in urbanized societies (Champion and Hugo 2003). However, there is opportunity for selection created by variation among individuals in lifetime reproductive success; the latter has remained significant and even tends to increase, being now the major contributor to natural selection in those populations. Actually, the balance between modifications in age-specific survival and fertility will not necessarily result in a decline in the total opportunity for selection. In the Utah population, for example, Moorad (2013) documented an increase in I over the course of the demographic transition that took place during the nineteenth century, despite a clear reduction in the component of I due to survival. Consequently, selection on traits related to mating and fertility gets stronger in modern populations. The trait thus far most consistently associated with that change in selection is age at maturity, where selection for earlier maturation has been driven by the decreased cost of earlier reproduction that results from lower mortality rates in infancy and childhood (Stearns et al. 2010).

However, culture is changing to favour later age at first birth: dual-career families, professional specialization, social networks less centred on familial bonds, and freedom of sexual habits (including birth control) influence individual reproductive decisions. The result is an increase in the period of life in which individuals experience a conflict between biology and culture. Whereas genes associated with earlier parental age at first birth and lower educational attainment are selected for in several post-industrial populations, both traits show opposed trends, i.e., toward a later reproductive onset and a longer education (Stearns et al. 2010; Beauchamp 2016; Kong et al. 2017). However, evolution can be cryptic, i.e., the genetic response to selection can be masked by environmentally induced (here culturally induced) phenotypic change in the opposite direction (Merilä et al. 2001); populations that experienced cryptic evolution could react differently to future environmental changes altering trait expression from those that did not.

16.4 Urban selection

Determining how cities shape selection will involve addressing several broad questions that we discuss in the next sections: Does the opportunity for selection in urbanized areas differ consistently from that in the countryside? Are there selective agents specific to urban areas and what traits do they target? Do the diversity and scale of urban habitats and ecological processes cause heterogeneity in selection? Do particularities exist in the genetic makeup of urbanites that modulate trait expression and the potential for responses to selection? Table 16.1 presents some traits on which the urban environment may exert specific selective pressures.

16.4.1 Opportunity for selection in cities

In analysing Demographic and Health Surveys (of the DHS Program/ICF, 2004–2017) from seventy-two countries, Hruschka and Burger (2016) found that the fertility component (I_f) of the opportunity for selection (I) increases with the decrease in mean fertility (μ_f) associated with the transition to modernity (I_f estimated by Crow's (1958) index: $I_f = \sigma_f^2 / \mu_f^2$, where σ_f^2 denotes the variance in fertility). But how does urbanization per se contribute to this trend? In Figure 16.1A–C we plotted Hruschka and Burger's fertility statistics as a function of urbanization, defined as the proportion of the population living in urban areas. We observe a steeper decline in μ_f than σ_f^2 (Figure 16.1A,B), resulting in a rather steep increase in I_f (Figure 16.1C), although certain urbanized populations have very little opportunity for selection because their variance in fertility is quite low. Consequently, more urbanized populations appear to exhibit, on average, a greater potential for selective changes in fertility traits than less urbanized ones.

The picture is, however, more complex than these summary results suggest. A regression model controlling for household wealth and education uncovered an overall increase in fertility variance with increasing urbanization, but the model with the best fit included a negative interaction between urban proportion and education: in poorly educated populations, σ_f^2 increases with urban proportion, while the trend is reversed (but weak) in areas where

Table 16.1 Some human traits on which the urban environment could exert selective pressures that would differ from the effects of modernization in general.*

Trait type	Trait	Potential or demonstrated effects of modernization (M) and urbanization (U) on trait expression or selection
Behavioural	Propensity to adopt unhealthy habits	M: behaviours that appeared or increased with modernization, that compromise health (e.g., smoking cigarettes, calorie-rich diet), and that may create selection for personality traits (e.g., lower willingness to take risks) associated with their avoidance. U: propensity to adopt unhealthy behaviours, and therefore the strength of selection against these behaviours, may vary with the perception of risk of extrinsic mortality across urban populations stratified by inequalities (e.g., poor vs rich neighbourhoods (Pepper and Nettle 2014)).
Physiological	Total blood cholesterol	M: selection for lower cholesterol levels caused by changes in diet, physical activity, and other factors (Byars et al. 2010; Stearns and Medzhitov 2016). U: heterogeneity in food availability and diet (e.g., 'food deserts'; see also previous trait) may modulate fitness costs associated with cholesterol levels in different urban areas.
Homeostastic	Sleep architecture	M: evolved sleep architecture is mismatched to modern lifestyle, which could prompt selection for different sleep patterns (Nunn et al. 2016). U: stressors such as artificial lighting, noise, and density further disturb sleep, impacting metabolism, immune response, cognitive functions, and sexual habits (Nunn et al. 2016), which could cause correlated selection on sleep and other traits.
Defence	Immune response	M: increase in immunoregulatory deficit and prevalence of autoimmune and other diseases (Rook 2012); possible reduction in acquired immunity (Klein et al. 2018). U: air pollution aggravates immune disorders (Campa and Castanas 2008) and could select for pollution-resistant alleles; genetic mixing in cities (Nalls et al. 2009) may increase genetic variance in immune response (Liston et al. 2016).
Life history	Fertility	M: fertility decline with the demographic transition (Hruschka and Burger 2016), but nevertheless an increase in opportunity for selection derived from fertility (Courtiol et al. 2013; Moorad 2013; Corbett et al. 2018). U: different socioeconomic trajectories may generate variation in fertility and in opportunity for selection among urban populations (Figure 16.1; Gries and Grundmann 2018); reproductive rate may scale negatively with city size (Burnside et al. 2012).
	Fetal growth rate	M: nutrition-rich environment *in utero* can cause infant overweight and higher risk of obesity at adult age (Stearns and Medzhitov 2016). U: pollution and maternal stress associated with urban living could slow down fetal growth (Huang et al. 2018; Wang and Yang 2019).
Vital rate	Juvenile survival	M: decrease in juvenile mortality to very low levels, thus decreasing the opportunity for selection (Courtiol et al. 2013; Corbett et al. 2018). U: urban stressors such as air pollution can generate new sources of mortality (WHO 2018), hence possibly increasing variance in fitness.

* This list is not meant to be exhaustive. It illustrates the diversity of traits and functions potentially affected by urbanization. We do not yet know how most of these effects explicitly translate into selective pressures.

most people have, at a minimum, an education through primary school (Figure 16.1D). When I_f is taken as the response variable instead of σ_f^2, a model excluding urban proportion as a predictor has a better fit than one including it (not shown). This suggests that the diversity of urbanization contexts results in a diversity of changes in I_f with no clear global trend. On this subject, Gries and Grundmann (2018) indicate that fertility changes can vary along local economic trajectories: some developing countries currently completing the transition to modernity are taking different paths from the one previously followed by western societies. As discussed by Stulp and Barrett (2016), local institutional and cultural factors can cause the relationship between fertility and factors such as income, wealth, and education to diverge from expectations based on evolutionary tradeoffs. Therefore, the effect of urbanization on fertility parameters determining the opportunity for selection is probably context-dependent.

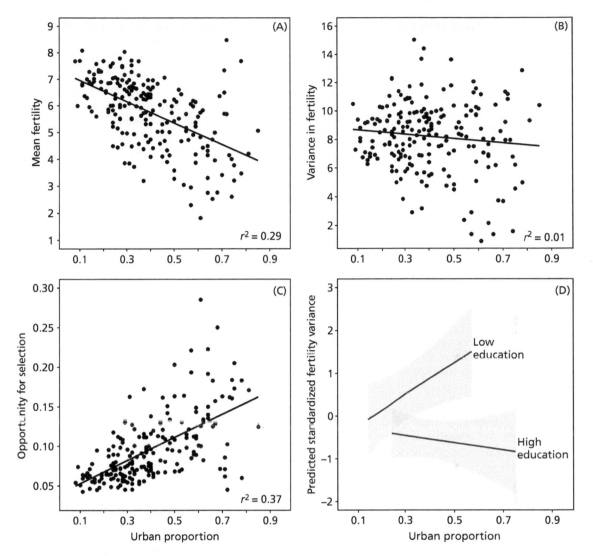

Figure 16.1 Human fertility and urbanization. (A–C) Relationship between female fertility statistics (mean, variance, or opportunity for selection) and proportion of population living in urban areas. Data points are means for fertility from 200 surveys conducted in 72 countries between 1990 and 2014 (one to eight surveys per country). Lines are slopes of least-squares regression with only the urban proportion (UP) entered as predictor. (D) Predicted variance in fertility (lines) ± 95 per cent confidence intervals (shaded areas) from a model controlling for mean household wealth *per capita* and proportion of population with at least a primary education. Only the most recent survey for each country was included in the analysis. Relationship between predicted fertility and urban residence is shown separately for the first ('low') and third ('high') terciles for education, which are composed of populations with respectively 21.9 per cent and 95.8 per cent of educated people, on average. Predictions were done using these average proportions and with the minimal wealth value observed for each tercile. Note that these regression analyses conducted on aggregated data should serve to stimulate further investigation, not to draw final conclusions. (Aggregated data and variance statistics from Hruschka and Burger (2016) kindly provided by the authors; original data from the Demographic and Health Surveys Program (ICF 2004–2017).)

Some caution is advised in interpreting the results in Figure 16.1. First, they do not tell us how much of the variation reflects interindividual differences in fertility rates vs random deviations of a Poisson process under a homogeneous rate (Hruschka and Burger 2016). Second, Crow's index is a crude measure of the opportunity for selection, since it does not account for age-specific vital rates that mould the covariance between fertility and relative fitness (Moorad 2013).

16.4.2 Urban agents of selection: stressors

Given that there is opportunity for selection, we may ask if human traits are targeted by selective agents specific to cities. It is practical to divide these agents into 'stressors' (e.g., chemicals) and 'sociocultural factors', although this separation is fuzzy (i.e., stressors are in part the products of cultural changes). While the opportunity for selection in modern societies is dominated by the variance in fertility traits, urban stressors can create new opportunities for selection though differences in survival, as we illustrate here with the example of air pollution.

Air pollution is composed of a diversity of toxic particles and molecules that enter the body through respiratory and digestive tracts, including oxides emitted by vehicles (NO_x, SO_2, CO), heavy metals, and oil-related compounds like naphthalene, styrene, and toluene that are contained in chemical solvents. Pollution increases oxidative stress in animals living in cities and can decrease their fitness (Chapter 13; Isaksson 2010). Similarly, it generates oxidative damage and triggers inflammatory responses in humans, causing or aggravating respiratory, cardiovascular, developmental, immune, and degenerative diseases (D'Amato 2002; Kampa and Castanas 2008). These health outcomes have consequences on fitness components. The World Health Organization (2018) estimated that 4.2 million persons died from ambient and 3.8 million from household air pollution in 2016, including tens of thousands of children. It is conceivable that individuals vary in their genetic capacity to endure high pollutant concentrations, prompting selection for pollution resistance. For instance, cytochrome P450 genes code for enzymes that can metabolize organic compounds emitted by chemical solvents and smoke. One of these enzymes, produced by the P450 2A13 gene, seems to be particularly efficient and is polymorphic (Fukami et al. 2008). Fitness costs of pollution can also vary with age. For instance, survival can be particularly compromised in children suffering from respiratory diseases, or alternatively in older individuals whenever costs depend on exposure time (e.g., accumulation of oxidative damage or of heavy metals (Kampa and Castanas 2008)). Therefore, adding an age-dependent layer of mortality increase could reshape selection on senescence rates in highly polluted areas.

Urban stressors can also affect growth, development, and fertility. For instance, Huang et al. (2018) report a negative correlation between body mass index in boys aged 9–15 and exposure to dioxides *in utero* or during childhood in Hong Kong. Pollution reduces fertility in both animals (e.g., clutch size in birds (Isaksson 2010)) and humans (Nieuwenhuijsen et al. 2014). Another example involves sleep patterns. Urban noise, artificial lighting, and density can disturb sleep, impacting metabolism, immune response, cognitive functions, sexual habits, and investment in reproduction (Nunn et al. 2016). This could select for sleep architectures (e.g., number and duration of sleep bouts) that are better adapted in urban areas (see also section 16.4.3). Sleep architecture exhibits genetic variation (Jones et al. 2013) and could respond to selection.

These examples of urban stressors that could feed human evolution are far from exhaustive. Others include road traffic increasing risks of accidental deaths or injuries for drivers, pedestrians, and cyclists, global warming accentuating heat islands, in which strokes during heat waves can be deadly (Tan et al. 2010), and pesticide-resistant parasites in densely populated city cores, such as lice, bed bugs, and rats that can be vectors of diseases (Johnson and Munshi-South 2017). Much remains to be explored in that matter.

16.4.3 Urban agents of selection: sociocultural factors

Cities superpose two heterogeneous layers—habitats and inhabitants—that make up the urban landscape of natural selection. City cores, suburbs, neighbourhoods, industrial zones, commercial sectors, parks, and other kinds of areas altogether form a mosaic of nested and interconnected habitats (Chapter 4) used by populations that are themselves stratified into social classes, ancestry groups, professional networks, and cultural assemblages (e.g., leisure, religion). It is not easy to extract from this complexity the sociocultural (including economic) factors that locally mould human traits and fitness. We can nevertheless identify broad types of factors—wealth, inequalities, social interactions, learning, lifestyle—that probably shape the urban quilt of selective pressures.

In developed countries, the spectacular growth of wealth increases the fitness payoff to parents of investing additional resources in offspring quality, for example in cognitive capital through higher education, at the expense of offspring number (Kaplan 1996; Gries and Grundmann 2018; but see Stulp and Barrett 2016). Individuals may trade their reproduction with investment in own education to gain socioeconomic returns in the market economy, as suggested by a study by Alvergne and Lummaa (2014) of Mongolian populations; this study found that such a strategy only paid off in urban areas. In a context of low juvenile mortality and enhanced contribution of reproductive traits to the opportunity for selection (section 16.4.1), mate competition may influence this greater investment in self or offspring capital (Mace 2008; Alvergne and Lummaa 2014) and boost the evolutionary significance of sexually selected characters in urban populations.

Inequalities in income, education, medical care, and access to a sanitary environment peak in urban areas and create disparities in health and survival of individuals from different socioeconomic strata (Bor et al. 2017). Individuals may consequently express behavioural and life-history strategies that reflect their resources/status, and underlying evolved tradeoffs. For example, one hypothesis proposed to explain the greater prevalence of unhealthy habits, such as smoking, in lower social classes is that people have less incentive to engage in healthy behaviours when they perceive the risk of extrinsic mortality to be higher (Pepper and Nettle 2014). It rests on the assumption that investing in healthy behaviour is costly because it diverts energy from other adaptive functions such as resource acquisition or mating (Pepper and Nettle 2014). Whether such modulations in health-related behaviours remain adaptive in contemporary cities should depend on how well the perception of mortality risks by citizens correlates with real risks (Adams et al. 2014). Plausibly, interactions between social class and urban microhabitats, which can be more or less healthy and safe, create heterogeneity in selection within cities. Unhealthy habits could also develop when people must trade-off among demanding vital activities. Fast-food consumption is a good example. Zagorsky and Smith (2017) found that Americans living in city cores are more frequent fast-food eaters. However, contrary to a common perception, fast-food consumption was not greater within lower socio-economic classes but, if anything, somewhat higher among middle-class people. Longer working hours and, to a lesser extent, fertility are among the factors that correlated positively with fast-food consumption. This suggests the hypothesis that people who work long hours and have limited leisure time could rear on average (slightly) larger families when relying on fast food on a regular basis. If so, the fitness payoff of this strategy should vary as a function of access to fast food, which increases along the rural–urban gradient (Zagorsky and Smith 2017).

Urbanization modifies social interactions as well. In particular, low fertility rates and genetic mixing (see section 16.4.5) in modern cities mean fewer close relatives to interact with and a lower return in inclusive fitness expected from kin-oriented behaviours. The multiplication of blended families can also change the fitness return of intrafamilial interactions (competition, cooperation) when half- or step-siblings are involved, hence the expression of (and selection on) life-history strategies. For example, Clutterbuck et al. (2015) found that English female teenagers who shared their homes with more half-/step-siblings benefitted from less familial support, experienced several residential relocations, and reached menarche younger.

As the influence of relatives declines, the horizontal transmission of culture increases at the expense of the vertical transmission along familial lineages (Creanza et al. 2017). Cultural norms acquired through social learning centred on non-kin (e.g., Internet communities, professional networks) sometimes promote healthy behaviours, such as physical activity, balanced nutrition, and vaccination, and sometimes unhealthy behaviours, such as drinking and risk-taking. They may also influence decisions that impact the reproduction of parents or the growth, survival, and health of their offspring. These include, for example, requesting genetic screening, avoiding products that contain harmful molecules (e.g., endocrine disruptors), or choosing to breast-feed. Cultural factors interact with local urban conditions. For instance, incentives to engage in active transport (i.e., walking or using self-propelled vehicles) for daily commuting may be higher in cities due to cultural norms and infrastructures (e.g., bike paths or availability of local trade facilities), while exposure to air pollution in cities where traffic is dense can mitigate the health benefits of active transport (Maizlish et al. 2013).

To better understand how cultural norms and learning modulate selection within different urban strata, we need to develop models connecting spatial and cultural variables at fine scale. It would be interesting, for example, to study how vaccine hesitancy, recently listed among the biggest challenges to human health by the World Health Organization, causes heterogeneity in infection risks and in selection for immune response, as a function of the spread of, and adherence to, anti-vaccine ideas in groups that exploit urban habitats differently, particularly when those groups are endogamous.

Lastly, lifestyles fashioned by sociocultural factors can impose new forms of selection. Returning to the example of sleep, Nunn et al. (2016) propose that evolutionary tradeoffs between sleep and other fitness-related functions (e.g., foraging) may be mismatched with urban life, inducing symptoms such as narcolepsy and seasonal affective disorder that previously may have had an adaptive component (e.g., tonic immobility associated with narcolepsy might have served to simulate death in response to approaching predators (Nunn et al. 2016)). Flexible work schedules in specialized jobs, norms against mother–child co-sleeping, and night-life socialization, among others, could select for sleep architectures that differ from traditional ones through their effects on metabolism, immunity, cognitive functions, juvenile survival, and other traits affecting fitness.

16.4.4 Scale of urban selection

Scaling effects that are important in biological and ecological systems (see also Chapter 2) include allometric relationships among phenotypic traits and habitat selection in hierarchically structured landscapes. Cities vary in scale, from small towns to huge megalopolises. Recent research has shown that major urban patterns and functions, such as infrastructure, *per capita* gross domestic product (GDP), income, economic specialization, traffic congestion, crowding, social interactions, innovation, and disease transmission scale non-linearly with urban population size (N) according to some basic

rules (Bettencourt et al. 2007; Bettencourt 2013). For example, social interactions, traffic congestion, crowding, and economic specialization increase faster than N, while the reverse is true for infrastructures. The scale effects could shape trait expression and selection. In particular, Burnside et al. (2012) state that resources and energy flow in urbanized societies enable 'people to extract and transport astronomically more resources and live at much higher population densities but at the cost of a slower life history and lower reproductive rate'. It takes a huge amount of extra-metabolic energy (e.g., fossil, hydroelectric, or nuclear power) to produce a child who will be competitive in a modern society. Thus, Burnside et al. (2012) suggest that urban scaling could extend the well-documented allometric scaling between metabolic (B) and fertility (F) rates in mammals, whereby $F \propto B^{-1/3}$: as city size increases, more extra-metabolic energy is used *per capita* (i.e., similar to an increase in B), so fertility decreases because each offspring is more costly to produce.

Another example of urban-scale effects on human traits is indirectly suggested by a recent study conducted on pregnant women from New Jersey. Wang and Yang (2019) found that longer commuting distances between women's home and their work are associated with poor fetal growth, higher maternal stress, and less access to prenatal care. If we think of urban sprawl, we can easily imagine how urban scaling in mobility could contribute to variation in these response variables, and, provided that they correlate with fitness (which was not investigated in this study), to selection. Therefore, linking urban-scaling effects on natural selection seems relevant for future research in urban evolutionary biology.

16.4.5 Urban disease genetics

A few studies suggest that urbanization can trigger changes in genetic parameters that shape the prevalence and penetrance of diseases and thus their evolutionary role. First, the mixing of ancestries and the large population sizes of urban areas can decrease homozygosity. The risk of developing several recessive disorders and complex diseases, including cancer and mental disorders, is known to increase with homozygosity. In their cohort study of North Americans of European descent, Nalls et al. (2009) found a decline during the twentieth century in autozygosity, the genomic signature of consanguinity, a pattern they suggest is explained by increased mobility and urbanization.

Second, mobility may favour adaptive introgression, as it did in our recent evolutionary history for alleles increasing infectious disease resistance (Johnson and Voight 2018). If specific alleles are advantageous across populations sharing similar urban conditions (see Chapter 3), they could be selected locally, spread with migration, and help to mitigate diseases in cities. Massive immigration and population admixture can also modify the frequency of beneficial or deleterious alleles already present in urban populations. For example, Aceves et al. (2006) documented a decline in the frequency of allele ε4 at the apolipoprotein E (*APOE*) gene, when going from rural to urban Mexican areas, presumably due to greater ethnic mixing in the latter. This allele, which is associated with higher cholesterol levels and involved in mismatches, is advantageous in calorie-restricted nutritional environments, where it improves the cognitive development of children by protecting them against frequent diarrhoea (Oriá et al. 2005). Polymorphism at *APOE* may also allow for the adjustment of cholesterol levels to the metabolic requirements imposed by external temperature, so that ε4 frequency, which ranges from 0 to ~ 0.4 across human populations, may reflect geographic adaptation to climate (Eisenberg et al. 2010). However, ε4 is harmful in our calorie-rich modern environments, where it increases the risk of developing heart disease, Alzheimer's, and arteriosclerosis (Eisenberg et al. 2010). Carriers of ε4 may be at greater risk of developing Alzheimer's in areas with greater air pollution (Calderón-Garcidueñas et al. 2015), suggesting that interactions between pollution and other factors, such as nutrition, modulate the fitness of ε4 carriers in urban populations.

Third, the frequency of alleles and polygenic architecture of complex traits and diseases may differ across rural–urban gradients and among urban populations (see Chapter 5). For example, Colodro-Conde et al. (2018) found that individuals with higher polygenic risk scores for schizophrenia tended to live in zones where population density is higher, possibly due in part to non-random migration of genes, although density explained a small amount in schizophrenia risk. Comparing additive genetic

(G) matrices for correlated health and fitness traits across populations is perhaps an important next step in determining how urbanity modulates evolutionary potential. Genomic datasets from large cohort studies are increasingly exploited to estimate quantitative genetic parameters (e.g., Tropf et al. 2015) and should offer growing opportunities to compare populations living in different habitats. Even though these datasets may not include urban descriptors, some can be gathered from open-data environments (Xu et al. 2017), provided that, at a minimum, the location of residence of participants is known.

Finally, changes in gene expression could be a fast track to adapting to urban life. Several factors that have been changing rapidly since the transition to modernity, including nutrition, pollution, and social interactions, are known to modulate gene expression, and their epigenetics varies across habitats (e.g., forest, agricultural, urban (Fagny et al. 2015)) and along rural–urban gradients (see Chapter 5). Johnson and Voight (2018) report that a large proportion of genetic variants that were under positive selection in recent human history appear to be in non-protein coding segments, and suggest that regulatory genes may have been hotspots for recent evolution. They identified quantitative trait loci associated with gene expression across several tissues. We now need research that reveals the mechanisms mediating changes that have actually occurred in humans living in cities.

16.5 Wrapping up: eco-evolutionary dynamics in the city

By modifying phenotypes, rapid evolution can alter population dynamics that in turn change the selection exerted on the evolving traits (Kokko and Lopez-Sepulcre 2007; Hendry 2017). Some evidence for such eco-evolutionary dynamics exists for at least one pre-industrial population (Box 16.3) and provides insight into how it could also occur in urban societies. It shows that rapid evolution in life history can inflate population growth, with potential feedbacks on selection on vital rates.

The abrupt shift in human ecology triggered by urbanization could spark eco-evolutionary feedbacks. Several traits whose expression changed with the transition to modernity, including fertility traits, are

Box 16.3 Eco-evolutionary dynamics in a human population

Pelletier et al. (2017) showed that the rapid evolution of maternal age at first reproduction (AFR) reported for women on Île aux Coudres (Milot et al. 2011) increased the population growth rate. Between 1772 and 1880, they estimated that the size of the population would have ended up about 12 per cent smaller without evolution (Figure 16.2A). Using the same data here, we observe an increase in the strength of selection on juvenile survival during the same period (Figure 16.2B). Therefore, as AFR shifted toward earlier ages, the contribution of juvenile survival to the total variance in fitness increased. While we did not measure the specific contribution of the genetic change in AFR to the trend depicted

in Figure 16.2B, it is expected that earlier reproductive onset will increase juvenile mortality (Stearns 1992). Therefore, it is plausible that the biological evolution of AFR partly explains the increased variance in juvenile survival on Île aux Coudres. This hypothesis should be formally tested using quantitative genetic methods. Importantly, the trend in Figure 16.2B suggests that mutations increasing juvenile mortality risk were more strongly counter-selected at the turn of the twentieth century than at the turn of the nineteenth century. This intensification of selection could produce positive eco-evolutionary feedbacks by improving juvenile survival, which would further increase selection for earlier AFR by reducing its cost.

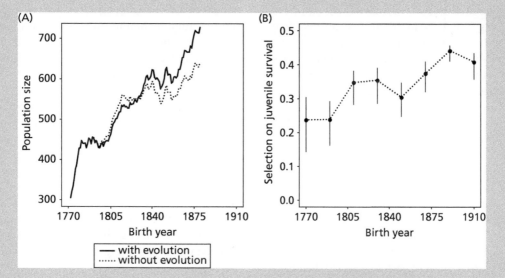

Figure 16.2 Elements of eco-evolutionary dynamics in the population of Île aux Coudres. (A) Observed population growth (solid line) compared to that expected without AFR evolution (dotted line; reproduced and adapted from Pelletier et al. (2017). Original source licensed under a Creative Commons Attribution 4.0 International Licence <https://creativecommons.org/licenses/by/4.0/>). (B) Selection on juvenile (0–15 years old) survival for women born 1770–1910. Values are covariances between survival and relative fitness (± 95 per cent CI) from regression models incorporating all vital rates (i.e., equivalent to Hamilton's sensitivities; see Moorad (2013) for methods). (Data from the Île aux Coudres Population Register provided by F.M. Mayer and E. Milot.)

---

both heritable and under selection in contemporary populations. The proportion of total phenotypic variation that is due to interindividual genetic differences could now be higher than before, augmenting evolutionary potential, as found by Bolund et al. (2015) in Finnish populations for heritable life-history traits. As these authors suggest, the transition to modernity ensures a better access to resources that may decrease the environmental component of phenotypic

variance, while the relaxation of social norms on reproductive decisions may further facilitate the expression of individual genetic differences. Moreover, the pace of human life (energy consumption, social interactions, movement, etc.) and of human-induced changes in ecosystems, accelerates with city size (Bettencourt et al. 2007; Alberti et al. 2016), helping to maintain evolutionary disequilibrium and, presumably, eco-evolutionary dynamics.

Rapid evolution may have practical, important implications for human societies. Burger and DeLong (2016) identified microevolution as one of several factors that could modify fertility in the future but are not yet considered in demographic forecasts. The contribution of evolution of age at first birth to the population growth rate of the pre-industrial population of Île aux Coudres caused dynamics to differ significantly from the forecasts that demographers would have made using standard projection analysis (Box 16.3; Pelletier et al. 2017). Even small changes in the rate of population growth or decline may have a major impact on social policy and anthropogenic change.

16.5.1 Eco-evolutionary dynamics of health

Health traits may be particularly susceptible to eco-evolutionary dynamics because the strong age component often found in their expression should make them sensitive to shifts in selection on vital rates, and hence on genetic variants affecting health. For instance, we have seen that the effect of air pollution on fitness components can vary with age. To take a hypothetical example, selection for faster life-history strategies (e.g., for earlier age at first reproduction) could result in demographic changes that strengthen or relax selection for pollution-resistant alleles in highly polluted areas, depending on the age-specific vital rates they alter. Conversely, a change in the frequency of an allele modulating the physiological response to pollution could have a feedback effect on life-history evolution. For example, pollution-induced child mortality or poor juvenile growth (e.g., see section 16.4.2) could trigger responses in life history, for example by delaying maturity, a trend that could be attenuated or modified as the beneficial allele spreads. Alternatively, when the severity of a pollution-related health problem depends on exposure time and thus on age, faster life-history strategies may be favoured because air pollution differentially increases the extrinsic mortality of older individuals. The quantitative impact that such processes would have on selection for pollution-resistant individuals will thus depend both on the genetic architecture of pollution resistance and on the age-specific fitness components most affected by pollution.

16.5.2 Implications for optimality models

Behavioural ecology and economics have sought to explain the (mal)adaptive value of life-history changes in contemporary societies, relying strongly on optimality models (Borgerhoff Mulder 1998; Shenka et al. 2013). Such models typically neglect the possibility of rapid evolution (Box 16.3). In addition, empirical studies testing the predictions of these models often use fitness proxies that mix the fitness of the parents with that of their descendants, such as the number of surviving children or of grandchildren, conflating selection and inheritance (Thomson and Hadfield 2017). These problems will aggravate the discrepancy between the realised response (e.g., of fertility) in populations where life history evolves rapidly and the response predicted by optimality models. The latter are nevertheless important to explain individual reproductive decisions in urban societies and to predict their effects on eco-evolutionary dynamics. For example, the continued selection for earlier AFR in several populations seems to contradict adaptive hypotheses proposed by anthropologists to explain low fertility in contemporary populations. However, selection on AFR is typically estimated using direct fitness that does not capture indirect benefits or costs to offspring that contribute to inclusive fitness. No single approach is likely, by itself, to explain the adaptive value of reproductive patterns in cities. Individual-based models incorporating realistic eco-evolutionary dynamics offer some hope.

16.6 Conclusions

16.6.1 Challenges for future research

In this chapter, we discussed the elements needed to understand the evolutionary potential of humans living in cities. These include:

- the major shift in the opportunity for selection that comes with the transition to modernity;
- the concomitant changes in our physical, social, economic, nutritional, and epidemiologic environment modifying the expression of evolutionary tradeoffs and genetic variance;
- the population and quantitative genetic consequences of urbanization;

- urban challenges to health and their fitness consequences;
- the heterogeneity and scaling of urban selection; and
- the non-equilibrium state of contemporary populations, which may fuel eco-evolutionary feedbacks.

We underlined the limitations of available data that could be used to quantify these influences on human evolution in urban populations. However, some of the large cohort datasets, collected for medical research, contain variables related to fitness that have been used to measure selection (e.g., Framingham (Byars et al. 2010)). Fortunately, the idea that evolutionary biology can contribute to medical research is gaining traction (Wells et al. 2017). This momentum should be used to promote a more systematic collection of data tailored towards evolutionary biology studies (e.g., fitness components, environmental factors that include urban descriptors). We also proposed approaches to assessing the role of urban habitats in shaping human evolution, including changes in selection along gradients of urbanization, as is done in animal and plant studies and as discussed in other chapters of this book.

A major gap in our knowledge of evolutionary processes in contemporary human populations is the lack of measures of selection in non-western societies. The study of human microevolution has not been characterized by the breadth of ethnocultural coverage that marks fields like anthropology and human genetics. Selection coefficients and genetic variance parameters are usually measured on populations with European backgrounds (with some exceptions (e.g., Courtiol et al. 2013)). Cities in the Middle East, Africa, and Asia are likely to differ in the factors that shape selection, such as population density, climate, pollution, and cultural norms.

Since the transition to modernity, human populations have certainly not been at genetic, demographic, or cultural equilibrium. Continuing urbanization will help to maintain that disequilibrium. In this context, we need the insights from both optimality and microevolutionary models to understand the widespread changes in life-history and health traits that are occurring in human populations. We must also consider how the switch from mainly vertical to more horizontal cultural transmission alters functional behavioural responses, in particular those pertaining to mating,

fertility, and health. Better integration of social learning, gene-culture coevolution, and evolutionary economics with evolutionary genetics is desirable.

Finally, little is known about how the connexion between cities and the countryside, via commuting and gene flow, modulates selection and the responses to it. To use the source-sink metaphor (Dias 1996), when cities first formed, they were presumably sinks for rural genes that were mismatched to the new environment (Cochran and Harpending 2009). As cities have grown to dominate the global population numerically, that relationship may have started to reverse, with cities becoming the sources and the countryside becoming the sink. How that will play out on evolutionary processes should be one direction for future research.

16.6.2 Transhumanism: the rise of a new selective force

We close this chapter with some thoughts on an issue that has received little attention from evolutionary biologists despite its potential major impacts on our future evolution: transhumanism, i.e., the editing of human genes and phenotype. The transhumanist agenda is ambitious: correcting our deficiencies, enhancing our capacities (e.g., computer–neuron interfaces), extending our lifespan, and directing our future evolution (Askland 2011; Groff 2015). We are not primarily interested in such Promethean dreams, which neglect the multitrait, ecological, stochastic, and non-teleological context in which evolution by natural selection takes place. Instead we focus on the immediate consequences that transhumanism might have on microevolutionary processes. There are several reasons why gene editing techniques, in particular those such as CRISPR-Cas9 that can target the germline, may become new agents of selection.

As for conventional health care, access to costly gene editing technologies is unlikely to be either universal or uniform along rural–urban gradients (e.g., Sibley and Weiner 2011). People living in cities will be more likely to have access to gene editing, including interventions using CRISPR-Cas9. People will also differ in their willingness to use these technologies. Therefore, any genetically or culturally heritable trait that influences the probability of using gene editing for the sake of improving survival or fertility, for self or offspring, will potentially translate into selection on that trait that depends on class and geography. In addition, that some genes modulating disease expression show balanced polymorphism in human populations (Kozubek 2018; Sato and Kawata 2018) suggests that manipulating them would have fitness consequences that vary across populations and environments.

People who will have recourse to gene editing could, consciously or not, trade-off different components of their fitness or that of their offspring. For instance, 'repairing' the *BRCA* alleles that increase cancer risks could affect fertility traits (section 16.3.1). Would the net effect on fitness favour or disfavour those desired alleles? People could likewise trade-off competitive advantage with disease risks. 'Gene doping' of physical or cognitive performance may be practicable soon (Kozubek 2018), and variants improving executive functions may become particularly attractive in urban societies where competition for specialized jobs, socioeconomic status, and attractiveness on the mating market is intense. For example, variation at the *COMT* gene modulates prefrontal dopamine catabolism and correlates with functions such as concentration and decision-making, but also with the expression of various diseases (Wishart et al. 2011; Kozubek 2018; Satterfield et al. 2018). The effect of *COMT* on cognitive function may further depend on sleep homeostasis (Satterfield et al. 2018), which, as we saw, is likely to be disturbed by urban stressors and lifestyle. Thus, editing pleiotropic genes like *COMT* would potentially alter the expression of focal and correlated traits in ways that vary along rural–urban gradients. How would the fitness of different life-history strategies be affected in modern populations dominated by sexual and fertility selection?

Transhumanist ideas deserve more attention from evolutionary biologists and ecologists. Gene editing with CRISPR-Cas9 is ready to jump into the evolutionary game and may not wait for public and legal approval, for it is surprisingly easy to use, especially for germ-line manipulations. We doubt that such technologies will annihilate selection. Rather, we should ask how they might interact with microevolutionary processes in urbanized societies. We have barely glimpsed what our biological future could be, but we can bet that natural selection will be a part of it, and the role of culture in determining selection in humans will only grow stronger.

Acknowledgements

We thank Daniel Hruschka and Oskar Burger for providing data and comments on Figure 16.1. We also thank Jacinthe Gosselin and Raphaëlle Mercier Gauthier for the selection analyses used in Figure 16.2B, as well as Jacob Moorad for advice on methods. We thank Marta Szulkin, Colin Garroway, Erik Postma, Anne Charmantier, and Jason Munshi-South for providing useful comments on the original manuscript. Part of this research was supported by the Natural Sciences and Engineering Research Council of Canada (Discovery Grant RGPIN-2015-03683) to E.M. and by the Québec Fonds de recherche Nature et Technologies (New Researchers Grant 2017-NC-197156) to E.M.

References

Aceves, D., Ruiz, B., Nun, P., et al. (2006). Heterogeneity of apolipoprotein E polymorphism in different Mexican populations. *Human Biology*, 78, 65–75.

Adams, J., Stamp, E., Nettle, D., Milne, E.M.G., and Jagger, C. (2014). Socioeconomic position and the association between anticipated and actual survival in older English adults. *Journal of Epidemiology and Community Health*, 68, 818–25.

Alberti, M., Marzluff, J., and Hunt, V.M. (2016). Urban driven phenotypic changes: empirical observations and theoretical implications for eco-evolutionary feedback. *Philosophical Transactions of the Royal Society B*, 372: 20160029.

Alvergne, A. and Lummaa, V. (2014). Ecological variation in wealth-fertility in Mongolia: the 'central theoretical problem of sociobiology' not a problem after all? *Proceedings of the Royal Society B: Biological Sciences*, 281: 20141733.

Askland, A.S. (2011). The misnomer of transhumanism as directed evolution. *International Journal of Emerging Technologies and Society*, 9, 71–8.

Balter, M. (2005). Are humans still evolving? *Science*, 309, 234–7.

Barnes, I., Duda, A., Pybus, O.G., and Thomas, M.G. (2010). Ancient urbanization predicts genetic resistance to tuberculosis. *Evolution*, 65, 842–8.

Beauchamp, J.P. (2016). Genetic evidence for natural selection in humans in the contemporary United States. *Proceedings of the National Academy of Sciences of the United States of America*, 113, 7774–9.

Bettencourt, L.M.A. (2013). The origins of scaling in cities. *Science*, 340, 1438–41.

Bettencourt, L.M.A., Lobo, J., Helbing, D., Kühnert, C., and West, G.B. (2007). Growth, innovation, scaling, and the pace of life in cities. *Proceedings of the National*

Academy of Sciences of the United States of America, 104, 7301–306.

Bolund, E., Hayward, A., Pettay, J.E., and Lummaa, V. (2015). Effects of the demographic transition on the genetic variances and covariances of human life-history traits. *Evolution*, 69, 747–55.

Bor, J., Cohen, G.H., and Galea, S. (2017). Population health in an era of rising income inequality: USA, 1980–2015. *Lancet*, 389, 1475–90.

Borgerhoff Mulder, M. (1998). The demographic transition: are we any closer to an evolutionary explanation? *Trends in Ecology and Evolution*, 13, 266–70.

Boyle, E.A., Li, Y.I., and Pritchard, J.K. (2017). An expanded view of complex traits: from polygenic to omnigenic. *Cell*, 169, 1177–86.

Burger, O. and DeLong, J.P. (2016). What if fertility decline is not permanent? The need for an evolutionarily informed approach to understanding low fertility. *Philosophical Transactions of the Royal Society B*, 371, 20150157.

Burnside, W.R., Brown, J.H., Burger, O., et al. (2012). Human macroecology: linking pattern and process in big-picture human ecology. *Biological Reviews*, 87, 194–208.

Byars, S.G., Ewbank, D., Govindaraju, D.R., Stearns, S.C., and Ellison, P.T. (2010). Natural selection in a contemporary human population. *Proceedings of the National Academy of Sciences of the United States of America*, 107, 1787–92.

Byars, S.G., Huang, Q.Q., Gray, L.-A., et al. (2017). Genetic loci associated with coronary artery disease harbor evidence of selection and antagonistic pleiotropy. *PLOS Genetics*, 13, e1006328.

Calderón-Garcidueñas, L., Mora-Tiscareño, A., Franco-Lira, M., et al. (2015). Decreases in short term memory, IQ, and altered brain metabolic ratios in urban apolipoprotein ε4 children exposed to air pollution. *Journal of Alzheimer's Disease*, 45, 757–70.

Champion, T. and Hugo, G. (2003). *New Forms of Urbanization: Beyond the Urban–Rural Dichotomy*. Ashgate Publishing, Aldershot.

Clutterbuck, S., Adams, J., and Nettle, D. (2015). Frequent residential relocations cumulatively accelerate menarcheal timing in a sample of English adolescent girls. *Journal of Biosocial Science*, 47, 188–202.

Cochran, G. and Harpending, H. (2009). *The 10,000 Year Explosion: How Civilization Accelerated Human Evolution*. Basic Books, New York.

Colodro-Conde, L., Couvy-Duchesne, B., Whitfield, J.B., et al. (2018). Association between population density and genetic risk for schizophrenia. *JAMA Psychiatry*, 75, 901–910.

Corbett, S., Courtiol, A., Lummaa, V., Moorad, J., and Stearns, S.C. (2018). The transition to modernity and chronic disease: mismatch and natural selection. *Nature Reviews Genetics*, 391, 1–12.

Courtiol, A., Rickard, I.J., Lummaa, V., et al. (2013). The demographic transition influences variance in fitness and

selection on height and BMI in rural Gambia. *Current Biology*, 23, 884–9.

Creanza, N., Kolodny, O., and Feldman, M.W. (2017). Cultural evolutionary theory: how culture evolves and why it matters. *Proceedings of the National Academy of Sciences of the United States of America*, 114, 7782–9.

Crow, J.F. (1958). Some possibilities for measuring selection intensities in man. *Human Biology*, 30, 1–13.

D'Amato, G. (2002). Environmental urban factors (air pollution and allergens) and the rising trends in allergic respiratory diseases. *Allergy*, 57(S72), 30–33.

Daub, J.T., Hofer, T., Cutivet, E., et al. (2013). Evidence for polygenic adaptation to pathogens in the human genome. *Molecular Biology and Evolution*, 30, 1544–58.

DeSantis, C.E., Ma, J., Sauer, A.G., Newman, L.A., and Jemal, A. (2017). Breast cancer statistics, 2017, racial disparity in mortality by state. *CA: A Cancer Journal for Clinicians*, 67, 439–48.

Dias, P.C. (1996). Sources and sinks in population biology. *Trends in Ecology & Evolution*, 11, 326–30.

Eisenberg, D.T.A., Kuzawa, C.W., and Hayes, M.G. (2010). Worldwide allele frequencies of the human apolipoprotein E gene: climate, local adaptations, and evolutionary history. *American Journal of Physical Anthropology*, 143, 100–111.

Fagny, M., Patin, E., MacIsaac, J.L., et al. (2015). The epigenomic landscape of African rainforest hunter-gatherers and farmers. *Nature Communications*, 6, 10047.

Field, Y., Boyle, E.A., Telis, N., et al. (2016). Detection of human adaptation during the past 2000 years. *Science*, 354, 760–64.

Fukami, T., Katoh, M., Yamazaki, H., Yokoi, T., and Nakajima, M. (2008). Human cytochrome P450 2A13 efficiently metabolizes chemicals in air pollutants: naphthalene, styrene, and toluene. *Chemical Research in Toxicology*, 21, 720–25.

Gagnon, A. and Mazan, R. (2009). Does exposure to infectious diseases in infancy affect old-age mortality? Evidence from a pre-industrial population. *Social Science & Medicine*, 68, 1609–16.

Gould, S.J. (2000). The spice of life. *Leader to Leader*, Winter 2000, 14–19.

Gries, T. and Grundmann, R. (2018). Fertility and modernization: the role of urbanization in developing countries. *Journal of International Development*, 30, 493–506.

Groff, L. (2015). Introduction to second special issue. *World Futures Review*, 7, 113–15.

Hendry, A.P. (2017). *Eco-Evolutionary Dynamics*. Princeton University Press, New Jersey.

Hruschka, D.J. and Burger, O. (2016). How does variance in fertility change over the demographic transition? *Philosophical Transactions of the Royal Society B*, 371, 20150155.

Huang, J.V., Leung, G.M., and Schooling, C.M. (2018). The association of air pollution with body mass index: evidence from Hong Kong's 'Children of 1997' birth cohort. *International Journal of Obesity* 43(Suppl 2).

Isaksson, C. (2010). Pollution and its impact on wild animals: a meta-analysis on oxidative stress. *EcoHealth*, 7, 342–50.

Jain, K. and Stephan, W. (2017). Modes of rapid polygenic adaptation. *Molecular Biology and Evolution*, 34, 3169–75.

Johnson, K.E. and Voight, B.F. (2018). Patterns of shared signatures of recent positive selection across human populations. *Nature Ecology and Evolution*, 2, 713–20.

Johnson, M.T.J. and Munshi-South, J. (2017). Evolution of life in urban environments. *Science* 358, eaam8327.

Jones, C.R., Huang, A.L., Ptáček, L.J., and Fu, Y.-H. (2013). Genetic basis of human circadian rhythm disorders. *Experimental Neurology*, 243, 28–33.

Kampa, M. and Castanas, E. (2008). Human health effects of air pollution. *Environmental Pollution*, 151, 362–7.

Kaplan, H.S. (1996). A theory of fertility and parental investment in traditional and modern human societies. *Yearbook of Physical Anthropology*, 39, 91–135.

Klein, L.D., Huang, J., Quinn, E.A., et al. (2018). Variation among populations in the immune protein composition of mother's milk reflects subsistence pattern. *Evolution, Medicine, and Public Health*, 2018, 230–45.

Kokko, H. and Lopez-Sepulcre, A. (2007). The ecogenetic link between demography and evolution: can we bridge the gap between theory and data? *Ecology Letters*, 10, 773–82.

Kong, A., Frigge, M.L., Thorleifsson, G., et al. (2017). Selection against variants in the genome associated with educational attainment. *Proceedings of the National Academy of Sciences of the United States of America*, 114, E727–32.

Kozubek, J. (2018). *Modern Prometheus: Editing the Human Genome with Crispr-Cas9*. Cambridge University Press, Cambridge.

Lawson, D.W. and Borgerhoff Mulder, M. (2016). The off-spring quantity–quality trade-off and human fertility variation. *Philosophical Transactions of the Royal Society B*, 371, 20150145.

Liston, A., Carr, E.J., and Linterman, M.A. (2016). Shaping variation in the human immune system. *Trends in Immunology*, 37, 637–46.

Lynch, M. (2016). Mutation and human exceptionalism: our future genetic load. *Genetics*, 202, 869–75.

Mace, G.M. (2008). Reproducing in cities. *Science*, 319, 764–6.

Mace, R. (2000). An adaptive model of human reproductive rate with wealth inheritance: why people have small families. In: Cronk, L., Chagnon, N., and Irons, W. (eds) *Adaptation and Human Behavior: An Anthropological Perspective*, pp. 261–81. Aldine de Gruyter, Hawthorne, NY.

Maizlish, N., Woodcock, J., Co, S., et al. (2013). Health cobenefits and transportation-related reductions in greenhouse gas emissions in the San Francisco Bay area. *American Journal of Public Health*, 103, 703–709.

Merilä, J., Kruuk, L.E.B., and Sheldon, B.C. (2001). Cryptic evolution in a wild bird population. *Nature*, 412, 76–9.

Milot, E., Mayer, F.M., Nussey, D.H., et al. (2011). Evidence for evolution in response to natural selection in a contemporary human population. *Proceedings of the National Academy of Sciences of the United States of America*, 108, 17040–45.

Milot, E., Moreau, C., Gagnon, A., et al. (2017). Mother's curse neutralizes natural selection against a human genetic disease over three centuries. *Nature Ecology and Evolution*, 1, 1400–406.

Monosson, E. (2014). *Unnatural Selection: How We Are Changing Life, Gene by Gene*. Island Press, Washington, DC.

Moorad, J.A. (2013). A demographic transition altered the strength of selection for fitness and age-specific survival and fertility in a 19th century American population. *Evolution*, 67, 1622–34.

Morel, B.A. (1857). *Traité des Dégénérescences Physiques, Intellectuelle et Morales de l'Espèce Humaine et des Causes qui Produisent ces Variétés Maladives*. J.B. Ballière, Paris.

Mostafavi, H., Berisa, T., Day, F.R., et al. (2017). Identifying genetic variants that affect viability in large cohorts. *PLOS Biology*, 15, e2002458.

Muller, H.J. (1950). Our load of mutations. *American Journal of Human Genetics*, 2, 111–76.

Nalls, M.A., Simon-Sanchez, J., Gibbs, J.R., et al. (2009). Measures of autozygosity in decline: globalization, urbanization, and its implications for medical genetics. *PLOS Genetics*, 5, e1000415.

Nieuwenhuijsen, M.J., Basagaña, X., Dadvand, P., et al. (2014). Air pollution and human fertility rates. *Environment International*, 70, 9–14.

Nunn, C.L., Samson, D.R., and Krystal, A.D. (2016). Shining evolutionary light on human sleep and sleep disorders. *Evolution, Medicine, and Public Health*, 2016, 227–43.

Oriá, R.B., Patrick, P.D., Zhang, H., et al. (2005). APOE4 protects the cognitive development in children with heavy diarrhea burdens in northeast Brazil. *Pediatric Research*, 57, 310–16.

Peischl, S., Dupanloup, I., Foucal, A., et al. (2018). Relaxed selection during a recent human expansion. *Genetics*, 208, 763–77.

Pelletier, F., Pigeon, G., Bergeron, P., et al. (2017). Eco-evolutionary dynamics in a contemporary human population. *Nature Communications*, 8, 15947.

Pepper, G.V. and Nettle, D. (2014). Socioeconomic disparities in health behaviour: an evolutionary perspective. In: Gibson, M.A. and Lawson, D.W. (eds) *Applied Evolutionary Anthropology*, pp. 225–43. Springer, New York.

Pettay, J.E., Kruuk, L., Jokela, J., and Lummaa, V. (2005). Heritability and genetic constraints of life-history trait evolution in preindustrial humans. *Proceedings of the National Academy of Sciences of the United States of America*, 102, 2838–43.

Pickrell, J.K., Coop, G., Novembre, J., et al. (2009). Signals of recent positive selection in a worldwide sample of human populations. *Genome Research*, 19, 826–37.

Pritchard, J.K. and Di Rienzo, A. (2010). Adaptation—not by sweeps alone. *Nature Reviews Genetics*, 11, 665–7.

Rickard, I. (2013). Sir David Attenborough is wrong: humans are still evolving. *The Guardian*, 10 September.

Rook, G.A.W. (2012). Hygiene hypothesis and autoimmune diseases. *Clinical Reviews in Allergy and Immunology*, 42, 5–15.

Ruiz-Calderon, J.F., Cavallin, H., Song, S.J., et al. (2016). Walls talk: microbial biogeography of homes spanning urbanization. *Science Advances*, 2, e1501061.

Sabeti, P.C., Schaffner, S.F., Fry, B., et al. (2006). Positive natural selection in the human lineage. *Science*, 312, 1614–20.

Santangelo, J.S., Rivkin, L.R., and Johnson, M.T.J. (2018). The evolution of city life. *Proceedings of the Royal Society B: Biological Sciences*, 285, 20181529.

Sato, D.X. and Kawata, M. (2018). Positive and balancing selection on SLC18A1 gene associated with psychiatric disorders and human-unique personality traits. *Evolution Letters*, 2, 499–510.

Satterfield, B.C., Hinson, J.M., Whitney, P., et al. (2018). Catechol-O-methyltransferase (COMT) genotype affects cognitive control during total sleep deprivation. *Cortex*, 99, 179–86.

Shenka, M.K., Towner, M.C., Kress, H.C., and Alam, N. (2013). A model comparison approach shows stronger support for economic models of fertility decline. *Proceedings of the National Academy of Sciences of the United States of America*, 110, 8045–50.

Sibley, L.M. and Weiner, J.P. (2011). An evaluation of access to health care services along the rural–urban continuum in Canada. *BMC Health Services Research*, 11, 20.

Solomon, S. (2016). *Future Humans: Inside the Science of our Continuing Evolution*. Yale University Press, New Haven, CN.

Stearns, S.C. (1992). *The Evolution of Life Histories*. Oxford University Press, Oxford.

Stearns, S.C. and Medzhitov, R. (2015). *Evolutionary Medicine*. Sinauer, Sunderland, MA.

Stearns, S.C., Byars, S.G., Govindaraju, D.R., and Ewbank, D. (2010). Measuring selection in contemporary human populations. *Nature Reviews Genetics*, 11, 611–22.

Stock, J.T. (2008) Are humans still evolving? *EMBO Reports*, 9, S51–4.

Stulp, G. and Barrett, L. (2016). Wealth, fertility and adaptive behaviour in industrial populations. *Philosophical Transactions of the Royal Society B*, 371, 20150153.

Stulp, G., Barrett, L., Tropf, F.C., and Mills, M. (2015). Does natural selection favour taller stature among the tallest

people on earth? *Proceedings of the Royal Society B: Biological Sciences*, 282, 20150211.

Sykes, B. (2004). *Adam's Curse: A Future Without Men*. Norton & Company, New York.

Tan, J., Zheng, Y., Tang, X., et al. (2010). The urban heat island and its impact on heat waves and human health in Shanghai. *International Journal of Biometeorology*, 54, 75–84.

Templeton, A.R. (2010). Has human evolution stopped? *Rambam Maimonides Medical Journal*, 1, e0006.

Thomson, C.E. and Hadfield, J.D. (2017). Measuring selection when parents and offspring interact. *Methods in Ecology and Evolution*, 8, 678–87.

Tishkoff, S.A., Reed, F.A., Ranciaro, A., et al. (2007). Convergent adaptation of human lactase persistence in Africa and Europe. *Nature Genetics*, 39, 31–40.

Tort, P. (1983). *La Pensée Hiérarchique de l'Évolution*. Aubier, Paris.

Tropf, F.C., Stulp, G., Barban, N., et al. (2015). Human fertility, molecular genetics, and natural selection in modern societies. *PLOS ONE*, 10, e0126821.

Walter, B.S. and DeWitte, S.N. (2017) Urban and rural mortality and survival in Medieval England. *Annals of Human Biology*, 44, 338–48.

Wang, X., Byars, S.G., and Stearns, S.C. (2013). Genetic links between post-reproductive lifespan and family size in Framingham. *Evolution, Medicine, and Public Health*, 2013, 241–53.

Wang, Y. and Yang, M. (2019). Long commutes to work during pregnancy and infant health at birth. *Economics and Human Biology*, 35, 1–17.

Wells, J.C.K., Nesse, R.M., Sear, R., Johnstone, R.A., and Stearns, S.C. (2017). Evolutionary public health: introducing the concept. *The Lancet*, 390, 500–509.

Wills, C. (1998). *Children of Prometheus: The Accelerating Pace of Human Evolution*. Perseus Books, Reading.

Wishart, H.A., Roth, R.M., Saykin, A.J., et al. (2011). COMT Val158Met genotype and individual differences in executive function in healthy adults. *Journal of the International Neuropsychological Society*, 17, 174–80.

World Health Organization (2018). Global Health Observatory data repository. http://www.who.int/gho/en/.

Xu, S., Ye, Y., and Xu, L. (2017). Complex power: an analytical approach to measuring the degree of urbanity of urban building complexes. *International Journal of High-Rise Buildings*, 6, 165–75.

Zagorsky, J.L. and Smith, P.K. (2017). The association between socioeconomic status and adult fast-food consumption in the U.S. *Economics and Human Biology*, 27, 12–25.

List of Glossary Terms Definition

Adaptation The process by which populations evolve towards higher mean fitness in an environment in response to natural selection. Also, the trait values that maximize fitness within an environment.

Adaptive evolution Change in the genetic composition (i.e., allele frequencies) of a population in response to selection.

Adaptive parallelism The repeated evolution of the same phenotypes or genotypes in response to similar selection pressures across populations or species. See also non-adaptive parallelism.

Adaptive peak Area of maximum fitness in an adaptive landscape, from which movement in any direction results in lower fitness.

Akaike Information Criterion (AIC) Estimator founded on information theory and widely used for statistical inference. For a given set of data, AIC compares the quality of a set of statistical models to each other, providing a means for model selection.

Antagonism An interaction between two species in which one benefits at the expense of the other, measured in terms of fitness and/or population growth.

Antagonistic pleiotropy When one gene influences more than one trait, where at least one of these traits is beneficial to the organism's fitness and at least one is detrimental to the organism's fitness.

Antioxidants Different compounds that inhibit oxidation of proteins, lipids, and DNA, commonly referred to as oxidative damage.

Artificial light at night (ALAN) Illumination by human-introduced light sources during the dark hours of the day. Also referred to as light pollution.

Biomechanics A field at the interface of biology and physics focused on understanding the processes that enable biological structures to move and how morphology and physiology influence motion and function.

Citizen science Contribution to scientific research from members of the general public that are not scientists, in particular through the collection and/or analysis of data.

City A large town or metropolitan area inhabited by humans, often used synonymously with 'urban area'.

Climbing morphology Organismal structures specialized for climbing vertical or branching surfaces. These traits vary widely across taxa and include claws, wet and dry adhesive structures, and adduction or grasping mechanisms.

Cognitive buffer Notion that cognition can protect individuals against maladaptation through learning and matching habitat choice.

Colinearity Correlation between predictor variables, such that they express a linear relationship in a statistical model.

Common garden experiment An experimental design used to separate the genetic and environmental components of phenotypic divergence between populations. In this design, individuals from separate natural populations are reared in controlled, identical environments. By eliminating the influence of environmental variation on development, the influence of fixed genetic differences on phenotypic divergence can be identified.

Community-weighted mean (CWM) trait value or abundance-weighted community trait value Community trait average calculated based on the relative abundances (as fraction) of all species in a community times their trait value. Species trait values can be obtained from the literature or can be based on measurements in the studied communities. In the latter case, one can use average trait values for each species across the meta-community, or use patch-specific values.

Contemporary evolution Biological evolution observable over contemporary timescales relative to human life-spans, typically less than a few hundred years.

Context dependency A change in the outcome of an inter-specfic interaction in response to the abiotic, biotic, or genetic background in which the interaction occurs.

Countergradient variation Evolutionary divergence in the sensitivity of populations to environmental variation (i.e., phenotypic plasticity) resulting in populations expressing similar phenotypes across an environmental gradient but divergent phenotypes when reared under identical conditions (i.e., a common garden).

De novo mutation A new germline mutation not inherited from either parent.

Decision-making The process of making a choice among available options, such as deciding where to breed or what to eat.

Demographic transition The transition from populations with high birth and death rates, dominated by the young, to populations with low birth and death rates, dominated by the old.

Detoxification Removal of toxic compounds from the body.

Dry adhesion A form of adhesion used by some climbing species involving specialized structures that enable clinging to substrates without the use of sticky or fluid adhesives. For example, some lizard species have specialized toe scales with hierarchical structures that enable microscopic surface interactions and adhesion via van der Waals forces.

Eco-evolutionary dynamics The dynamics resulting from interactions between ecological and evolutionary processes. It is increasingly recognized that evolutionary and ecological processes can operate at the same temporal and spatial scales, and that interactions between these processes may lead to important deviations from expectations from pure ecological or pure evolutionary processes.

Eco-evolutionary feedbacks Refer to a full feedback loop between ecological and evolutionary change. In its narrow-sense definition, evolutionary trait change in a population impacts ecological change, which subsequently impacts the evolution of the same trait in that same population. In a community or metacommunity context, however, feedback loops can involve multiple intermediate steps and can become indirect. In a broad-sense definition, an eco-evolutionary feedback refers to the feedback between ecological and evolutionary change itself.

Eco-evolutionary partitioning metrics Metrics to estimate the relative contributions or effect sizes of ecological and evolutionary drivers of ecological responses. These ecological responses can relate to demography, community trait values and community composition, or ecosystem characteristics and functions. For trait change, eco-evolutionary partitioning metrics can differentiate between phenotypic plasticity, genetic change, community trait change mediated by changes in species composition, and the interactions among these contributions. Most eco-evolutionary partitioning metrics are based on the Price equation, on regression analyses, or on reaction norm-based approaches.

Effective population size (Ne) The size of an idealized population exhibiting the same rate of loss of genetic diversity as the focal population of interest. In practice, Ne is often approximated to the number of breeding individuals in the population.

Eigenvector Concept used in numerical data analysis, particularly in the context of PCA or canonical analysis; property of a data matrix, which specifies directions in which a system expands or shrinks.

Endocrine-disrupting compounds Chemical compounds (usually man-made) that interfere with the endocrine (hormone) system and generate adverse effects in development, reproduction, brain function, and immune system. Common examples are polychlorinated biphenyls (PCBs), which affect brain and reproductive development and have entered the environment from various manufactured materials.

Endocrinology The study of hormonal (endocrine) regulation of biological processes, such as behaviour, growth, metabolism, reproduction, and stress responses.

Epibiota Plants and animals that live on the surface of another species or a firm substrate. Attached species like barnacles and fungi fall into this category.

Eutrophication An overabundance of nutrients and their effects on aquatic ecosystems. When water bodies receive increased nutrients like nitrogen and phosphorus, they experience a bloom in algae growth, which is followed by a decrease in dissolved oxygen as bacteria use oxygen during decomposition when the algae and other plant life dies.

Evolution Change in the distribution of phenotypes attributable to underlying heritable genetic variation.

Evolution-mediated priority effect (or monopolization effect) A priority effect that is mediated by evolution. Through local adaptation to the conditions that prevail in a patch, an early colonizer of that patch gains a fitness advantage over later immigrants. This fitness advantage adds to the numerical advantage of the early colonizer and can make priority effects longer-lasting or permanent.

Evolutionary rescue A situation in which adaptive genetic change allows a population to survive an adverse environmental change. In the absence of evolution, the population would have gone extinct.

Evolving metacommunity ecology The integrated analysis of community ecology and evolution in spatially explicit landscapes. Evolving metacommunity ecology considers how communities are structured by environmental and spatial gradients (cf. metacommunity ecology), how species genetically respond to those same environmental and spatial gradients, and how both influence each other.

Fitness Number of descendants produced by an individual relative to the average produced by other individuals in the population.

Flight initiation distance (FID) The distance at which an organism flees from a perceived approaching predator. Often used to assess perception of risk and habituation in human-dominated habitats.

Gene flow The exchange of genes and alleles between populations/areas that subsequently contributes to the future gene pool of the recipient population/area; a result of successful reproduction by dispersing individuals.

Genetic drift Stochastic changes in allele frequencies across generations due to random sampling effects of parental genotypes, rather than allele frequency changes due to natural selection or mutation.

Genetic quality Breeding value of an individual for total fitness in a specific environment.

Genetic variation Degree to which DNA sequences differ across organisms. Genetic variation in combination with *environmental variation* explains the total *phenotypic variation* observed in a population.

Genotyping by sequencing (*sensu lato*) Methods that reduce the complexity of multiple DNA samples before sequencing them and eventually calling SNP.

Glucocorticoids Family of steroid hormones found across all vertebrates. Glucocorticoids are involved in regulation of numerous biological processes, including metabolism, reproduction, growth, immune function, and responses to acute stressors. Includes corticosterone (primary glucocorticoid in birds, reptiles, amphibians, and some rodents) and cortisol (primary glucocorticoid in fish as well as humans and most other mammals).

Grasping/adduction A form of adhesion used in climbing that does not involve specialized adhesive mechanisms but instead relies on friction and/or claws interlocking with surface asperities, and forces generated by muscle contraction. Adduction is the motion of moving a limb towards the midline of the body, which enhances grasping, particularly on flatter substrates.

Habituation Cognitive process that involves the reduction of a pre-programmed response as a result of repeated exposure to a normally eliciting stimulus.

Heritability Fraction of phenotypic variability that can be attributed to genetic variation. Broad-sense heritability is $H^2 = V_G/V_p$, where V_G is the trait genetic variance, and V_p the total phenotypic variance. Narrow-sense heritability is $h^2 = V_A/V_p$, where V_A is the additive genetic variance. Heritability estimates range in value from 0 (no genetic variation) to 1 (all variation in a population is due to differences between genotypes).

Hydrology The field of study concerned with how water moves through the atmosphere, oceans, rivers, and ground.

Immunoregulatory deficit The physiological mechanisms, such as cell–cell interactions, that regulate the immune system to trigger appropriate responses to pathogens at the organismal level.

Imprinting Any kind of rapid, phase-sensitive learning that is independent of the consequences of behaviour. Imprinting is hypothesized to have a critical period. Habitat imprinting is, for example, a learned habitat preference acquired during a specific stage (often juvenile or immature stages) through exposure to a specific habitat type during this sensitive stage.

Individual diet specialization Within a single population, different individuals may specialize in eating different things. This might reduce resource competition.

Intraspecific trait variation Trait variation within and among populations of a given species. Intraspecific trait variation can involve genetic differences, phenotypic plasticity (including epigenetic effects), or ontogenetic trait shifts.

Intrinsic water efficiency The ratio of carbon assimilated via photosynthesis to water lost via stomatal conductance, as indicated by stable carbon isotope analysis.

Isolation by distance (IBD) A general spatial genetic pattern where pairs of individuals or populations become more genetically dissimilar the further apart they are geographically. This is a consequence of limited dispersal, and can be thought of as a clinal statistical relationship between genetic and geographic distance. IBD is the null expectation in most spatial genetics studies, and the most commonly observed empirically.

Isolation by resistance (IBR) A spatial genetic pattern where the genetic similarity of pairs of individuals or populations is associated with the amount of resistance they would encounter when moving across the landscape, rather than simple Euclidean distance separating them (as in IBD). Resistance is typically quantified using the least-cost path or circuit-based resistance models.

Landscape genetics A set of approaches or tools to study how geography and environments structure genetic variation across space. Early development drew from the fields of population genetics and landscape ecology, but now landscape genetics is a subdiscipline developing separate methods and analytical tools to address landscape-level questions about movement and adaptation across space.

Life-history theory A framework to explain how adaptive evolution shapes species' life-history traits. The theory assumes fundamental tradeoffs between survival, growth, and reproduction.

Life-history traits Traits that characterize how and when a species transitions through its life cycle (e.g., age at first reproduction). These traits are closely related to survival and fecundity.

Local adaptation Evolution of a higher mean fitness of a local population in its native environment compared to other environments, and compared to other populations in this environment.

Locomotor performance Organismal function related to moving from one place to another in a given environment. Locomotor performance may be measured in many ways: for example, speed or agility.

Matching habitat choice The process by which an animal selects those habitats or resources that better match its phenotype with the environment. By making these choices, animals maximize their fitness in the environmental conditions they live in.

Metabolism Collective term for three life-sustaining chemical reactions—catabolism (breakdown of food to generate energy), anabolism (conversion of food to cellular building-blocks), and excretion (elimination of waste products, especially nitrogenous wastes).

Metacommunity A set of local communities that are connected via dispersal of their member species. Local community composition and other community features can be determined by both local processes (e.g., environmental filters) and regional processes (i.e., dispersal).

Microbiota The community of bacteria, fungi, viruses, and parasites that inhabit the gut, skin, nose, and other tissues and organs of the body.

Mismatch The states of traits expressed in one environment are not optimal in that environment because they evolved in a different one, either in space or in time.

Modernization The change in human ecology sparked by the industrial revolution starting around the turn of the nineteenth century and involving demographic and epidemiological transitions, major changes in work and lifestyle, and ongoing urbanization.

Mutualism An interaction between two species in which both partners experience a net benefit, measured in terms of fitness and/or population growth.

Natural selection A sorting process producing a change in phenotype mean or variance in a population through differential birth and death resulting from some individuals being phenotypically better equipped than others to survive and/or reproduce in their current environment.

Neutral evolution Change in the genetic composition of a population that occurs due to the random processes of mutation and/or genetic drift.

Non-adaptive parallelism The repeated evolution of similar patterns of genomic diversity due to landscape features of cities driving consistent changes in non-adaptive evolutionary processes (e.g., drift and gene flow).

Nutrition Link between diet type and performance such as growth, maintenance, reproduction, and health.

Nutritional mutualism A mutually beneficial interaction between two species involving the provision of one or more limiting nutrients by each of the partners.

Opportunity for selection Defined by $I = V_W/W^2$, where V_W is the variance in absolute fitness and W is the mean fitness in a population. It is equivalent to the variance in relative fitness, and provides an empirical estimate of the maximum strength of selection acting on a particular population.

Outreach Popularizing science to promote science education and awareness in the general public.

Oxidative stress Imbalance between the pro-oxidants and antioxidants, with a surplus of the former. Oxidative stress is often manifested in the increased generation of oxidative damages.

Phenology Timing of periodic biological events; for example, spring flowering, autumn migration, and mating season.

Phenotype–environment match Extent to which the morphology, physiology, behaviour, and life history of an organism will allow it to survive and reproduce in a specific environment.

Phenotypic change vector analysis A method that calculates the mean multivariate trait value (i.e., 'centroid') and connects the centroids for each city by a line forming a vector. The angles (Θ) and differences in lengths (ΔL) among the vectors measure the direction and magnitude of trait divergence across population pairs (cities in this case), respectively.

Phenotypic plasticity The ability of one genotype to produce multiple phenotypes in response to the environment.

Philopatry The tendency of an organism or species to stay within or return to a particular area, such as its place of birth or a breeding area.

Phytoplankton Small, typically single-celled, aquatic organisms that photosynthesize.

Plastic rescue Population recovery before extinction mediated by plastic phenotypic responses, with no need of evolutionary change.

Plasticity Change in the phenotypic expression of a single genotype in response to environmental variation.

Pleiotropy One gene controls multiple phenotypes.

Polygenic trait Multiple genes control one phenotype.

Population persistence The phenomenon by which a population of organisms continues through generations (in opposition to population extinction).

Principal component analysis (PCA) A multivariate technique analysing a data table where observations are described by several variables. PCA reduces the large number of often collinear variables to a smaller number of orthogonal and independent variables (principal component axes) that can be visualized in two- or three-dimensional space.

Problem-solving The process by which animals use generic or *ad hoc* methods to find solutions to problems. In cognitive science, problem-solving tasks (e.g., opening a box or removing an obstacle to obtain a reward) are used to estimate animals' behavioural and cognitive performances, and to characterize cognitive processes.

Protection mutualism A mutually beneficial interaction between two species involving the provision of defence against the biotic and abiotic environment by a partner in return for a reward, typically food or shelter.

Reciprocal transplant experiment An experimental design used to test whether evolved shifts in trait values between populations are adaptive. In this design, individuals from two populations are placed into their 'home' environments and their 'away' environments, and the fitness of each population in its home and away environment is assessed. The expectation for local adaptation is for each population to have its highest fitness in its home environment compared with the away environment.

Reproductive physiology Collective term that encompasses the processes that regulate mating as well as production of and investment in offspring.

Reproductive strategies Different approaches through which members of a population can achieve maximal fitness.

Scansorial Terrestrial organisms specialized for both climbing and terrestrial movement.

Selection differential (S) A measure of the strength of selection that is equal to the within-generational difference between population means before and after selection. Selection differentials measure total (both direct and indirect) selection on a trait. For directional selection, it is equivalent to the regression of relative fitness on phenotype.

Selection gradient (β) A measure of the strength of selection acting directly on the trait of interest (as opposed to selection differential estimating total direct and indirect selection via other correlated traits) that is equal to the partial regression coefficients of relative fitness on a given phenotype (trait).

Selection pressure Factors that affect an organism's ability to survive and reproduce.

Sequencing Determining the succession of nucleotides along DNA segments.

Sexual selection Competition between individuals over access to copulations and fertilizations in sexually reproducing organisms; responsible for many of the more extravagant traits found in nature.

Single nucleotide polymorphism (SNP) Recurrent variation of a single nucleotide at a specific position in the genome.

Spatial autocorrelation Increasing similarity in measures occurring with decreasing geographical distance. Such similarity can arise due to the fact that individuals occupying the same environment will resemble each other to a greater degree than they otherwise might if exposed to different habitats.

Species sorting Refers to the process by which local community composition in a metacommunity is determined by niche differences among species, such that species composition is driven by the environmental conditions prevailing in the local patches. Species sorting can only be effective if dispersal is sufficiently high (such that all species can reach their preferred habitats) but not too high (so that no source-sink dynamics occur, in which species also are found in habitats where they cannot sustain a viable population).

Standing genetic variation Allelic variation that has been segregating within a population.

Stream dewatering Disappearance of water from a stream. This can be caused by dam construction, water withdrawal for agricultural use, underground mining, or drought.

Structural variants Variation in the structure of the genome, within chromosomes, such as deletions, insertions, duplications, copy-number variations, inversions, and translocations.

Sustainable cities A city is said to be sustainable when it meets the biological, social, and environmental needs of its current inhabitants without degrading the ability of future generations to meet their needs. Resources involved in sustainability include water, energy, food, and biodiversity, among others. The rate at which these are used or damaged by waste, pollution, overdevelopment, or other factors determines whether the resources are used sustainably.

Terrestrial morphology Organismal structures specialized for overground movement such as walking, running, and jumping, as opposed to specializations for climbing, digging, or swimming.

Thigmotaxis An innate avoidance of open environments and affinity for dark or enclosed areas.

Transhumanism A movement that advocates the enhancement of human condition with technologies, such as gene editing to repair disease-causing alleles or to improve phenotypes, prostheses to augment physical performance, or computer–neuron interfaces to enhance cognitive functions.

Transport mutualism A mutually beneficial interaction between two species involving the movement of partners themselves, or movement of partners' gametes, in return for a reward, typically food.

Urban eco-evolutionary dynamics Interactions between evolutionary and ecological processes in urban areas. These interactions can be reciprocal, with urban ecological dynamics influencing evolutionary trajectories, and urban evolution impacting population, community, or ecosystem ecology.

Urban heat island Generation of warmer environmental temperatures within human settlement areas compared with nearby non-urban areas.

Urban mosaic/urban matrix Cities are characterized by a high proportion of built-up and paved areas such as buildings, roads, and pavements. These are interspersed with green spaces such as parks, road verges, forest remnants, small gardens, and avenue trees. The resulting patchiness of highly contrasted habitat types is referred to as the urban mosaic/urban matrix.

Urbanization gradient A heuristic tool used to characterize spatial and/or temporal patterns of human development, ranging from high-density areas dominated by built structures (urban) to low-density areas dominated by natural, agricultural, or pastoral open space (rural); also called 'urban-rural gradient'.

Urbanization Process related to the growth of cities and associated human activities; involves a human population shift from rural to urban areas.

Variable—continuous, ordinal, nominal A continuous variable is one that has an infinite number of possible values. While population counts have values that vary in discrete increments, a variable such as population density is a ratio and consequently has values that vary continuously. Continuous variables can be reclassified into categorical variables, which include ordinal variables (e.g., categories that can be ordered: high, medium, and low density) and nominal variables (e.g., land type: forest/grassland/urban). A nominal variable that only has two categories is referred to as dichotomous (e.g., urban/rural).

Variance partitioning Quantifies the percentage of variation that could be explained by the variables/factors (e.g., urban versus non-urban), while accounting for variation due to other factors (depending on models used).

Vital rate Either fertility or mortality rate for a specific age range.

Wet adhesion A form of adhesion used by some climbing species involving the secretion of a sticky or fluid adhesive substance to create a bond with the surface.

Whole genome sequencing (WGS) Determining the complete DNA sequence of an organism's genome.

Zooplankton Small aquatic animals, such as copepods, the larvae of larger organisms, and krill that inhabit open-water regions of aquatic environments.

Index

Tables, figures, glossary terms and boxes are indicated by an italic *t*, *f*, *g*, and *b* following the page number.

A

Abax ater 40*t*
abundance-weighted community trait
    value *see* community-weighted
    mean (CWM) trait values
Accipiter trivirgatus 241*b*
acorn ant (*Temnothorax curvispinosus*)
    8, 40*t*, 43, 77*t*, 78, 98–9, 102, 102*f*,
    104*f*, 106*f*, 237*t*
Acridotheres tristis 260*t*, 261
adaptation
    defined 289*g*
    sources of urban 8
    see also local adaptation
adaptive evolution, defined 289*g*
adaptive introgression 37*b*, 46,
    50*t*, 281
adaptive parallelism 289*g*
adaptive peak 255, 257, 258, 289*g*
Aedes aegypti 40*t*, 44
Aegean wall lizard (*Podarcis*
    erhardii) 204
age at first reproduction 273*b*, 282*b*, 283
Ailanthus altissima 20, 56
air pollution 43
    birds and 221–2
    human health and 278, 280,
    281, 283
    melanism and 97–8
    mutualism and 115
Akaike Information Criterion (AIC)
    25, 26, 289*g*
algal blooms 168, 190
Allen's rule 97
allergies 274
allozymes 57
alpine mustard (*Erysimum*
    mediohispanicum) 118
Alzheimer's disease 274, 281
Ambrosia artemisiifolia 78–9, 101, 103,
    104*f*, 106*f*, 148–50, 177
ambush predation 201, 202–3

Ambystoma
    maculatum 261–2
    tigrinum 61
American kestrel (*Falco sparverius*) 227
amphipod (*Hyalella azteca*) 166
annual meadow grass (*Poa annua*) 132
Anolis 200–1, 202, 203, 205–6, 209,
    210, 259*f*
    cristatellus 39, 40*t*, 97, 104*f*, 106*f*,
    203–4, 205*f*, 206, 209, 210, 256
    sagrei 200–1
    stratulus 210, 256
anonymous markers 80
Anopheles gambiae 166
antagonism 112, 124, 289*g*
    shifts from mutualism to
    antagonism 113, 116*t*
antagonistic pleiotropy 274, 289*g*
anthrodependent species 20, 55
anthropogenic food sources 16*t*, 124,
    203, 225, 226, 229, 244
antibiotic resistance 166–7
antioxidants 218–19, 219*b*, 220, 222,
    228, 289*g*
ants 104*f*
    limb length 97, 213*b*
    parallel evolution 40*t*
    protection mutualism 120–1,
    121*f*, 124
    seed dispersal and 119–20
    thermal tolerance 8, 43, 77*t*, 78, 97,
    98–9, 102, 102*f*, 106*f*, 213*b*, 237*t*
Aphelacoma californica 224
Aphelocoma coerulescens 225, 226
aphids 123
Apodemus 20, 62
    agrarius 62
    flavicollis 62
    sylvaticus 67
APOE (apolipoprotein E) gene 281
appendage size, urban heat island
    effects 97

aquatic environments 157–69, 159*t*
    body size 96–7, 106*f*, 164, 178*b*,
    179*f*, 182–3
    chemical pollution 43–4, 46, 159*t*,
    165–7, 168–9
    competition 159*t*, 160–1
    diet 159*t*, 161
    habitat fragmentation 44, 159*t*,
    161–2
    light pollution 159*t*, 160, 161, 167–8
    morphology 96–7, 164–5
    mutualism 112
    noise pollution 159*t*, 168
    pace-of-life 43, 102, 164–5
    phenology 163–4, 167
    predation 159*t*, 160, 167
    sex determination 165, 166
    temperature 96–7, 106*f*, 159*t*,
    163–5, 177
    turbidity 169
    urban stream flow 159*t*, 162–3
aquatic species, morphology 163
Arabidopsis 137
Argentine ant (*Linepithema*
    humile) 119
Aristelliger praesignis 209
arroyo chub (*Gila orcuttii*) 40*t*, 44
artificial light at night (ALAN)
    16*t*, 289*g*
    reproductive endocrinology and
    167, 223–4
    see also light pollution
aryl hydrocarbon receptors
    (AHRs) 43–4, 222
associative learning 259*b*, 259*f*
assortative mating 258
Athene cunicularia 19, 40*t*, 63, 67,
    237*t*, 244
Atlantic killifish (*Fundulus*
    heteroclitus) 40*t*, 42, 43–4, 46, 49,
    77*t*, 81, 82, 166, 169
Atta sexdens rubropilosa 98

autoimmune diseases 274, 276t
autozygosity 280–1

B

bacteria
  antibiotic resistance 166–7
  gut microbiomes 9, 274
  rhizobial 121–3, 123f, 152
Bahamas mosquitofish (*Gambusia hubbsia*) 38, 160–1, 162, 169
Baldwin effect 257, 258
ballistic interception predation 201, 202–3
Barbados bullfinch (*Loxigilla barbadensis*) 260t
barn swallow (*Hirundo rustica*) 226
basal metabolic rate (BMR) 228, 229
Bayesian coalescent approach 58
beach mouse (*Peromyscus polionotus*) 38
bedbugs (*Cimex*) 20, 44, 189, 278
behaviourally mediated habitat use 199–201, 200f, 212f
Bergmann's rule 96–7
biomechanics 289g
birds
  air pollution and 221–2
  anthropogenic food 124, 225, 226, 229, 244
  body size 96
  candidate gene studies 77t, 80–1, 85–6
  endocrine responses to challenges 226, 227
  epigenetic variation 85–6
  flight initiation distance (FID) 200–1
  human disturbance 227–8
  light pollution and 224
  mating strategies 244–5
  metabolic responses 229
  niche use changes 191
  noise pollution and 237t, 243, 246
  parasite loads 124
  phenological change 101
  predation rates 124
  reproductive endocrinology 223, 224, 225
  seed dispersal and 119, 120
  sexual selection 236, 237t, 239–40, 241b, 242, 243, 244–5, 246
Biston betularia 45, 77t, 82, 97–8
black swan (*Cygnus atratus*) 77t, 81, 262
black-capped chickadee (*Poecile atricapillus*) 260, 260t
Blattella germanica 20
blue tit (*Cyanistes caeruleus*) 25, 224
body size

aquatic species 96–7, 106f, 164, 178b, 179f, 182–3
  climbing and 207
  habitat fragmentation and 176
  urban heat island effects 96–7, 103, 105, 106f, 164, 176
Bombus
  lapidarius 39–42, 40t, 43, 49, 63, 77t, 81, 83, 189
  vosnesenskii 66–7
BRCA genes 274, 285
breast-feeding 274
brown rat (*Rattus norvegicus*) 4, 20, 40t, 55, 56f, 59f, 62, 63, 65b, 67
brushtail possum (*Trichosurus cunninghami*) 244
Buchnera 123
Burkholderia 123
burrowing owl (*Athene cunicularia*) 19, 40t, 63, 67, 237t, 244

C

Calamospiza melanocorys 236
Calomys musculinus 67
canary (*Serinus canaria domestica*) 237t, 243
cancer 274, 280, 285
candidate genes 49, 77t, 80–1, 85–6, 219b
Cape glossy starling (*Lamprotornis nitens*) 222
Cardinalis cardinalis 97, 237t, 240
carp (*Catla catla*) 167
Cataglyphis bombycina 97
Catla catla 167
Centaurea solstitialis 117–18
Centaurium erythraea 131, 139
Cepaea
  hortensis 4, 25
  nemoralis 4, 25
Cervus elaphus 26
Chamaecrista fasciculata 121, 121f
chemical pollution
  aquatic environments 43–4, 46, 77t, 81, 82, 159t, 165–7, 168–9
  birds and 221–2
  detoxification responses 221–2
  human health and 278, 280, 281, 283
  parallel evolution and 43–4, 46
  physiology and 221–2
  sexual selection and 244
chickweed (*Stellaria media*) 132
chitinolytic fungus 100, 104f
cholesterol levels 276t, 281
Chorthippus brunneus 97, 106f
Cimex 20, 44, 189, 278
CIRCUITSCAPE package 60, 64f
cities 289g

citizen science 4, 25, 86–7, 289g
claw morphology 207f, 208, 210
climate change 136, 137, 139, 143, 190, 244
climbing behaviour 205–6, 205f
climbing morphology 206–11, 207f, 212f, 289g
coalescent modelling approaches 7, 58
cockroaches 20, 44, 189
Coenagrion puella 100, 106f, 162, 237t, 242, 243, 246, 246f
cognition 253–63
  adaptive evolution and 255f, 256–8, 257f
  evolution of 258–61, 259b, 260t
  experimental approaches 258–61, 259b, 260t
  fitness and 256, 261–2
  future directions 261–2
  learning 255f, 256, 257, 258, 259b, 259f, 260t
  matching habitat choice 255f, 256, 257–8, 292g
  phenotype–environment mismatch and 255–6, 255f
  plastic rescue 255f, 256, 292g
cognitive buffer 256–8, 257f, 289g
cold tolerance 43, 98, 98f, 99, 106f, 138, 145
colinearity 289g
Columba livia 243, 257–8, 257f
Commelina communis 117, 117f
common centaury (*Centaurium erythraea*) 131, 139
common frog (*Rana temporaria*) 57, 80
common garden experiments 8, 10, 24f, 25, 147t, 176–7, 289g
  evolving metacommunities 187–9, 188f, 190
  physiological traits 219, 219b, 223, 228, 229
  plant life-history adaptation 101, 148
  sexual selection 246, 246f
  urban heat island effects 95, 96–7, 98–9, 101, 102f
common groundsel (*Senecio vulgaris*) 132
common myna (*Acridotheres tristis*) 260t, 261
common ragweed (*Ambrosia artemisiifolia*) 78–9, 101, 103, 104f, 106f, 148–50, 177
common wall lizard (*Podarcis muralis*) 55, 67, 237t
community-weighted mean (CWM) trait values 184f, 185–7, 186f, 289g
COMT gene 285

contemporary evolution 289*g*
context dependency 289*g*
convergence 37*b*
Copsychus saularis 237*t*
Cornu asperum 56
coronary artery disease 272, 273*b*, 274
corticosterone 226, 227–8
countergradient variation 100, 148–9,
    149*f*, 150, 289*g*
Crepis sancta 39, 77*t*, 87, 101, 104*f*,
    106*f*, 117, 119, 132*f*, 133–4
    see also sidewalk plants
crested goshawk (*Accipiter trivirgatus*)
    241*b*
CRISPR/Cas9: 4, 284, 295
Culex mosquitoes 163, 246
curve-billed thrashers (*Toxostoma
    curvirostre*) 227–8
Cyanistes caeruleus 25, 224
cyanogenesis 39, 99, 103, 138, 145, 151
Cygnus atratus 77*t*, 81, 262

D
dams 44
damselfly (*Coenagrion puella*) 100,
    106*f*, 162, 237*t*, 242, 243, 246, 246*f*
Danio rerio 167
Daphnia magna 43, 87, 96–7, 98, 100,
    101, 102, 104*f*, 106*f*, 163, 164,
    165–6, 167, 177, 178*b*, 179*f*, 180,
    182–3, 190
dark-eyed junco (*Junco hyemalis*) 43,
    223, 236, 237*t*, 240, 257
Darwin's finches 77*t*, 86, 191, 245
data replication 7
dayflower (*Commelina communis*)
    117, 117*f*
de novo mutation 79, 80, 131, 290*g*
    parallel evolution and 37*b*, 40*t*, 44
decision-making 290*g*
degeneration arguments 270*b*
degeneration theory 269*b*
degus (*Octodon degus*) 202
delicate skink (*Lampropholis delicata*)
    260*t*
demographic transition 273–5, 290*g*
detoxification 221–2, 290*g*
diabetes 272, 274
Dipodomys merriami 203
dispersal
    seed dispersal mutualisms 116*t*,
        119–20, 124
    sidewalk plants 119, 134–6, 135*f*,
        138, 147–8
    thermal barriers to 107
DNA methylation 8, 85–6
domestication 9, 213*b*
dose-response relationships 220–1, 220*f*

DRD4 (dopamine receptor) gene 77*t*,
    80–1, 85, 262
dry adhesion 207*f*, 208, 210, 290*g*

E
eastern grey squirrel (*Sciurius
    carolinensis*) 17–18, 200–1
eastern mosquitofish (*Gambusia
    holbrooki*) 164
eastern tiger salamander (*Ambystoma
    tigrinum*) 61
eco-evolutionary dynamics 180,
    189–90, 282*b*, 282*f*, 283, 290*g*
eco-evolutionary feedbacks 180,
    189–90, 290*g*
eco-evolutionary partitioning metrics
    184*f*, 185–7, 186*f*, 189, 190, 290*g*
ecological niche models *see* habitat
    suitability models (HSM)
education and outreach 3–4
effective population size (Ne) 44,
    46, 57, 67, 80, 131, 134, 139, 150,
    162, 290*g*
eigenvectors 26, 58, 290*g*
empirical approaches 146, 147*t*
endocrine-disrupting compounds
    165, 166, 290*g*
endocrinology 290*g*
    food availability/quality and 224–5,
        226–7
    HPA axis 225–8
    human disturbance and 227–8
    light pollution and 167, 223–4
    reproductive 167, 222–5
    responses to challenges 225–8
Engystomops pustulosus 237*t*, 240,
    241–2, 246
environmental heterogeneity 15–17
    parallel evolution and 20, 39, 44–5,
        48–9, 50*t*
    scale 23, 27, 27*f*
environmental metrics 21*t*
epibiota 161, 290*g*
epidemiological transitions 271–2,
    273–4
epigenetic processes 2, 8, 75, 77*t*,
    85–6, 87
Episyrphus balteatus 117*f*
Erithacus rubecula 43, 237*t*
Erysimum mediohispanicum 118
estimated effective migration surfaces
    (EEMS) analysis 58, 63
European blackbird (*Turdus merula*)
    19, 40*t*, 43, 49, 68, 80–1, 223, 224,
    228, 240, 262
European robin (*Erithacus rubecula*)
    43, 237*t*
Eurycea bislineata 65*b*, 65*f*

eutrophication 168, 290*g*
evolution, defined 290*g*
evolutionary clustering approaches
    57–8, 68
evolutionary rescue 8, 166, 182, 183,
    184, 290*g*
evolution-mediated priority
    effect 290*g*
evolving metacommunities
    175–93
    common garden experiments
        187–9, 188*f*, 190
    eco-evolutionary dynamics 180,
        189–90
    eco-evolutionary feedbacks 180,
        189–90
    eco-evolutionary partitioning
        metrics 184*f*, 185–7, 186*f*,
        189, 190
    framework 177–81
    future directions 190–2
    hypothetical example 181–4,
        182*f*, 184*f*
    metacommunity ecology 177, 180
    multispecies approach 190
    niche use changes 191
    zooplankton 178–80, 178*b*, 179*f*,
        182–3
evolving metacommunity
    ecology 177, 180, 290*g*
extinction debt 19

F
Falco sparverius 227
fast-food consumption 279
feral rock pigeons (*Columba livia*) 243,
    257–8, 257*f*
fertility, human 269*b*, 274, 275–7,
    276*t*, 277*f*, 278, 279, 280, 281–3,
    284–5
fetal growth 276*t*, 280
fish
    chemical pollution and 43–4, 46,
        77*t*, 81, 82, 166, 169
    competition 160–1
    dietary shifts 161
    habitat fragmentation and 162
    light pollution and 167, 168
    morphology 163, 164
    noise pollution and 168
    parallel evolution 38, 40*t*, 43–4,
        46, 49
    phenology 163–4
    predation 160
    sex determination 165
    sexual selection 167, 169
    turbidity and 169
    urban stream flow and 162–3

fitness 291*g*
  cognition and 256, 261–2
  sexual selection and 236–9
  urban heat island effects 102–3, 102*f*
flight initiation distance (FID)
  199–201, 200*f*, 291*g*
Florida scrub-jay (*Aphelocoma
  coerulescens*) 225, 226
food availability/quality
  anthropogenic food sources 16*t*,
    124, 203, 225, 226, 229, 244
  HPA responses 226–7
  metabolic responses 229
  reproductive endocrinology and
    224–5
fragmentation *see* habitat
  fragmentation
Fundulus
  grandis 82
  heteroclitus 40*t*, 42, 43–4, 46, 49, 77*t*,
    81, 82, 166, 169

G

Gambusia 160
  holbrooki 164
  hubbsia 38, 160–1, 162, 169
garden skink (*Lampropholis
  guichenoti*) 201
Gasterosteus aculeatus 38
geckos 207*f*, 208, 209
Gelsemium sempervirens 118
gene editing 284–5
gene flow 7–8, 54–69, 131, 291*g*
  analytical approaches 57–62, 59*f*
  aquatic environments 162
  cognition and 255*f*, 258
  empirical studies 62–8, 64*f*
  future directions 68–9
  habitat fragmentation and 44, 119,
    134, 150–1, 162
  landscape genetics 60–1, 67
  landscape genomics 61–2, 68
  molecular markers 57–8
  mutualism and 119
  parallel evolution and 37*b*, 40*t*, 44,
    45–6, 49, 50*t*
  plant life-history traits and 150–1,
    152
  sampling strategies 63, 65*b*, 65*f*
  spatial population genomics 58–60,
    59*f*, 63–6
genetic architecture 19, 139
  parallel evolution and 37*b*, 46
genetic drift 7–8, 57–60, 61, 62–3,
  67–8, 131, 291*g*
  aquatic environments 162
  habitat fragmentation and 44, 46,
    134, 144, 150–1, 162

mutualism and 119
  parallel evolution and 37*b*, 40*t*, 44,
    46, 50*t*
  plant life-history traits 144, 150–1
genetic quality 291*g*
genetic variation 7–8, 37*b*, 79, 131,
  291*g*
genotype by environment association
  (GEA) tests 7, 62, 68
genotyping by sequencing (*sensu lato*)
  81, 291*g*
Geospiza
  fortis 77*t*, 86
  fuliginosa 77*t*, 86
German cockroach (*Blattella
  germanica*) 20
germination traits 151–2
gibbons 202
Gila orcuttii 40*t*, 44
gliding 206, 210–11
glucocorticoids 218–19, 223, 224, 226,
  227, 291*g*
golden shiner (*Notemigonus crysoleucas*)
  166
gonadotropins 167, 223
grasping/adduction 207–8, 207*f*, 291*g*
grasshoppers (*Chorthippus*) 97,
  106*f*, 237*t*
great tit (*Parus major*) 20, 26, 77*t*, 78,
  83, 85–6, 131, 224, 228, 229, 237*t*,
  260*t*, 261
green spaces, urban 56–7, 65*b*, 199, 199*f*
grey snapper (*Lutjanus griseus*) 161
grey treefrog (*Hyla versicolor*) 167
Greya moths 112
growth rates, urban heat island
  effects 100
Gulf killifish (*Fundulus grandis*) 82
gut microbiomes 9, 274

H

habitat constraint hypothesis 206
habitat fragmentation 199, 199*f*
  aquatic environments 44, 159*t*, 161–2
  behaviourally mediated habitat use
    and 199–201, 200*f*
  body size and 176
  genetic drift and 44, 46, 134, 144,
    150–1, 162
  mating strategies and 244
  parallel evolution and 44, 46
  plant life-history traits and 144,
    150–1
  seed dispersal and 119, 134–6, 135*f*,
    138, 144, 147–8
  sexual selection and 162, 244
  sidewalk plants 119, 134–6, 134*t*,
    135*f*, 138, 147–8

habitat suitability models (HSM) 56*f*,
  60–1, 67
habituation 200–1, 200*f*, 202, 258,
  259*b*, 260*t*, 291*g*
Haemorhous mexicanus 237*t*, 245,
  258, 260*t*
hawksbeard (*Crepis sancta*) 39, 77*t*, 87,
  101, 104*f*, 106*f*, 117, 119, 132*f*, 133–4
  see also sidewalk plants
health, human
  eco-evolutionary dynamics of 283
  epidemiological transitions 271–2,
    273–4
  gene editing and 284–5
  immunoregulatory deficit 274,
    276*t*, 291*g*
  pollution and 278, 280, 281, 283
  urban disease genetics 280–1
heat tolerance
  mutualism and 123
  see also urban heat island effects
heavy metals 122, 165–6, 221, 278
Hemidactlyus 209
herbivory 124, 145
  cyanogenesis and 99, 145
  protection mutualism 116*t*, 120–1,
    121*f*, 124
heritability 26, 95, 136, 201, 261, 291*g*
herring gull (*Larus argentatus*) 8, 40*t*
Hirundo rustica 226
histone modification 85
homozygosity 280–1
hormetic responses 220–1, 220*f*,
  229, 230
house finch (*Haemorhous mexicanus*)
  237*t*, 245, 258, 260*t*
house mouse (*Mus musculus*) 8, 68
house sparrow (*Passer domesticus*) 9,
  49, 222, 226, 243, 260*t*
HPA axis 225–8
HPG axis 223, 224, 225, 226
human disturbance 227–8
human evolution 268–85
  antagonistic pleiotropy 274
  demographic transition and 273–5
  dietary shifts 229, 271
  eco-evolutionary dynamics 282*b*,
    282*f*, 283
  end of natural selection 269*b*
  epidemiological transitions 271–2,
    273–4
  fertility 269*b*, 274, 275–7, 276*t*, 277*f*,
    278, 279, 280, 281–3, 284–5
  future directions 283–4
  modernization and 269, 269*b*,
    273–5, 276*t*
  sexual selection 243
  sociocultural factors 278–80

timescales for genetic responses 272–3, 273*b*
transhumanism 284–5, 293*g*
transition to agriculture 271–3, 273*b*
urban disease genetics 280–1
urban stressors 278
urbanization and 269–71, 275–81, 276*t*, 277*f*
Hyalella azteca 166
hydrogen cyanide (HCN) *see* cyanogenesis
hydrology 291*g*
Hyla
  ebraccata 243
  versicolor 167
hypermethylation 86
hypothalamic–pituitary–gonadal (HPG) axis 223, 224, 225, 226
hypoxia 168–9

I

immunoregulatory deficit 274, 276*t*, 291*g*
impervious surface area 14, 56, 93–4, 93*f*
imprinting 255, 255*f*, 258, 291*g*
inbreeding depression 139
Indian rock agamas (*Psammophilus dorsalis*) 201, 259*b*
individual diet specialization 161, 291*g*
infectious diseases 271–2, 274, 281
insecticide resistance 44, 166
interception predation 201, 202–3
intraspecific trait variation 291*g*
intrinsic water efficiency 291*g*
introduced species 20
isolation by distance (IBD) 26, 58, 60, 61, 63, 66, 291*g*
isolation by environment (IBE) 61, 66
isolation by resistance (IBR) 60, 61, 64*f*, 66–7, 291*g*

J

jumping and landing 206, 210, 211
Junco hyemalis 43, 223, 236, 237*t*, 240, 257
juvenile mortality, human 275, 276*t*, 279, 282*b*, 282*f*, 283

K

kangaroo rat (*Dipodomys merriami*) 203

L

Lampropholis
  delicata 260*t*
  guichenoti 201

Lamprotornis nitens 222
landscape genetics 60–1, 67, 177–8, 291*g*
landscape genomics 61–2, 68
lark bunting (*Calamospiza melanocorys*) 236
Larus argentatus 8, 40*t*
latent factor mixed models (LFMM) 68
Latrodectus hesperus 40*t*, 45, 63
laughing dove (*Streptopelia senegalensis*) 222
leaf cutter ant (*Atta sexdens rubropilosa*) 98
learning 255*f*, 256, 257, 258, 259*b*, 259*f*, 260*t*
leopard frog 246
Lepidium virginicum 40*t*, 43, 46, 101, 104*f*, 106*f*, 131, 138, 148, 150
leprosy 19–20
lice 44, 189, 278
life-history theory 146, 164, 291*g*
life-history traits 292*g*
  human 269*b*, 271, 273–5, 273*b*, 276*t*, 277*f*, 278, 279, 281–3, 282*b*, 284–5
  pace-of-life syndrome 43, 102, 106, 164–5, 228, 240–1
  rapid evolution of 139
  trait covariances and tradeoffs 146
  urban heat island effects 100–1, 103, 106, 106*f*, 136–8, 137*t*, 144–5, 152–3, 163–4
  see also plant life-history traits
light pollution 16*t*, 43
  aquatic environments 159*t*, 160, 161, 167–8
  birds and 224
  plant life-history traits 151
  reproductive endocrinology and 167, 223–4
  sexual selection and 237*t*
limb morphology 39, 97, 203–4, 204*f*, 207–8, 207*f*, 209, 210
Limitative arguments 269*b*, 270*b*
Linepithema humile 119
Lithophragma parviflorum 112
little brown bat (*Myotis lucifugus*) 245
lizards 20, 55, 201, 203–4, 208, 209, 210, 237*t*
  see also *Anolis*
local adaptation 5, 27, 49, 78–9, 80, 81, 95, 102*f*, 103, 254, 292*g*
  empirical approaches 146, 147*t*
  gene flow and 45, 134, 162
  landscape genomics 61–2, 68
  plant life-history traits 143, 147–50, 149*f*, 150*f*

locomotor evolution *see* terrestrial locomotor evolution
locomotor performance 202–3, 209, 292*g*
long-term inference 7
Loxigilla barbadensis 260*t*
luteinizing hormone (LH) 223
Lutjanus griseus 161
Lyme disease 4
lynx spider (*Oxyopes salticus*) 244
Lythrum salicaria 137

M

Macaca mulatta 261
Mamestra brassicae 237*t*
matching habitat choice 255*f*, 256, 257–8, 292*g*
maternal age at first reproduction (AFR) 273*b*, 275, 282*b*, 282*f*, 283
mating strategies 244–5
meadow vole (*Microtus pennsylvanicus*) 262
medical research 284
melanism 77*t*, 82, 97–8
melatonin 167, 168, 223, 224
Melospiza melodia 55, 237*t*, 242
MEMGENE package 58
metabolism 228–9, 292*g*
metabolites 228
metacommunities 292*g*
  see also evolving metacommunities
metacommunity ecology 177, 180, 290*g*
metagenomics 9, 84–5, 87
metal pollutants 122, 165–6, 221, 278
mice 4, 8, 20, 38, 57, 62, 63, 64*f*, 65*b*, 66, 67, 189, 213*b*
microbiota 9, 274, 292*g*
microplastic pollution 166
microsatellite markers 57–8, 61, 62–3, 80, 134, 134*t*, 245
Microtus pennsylvanicus 262
mismatch 292*g*
mitochondrial haplotypes 57, 68
model averaging 25
modernization 269, 269*b*, 273–5, 276*t*, 292*g*
molecular markers 57–8, 80
monopolization effect *see* evolution-mediated priority effect
Moran's eigenvector maps (MEM) 26, 58
morphology
  aquatic species 96–7, 163, 164–5
  climbing 206–11, 207*f*, 212*f*, 289*g*

morphology (*cont.*)
  limb 39, 97, 203–4, 204*f*, 207–8, 207*f*,
    209, 210
  terrestrial 202–5, 204*f*, 293*g*
  urban heat island effects 96–8, 103,
    105, 106*f*, 164–5, 176
mosquitos 40*t*, 44, 163, 164–5, 166,
  167, 189, 246
moths 45, 77*t*, 82, 97–8, 112
mountain chickadee (*Poecile gambeli*)
  260*t*
movement and gene flow 54–69
  analytical approaches 57–62, 59*f*
  empirical studies 62–8, 64*f*
  future directions 68–9
  landscape genetics 60–1, 67
  landscape genomics 61–2, 68
  molecular markers 57–8
  sampling strategies 63, 65*b*, 65*f*
  spatial population genomics 58–60,
    59*f*, 63–6
movement of organisms
  climbing behaviour 205–6, 205*f*
  climbing morphology 206–11, 207*f*,
    212*f*, 289*g*
  gliding 206, 210–11
  jumping and landing 206, 210, 211
  vertical migration in zooplankton
    160, 167
  see also movement and gene flow;
    terrestrial locomotor evolution
multimodel inference 25
Mus musculus 8, 68
muscular morphology 204, 204*f*
mutation rates 8, 57, 75, 80,
  85, 107
mutualism 111–25, 152, 292*g*
  future directions 123–5
  mechanistic perspective 113–15,
    114*f*, 116*t*
  nutritional mutualism 115, 116*t*,
    121–3, 123*f*, 292*g*
  partner behaviour changes 115
  partner loss 115, 116*t*, 119, 122
  partner switching 114, 116*t*, 118,
    119, 120, 121, 122
  pollination mutualisms 114,
    115–19, 116*t*, 117*f*, 124, 125
  protection mutualism 116*t*, 120–1,
    121*f*, 124, 293*g*
  seed dispersal mutualisms 116*t*,
    119–20, 124
  shifts from mutualism to
    antagonism 113, 116*t*
  trait–fitness relationship
    changes 113–14
  transport mutualism 115–20, 116*t*,
    117*f*, 124, 125, 293*g*

mycorrhizal fungi 121–3, 152
Myotis lucifugus 245

N
natural selection 8, 292*g*
neuromuscular changes 204, 204*f*
neutral evolution 292*g*
niche use changes 162, 191
night lighting *see* artificial light at
  night (ALAN); light pollution
nitrogen pollution 168
noise pollution 43, 131
  aquatic environments 159*t*, 168
  sexual selection and 168, 237*t*, 239,
    243, 246
non-adaptive parallelism 42, 292*g*
northern cardinal (*Cardinalis
  cardinalis*) 97, 237*t*, 240
northern two-lined salamander
  (*Eurycea bislineata*) 65*b*, 65*f*
Notemigonus crysoleucas 166
nutrient pollution 159*t*, 168–9
nutrition 292*g*
nutritional mutualism 115, 116*t*,
  121–3, 123*f*, 292*g*

O
Octodon degus 202
'old friends' hypothesis 271, 274
oligogenic traits 50*t*
open-field anxiety 199–201
operant conditioning 259*b*
opportunity for selection 292*g*
Oriental magpie-robin (*Copsychus
  saularis*) 237*t*
Ontological arguments 269*b*, 270*b*
outreach 292*g*
oxidative stress 39–42, 49, 222, 228,
  278, 292*g*
Oxyopes salticus 244

P
pace-of-life syndrome 43, 102, 106,
  164–5, 228, 240–1
Papio cynocephalus 229
parallel evolution 20, 36–50, 37*b*,
  50*t*, 178
  agents driving 42–4
  causes of non-parallel responses to
    urbanization 44–6, 48–9
  environmental heterogeneity
    and 20, 39, 44–5, 48–9, 50*t*
  frequency of parallel responses to
    urbanization 39–42, 40*t*
  future directions 47–9, 47*b*, 48*f*
  gene flow and 37*b*, 40*t*, 44, 45–6,
    49, 50*t*
  genetic architecture and 37*b*, 46

genetic drift and 37*b*, 40*t*, 44, 46, 50*t*
  habitat fragmentation and 44, 46
  pollution and 43–4, 46
  urban heat island effects 38–9,
    43, 105
parasitism 124
Parus major 20, 26, 77*t*, 78, 83, 85–6,
  131, 224, 228, 229, 237*t*, 260*t*, 261
Passer domesticus 9, 49, 222, 226,
  243, 260*t*
Passer montanus 227
patch dynamics 15–17
Penicillium bilaiae 106*f*
Penicillium lagena 106*f*
peppered moth (*Biston betularia*) 45,
  77*t*, 82, 97–8
Perca fluviatilis 167
perch (*Perca fluviatilis*) 167
Peromyscus 20
  leucopus 4, 57, 62, 63, 64*f*, 65*b*, 66,
    67, 77*t*, 83, 213*b*, 262
  polionotus 38
pesticide resistance 44, 123, 166,
  189, 221
phenology 143, 292*g*
  aquatic species 163–4, 167
  sidewalk plants 136–8, 137*t*
  urban heat island effects 100–1,
    103, 106, 106*f*, 136–8, 137*t*,
    163–4
phenotype–environment match 292*g*
phenotypic change vector analysis
  (PCVA) 47*b*, 48*f*, 292*g*
phenotypic plasticity 2, 8, 75, 76*f*,
  78–9, 143, 292*g*
  cognition and 255*f*, 256, 262
  community-wide trait changes 176,
    178*b*, 181, 183, 186*f*, 187
  epigenetic variation and 75, 85
  gut microbiomes 9
  locomotor morphology 204
  mating strategies 245
  physiology 218–19, 223, 224, 225,
    228, 229, 230
  plant life-history traits 101, 149,
    149*f*, 150, 152
  sexual selection 239, 245, 246, 247
  urban heat island effects 94–5, 94*f*,
    98–100, 99*f*, 101, 136
pheromones 167, 237*t*
philopatry 292*g*
physiology 217–30, 291*g*
  chemical pollution and 221–2
  detoxification 221–2, 290*g*
  endocrine responses to challenges
    225–8
  food availability/quality and 224–5,
    226–7, 229

hormetic responses 220–1, 220*f*, 229, 230
HPA axis regulation 225–8
human disturbance and 227–8
light pollution and 167, 223–4
measuring selection on 219*b*
metabolic responses 228–9, 292*g*
plasticity 218–19, 223, 224, 225, 228, 229, 230
reproductive endocrinology 167, 222–5
sidewalk plants 136–8, 137*t*, 139
stressors and stimulators 220–1, 220*f*
urban heat island effects 98–100, 98*f*, 99*f*, 103–6, 106*f*, 133*f*, 136–8, 137*t*, 139
phytoplankton 292*g*
phytoplankton blooms 168, 190
Pipilo maculatus 244
plant life-history traits 142–53
empirical evidence of adaptive evolution 147–50, 149*f*, 150*f*
future directions 151–3
gene flow and 150–1, 152
non-adaptive evolution 144, 150–1
phenotypic plasticity 101, 149, 149*f*, 150, 152
trait covariances and tradeoffs 146
urban heat island effects 100–1, 103, 106*f*, 136–8, 137*t*, 144–5, 152–3
plants
cyanogenesis 39, 99, 103, 138, 145, 151
nutritional mutualism 115, 116*t*, 121–3, 123*f*
phenological change 100–1
pollination mutualisms 114, 115–19, 116*t*, 117*f*, 124, 125
protection mutualism 116*t*, 120–1, 121*f*, 124
seed dispersal mutualisms 116*t*, 119–20, 124
size 97
see also plant life-history traits; sidewalk plants
plastic rescue 255*f*, 256, 292*g*
plasticity 292*g*
see also phenotypic plasticity
pleiotropy 46, 292*g*
antagonistic 274, 289*g*
Poa annua 132
Podarcis
erhardii 204
muralis 40*t*, 55, 67
Poecile
atricapillus 260, 260*t*
gambeli 260*t*

pollen limitation 114, 116–17, 116*t*
pollination mutualisms 114, 115–19, 116*t*, 117*f*, 124, 125
pollinators 131, 139, 145
pollution
aquatic environments 43–4, 46, 159*t*, 165–9
birds and 221–2
detoxification responses 221–2
human health and 278, 280, 281, 283
melanism and 97–8
mutualism and 115
parallel evolution and 43–4, 46
physiology and 221–2
sexual selection and 244
see also light pollution; noise pollution
pollution metrics 21*t*
polychlorinated biphenyls (PCBs) 43–4, 166
polygenic traits 46, 50*t*, 81, 82–3, 178, 272, 273*b*, 281, 292*g*
polymerase chain reaction (PCR) 57
pooled sequencing 83–4, 84*f*
population genomics 75, 76*f*, 77*t*, 79–85, 84*f*, 87
anonymous markers 80
candidate genes 77*t*, 80–1
methodological and taxonomic perspectives 83–5, 84*f*
polygenic adaptation 81, 82–3
spatial 7, 58–60, 59*f*, 63–6
whole genome sequencing (WGS) 77*t*, 82
population persistence 292*g*
pounce-pursuit predation 202–3
predation 97, 124
aquatic environments 159*t*, 160, 167
behaviourally mediated habitat use and 199–201, 200*f*
climbing behaviour and 205–6
flight initiation distance (FID) 199–201, 200*f*, 291*g*
open-field anxiety 199–201
protean behaviour 201–2
terrestrial locomotion and 201–5, 204*f*
principal component analysis (PCA) 18*b*, 18*f*, 23–4, 24*f*, 34–5, 292*g*
problem-solving 293*g*
Procyon lotor 229, 244
Promethean: 269*b*, 270*b*, 284
protean behaviour 201–2
protection mutualism 116*t*, 120–1, 121*f*, 124, 293*g*
Psammophilus dorsalis 201, 259*b*
Pseudogymnoascus pannorum 106*f*
Pterostichus madidus 40*t*

Puerto Rican crested anole (*Anolis cristatellus*) 39, 40*t*, 97, 104*f*, 106*f*, 203–4, 205*f*, 206, 209, 210, 256
purple loosestrife (*Lythrum salicaria*) 137
pursuit predation 201, 202–3
pyrethroid insecticides 44, 166

Q
quantifing urbanization 13–29, 16*t*, 28*t*
classic urban ecology frameworks 15–17, 18*b*, 18*f*
literature review 24–5
model selection and variable fitting 25–6
parallel urban evolution framework 20
principal component analysis (PCA) 18*b*, 18*f*, 23–4, 24*f*, 34–5, 292*g*
scale 26–8, 27*f*
spatial autocorrelation control 26
temporal aspects 17–20
univariate versus multivariate approaches 23–4
urban metrics 20–3, 21*t*
quantitative genetics 75, 76*f*, 77–9, 77*t*, 86–7

R
raccoon (*Procyon lotor*) 229, 244
Rana temporaria 57, 80
rats 4, 20, 40*t*, 44, 55, 56*f*, 59*f*, 62, 63, 65*b*, 67, 189, 203, 278
Rattus norvegicus 4, 20, 40*t*, 55, 56*f*, 59*f*, 62, 63, 65*b*, 67
reciprocal transplant experiments 10, 147*t*, 176–7, 293*g*
evolving metacommunities 188*f*, 189, 190
physiological traits 219*b*
plant life-history adaptation 101, 148–50, 152
urban heat island effects 95, 97, 101, 102–3
red deer (*Cervus elaphus*) 26
red fox (*Vulpes vulpes*) 56
red-tailed bumble bee (*Bombus lapidarius*) 39–42, 40*t*, 43, 49, 63, 77*t*, 81, 83, 189
replication 20
reproductive physiology 167, 222–5, 293*g*
reproductive strategies 244–5, 293*g*
resurrection ecology 191–2
Reticulitermes flavipes 67
reversal learning 258, 259*b*, 260*t*
rhesus macaque (*Macaca mulatta*) 261

rhizobial bacteria 121–3, 123*f*, 152
road salting 153, 160, 165, 166
root symbionts 113, 121–3, 123*f*, 152

S

salamanders 40*t*, 61, 65*b*, 65*f*, 160, 166, 261–2
salt tolerance 153, 160, 166
satellite sensor data 20, 21*t*
scale 23, 26–8, 27*f*
scale insects 101
scansorial species 198, 199, 202, 204, 293*g*
climbing behaviour 205–6, 205*f*
climbing morphology 206–11, 207*f*, 212*f*, 289*g*
Sceloporus occidentalis 201
schizophrenia 281
Sciurus carolinensis 17–18, 200–1
seed dispersal
habitat fragmentation and 119, 134–6, 135*f*, 138, 144, 147–8
mutualisms 116*t*, 119–20, 124
sidewalk plants 119, 134–6, 135*f*, 138, 147–8
seed dormancy 151, 152
selection differential 293*g*
selection gradient 293*g*
selection pressure 293*g*
temporal aspects of impacts 17–20
selective harvesting 244
self-fertilization 139
self-pollination 117
Semotilus atromaculatus 40*t*
Senecio vulgaris 132
sequencing 293*g*
Serinus canaria domestica 237*t*, 243
SERT (serotonin transporter) gene 49, 77*t*, 80–1, 85–6, 219*b*, 262
sex determination 165, 166
sexual imprinting 255*f*, 258
sexual selection 8, 234–48, 235*f*, 237*t*, 293*g*
aquatic species 162, 167, 168, 169
chemical pollution and 244
climate change and 244
evidence for evolutionary changes 245–6, 246*f*
fitness and 236–9
future directions 247–8
habitat fragmentation and 162, 244
light pollution and 167
mating and reproductive strategies 244–5
noise pollution and 168, 237*t*, 239, 243, 246

phenotypic plasticity 239, 245, 246, 247
responses of signal receivers 243
responses of signal senders 242–3
selection pressure changes 235–6, 235*f*, 239–42, 241*b*, 241*f*
selective harvesting and 244
speciation 246–7
stages of urban colonization and 241*b*, 241*f*, 242
turbidity and 169
sidewalk plants 77*t*, 130–40, 132*f*, 133*f*
habitat fragmentation 119, 134–6, 134*t*, 135*f*, 138, 147–8
phenology 136–8, 137*t*
physiology 136–8, 137*t*, 139
seed dispersal 119, 134–6, 135*f*, 138, 147–8
urban heat island effects 133*f*, 136–8, 137*t*, 139
water-use efficiency 136–7, 137*t*, 138
single nucleotide polymorphisms (SNPs) 19, 49, 57–8, 61, 65*b*, 68, 77*t*, 80–1, 83, 84, 245, 293*g*
sink populations 190
sleep architecture 276*t*, 278, 280
smart cities 4
snails 4, 25, 56
snow cover 99, 137, 138, 145
sociodemographic metrics 21*t*
song sparrow (*Melospiza melodia*) 55, 237*t*, 242
spatial autocorrelation 26, 293*g*
spatial genetic diversity (sGD) approach 58
spatial population genomics 7, 58–60, 59*f*, 63–6
speciation 246–7
species sorting 177, 178*b*, 181, 293*g*
spiders 40*t*, 45, 63, 176, 177, 180, 244
spotted salamander (*Ambystoma maculatum*) 261–2
spotted towhee (*Pipilo maculatus*) 244
standing genetic variation 293*g*
Stellaria media 132
stickleback (*Gasterosteus aculeatus*) 38
stinkbugs 123
stream dewatering 161, 293*g*
stream flow 159*t*, 162–3
Streptopelia senegalensis 222
stress physiology 100
striped field mice (*Apodemus agrarius*) 62
structural variants 293*g*
sustainable cities 4–5, 293*g*
synthetic organic compounds 166

T

Taeniopygia guttata 227
Teleogryllus commodus 237*t*
Temnothorax curvispinosus 8, 40*t*, 43, 77*t*, 78, 98–9, 102, 102*f*, 104*f*, 106*f*, 237*t*
temperature *see* urban heat island effects
temperature–size rule 96–7, 176
temporal metrics 21*t*
tendons 204, 204*f*, 210
termite (*Reticulitermes flavipes*) 67
terrestrial locomotor evolution 197–213
behaviourally mediated habitat use 199–201, 200*f*, 212*f*
climbing behaviour 205–6, 205*f*
climbing morphology 206–11, 207*f*, 212*f*, 289*g*
future directions 212–13
indirect effects 213*b*
spatial organization of habitats and 199–205, 199*f*, 200*f*, 204*f*, 212*f*
substrate properties and 205–11, 205*f*, 207*f*, 212*f*
terrestrial locomotion 201–5, 204*f*, 212*f*
terrestrial morphology 202–5, 204*f*, 293*g*
terrestrial morphology 202–5, 204*f*, 293*g*
thermal physiology 98–100, 98*f*, 99*f*, 103–6, 106*f*, 133*f*, 136–8, 137*t*, 139
thermal tolerance *see* urban heat island effects
thiamine deficiencies 229
thigmotaxis 199–201, 293*g*
Toxostoma curvirostre 227–8
transhumanism 284–5, 293*g*
transport mutualism 115–20, 116*t*, 117*f*, 124, 125, 293*g*
transposable elements 45
tree frogs 167, 206, 243
tree of heaven (*Ailanthus altissima*) 20, 56
tree sparrow (*Passer montanus*) 227
Trichoderma koningii 106*f*
Trichosurus cunninghami 244
Trifolium repens 25, 39, 40*t*, 87, 99, 103, 104*f*, 106*f*, 119, 122, 123*f*, 138, 145, 149, 151
trophobionts 120–1
tuberculosis 19–20
túngara frog (*Engystomops pustulosus*) 237*t*, 240, 241–2, 246
turbidity 169

Turdus merula 19, 40*t*, 42, 43, 68, 80–1, 223, 224, 228, 240, 262
turtles 161, 165

U

UHI *see* urban heat island effects
urban areas, definitions of 14
urban disease genetics 280–1
urban eco-evolutionary dynamics 180, 189–90, 282*b*, 282*f*, 283, 294*g*
urban exploiters 20, 55
urban green spaces 56–7, 65*b*, 199, 199*f*
urban heat island effects 91–107, 92*f*, 93*f*, 190, 237*t*, 294*g*
  aquatic environments 96–7, 106*f*, 159*t*, 163–5, 177
  body size 96–7, 103, 105, 106*f*, 164, 176
  evolutionary responses 94–5, 94*f*
  fitness 102–3, 102*f*
  future directions 107
  life-history traits 100–1, 103, 106, 106*f*, 136–8, 137*t*, 144–5, 152–3, 163–4, 177
  meta-analysis 103–7, 104*f*, 105*f*, 106*f*
  morphology 96–8, 103, 105, 106*f*, 164–5, 176
  parallel evolution and 38–9, 43, 105
  phenotypic plasticity and 94–5, 94*f*, 98–100, 99*f*, 101, 136
  physiology 98–100, 98*f*, 99*f*, 103–6, 106*f*, 133*f*, 136–8, 137*t*, 139
  sex determination and 165
  sidewalk plants 133*f*, 136–8, 137*t*, 139

urban metrics 20–3, 21*t*
urban mosaic/urban matrix 27, 27*f*, 55–7, 56*f*, 294*g*
urban starthistle (*Centaurea solstitialis*) 117–18
urbanization 74, 176, 294*g*
  see also quantifing urbanization
urbanization gradient 294*g*
urban–rural gradients 15–17, 18*b*, 24*f*, 25

V

variables
  continuous 294*g*
  nominal 294*g*
  ordinal 294*g*
variance partitioning 47*b*, 48*f*, 294*g*
Viola pubescens 120
Virginia pepperweed (*Lepidium virginicum*) 40*t*, 43, 46, 101, 104*f*, 106*f*, 131, 138, 148, 150
vital rate 294*g*
voltage-gate sodium channel gene (vgsc) 44
Vulpes vulpes 56

W

warming *see* urban heat island effects
water flea (*Daphnia magna*) 43, 87, 96–7, 98, 100, 101, 102, 104*f*, 106*f*, 163, 164, 165–6, 167, 177, 178*b*, 179*f*, 180, 182–3, 190
water-use efficiency, sidewalk plants 136–7, 137*t*, 138

western black widow spider (*Latrodectus hesperus*) 40*t*, 45, 63
western fence lizard (*Sceloporus occidentalis*) 201
western scrub-jay (*Aphelacoma californica*) 224
wet adhesion 207*f*, 208, 210, 294*g*
white clover (*Trifolium repens*) 25, 39, 40*t*, 87, 99, 103, 104*f*, 106*f*, 119, 122, 123*f*, 138, 145, 149, 151
white-crowned sparrow (*Zonotrichia leucophrys*) 237*t*, 240
white-footed mouse (*Peromyscus leucopus*) 4, 57, 62, 63, 64*f*, 65*b*, 66, 67, 77*t*, 83, 213*b*, 262
whole genome sequencing (WGS) 77*t*, 82, 245, 294*g*
wood mouse (*Apodemus sylvaticus*) 67

Y

yellow baboon (*Papio cynocephalus*) 229
yellow-faced bumble bee (*Bombus vosnesenskii*) 66–7
yellownecked mouse (*Apodemus flavicollis*) 62

Z

zebra finch (*Taeniopygia guttata*) 227
zebrafish (*Danio rerio*) 167
zenaida dove (*Zenaida aurita*) 260*t*
Zonotrichia leucophrys 237*t*, 240
zooplankton 160, 167, 168, 178–80, 178*b*, 179*f*, 182–3, 190, 294*g*
  see also *Daphnia magna*